Physical Chemistry of Iron and Steel Manufacture

Physical Chemistry of Iron and Steel Manufacture

C. BODSWORTH

CBS PUBLISHERS & DISTRIBUTORS PVT. LTD.

New Delhi • Bengaluru • Chennai • Kochi • Mumbai • Pune

ISBN : 978-81-239-2434-2

First Indian Edition : 1988

This Edition has been reprinted in India by arrangement with Longmans, Green & Company Limited

Published by Satish Kumar Jain and produced by V.K. Jain for
CBS Publishers & Distributors Pvt. Ltd.,
CBS Plaza, 4819/XI Prahlad Street, 24 Ansari Road, Daryaganj,
New Delhi - 110002, India. • Website: www.cbspd.com
e-mail: delhi@cbspd.com, cbspubs@airtelmail.in
Ph.: 23289259, 23266861, 23266867 • Fax: 011-23243014

Branches:

- ***Bengaluru:*** Seema House, 2975, 17th Cross, K.R. Road, Bansankari 2nd Stage, Bengaluru - 560070
  • Ph.: +91-80-26771678/79 • Fax: +91-80-26771680
  • E-mail: cbsbng@gmail.com, bangalore@cbspd.com
- ***Pune:*** Bhuruk Prestige, Sr. No. 52/12/2+1+3/2, Narhe, Haveli (Near Katraj-Dehu Road By-pass), Pune - 411041
  • Ph.: +91-20-64704058/59, 020-32392277 • E-mail: pune@cbspd.com
- ***Kochi:*** 36/14, Kalluvilakam, Lissie Hospital Road, Kochi - 682018, Kerala • Ph.: +91-484-4059061-65
  • Fax: +91-484-4059065 • E-mail: cochin@cbspd.com
- ***Chennai:*** 20, West Park Road, Shenoy Nagar, Chennai - 600030
  Ph.: +91-44-26260666, 26208620 • Fax: +91-44-42032115
  • E-mail: chennai@cbspd.com
- ***Mumbai:*** 83-C, Ist Floor, Dr. E. Moses Road, Worli, Mumbai-400 018, Maharashtra Ph.: +91-9833017933, 022-24902340/24902341
  • E-mail: mumbai@cbspd.com

Printed at :
Nikunj Print Process, Delhi

Preface

The development of the iron and steel manufacturing processes to their present standard of efficiency has been brought about largely by trial and error and by experience gained under production conditions. In comparison, our knowledge of the physico-chemical factors which determine the rate and extent of the various refining reactions is still at a comparatively early stage of development. However, progress in this field has been rapid during the past twenty-five years or so and it is now possible to define precisely the equilibrium conditions and the factors affecting the kinetics of many of these reactions.

The art, as distinct from the science, of ferrous extraction and refining operations has been fully described in many textbooks, but no recent publications have brought together the wealth of physico-chemical data pertinent to this field which is now available, or have shown how these data can be used to further the understanding and control of the existing processes and assist in the development of new processes. This text is concerned mainly with the application of the physico-chemical data and only brief reference is made to processing operations, although more such detail is given for some of the newer methods of production of iron and steel.

The text has been prepared primarily for usc by students studying ferrous extraction and refining operations, the research worker engaged in obtaining fundamental information in this field and the process metallurgist who wishes to apply the rapidly developing theoretical treatment to the process he is operating. To meet individual requirements the text is divided into four parts.

Part one contains an explanation of those aspects of chemical thermodynamics, kinetic theory and the structure and properties of liquid slags which are subsequently applied to the extraction and refining processes in the remainder of the book. Thermodynamic theory is covered in sufficient detail to meet the needs of those process metallurgists with only a limited knowledge of the subject and for use in connection with lecture courses dealing with the application of chemical thermodynamics to metallurgical problems. Worked examples are given to illustrate the evaluation and application of the thermodynamic relations. The mathematical treatment has been simplified so far as possible without undue approximations. Some topics, such as the ternary Gibbs–Duhem relations, are dealt with in more detail than is subsequently necessary in the text to enable the reader to understand the theoretical background of any relevant thermodynamic papers in the scientific and technological journals. The process metallurgist may choose to omit part of these chapters, but it is important to become familiar with the relations between the concepts of activity, activity coefficient, free energy and the equilibrium constant (Sections 1.7, 1.8 and 1.10) and the methods used to assess the activities of solutes in multi-component solutions (Section 2.5).

Thermodynamics deals only with equilibrium conditions and these conditions may not be achieved in a reasonable period of time in an industrial process, so it is necessary to examine the factors affecting the kinetics of the reactions. The application of kinetic theory to ferrous extraction and refining operations has so far produced only a little quantitative data, but qualitative explanations can be advanced for most of the non-equilibrium conditions which are encountered. A brief introduction is given, therefore, to those aspects of the theory which have been applied successfully to this field.

The theoretical approach is explained in the second part in relation to the gas–solid, gas–liquid and slag–metal reactions in the blast furnace and, in the third part, in relation to the Bessemer converter, open-hearth and electric-arc furnace processes, together with some of the newer processes which have been proposed for the production of iron and steel. Some cross-reference between chapters is unavoidable but, so far as possible without introducing excessive repetition, the chapters dealing with each of the processes are self-contained and, for example, the reader who is interested only in steel-making can, if he so chooses, omit the chapters dealing with iron-making.

Part four contains more advanced chapters in which detailed consideration is given to the factors affecting the activities of various oxides in liquid slags and the distribution of carbon, oxygen, phosphorus, sulphur and other elements between gas, metal and slag. Available data are critically assessed and methods are described for calculating equilibrium conditions. Tables of interaction coefficients are given in an appendix.

As the contents of the book were planned originally it was intended to include chapters dealing with dissolved gases and with vacuum melting and casting. However, as the writing proceeded it became evident that the limited quantitative data available for these topics were better introduced in relation to the various refining processes and the distribution of specific elements than in isolated chapters.

Contrary to usual practice, the names of authors are largely excluded from the text. This is not intended in any way as disrespect of their contributions, but if the normal practice had been adopted certain names such as Chipman, Richardson, Schenck and their associates would appear several times on most of the pages. Omission of names makes the text easier to read, but requires more effort in consulting the references on the part of the reader who wishes to distinguish between the contributions made by the various research and development groups. However, the author believes that the improved continuity of the text fully justifies this additional effort. References are given at the end of each chapter. More than 500 references are included, which have been selected to draw attention to the various types of approach to each of the problems considered and also to provide the process metallurgist with further reading on the practical applications of some of the more recent refining techniques.

Acknowledgements

The author is indebted to the many friends and colleagues who have assisted in the preparation of this book and who have offered valuable criticisms of the contents and the presentation of the manuscript.

The data for open-hearth furnace heats given in Chapter 9 and Appendix 2 were kindly supplied by the Lancashire Steel Co. Ltd., the Steel Company of Wales Ltd., and the United Steel Cos. Ltd. The following publishers and learned societies have granted permission to reproduce diagrams from published work: the American Chemical Society, the *American Journal of Science*, the Faraday Society, the Institution of Mining and Metallurgy, the Iron and Steel Institute, the Metallurgical Society of A.I.M.E., the Society of Chemical Industry, the Royal Society, Verein Deutscher Eisenhüttenleute and John Wiley & Sons, Inc.

Contents

PART 3 STEEL-MAKING

PART 4 SLAG–METAL REACTIONS

Notation

The notation selected for the various quantities are those in most common use and are largely based upon the recommendations of the International Union of Pure and Applied Physics (1955) and the International Union of Pure and Applied Chemistry (1955).

Square and round brackets surrounding a molar quantity denote components of the metal and slag respectively.

The meaning of all the symbols used are listed in the following Table for ready reference.

Symbol	*Meaning*	*Common Units*
A	Helmholtz function	cal mol^{-1}
a	Activity	
c	Concentration	
C_P	Molar heat capacity at constant pressure	cal deg^{-1} mol^{-1}
C_V	Molar heat capacity at constant volume	cal deg^{-1} mol^{-1}
D	Diffusion coefficient	cm^2 sec^{-1}
E	Activation energy for diffusion	cal mol^{-1}
f	Activity coefficient (Henry)	
f_Y^X	Interaction coefficient; the effect of element X upon the activity coefficient of element Y	
$\mathscr{F}$	Fugacity	
G	Gibbs function or Gibbs free-energy	cal mol^{-1}
G^0	Standard free-energy	cal mol^{-1}
$\bar{G}$	Partial molar free-energy	cal mol^{-1}
G^M	Integral molar free-energy of mixing	cal mol^{-1}
G_A^M	Relative partial molar free-energy of a component A	cal mol^{-1}
G_A^E	Excess partial molar free-energy of a component A	cal mol^{-1}
H	Enthalpy, heat content	cal mol^{-1}
H^0	Standard enthalpy	cal mol^{-1}
$\bar{H}$	Partial molar enthalpy	cal mol^{-1}
H^M	Integral molar enthalpy (or heat) of mixing	cal mol^{-1}
H_A^M	Relative partial molar enthalpy of a component A	cal mol^{-1}
k	Boltzmann's constant	$1{\cdot}38 \times 10^{-16}$ erg deg^{-1} $3{\cdot}30 \times 10^{-22}$ cal deg^{-1}
K	Equilibrium constant	
K'	Apparent equilibrium constant	
κ	Henry's law constant	
ln	Natural logarithm; $\log_e$	

Symbol	*Meaning*	*Common Units*
log	Common logarithm; $\log_{10}$	
L_{tr}	Latent heat of transformation	cal mol^{-1}
M	Molecular weight	grams
N	Mole fraction	
N_0	Avogadro's number	$6{\cdot}02 \times 10^{23}$ mol^{-1}
n	Number of moles	
P	Pressure	atmospheres
p	Partial pressure	atmospheres
Q	Arrhenius activation energy	cal mol^{-1}
q	Heat absorbed	cal mol^{-1}
r	Radius	cm mol^{-1}
R	Gas constant	8·3147 Joules deg^{-1} 1·986 cal deg^{-1}
S	Entropy	cal deg^{-1}
S^0	Standard entropy	cal deg^{-1}
$\bar{S}$	Partial molar entropy	cal deg^{-1} mol^{-1}
S_A^M	Relative partial molar entropy of a component A	cal deg^{-1} mol^{-1}
S_A^E	Excess partial molar entropy of a component A	cal deg^{-1} mol^{-1}
T	Temperature	deg Kelvin
U	Internal energy	cal mol^{-1}
V	Volume	mol; cm^3
W	Work done by a system	cal
γ	Activity coefficient (Raoult)	
γ^0	Activity coefficient at infinite dilution	
ϵ_X^Y	Interaction parameter $\dfrac{\partial \ln f_X}{\partial N_Y}$	
e_X^Y	Interaction parameter $\dfrac{\partial \log f_X}{\partial \mathrm{wt}\%\ Y}$	
ζ	Viscosity	dynes sec^{-1} cm^{-2}
κ	Rate constant	
μ	Chemical potential	cal mol^{-1}
ρ	Density	grams ml^{-1}
σ	Surface tension	dynes cm^{-1}
ω	Probability factor; number of distributions	
(g)	Gas	
(s)	Solid	
(l)	Liquid	

PART I

THEORY

CHAPTER 1

Thermodynamic Functions

1.1 Internal Energy

The classical theory of thermodynamics is based on two fundamental laws which are simply statements of the results of practical observations regarding the interconversion of different forms of energy. A corollary to the second law is now classed as a third law of thermodynamics. By the application of certain well established mathematical procedures it is possible to derive from these laws a large number of relations which can be employed for the solution of many different types of problems, but in this book consideration will be restricted mainly to those relations which determine the conditions for chemical equilibrium and whether or not, under specified conditions, a reaction can occur spontaneously.

A chemical *reaction* can be defined as a process whereby a system is transferred from one state to another. The *system* is here defined as that part of the universe selected for consideration. It may contain any number of atoms, chemical substances or other entities but, for convenience, it is customary to select the boundaries of the system so that the various parts are in some way interconnected. The system may be separated by a real or an imaginary boundary, through which energy can be exchanged with the surroundings. In some instances it is desirable to include the immediate surroundings in the part of the universe considered, in which case it is referred to as an *isolated system.*

A *homogeneous system* contains only one phase, such as a gas, a mixture of gases or a liquid or solid solution, whereas a *heterogeneous system* contains more than one phase. In relation to iron and steel refining, the gas, liquid slag and liquid metal may usually be considered as homogeneous systems, whilst mixtures of two or more of these are heterogeneous, but it is important to recognise that the slag, or the metal, is also heterogeneous if it contains more than one phase.

If we exclude from consideration such effects as electrical, magnetic and gravitational forces (which do not enter into the refining reactions), the thermodynamic *state* of a system is defined in terms of four characteristic properties: temperature, pressure, volume and composition. If one of these variables is changed the state of the system is also changed, but in order to define the state of the system it is not necessary to assign values to all four variables as the first three are interrelated through an equation of state. For example, if the temperature and pressure are each assigned specific values then the volume is also fixed.

The total energy of a system may be subdivided into the kinetic and characteristic or *internal energy* components and the latter is represented by the symbol U. In chemical thermodynamics the changes in kinetic energy accompanying a reaction are either zero or very small and this treatise will be concerned only with the internal energy component. The absolute energy of a system cannot be measured, for all energy levels are relative to some arbitrary reference state and, in regard to thermodynamics, only the difference in the internal energy (ΔU) of a system when it is taken from one state to another is measurable.

The word thermodynamics suggests a relation between heat and work and these two quantities are related to the internal energy through the *Law of Conservation of Energy*, or the *First Law of Thermodynamics*, which can be stated simply in the form: '*energy can be neither created nor destroyed*'. If energy of one kind is consumed then an equivalent amount of some other kind of energy must be formed. In mathematical form this statement is expressed as:

$$\Delta U = q - w \qquad 1.1$$

where, by convention, q is the heat *absorbed* by the system from the surroundings and w is the work *done* by the system on the surroundings. The internal energy must therefore remain constant in an isolated system.

A system can be taken through a complete cycle from some state A to another state B by altering only one state variable (e.g., by adding heat to or removing heat from the surroundings but without performing work) and then restored to state A by suitably changing another variable (e.g., by performing work whilst holding the heat content constant). It follows from the first law of thermodynamics that the change in the internal energy over the complete cycle is zero; otherwise energy must be either created or destroyed. Hence the change in the internal energy of the system is dependent only on the initial and final states and is independent of the path followed to effect the change in state. In consequence the internal energy is called a *state property* or a *characteristic function* of the state of the system which may be treated mathematically in a similar manner to the other state functions such as temperature, pressure and volume. Thus the change in internal energy during a reaction which transfers a system from a state A to another state B is a single-valued function of the difference between the values of, for example, the variables volume, temperature and composition in the two states, i.e.,

$$\Delta U = U(V_B.T_B.N_B) - U(V_A.T_A.N_A) \qquad 1.2$$

From the fundamental theorem of partial differentiation, the change in internal energy accompanying a very small change in the state of the system is given by:

$$dU = \left(\frac{\partial U}{\partial V}\right)_{TN}.dV + \left(\frac{\partial U}{\partial T}\right)_{VN}.dT + \left(\frac{\partial U}{\partial N}\right)_{VT}.dN \qquad 1.3$$

where each of the bracketed differential terms expresses the rate of change of U with the variable in the denominator when the other variables are held constant. Any differential quantity of a property which can be expressed in this way is called a complete or an exact differential of the property.* Because the value of the internal energy change is dependent on the amount of the system, it is also called an *extensive* property and is usually measured in units of calories (or kilocalories, 1 kcal = 1000 cal) per gram-atom or per gram-mole. This is abbreviated to cal g-atom^{-1} or cal mol^{-1}.

In contrast to the internal energy the heat and work terms in equation 1.1 are not characteristic functions of the system for, as stated in the previous paragraph, the change from state A to state B can be accomplished by changing only one of these variables whilst holding the other constant. Thus the heat and work terms *are* dependent on the path followed between the initial and final states. They cannot be expressed as exact, but only as incomplete, differentials and for a very small change in state:

$$dU = \partial q - \partial w \qquad 1.4$$

For mathematical convenience it is preferable that equation 1.1 be expressed in terms of exact differential quantities. This is possible for certain specific cases which form the fundamental basis for the development of the classical thermodynamic approach.

1.2 Enthalpy

The work term can be expressed as a function of the pressure, the volume and various forces, such as electrical and magnetic, which are imposed upon the system. However, in the chemical reactions considered in this

* Because, in general, each of the thermodynamic variables does not vary linearly with the other variables, it is necessary to evaluate the variables by integration of a differential relation. An exact differential quantity can usually be integrated without difficulty, but an incomplete differential is less readily manipulated.

treatise, it is reasonable to assume that these other forces do not affect the reaction and that the only work performed is that due to a change in volume. In this case the work change is zero for a process which occurs at constant volume and

$$\Delta U_V = q_V \quad 1.5$$

where the subscript V indicates constant volume.

More commonly in practice a reaction occurs at constant pressure, usually atmospheric, in which case the volume of the system is increased or decreased during the reaction by an amount ΔV and:

$$w_P = P\Delta V \quad 1.6$$

Also for a very small change:

$$\partial w_P = P\,dV \quad 1.7$$

Substituting equation 1.7 for ∂w in equation 1.4 and rearranging terms:

$$\partial q_P = dU + P\,dV \quad 1.8$$

The heat change is now expressed in terms of the state variables U, P and V and it follows from the previous definition that the heat change is a characteristic function of the state of a system which undergoes a change at constant pressure, i.e., the value is dependent only on the initial and final states and is independent of the reaction path.

The quantity $U+PV$ occurs frequently in thermodynamic equations and it is convenient to represent it by a single symbol. The quantity is called accordingly the *heat content* or *enthalpy* of the system and is represented by the symbol H. That is:

$$H = U + PV \quad 1.9$$

or for an infinitesimal change at constant pressure:

$$\partial q_P = dH = dU + P\,dV \quad 1.10$$

The enthalpy is an extensive property and the units are usually the same as those used to measure the internal energy, i.e., cal g-atom^{-1} or cal mol^{-1}.

HEATS OF REACTION AND FORMATION

The heat change associated with a chemical reaction at constant pressure and temperature is equal to the difference between the total heat contents of the products and of the reactants at that temperature, i.e.,

$$\Delta q_P = \Delta H = \sum H_{(\text{products})} - \sum H_{(\text{reactants})} \quad 1.11$$

and ΔH is called the *heat of reaction*. For a reaction such as:

$$A_x B_y + C_z \rightleftharpoons A_x C_z + B_y$$

the heat of reaction is given by:

$$\Delta H = H_{A_x C_z} + H_{B_y} - H_{A_x B_y} - H_{C_z} \quad 1.12$$

When ΔH is positive (i.e., $\Delta H_{\text{(products)}}$ is more positive than $\Delta H_{\text{(reactants)}}$) heat is absorbed and the reaction is endothermic. Conversely, ΔH is negative and heat is evolved for an exothermic reaction.

The enthalpy change accompanying the formation at constant temperature and pressure of a compound AB from the elements A and B:

$$A + B \rightleftharpoons AB$$

is given by:

$$\Delta H = H_{AB} - H_A - H_B$$

and this quantity is called the *heat of formation* of the compound.

The physical state and condition of each of the substances taking part in a reaction are dependent on the reaction temperature. For example, a pure substance changes from solid to liquid when heated at the melting point and from liquid to gas at the boiling point. One or more changes in the atomic configuration (allotropic or solid-state phase changes) may also occur during the heating of a substance below the melting point. Each of these changes is accompanied by a change in the enthalpy of the substance and ΔH for the reaction is dependent on the physical state of each of the components at the reaction temperature. In order to compare the enthalpies of different substances it is necessary to select some *reference state* and this is usually chosen as the stable form which exists at atmospheric temperature (25°C or 298·16°K*) and one atmosphere pressure.

Because the zero level for the internal energy is chosen arbitrarily, it is evident from equation 1.9 that there is no absolute zero to which the

* The zero point on the thermodynamic temperature scale was determined by Kelvin from consideration of the efficiency of a reversible gas engine and coincides with the so called absolute zero on the absolute-ideal-gas scale. For convenience the size of the degree on the Kelvin scale (°K) is taken to be the same as on the centigrade scale and it is then found that 0°C is equal to 273·16°K. Whenever the temperature T appears in a thermodynamic statement or equation it refers to the Kelvin scale, and temperatures measured in degrees centigrade must be converted by adding 273·16° before insertion into an equation. For example, if it is required to evaluate a thermodynamic relation at 1,600°C the temperature used in the evaluation should be 1,873·16°K. However, at elevated temperatures the accuracy of the thermodynamic data and temperature measurement rarely justify inclusion of fractions of a degree, and for most purposes it is sufficient to add 273° to the centigrade temperature. This practice is adopted for all temperatures referred to subsequently in the text.

enthalpy of a substance can be related. Consequently it is permissible to select any fixed point as an arbitrary zero and by convention the heat contents of the elements are taken as zero when they are in their reference states. It follows that at the reference temperature the enthalpy of a compound is equal to its heat of formation. Useful compilations of reference (sometimes called 'standard') enthalpies and other thermodynamic functions have been prepared, for example, by Kubaschewski and Evans [1], Rossini [2] and by Elliott and Gleiser [3]. The following example illustrates the evaluation of standard heats of reaction from the tabulated data.

The reduction of solid ferric oxide by solid carbon to form iron and carbon monoxide is represented by the equation:

$$Fe_2O_{3(s)} + 3C_{(g)} \rightleftharpoons 2Fe_{(s)} + 3CO_{(g)} \tag{1.13}$$

where the subscripts (s) and (g) signify respectively solid and gas. At 298°K

$$H_{Fe_2O_3} = -196{\cdot}3 \text{ kcal mol}^{-1}$$

$$H_{CO} = -26{\cdot}4 \text{ kcal mol}^{-1}$$

$$H_{Fe} = H_C = 0 \text{ kcal mol}^{-1}$$

and the heat of reaction is:

$$\Delta H_{298°K} = 3(-26{\cdot}4) - (-196{\cdot}3) = +117{\cdot}1 \text{ kcal mol}^{-1}$$

The enthalpy change for the reaction could also be obtained by combining the heats of reaction for the following two equations:

$$Fe_2O_{3(s)} + 3CO_{(g)} \rightleftharpoons 2Fe_{(s)} + 3CO_{2(g)} \tag{1.14}$$

$$CO_{2(g)} + C_{(s)} \rightleftharpoons 2CO_{(g)} \tag{1.15}$$

Given that at 298°K

$$H_{CO_2} = -94{\cdot}5 \text{ kcal mol}^{-1}$$

and proceeding as before, the heats of reaction are $-6{\cdot}65$ kcal for reaction 1.14 and $+41{\cdot}25$ kcal for 1.15. If equation 1.15 is multiplied by three and added to 1.14 the product is identical with 1.13. Combining similarly the heats of reaction for the two processes:

$$\Delta H_{298°K(1.14)} + 3\Delta H_{298°K(1.15)} = \Delta H_{298°K(1.13)} \tag{1.16}$$

$$(-6{\cdot}65) + (+123{\cdot}75) = (+117{\cdot}1) \text{ kcal mol}^{-1}$$

The addition of heats of reaction in this example is an illustration of the application of *Hess's Law* which states that '*the enthalpy change for a reaction is the same whether it takes place in one or in a number of steps*'. This statement follows logically from the fact that the enthalpy change is dependent only on the initial and final states of the system. The additive relation only holds, however, when the thermodynamic state of each substance is the same in each of the reaction steps in which it appears. In the above example the carbon monoxide consumed in reaction 1.15 must be at the same temperature, pressure and concentration as that produced in reaction 1.14. The changes in any other extensive state property are also additive when this requirement of corresponding states is satisfied.

1.3 Heat Capacity

In calculating the heat of a reaction at other than the reference temperature, consideration must be given to the variation with temperature of the enthalpies of the substances taking part. Some tabulations give mean values over small intervals of temperature of the enthalpies of substances, but it is usually more precise to calculate the enthalpy at a chosen temperature from the reference enthalpy of the substance at 298°K and a quantity which is called the *heat capacity* (C) of the substance. The latter is defined as the ratio of heat absorbed by the substance when heated through a range of temperature to the resultant temperature increment: i.e.,

$$C = \frac{q_{T_2} - q_{T_1}}{T_2 - T_1} \qquad 1.17$$

Generally the heat capacity of a substance is a non-linear function of the temperature, and in order to define precisely the heat capacity at any given temperature it is necessary to reduce the temperature interval to an infinitesimal amount, i.e.,

$$C = \lim_{T_2 \to T_1} \left(\frac{q_{T_2} - q_{T_1}}{T_2 - T_1} \right)$$

$$= \frac{\partial q}{\partial T} \qquad 1.18$$

When q is measured in calories the units of heat capacity are cal deg^{-1} g-atom^{-1} or cal deg^{-1} mol^{-1}.

Because q is not a state property of the system the heat capacity is dependent on the path chosen for the reaction unless conditions are specified to define the path. For example, equation 1.5 shows that when the volume of the system is held constant the heat absorbed is equal to the change in the internal energy. Hence:

$$\left(\frac{\partial q}{\partial T} \right)_V = C_V = \left(\frac{\partial U}{\partial T} \right)_V \qquad 1.19$$

That is, the heat capacity of a substance at constant volume (C_v) is equal to the rate of change with temperature of the internal energy at constant volume.

Similarly, from equation 1.10 for a reaction at constant pressure,

$$\left(\frac{\partial q}{\partial T}\right)_P = C_P = \left(\frac{\partial H}{\partial T}\right)_P \qquad 1.20$$

and the heat capacity at constant pressure (C_P) is equal to the rate of change of the enthalpy with temperature at constant pressure.

Differentiating equation 1.11 with respect to temperature we obtain:

$$\left(\frac{\partial \Delta q}{\partial T}\right)_P = \left(\frac{\partial \Delta H}{\partial T}\right)_P = \left(\frac{\partial \sum H_{(\text{products})}}{\partial T}\right)_P - \left(\frac{\partial \sum H_{(\text{reactants})}}{\partial T}\right)_P$$

and substituting $\sum C_{P(\text{products})}$ and $\sum C_{P(\text{reactants})}$ for the bracketed terms on the right-hand side gives:

$$\left(\frac{\partial \Delta H}{\partial T}\right)_P = \Delta C_P = \sum C_{P(\text{products})} - \sum C_{P(\text{reactants})} \qquad 1.21$$

This is one way of expressing *Kirchoff's equation*, which shows that '*the rate of change with temperature of the heat of a reaction at constant pressure is equal to the change in the heat capacity accompanying the reaction.*'

Integration of equation 1.21 between temperature limits T_1 and T_2 gives

$$\int_{T_1}^{T_2} \left(\frac{\partial \Delta H}{\partial T}\right)_P = \Delta H_{T_2} - \Delta H_{T_1} = \int_{T_1}^{T_2} \Delta C_P dT \qquad 1.22$$

which can be evaluated for ΔH_{T_2} if ΔH_{T_1} and the variation of C_P with temperature are known. For most substances the heat capacity can be expressed as a function of temperature by a power series relation of the general form:

$$C_P = a + bT + cT^2 + dT^{-2} + eT^{-1/2} \ldots \qquad 1.23$$

where a, b, c, etc., are constants which are derived from experimental heat capacity data. (Not all the terms are required to define the heat capacity-temperature relation for one substance.) On integration this equation becomes:

$$\int_{T_1}^{T_2} C_P dT = a(T_2 - T_1) + \tfrac{1}{2}b(T_2^2 - T_1^2) + \tfrac{1}{3}c(T_2^3 - T_1^3)$$

$$- d(T_2^{-1} - T_1^{-1}) + 2e(T_2^{1/2} - T_1^{1/2}) \ldots \qquad 1.24$$

It is sometimes convenient to express equation 1.22 in general form by integrating between the limits $T_0(=0°K)$ and T to obtain:

$$\Delta H_T = \Delta H_0 + \int_{T_0}^{T} \Delta C_P dT \qquad 1.25$$

The quantity ΔH_0 has no fundamental meaning, but may be regarded as a value of ΔH for the reaction at 0°K if the reactants and products could co-exist in the same physical condition at 0°K as at T°K and if the heat capacity data are valid for temperatures down to the absolute zero. It can be evaluated by substituting for ΔH_T the change in the reference enthalpy at 298°K and integrating between the limits 0 and 298°K. Substituting the value obtained for ΔH_0, the equation can then be solved for any temperature within the range for which the heat capacity constants a, b, c, etc., are valid for each of the substances in the reaction.

When one of the components of a reaction undergoes allotropic transformation, fusion or evaporation within the temperature range considered, the latent heat (L_T) of the transformation must be included in the heat capacity relation at the transformation temperature (T_T). This latent heat is added to the enthalpy change for the transformation of a reactant and subtracted for the transformation of a product. Equation 1.25 is then expanded into the form:

$$\Delta H_T = \Delta H_0 + \int_{T_0}^{T_T} \Delta C_P . dT \pm L_T + \int_{T_T}^{T} \Delta C_P dT \ldots \qquad 1.26$$

The application of this equation is illustrated by the following example.

Pure calcium undergoes an allotropic change from the α to the β form at 440°C (713°K) and melts at 850°C (1,123°K). The change in the enthalpy for the reaction between pure calcium and gaseous oxygen to form pure calcium oxide at any temperature within the range 440 to 850°C is calculated as follows: The enthalpy of the reaction at 298°K is given as [1]:

$$Ca_{(\alpha)} + \tfrac{1}{2}O_{2(g)} \rightleftharpoons CaO_{(s)};\ \Delta H^0_{298°K} = -151{,}500 \text{ cal mol}^{-1}$$

and the heat capacities of the components are given by the following equations:

$$C_{P,Ca_{(\alpha)}} = 5{\cdot}31 + 3{\cdot}33 \times 10^{-3}T$$

$$C_{P,O_{2(g)}} = 7{\cdot}16 + 1{\cdot}00 \times 10^{-3}T - 0{\cdot}40 \times 10^5 T^{-2}$$

$$C_{P,CaO_{(s)}} = 11{\cdot}86 + 1{\cdot}08 \times 10^{-3}T - 1{\cdot}66 \times 10^5 T^{-2}$$

Hence

$$\Delta C_P = C_{P,CaO} - C_{P,Ca} - \tfrac{1}{2}C_{P,O_2}$$

$$= 2{\cdot}97 - 2{\cdot}75 \times 10^{-3}T - 1{\cdot}46 \times 10^5 T^{-2}$$

Substituting for C_P in equation 1.25 and integrating between the limits 0 and 298°K give:

$$\Delta H_2 = \Delta H_0 + 2{\cdot}97T - 1{\cdot}38\times10^{-3}T^2 + 1{\cdot}46\times10^5 T^{-1}$$

At 298°K the value of ΔH_2 is $-151{,}500$ cal mol^{-1}; hence ΔH_0 is equal to $-152{,}753$ cal mol^{-1}. Inserting this value in the last equation and solving for $T = 713$°K, the value of ΔH_2 at the temperature of the α-β transformation in pure calcium is found to be $-151{,}133$ cal mol^{-1}.

The latent heat of the transformation $Ca_{(\alpha)}$ to $Ca_{(\beta)}$ is 240 cal mol^{-1}. Therefore, for the reactions:

$$Ca_{(\alpha)} + \tfrac{1}{2}O_{2(g)} \rightleftharpoons CaO_{(s)}, \quad \Delta H_{713°K} = -151{,}133 \text{ cal mol}^{-1}$$

$$Ca_{(\beta)} \rightleftharpoons Ca_{(\alpha)}; \quad L_{tranf.} = -240 \text{ cal mol}^{-1}$$

Adding: $Ca_{(\beta)} + \tfrac{1}{2}O_{2(g)} \rightleftharpoons CaO_{(s)}; \quad \Delta H_{713°K} = -151{,}373 \text{ cal mol}^{-1}$

The heat capacity of $Ca_{(\beta)}$ is given by:

$$C_{P,Ca_{(\beta)}} = 1{\cdot}50 + 7{\cdot}74\times10^{-3}T + 2{\cdot}50\times10^5 T^{-2}$$

Taking the values for calcium oxide and oxygen as before:

$$\Delta C_P = 6{\cdot}78 - 7{\cdot}16\times10^{-3}T - 3{\cdot}96\times10^5 T^{-2}$$

The equation is set up as previously but using the new values of $\Delta H_{713°K}$ and ΔC_P, and this is evaluated for ΔH_0, which is found to be $-154{,}944$ cal mol^{-1}. Hence $\Delta H_{440-850°C} = -154{,}944 + 6{\cdot}78T - 3{\cdot}58\times10^{-3}T^2 + 3{\cdot}96\times10^5\ T^{-1}$ cal mol^{-1}. Pure β calcium melts at 850°C and, if the enthalpy change is required at a higher temperature, the latent heat of fusion of calcium must now be added and the procedure is repeated to obtain the enthalpy of formation of solid lime from liquid calcium and gaseous oxygen.

1.4 Entropy

The first law of thermodynamics indicates the energy change during a reaction, but it gives no information regarding whether or not a reaction is likely to occur under a given set of conditions; neither does it indicate the precise conditions under which heat can be converted into work, and vice versa. This information is obtained from consideration of the Second Law of Thermodynamics but, before this information can be obtained, it is first necessary to define certain conditions of a reaction.

A reaction is in *thermodynamic equilibrium* when the state functions temperature, pressure, composition, etc. are not changing with time. If the conditions are such that the system is not in mechanical, thermal and chemical equilibrium, then it is not in thermodynamic equilibrium.

A *spontaneous process* is one in which the initial properties of a system differ from the equilibrium values, and which proceeds naturally towards equilibrium without the application of an external imposed force. The process of combustion and the reaction between an acid and a base to form a salt are obvious examples. Such a reaction cannot be reversed and restored to its original state without some permanent energy change in the surroundings.

Now consider a system which is displaced from equilibrium only to an infinitesimal extent. For example, if a volume of gas enclosed in a cylinder by a frictionless piston is placed in a thermostat and the pressure on the piston is reduced below that of the gas by a very small amount, the gas expands and heat is absorbed from the thermostat to compensate for the energy loss. However, the expansion and heat transfer proceed at such a low rate that the temperature of the gas is always equal to that of the thermostat. If the pressure on the piston is now increased by a very small amount to restore the gas to the original pressure, the direction of the process is reversed and heat is returned to the reservoir. At all times during the cycle the system is virtually in thermodynamic equilibrium with the surroundings, and the state of the gas can be defined completely in terms of two variables such as the temperature and the pressure. Such a process is said to be *thermodynamically reversible* and may be regarded as the hypothetical passage of a system from one state to another through a series of equilibrium states. If, on the other hand, the volume of the gas is increased rapidly by making a marked reduction in the pressure on the piston, both temperature and pressure gradients are set up in the gas and the condition of the gas differs from that of the surroundings. The expansion is spontaneous and is not thermodynamically reversible. Similarly, if an insulated bar of material is initially at uniform temperature and a small amount of heat is supplied to raise the temperature of one end through an infinitesimal temperature interval, heat will flow along the bar towards the cooler end. If the temperature at the hot end is now lowered by withdrawing an equal quantity of heat to that supplied in the first step, the direction of heat flow is reversed and the temperature distribution along the bar can be restored reversibly to the initial condition. However, when one end of the bar is initially at a significantly higher temperature than the other, heat flows spontaneously from the hot to the cold end. The process cannot be reversed to make heat flow from the cold to the hot end and restore the initial condition.

The examples considered illustrate a general statement that spontaneous processes are not thermodynamically reversible. This concept of reversibility is the basis of the *Second Law of Thermodynamics*. The law can be stated in various forms, but the most useful for the discussion which follows is '*Heat cannot be transferred from a system at a lower temperature to one at a higher temperature without a permanent change in the surroundings*'.

With the aid of the second law it is possible to define the conditions under which the quantity ∂q can be expressed as a state function and a complete integral. This is achieved by the application of a Carnot reversible cycle to a heat engine.

THE CARNOT CYCLE

The Carnot cycle consists of two isothermal (i.e., constant temperature) and two adiabatic (i.e., change in volume and pressure without loss or gain of heat) steps, each of which is reversible. The process is most readily considered with regard to the cycling of an ideal gas, but the principle is applicable to any thermodynamic substance.

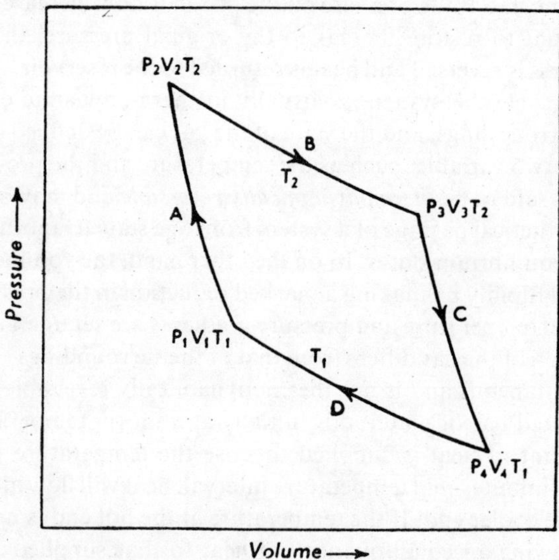

FIG. 1.1. Pressure–volume relation in the Carnot cycle.

Consider one mole of ideal gas which is contained in a cylinder closed by a frictionless piston and immersed in a large thermostat T_1, where it assumes the volume V_1 and pressure P_1 (Figure 1.1).

Step A. The temperature is increased to T_2 and the volume and pressure are changed to P_2 and V_2. From the Second Law this change can only be accomplished by the application of work or heat. In the present case it is assumed that the cylinder is removed from the thermostat and surrounded by a perfect insulator so that no heat is exchanged with the surroundings. The temperature of the gas is then raised to T_2 by an adiabatic and reversible compression.

Since no heat is exchanged with the surroundings, it follows from equation 1.1 that the change in the internal energy of the gas is equal to the work done in compression. It can be shown that the internal energy of an ideal gas at constant temperature is independent of the volume and hence, from equation 1.19, the work performed during the adiabatic compression is given by:

$$w_A = -C_V(T_1 - T_2) \qquad 1.27$$

Step B. The insulation is removed, the cylinder is placed in a large thermostat at temperature T_2 and the gas is allowed to expand isothermally and reversibly to P_3 V_3. Because the only work performed is an expansion, it is equal to PdV or, substituting for P from the equation of state for one mole of an ideal gas:

$$PV = RT \qquad 1.28$$

where R is the gas constant (1·986 cal deg^{-1}), the work performed in the second step is given by:

$$w_B = RT_2 \int_{V_2}^{V_3} \frac{dV}{V} = RT_2 \ln \frac{V_3}{V_2} \qquad 1.29$$

where ln signifies the logarithm to the base e (i.e. natural logarithm).

As the internal energy of a perfect gas at constant temperature is independent of the volume, it follows that the quantity of heat q_2 absorbed by the gas from the thermostat is equal to w_B and

$$q_2 = RT_2 \ln \frac{V_3}{V_2} \qquad 1.30$$

Step C. The cylinder is removed from the reservoir, insulated and expanded adiabatically and reversibly until the temperature of the gas is restored to T_1, the pressure and volume changing to P_4 and V_4. The work performed by the gas is given by:

$$w_C = C_V(T_1 - T_2) \qquad 1.31$$

where C_V has the same value as in stage A, since for an ideal gas it is independent of pressure or volume.

Step D. The insulation is removed, the cylinder is replaced in the thermostat at T_1 and the gas is compressed isothermally and reversibly back to P_1 and V_1. A quantity of heat q_1 is transferred to the thermostat; hence

$$w_D = q_1 = RT_1 \ln \frac{V_1}{V_4} \qquad 1.32$$

The total work performed in the cycle is the sum of the work terms in each step, but $w_A = -w_C$ and the total work is

$$w = q_1 + q_2 = RT_2 \ln \frac{V_3}{V_2} + RT_1 \ln \frac{V_1}{V_4} \qquad 1.33$$

For a very small step in the adiabatic process:

$$dU = -\partial w$$

which is equal to the first two terms in equation 1.3, because the state of a homogeneous system of constant composition is independent of the composition. Hence:

$$-\partial w = \left(\frac{\partial U}{\partial V}\right)_T dV + \left(\frac{\partial U}{\partial T}\right)_V dT \qquad 1.34$$

Equating $(\partial U/\partial T)_V$ to C_V from equation 1.19, ∂w to $[(RT)/V].dV$ from equation 1.29 and setting $(\partial U/\partial V)_T = 0$ for an ideal gas, this equation becomes:

$$C_V dT = -\frac{RT}{V} dV \qquad 1.35$$

Dividing through by RT and setting $dx/x = \ln x$, we obtain:

$$\frac{C_V}{R} \ln T = -d \ln V \qquad 1.36$$

Integrating between the limits $V_1 T_1$ and $V_2 T_2$ gives:

$$\ln \frac{V_2}{V_1} = \frac{C_V}{R} \ln \frac{T_2}{T_1} \qquad 1.37$$

Similarly
$$\ln \frac{V_3}{V_4} = \frac{C_V}{R} \ln \frac{T_2}{T_1} \qquad 1.38$$

whence
$$\frac{V_3}{V_4} = \frac{V_2}{V_1}$$

or
$$\frac{V_3}{V_2} = \frac{V_4}{V_1} \qquad 1.39$$

Substituting this equality in equation 1.33 gives:

$$q_1 + q_2 = w = R(T_2 - T_1) \ln \frac{V_3}{V_2} \qquad 1.40$$

and equating $\ln(V_3/V_2)$ to q_2/RT_2 by equation 1.30 and rearranging terms, we obtain

$$\frac{q_1 + q_2}{q_2} = \frac{T_2 - T_1}{T_2} \qquad 1.41$$

Because $q_1 + q_2$ is the work done during the cycle, the term on the left-hand side of the equation is equal to the ratio of the work done to the heat absorbed, which is usually defined as the efficiency of the engine.

The extension of the Carnot cycle to any thermodynamic substance can be examined by consideration of a coupled system in which an ideal gas engine drives a second engine containing the other substance through one complete reversible cycle, in forward and then in reverse motion, between the same temperature limits. In the forward cycle the ideal gas engine absorbs a quantity of heat q_2 at the higher temperature, performs a fixed amount of work w, and transfers a quantity of work q_1 to the lower temperature thermostat. The work

w is used to drive the second engine which absorbs heat q_1^1 at the lower temperature and releases heat q_2^1 at the higher temperature. The signs of each quantity are reversed during the return cycle. Since the net quantity of work performed in the complete cycle is zero, it is evident that the efficiency of both engines must be the same; otherwise in either the forward or the reverse cycle heat would be transferred from a low temperature to a higher temperature without any other change occurring in the system, and this is contrary to the requirements of the Second Law. It follows, therefore, that

$$\frac{q_1+q_2}{q_2} = \frac{q_1^1+q_2^1}{q_2^1} = \frac{T_2-T_1}{T_2} \qquad 1.42$$

Equation 1.41 can be arranged in the form:

$$\frac{q_1}{T_1}+\frac{q_2}{T_2} = 0 \qquad 1.43$$

and for a complete cycle:

$$\sum_{\text{cycle}} \frac{q_r}{T} = \frac{\partial q_r}{T} = 0 \qquad 1.44$$

where q_r signifies the heat content in a reversible process.

Any reversible cycle may be divided up into a number of Carnot cycles, each of which obeys equation 1.44 and, if the thermal steps of the Carnot cycles are infinitely small, the envelope of all the Carnot cycles approaches the smooth curve of the cyclic process under consideration. Thus, for a reversible process in which a system is taken from state A to state B and then back to A, the summation of all the reversible Carnot cycles can be written as:

$$\oint \frac{\partial q_r}{T} = \int_{A\ (\text{forward})}^{B} \frac{\partial q_r}{T} + \int_{B\ (\text{reverse})}^{A} \frac{\partial q_r}{T} = 0 \qquad 1.45$$

where the symbol $\oint$ signifies integration over one complete cycle. The path from A to B may be different from the B to A path, but the integration over the complete cycle is always equal to zero.

A state property or state function is defined on page 4 as one whose value is determined only by the initial and final states, and is independent of the reaction path followed. The quantity $\partial q_r/T$ fulfils these requirements and so it is an exact differential of a state property which is represented by the symbol S, i.e.

$$\frac{\partial q_r}{T} = dS \qquad 1.46$$

The change in S on going from state A to state B is given by:

$$\Delta S = S_B - S_A = \int_A^B \frac{\partial q_r}{T} \tag{1.47}$$

This property S of the system is called the *entropy* and, as its value depends on the amount of substance considered (i.e. it is an extensive property), it is usually measured in units of either cal deg^{-1} g-$atom^{-1}$ or cal deg^{-1} mol^{-1}, whichever is appropriate.

ENTROPY CHANGES IN REVERSIBLE AND NON-REVERSIBLE PROCESSES

The entropy change for a system which is taken reversibly through a complete cycle must be zero because the system is returned to its original state. That is:

$$dS = \oint \frac{\partial q_r}{T} = 0 \tag{1.48}$$

At every stage during the cycle the system is virtually in equilibrium with the surroundings (e.g., a thermostat in the Carnot cycle) and the quantity of heat gained or lost by the system is equal, but of opposite sign, to the heat change in the surroundings. Therefore, at any stage the entropy change of the system is also equal, but of opposite sign, to the entropy change of the surroundings and hence, for an isolated system, the entropy change of both the system *and* the surroundings is zero in a complete reversible cycle.

Now consider a system which is taken from state A to state B by an irreversible process. For example, the system may be heated from temperature T_1 to T_2 solely by performing work and without supplying heat. In this case $\partial q_{(\text{irreversible})}/T = 0$ but, because the entropy is a function only of the initial and final states, the entropy change for the irreversible process is identical with that for the reversible process. It follows that the entropy change is greater than $\partial q/T$ for spontaneous processes or is equal to $\partial q_r/T$ for a reversible process. That is

$$\Delta S \geqslant \int \frac{\partial q}{T} \tag{1.49}$$

and in an isolated system the entropy must either increase for a spontaneous process or remain constant at equilibrium.

From equation 1.19, for a reaction which occurs at constant volume:

$$\partial q = C_V . dT$$

and so for a reversible process at constant volume:

$$dS_V = \frac{C_V.dT}{T} = C_V.d\ln T \tag{1.50}$$

Similarly, from equation 1.20, for a reversible process at constant pressure:

$$dS_P = \frac{C_P.dT}{T} = C_P.d\ln T \tag{1.51}$$

and the entropy change accompanying the reversible heating of tne system and its surroundings from temperature T_1 to T_2 at constant pressure is given by:

$$\Delta S = S_{T_2} - S_{T_1} = \int_{T_1}^{T_2} C_P.d\ln T \tag{1.52}$$

For a small range of temperature C_P can be regarded as constant, in which case the integral is equal to $C_P\ln(T_2/T_1)$.

The entropy change for a process occurring at any temperature can thus be evaluated from the appropriate heat capacity data in similar manner to the method demonstrated (see p. 11) for the calculation of the enthalpy change. Where a phase transformation occurs in the temperature range considered, the entropy change accompanying the transformation ($\Delta S_T = L_T/T_T$) must be added or subtracted at the transformation temperature and the general form of equation 1.52 is:

$$\Delta S = S_{T_2} - S_{T_1} = \int_{T_1}^{T_T} \frac{C_P}{T}.dT \pm \frac{L_T}{T_T} + \int_{T_T}^{T_2} \frac{C_P}{T}.dT \ldots \tag{1.53}$$

For example, on page 11 the heat capacity for the reaction

$$Ca_{(\alpha)} + \tfrac{1}{2}O_{2(g)} \rightleftharpoons CaO_{(s)}$$

is given as

$$\Delta C_{P(298-713°K)} = 2{\cdot}97 - 2{\cdot}75\times10^{-3}T - 1{\cdot}46\times10^5 T^{-2}$$

Dividing by T and integrating, we obtain

$$\int \frac{\Delta C_P}{T} dT = 2{\cdot}97\ln T - 2{\cdot}75\times10^{-3}T + 0{\cdot}73\times10^5 T^{-2}$$

Hence, at 713°K:

$$\Delta S_{713°K} = \Delta S_{298°K} + \int_{298}^{713} \frac{\Delta C_P}{T}.dT$$

$$= \Delta S_{298°K} + 2{\cdot}97(\ln 713 - \ln 298) - 2{\cdot}75\times10^{-3}(713-298) + $$
$$+ 0{\cdot}73\times10^5(713^{-2} - 298^{-2})$$

which can be evaluated if $\Delta S_{298°K}$ is known.

It will be recalled that the choice of a zero level for the enthalpy of a substance is arbitrary and the statement of the Second Law gives no indication that there is a fixed zero level for entropy. However, the *Third Law of Thermodynamics* which originated from the work of W. Nernst and M. Planck, may be stated in the form: '*the entropy of any substance which is in complete internal equilibrium is zero at the absolute zero of temperature*'. The entropy of a substance at a temperature T is given accordingly by the relation:

$$S_T = S_0 + \int_{T_0}^{T_T} \frac{C_P}{T}.dT + \frac{L_T}{T_T} + \int_{T_T}^{T} \frac{C_P}{T} dT + \ldots \qquad 1.54$$

where, according to the Third Law, S_0 is zero. In practice it is very difficult to measure heat capacity data at temperatures in the vicinity of 0°K; reference (or standard) entropies are usually tabulated for a temperature of 25°C (298°K) but, in contrast to reference enthalpies, the entropies of the elements are finite at this temperature. These reference entropies would normally be used to evaluate $\Delta S_{298°K}$ in the last example. This value, together with the appropriate heat capacity data, could be used to derive a general equation for integration between the limits $T_0(=0°K)$ and the temperature considered. However, the heat capacity data used in the example are not valid at very low temperatures and the value of S_0 obtained by this method is fictitious.

Because entropy is an extensive state property the values are additive, and the entropy change for a reaction can be obtained by summation of the entropies of the reactants and products in similar manner to the method previously described to determine the heat of reaction. Thus, for the general case

$$A_xB_y + C_z \rightleftharpoons A_xC_z + B_y$$

$$\Delta S = S_{A_xC_z} + S_{B_y} - S_{A_xB_y} - S_{C_z} \qquad 1.55$$

when the entropies of all the substances taking part in the reaction are measured at the same temperature.

ENTROPY AND STATISTICAL MECHANICS

The definition of entropy in terms of the heat change during a reversible Carnot cycle is adequate for most purposes, but a clearer understanding of the physical significance of entropy and the requirement of '*complete internal equilibrium at 0°K*' in the statement of the Third Law can be obtained from statistical mechanical considerations.

It has long been recognised that the randomness of the system is increased in all spontaneous processes. For example, if two miscible liquids or gases are placed in contact it is highly improbable that the atoms or molecules of each species will remain separated, but with time they will intermix in a purely random manner. The probability that any one atom or molecule of one species will have only atoms or molecules of the same species surrounding it is very much less than the probability of both species being present in the immediate neighbourhood. This can be demonstrated simply by consideration of the following mechanical analogy. If an even number of coins arranged with heads uppermost are allowed to fall onto the floor, the probability of all the coins coming to rest with heads uppermost is very small. There is a greater probability that one of the coins will be reversed, a still greater probability that two coins are reversed, and so on up to the limit where half the coins show heads and half are reversed. The probability of any one arrangement occurring can be calculated statistically.

In similar manner the thermodynamic probability of any one thermodynamic arrangement of the component parts of a system can be calculated. If the symbol ω is used to represent this thermodynamic probability, it can be related to a quantity S by the relation

$$S = k \ln \omega \tag{1.56}$$

where k is a constant. It was shown by Boltzmann and Planck that this quantity S varies in like manner to the quantity called the entropy when k is equal to $1{\cdot}3808 \times 10^{-16}$ ergs deg^{-1}, which is equal to the ratio of the gas constant (R) to Avogadro's Number ($N_0 = 6{\cdot}023 \times 10^{23}$). From this equation it is evident that any increase in the thermodynamic probability or randomness of the system is accompanied by an increase in the entropy and the Second Law of Thermodynamics can be re-stated accordingly in the form '*the entropy increases when a spontaneous change occurs in an isolated system; at equilibrium the entropy is a maximum*'.

In order to satisfy the Third Law and make the entropy zero at 0°K the system must be in the state of lowest probability. The randomness of a pure substance is increased as it is changed from solid to liquid or from liquid to gas, and it follows that the state of lowest probability is the perfect crystalline substance. The state of lowest probability for equiatomic proportions of elements A and B is the condition where each A atom is surrounded only by B atoms and vice versa; any deviation from this condition introduces some degree of randomness and the entropy is increased.

1.5 Free Energy

Having defined the heat content of a system in terms of a single-valued function of the state of the system, for a reversible process in which the system performs no work other than that done against pressure, equation 1.1 can now be written in the form:

$$dU = TdS - PdV \qquad 1.57$$

and if the volume is held constant:

$$dU - TdS = 0 \qquad 1.58$$

The state of the system is then defined in terms of the state variables, internal energy, entropy and temperature.

HELMHOLTZ FUNCTION

It is convenient to introduce a new function A which is defined by the relation:

$$A = U - TS \qquad 1.59$$

Because it is defined completely in terms of state variables, like the internal energy it is also a characteristic function of the state of the system.

Differentiating equation 1.59 we have:

$$dA = dU - TdS - SdT$$

and on adding equation 1.57, we obtain the equation:

$$dA = -SdT - PdV \qquad 1.60$$

If the reaction proceeds reversibly at constant temperature:

$$dA = -PdV = -\partial w \qquad 1.61$$

which shows that the decrease in A, in a process which occurs at constant temperature, is equal to the reversible work performed by the system. The function A is therefore a measure of the maximum amount of energy which can be converted into useful work, and accordingly it is called the *maximum work function*. It is also called the *Helmholtz function* or the *Helmholtz free energy*.

GIBBS FREE ENERGY

The Helmholtz function is defined in terms of the changes in temperature and volume during the reaction, but experimentally and in industrial practice it is commonly found that the temperature and pressure are more readily controlled. These conditions are met by using the *Gibbs free energy G***, which is defined by the relation:

$$G = U+PV-TS \tag{1.62}$$

Since G is defined completely in terms of state variables, it follows that the Gibbs free energy is also a characteristic state property.

The Gibbs free energy can be related to the enthalpy by substituting equation 1.9 into 1.62 to obtain:

$$G = H-TS \tag{1.63}$$

and to the Helmholtz function by substituting equation 1.59

$$G = A+PV \tag{1.64}$$

Differentiating equation 1.62 we obtain:

$$dG = dU+P\,dV+V\,dP-T\,dS-S\,dT \tag{1.65}$$

and adding equation 1.57 gives:

$$dG = V\,dP-S\,dT \tag{1.66}$$

where the derivative of G is expressed completely in terms of P and T. Hence for a reversible process at constant pressure and temperature (i.e. at equilibrium) $dG = 0$.

Since for an irreversible process the change in the entropy is greater than the sum of all the q/T terms for the isothermal steps, the entropy change is greater than the sum of $(dU+P\,dV)/T$, i.e.

$$T\,dS > dU+P\,dV$$

If this relation is combined with equation 1.65, for irreversibility we obtain the relation:

$$dG < V\,dP-S\,dT$$

and so, for a reaction at constant temperature and pressure,

$$dG \leqslant 0 \tag{1.67}$$

which signifies that the change in free energy is either zero for a reversible process or less than zero (i.e. a negative quantity) for a spontaneous process. This means that all spontaneous processes which occur at constant pressure and temperature are accompanied by a decrease in the free energy of the system, and at equilibrium the free energy must

* The symbols A and G are used in this text to represent the Helmholtz function and Gibbs free energy respectively. Many texts employ the symbol F for the latter, particularly those published in the U.S.A., but the same symbol is [illegible]

therefore be a minimum. It follows, also, that a reaction cannot proceed spontaneously if the accompanying change in free energy is positive.

By similar reasoning it can be shown that a spontaneous reaction which occurs isothermally and reversibly at constant volume is accompanied by a decrease in the Helmholtz function and that this function is a minimum at equilibrium.

The Helmholtz function and Gibbs free energy can be used, therefore, to determine whether or not a reaction can proceed spontaneously under any chosen conditions which are imposed upon the system and also to determine the conditions under which the system is at equilibrium. However, a negative free-energy value does not necessarily mean that a reaction *will* proceed at a measurable rate under a given set of conditions, but indicates only that the reaction is possible. There are many cases where the free-energy change is negative but the reaction proceeds so slowly that the rate of change in the system cannot readily be measured. Whether or not the reaction will proceed to the equilibrium condition in a reasonable period of time depends on the kinetics of the reaction, which cannot be determined from thermodynamic considerations.

This can be illustrated by the following mechanical analogy. If two tanks containing water at different levels are connected by a pipe, water will flow through the pipe until a condition of equilibrium is established and the water in both tanks is at the same level. Provided the pipe is unobstructed, equilibrium is reached in a finite time. However, if a constriction is placed in the tube, the time taken to reach equilibrium is increased as the effective bore of the tube is reduced and tends towards infinity as the size of the obstruction approaches the diameter of the tube. The constriction does not alter the equilibrium but only the rate of approach to it. In order to determine when the water levels in the two tanks will cease to change it is necessary to know both the equilibrium condition and the rate of flow of water through the pipe. Similarly, to determine when the concentrations of reactants and products will cease to change in a chemical reaction a knowledge is required of both the equilibrium condition and the reaction rate. The free energy of a reaction is a measure only of the first of these.

The Helmholtz function and Gibbs free energy are extensive properties of the system, which are usually measured in units of cal g-atom^{-1} or cal mol^{-1}. Because they are extensive properties the values are additive, i.e. for the reaction

$$A+B \rightleftharpoons AB$$

the change in the Gibbs free energy is given by:

$$\begin{aligned} \Delta G &= G_{AB}-G_A-G_B \\ &= (H_{AB}-H_A-H_B)-T(S_{AB}-S_A-S_B) \\ &= \Delta H - T\Delta S \end{aligned} \quad 1.68$$

VARIATION OF FREE ENERGY WITH TEMPERATURE

Since ΔH and ΔS values are usually tabulated for room temperature, the free energy of a substance or the free energy of formation or reaction can be calculated directly at that temperature. For any other temperature the variation of ΔH and ΔS with temperature must be considered and the free energy at any temperature T can be obtained as the sum of the values of equations 1.25 and 1.54 at the same temperature, i.e.

$$\Delta G_T = \Delta H_0 + \int_0^T \Delta C_P . dT - \left[T\Delta S_0 + \int_0^T \frac{\Delta C_P}{T} . dT \right] \quad 1.69$$

Alternatively ΔH_0 and ΔS_0 may be replaced by the values of these functions at 298°K and the integration carried out over the temperature range 298° to T°K. Where one or more phase transformations occur in the temperature range considered, the expressions for the enthalpy and entropy must be expanded to include the appropriate enthalpies and entropies of transformation as indicated in the earlier sections.

When the heat capacity equations are integrated and like terms are added, the free energy as a function of temperature is given as a constant plus a number of terms which are raised to various powers of the temperature. However, it is rarely found that the accuracy of the data, and particularly the accuracy of the heats of formation at room temperature, justify the use of the extended equation. If the free-energy/temperature relation shows appreciable curvature and if in addition the reference enthalpies, entropies and heat capacities are known precisely for each of the components in the reaction, the variation of the free energy with temperature can usually be expressed with sufficient accuracy by use of an equation of the form:

$$\Delta G_T = \Delta H + bT \log T - T\Delta S \quad 1.70$$

where b is a constant and term $bT \log T$ represents the non-linear $\Delta G/T$ relation. The ΔH and ΔS values may differ from the reference values at 298°K and are obtained empirically by fitting the curve represented by equation 1.70 to the values obtained by solving equation 1.69 over the appropriate temperature range.

For many reactions the error introduced by assuming a non-temperature dependence of the reference enthalpy and entropy changes is well within the limits of the accuracy of the experimental evaluations at high temperatures, and the free energy as a function of temperature is adequately represented by

$$\Delta G_T = \Delta H_{298°K} - T\Delta S_{298°K} \quad 1.71$$

The heats and entropies of transformation are usually too large to be omitted and equations 1.70 or 1.71 are a reasonable approximation only up to the lowest temperature at which one of the substances in the reaction undergoes a change in atomic arrangement (allotropic or phase change) in the solid, or fusion or evaporation. If equation 1.69 is solved for the variation of the free energy with temperature between two adjacent transformation temperatures, it is generally found that the variation can again be expressed with sufficient accuracy for most practical purposes by an equation similar to 1.70 or 1.71, but which is now applicable only over the temperature range between the two transformation points. Thus the free-energy/temperature relation for a substance or for a reaction can often be represented adequately by a family of equations similar to the following series which has been suggested [4] for the free energy of formation of calcium oxide:

$$Ca_{(s)} + \tfrac{1}{2}O_{2(g)} \rightleftharpoons CaO_{(s)}$$

$$\Delta G_{(298-1,123°K)} = -151{,}325 + 23{\cdot}66T \text{ cal mol}^{-1} \qquad 1.72$$

Pure calcium melts at 1,123°K; above this temperature the reaction is

$$Ca_{(l)} + \tfrac{1}{2}O_{2(g)} \rightleftharpoons CaO_{(s)}$$

and the free-energy change is

$$\Delta G_{(1,123-1,760°K)} = -153{,}550 + 25{\cdot}64T \text{ cal mol}^{-1} \qquad 1.73$$

The boiling point of calcium is 1,760°K

$$Ca_{(g)} + \tfrac{1}{2}O_{2(g)} \rightleftharpoons CaO_{(s)}$$

and above this temperature the free-energy data are fitted by the relation:

$$\Delta G_{(1,760-2,500°K)} = -190{,}100 + 46{\cdot}62T \text{ cal mol}^{-1} \qquad 1.74$$

Calcium oxide melts at 2,870°K, but the accuracy of the available heat capacity data does not justify the use of the last equation beyond about 2,500°K.

The first equation in the above series is recommended for use over the temperature range 298–1,123°K. Pure calcium transforms from the α to the β form at 713°K, but the worked example in section 1.3 shows that this transformation is accompanied by a change in the enthalpy of only 250 calories. The enthalpy of formation of calcium oxide from oxygen and pure β-calcium at 1,123°K differs by only a few calories from the standard enthalpy of formation from α-calcium at room temperature, which is considerably less than the uncertainty in the standard enthalpy of the reaction. In this case the error incurred by neglecting the α-β

transformation is negligible. However, comparison of the recommended equations shows that there is a more marked change in the enthalpy accompanying the change from solid to liquid calcium, whilst the change from liquid to gaseous calcium is associated with a large increase in the enthalpy. The corresponding entropy changes are even more pronounced and reflect the increase in the number of possible thermodynamic arrangements (cf. equation 1.56) of the calcium atoms as the physical state is changed from solid to liquid and from liquid to gas.

The free energy of formation of calcium oxide as calculated from equation 1.72 at 1,123°K is equal to that calculated from equation 1.73 at the same temperature. This follows from the fact that the pure α and β forms of calcium can co-exist in equilibrium at 1,123°K and the free energy of transformation is zero at the transformation temperature.

1.6 Oxygen-Potential Diagrams

By comparison of free-energy equations similar to those given in equations 1.71 to 1.73 for the formation of various pure oxides from the pure elements, it is possible to decide the relative order of stability of the oxides at any temperature in terms of the free energies of formation at that temperature. However, this information is of limited use because the free-energy data are obtained by use of reference entropies and enthalpies; this restricts consideration, using the treatment which has been developed up to present, to pure metals and their oxides in the stable physical state (solid, liquid or gas) at the temperature considered and to gases at one atmosphere pressure. The treatment can, however, be extended simply in the following manner to allow the determination from the free-energy data of the actual partial pressure of oxygen (p_{O_2}) in equilibrium with a pure metal and its pure oxide as a function of the temperature.

For a reaction between a pure solid metal M and gaseous oxygen to form the pure solid oxide MO:

$$2M_{(s)} + O_{2(g)} \rightleftharpoons 2MO_{(s)} \qquad 1.75$$

the change in the free energy is given by:

$$\Delta G = 2G_{MO} - 2G_{M} - G_{O_2} \qquad 1.76$$

At any temperature the three phases can co-exist (i.e., are in equilibrium) at only one partial pressure of oxygen. At higher partial pressures the metal is converted to the oxide and, conversely, at lower partial pressures the oxide is reduced to the metal. When equilibrium is obtained by holding the metal and oxide at the partial pressure of oxygen appropriate to the temperature then, by definition, $\Delta G = 0$.

Because, in general, the equilibrium partial pressure of oxygen is greater or less than one atmosphere the free energies of the products and reactants differ from those in the standard states. We designate by a superscript zero (i.e. G^0) the free energies of the pure solids at one atmosphere pressure and of a gas at one atmosphere pressure. For the reaction where all the components are in their standard states the free-energy change is then written as:

$$\Delta G^0 = 2G^0_{MO} - 2G^0_M - G^0_{O_2} \qquad 1.77$$

ΔG^0 is called the *standard free-energy change* for the reaction at the temperature considered, and it is equal to zero only in the exceptional case where the standard oxygen pressure is in equilibrium with the two condensed phases.

Subtracting equation 1.77 from 1.76 since $\Delta G = 0$, we obtain for the free-energy change accompanying the change in the oxygen partial pressure from one atmosphere to the equilibrium partial pressure:

$$-\Delta G^0 = 2(G_{MO} - G^0_{MO}) - 2(G_M - G^0_M) - (G_{O_2} - G^0_{O_2}) \qquad 1.78$$

If this change is effected isothermally and reversibly then, from equation 1.66, $dG = VdP$. The volume of a solid is not changed significantly by small changes in the pressure; hence the first two terms on the right-hand side of equation 1.78 can be set at zero and the equation simplifies to:

$$-\Delta G^0 = (G_{O_2} - G^0_{O_2}) = -\int_{p_{O_2}=1}^{p_{O_2}(\text{equilibrium})} (VdP) \qquad 1.79$$

The liquefaction temperature of oxygen gas at one atmosphere pressure is $-182{\cdot}97°C$ ($90{\cdot}19°K$), and it is reasonable to assume that the gas approximates to the behaviour of an ideal gas at elevated temperatures, in which case it is permissible to substitute RT/P for V and so obtain:

$$\Delta G^0 = RT \int_{p_{O_2}=1}^{p_{O_2}(\text{equilibrium})} \frac{dp}{p} = RT \int_{p_{O_2}=1}^{p_{O_2}(\text{equilibrium})} d\ln p \qquad 1.80$$

For a small pressure change:

$$\Delta G^0 = RT \ln \frac{p_{O_2}(\text{equilibrium})}{p_{O_2}(1\text{ atmosphere})}$$

$$= RT \ln p_{O_2}(\text{equilibrium}) \qquad 1.81$$

since $\ln 1 = 0$

and $$\Delta G^0 = 4{\cdot}575T \log_{10} p_{O_2(\text{equilibrium})} \qquad 1.82$$

where the constant 4·575 is the product of the gas constant in calories and the factor for the conversion of logarithms base e to logarithms base 10. The quantity $RT \ln p_{O_2(\text{equilibrium})}$ is often called the *oxygen potential* of the system.

We have now defined the standard free energy G^0 for a solid or a gas. In general terms the standard free energy of any substance can be defined as the free energy when the substance is present in the standard state, which is usually selected as the pure solid, liquid or gas, whichever of these forms is stable at the temperature considered and at one atmosphere pressure. The standard free-energy change for a reaction can be evaluated from the free energies of the components of the reaction when each of these is in the appropriate standard state. In the same way the *standard enthalpy* H^0 and *standard entropy* S^0 can be defined as the values of these quantities for a substance which is in the standard physical state at the temperature considered. (It is important to distinguish the reference values of these quantities, which are usually tabulated for a temperature of 298·16°K and are designated $\Delta H_{298°K}$ etc., from those which apply at the temperature considered and are designated ΔH_T^0 etc.) The standard enthalpy and entropy changes are related to the standard free-energy change for a reaction by the equation:

$$\Delta G^0 = \Delta H^0 - T\Delta S^0 \qquad 1.83$$

The choice of the standard state is arbitrary and in Chapter 2 conditions are discussed under which it is convenient to select a standard state which differs from those considered here. This does not affect the application of the thermodynamic data except that the values of G^0, H^0 and S^0 for a substance (and hence for a reaction) are dependent on the way in which the standard states are defined.

GRAPHICAL REPRESENTATION OF FREE-ENERGY DATA

When the standard free-energy–temperature relation for a metal–oxygen–metal oxide system can be represented by a family of linear equations similar to equations 1.72 to 1.74, the graphical representation consists of a continuous line which changes slope at each transition temperature for the metal or the oxide and which is linear between adjacent transition points. The intercept of the line upon the temperature axis at 298°K and the slope of each linear portion is dependent respectively on the standard enthalpy change at 298°K and on the standard entropy change for the phases present in the relevant temperature range but, in consequence of the linear approximations, they are not necessarily equal to these quantities.

This method of plotting was used by Ellingham [5] and by Richardson and others [4, 6, 7, 8] to summarise equilibrium data for metal–oxygen–metal-oxide systems and also for metal–sulphur–metal sulphide, metal–carbon–metal carbide and similar systems. Figure 1.2 illustrates the data for the oxide systems of interest in ferrous extraction, the data being mainly those given by Richardson, with some amendments and additions. Each line represents the change in the oxygen potential or standard free energy when one mole of oxygen at one atmosphere pressure reacts with a pure element to form the pure oxide as shown on the diagram. The standard free-energy changes accompanying the oxidation of ferrous oxide (FeO) to magnetite (Fe_3O_4) and of magnetite to ferric oxide (Fe_2O_3) are also included.

All the lines for the metal-oxide systems have a positive slope, the initial slopes of which are approximately constant, since one mole of gaseous oxygen is consumed in the formation of the oxide and the contribution of the oxygen gas to the overall entropy change is very large in relation to the entropy difference between a solid metal and its solid oxide. The standard entropy change for a reaction such as that represented by equation 1.75 is given by the equation:

$$\Delta S^0 = (2S^0_{MO} - 2S^0_M \simeq 0) - S^0_{O_2} \qquad 1.84$$

and as the slope is equal to $-\Delta S^0$, the slope must be positive.

In contrast to this are the standard free-energy—temperature relations for the reactions between solid carbon and one mole of oxygen to form carbon dioxide or carbon monoxide:

$$C_{(s)} + O_{2(g)} \rightleftharpoons CO_{2(g)} \qquad 1.85$$

and

$$2C_{(s)} + O_{2(g)} \rightleftharpoons 2CO_{(g)} \qquad 1.86$$

In the first reaction one mole of oxygen is consumed to form one mole of carbon dioxide and, because the entropy contributions from the two gaseous components are almost equal, the plot is approximately horizontal, (i.e., $-\Delta S^0 = 0$). Two molecules of carbon monoxide gas are produced from one mole of oxygen by the second reaction, so the overall entropy change is positive and the slope of the plot is negative.

When a metal melts its entropy is increased, with the result that the total entropy of the reactants is raised, the entropy change for the reaction becomes more negative and the slope is increased. Conversely, when the oxide melts the overall entropy change is decreased and the slope is

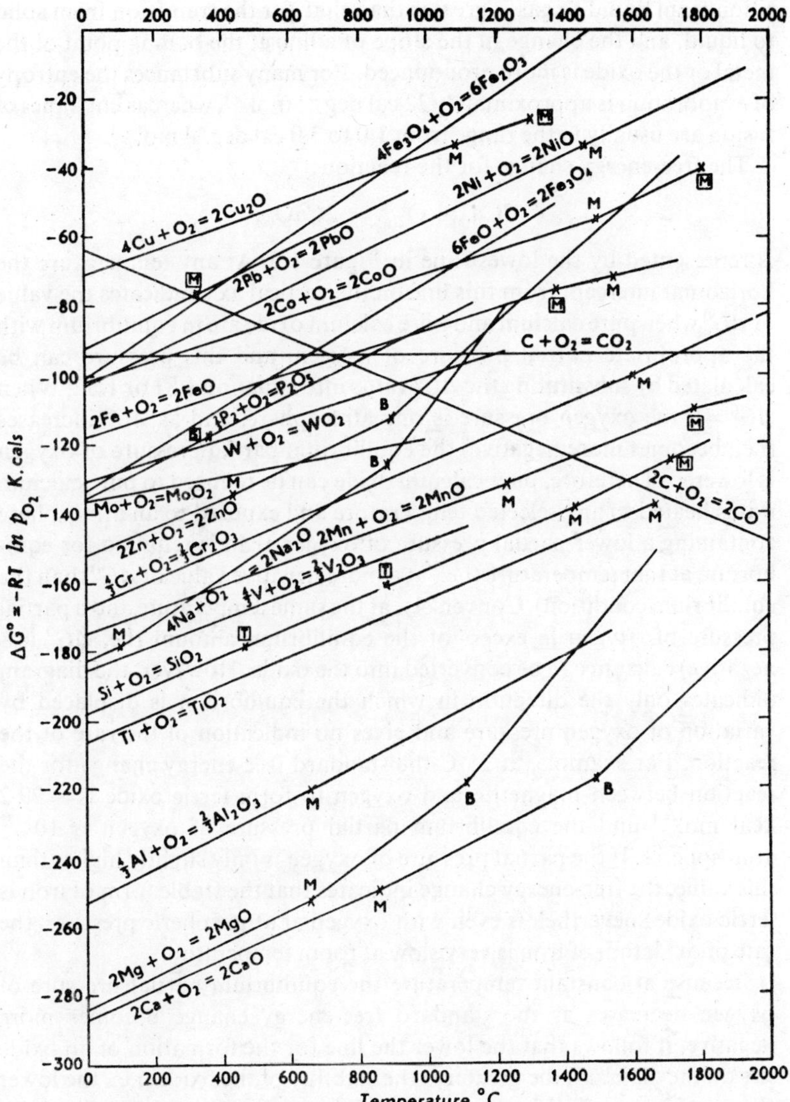

FIG. 1.2. Standard free energies of formation of metal oxides. (M, B and T are melting, boiling and transformation points respectively. Open letters refer to the metal and boxed letters to the oxide.)

lowered correspondingly. The entropy change accompanying the transition from liquid to gas is greater than that for the transition from solid to liquid, and the change in the slope of a line at the boiling point of the metal or the oxide is more pronounced. For many substances the entropy of evaporation is approximately 22 cal deg^{-1} mol^{-1}, whereas entropies of fusion are usually in the range from 1·0 to 3·0 cal deg^{-1} mol^{-1}.

The free-energy change for the reaction:

$$2Ca_{(s)} + O_{2(g)} \rightleftharpoons 2CaO_{(s)}$$

is represented by the lowest line in Figure 1.2. At any temperature the horizontal intercept from this line on the vertical axis indicates the value of ΔG^0 when pure calcium and pure calcium oxide are in equilibrium with the appropriate partial pressure of oxygen, and this pressure can be calculated by substituting the ΔG^0 value into equation 1.81 or 1.82. When $\Delta G^0 = 0$ the oxygen pressure is one atmosphere, and as ΔG^0 increases (i.e., becomes more negative) the equilibrium partial pressure of oxygen is lowered. Therefore, pure calcium oxide can be reduced to pure calcium if it is heated to any selected temperature and exposed to an atmosphere containing a lower partial pressure of oxygen than is required for equilibrium at that temperature (i.e., at a more negative value of ΔG^0 than the equilibrium condition). Conversely, at the same temperature and a partial pressure of oxygen in excess of the equilibrium amount (i.e. ΔG^0 less negative) calcium can be converted into the oxide. However, the diagram indicates only the direction in which the equilibrium is displaced by variation of oxygen pressure and gives no indication of the rate of the reaction. For example, at 25°C the standard free-energy change for the reaction between magnetite and oxygen to form ferric oxide is −99·2 kcal mol^{-1} and the equilibrium partial pressure of oxygen is 10^{-73} atmospheres. If the partial pressure of oxygen is only slightly higher than this value, the free-energy change indicates that the stable form of iron is ferric oxide; nevertheless even with oxygen at atmospheric pressure, the rate of oxidation of iron is very slow at room temperature.

Because at constant temperature the equilibrium partial pressure of oxygen decreases as the standard free-energy change becomes more negative, it follows that the lower the line for the formation of an oxide lies on the diagram, the greater is the stability of the oxide (i.e., the lower the partial pressure of oxygen at which it can be reduced to the metal). Hence a pure element is able to reduce the oxide of any other element whose ΔG^0 curve lies higher on the diagram. For example, the C–CO and C–CO_2 curves lie below the curve for the Pb–PbO reaction and lead oxide is unstable in the presence of carbon at all temperatures, whereas carbon reacts with magnesia to form the metal and carbon monoxide only at temperatures well above the boiling point of magnesium. Similarly, if pure chromic oxide is reduced to pure chromium metal any oxides of zinc, molybdenum, tungsten, iron, cobalt, nickel, lead and copper which are present will be reduced simultaneously. For a given temperature and partial pressure of oxygen, all oxides lying below the corresponding point on the diagram are stable and cannot be reduced.

EFFECT OF VARYING COMPOSITION

The equilibrium oxygen potentials may be changed appreciably when the oxides are not pure. Figure 1.3, which is taken from the data of Richardson, Jeffes and Withers [7] with some modifications, shows the standard free energies of formation of some silicate compounds. Comparison with Figure 1.2 shows that reduction to the metallic state occurs at a much lower partial pressure of oxygen than is required for the reduction of the corresponding metal oxide. In general, when the oxides or the elements are present as compounds or in solutions, they are not in their standard states as previously defined and the dissociation pressures are no longer indicated by the lines in Figure 1.2. This feature will be considered later (see p. 52).

Some oxides exist over a fairly wide composition range and the stoichiometric ratio may not be included in this range. For example, the lowest oxide of iron, wüstite, contains less iron than corresponds to the stoichiometric composition of ferrous oxide (FeO). Moreover, the composition of wüstite is not constant but varies over a fairly wide range. In equilibrium with pure iron the ratio of iron to oxygen atoms is close to unity and varies from 0·945 at 560°C to 0·950 at 1,371°C, but the range of compositions of wüstite in equilibrium with the next lowest oxide of iron, magnetite, is very much wider and varies from 0·945 at 560°C to 0·833 at 1,376°C [9] (Figure 6.5). The oxygen potentials for the formation of the oxide may be quoted with reference to:

(*a*) the stoichiometric composition,
(*b*) the actual phase at one or other of the saturation surfaces,
(*c*) some arbitrary composition within the range of existence of the actual phase.

and it is often necessary to refer back to the original data to determine which of these conditions is applicable.

It will be noticed that the iron/oxygen atom ratio of wüstite in equilibrium with iron at 560°C is equal to that in equilibrium with magnetite at the same temperature. This is because the three phases can co-exist at 560°C; at lower temperatures wüstite is unstable and decomposes into a mixture of iron and magnetite. This is shown in Figure 1.2 by the intersection of the Fe–FeO and the FeO–Fe_3O_4 lines at 560°C.

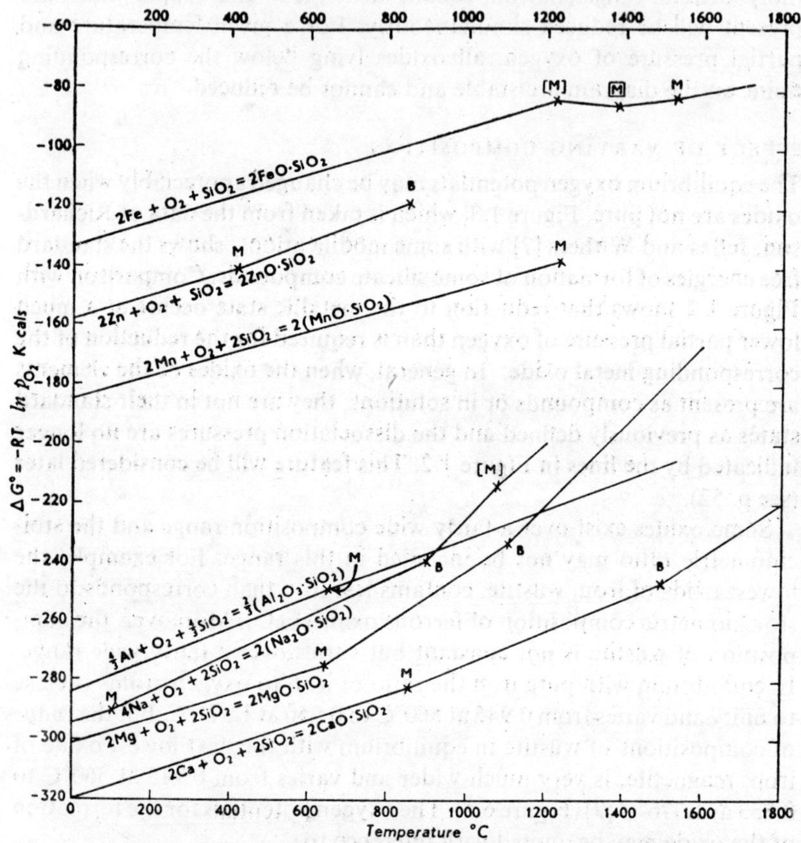

FIG. 1.3. Standard free energies of formation of metal silicates.

M = melting point of metal; $\boxed{M}$ = melting point of oxide

B = boiling point of metal; [M] = melting point of silicate

DETERMINATION OF VERY LOW OXYGEN PRESSURES

The oxygen potentials of interest in iron- and steel-making processes are usually too low for accurate measurements of the partial pressure of oxygen. The pressures can be measured indirectly, however, by observation of the equilibrium between the reaction considered and a reaction between gases in which oxygen is one of the participants. For example, if the reactions

$$2M_{(s)}+O_{2(g)} \rightleftharpoons 2MO_{(s)}$$

where M is a metal, and

$$2C_{(s)}+O_{2(g)} \rightleftharpoons 2CO_{(g}$$

are brought to equilibrium at the same partial pressure of oxygen, this pressure can be determined from the free-energy change for the gas reaction, the temperature, and the partial pressures of the carbon monoxide and carbon dioxide in the equilibrium gas mixture.

The equilibrium partial pressures of the gases in the C–CO reaction can be related to the standard free-energy change for the reaction by a similar method to that described at the beginning of this section. Thus, subtracting the standard free-energy change for reaction when all the gases are present at one atmosphere pressure from the free-energy change at equilibrium, we obtain

$$-\Delta G^0 = 2(G_{CO_2}-G^0_{CO_2})-2(G_{CO}-G^0_{CO})-(G_{O_2}-G^0_{O_2}) \qquad 1.87$$

Assuming that carbon monoxide, carbon dioxide and oxygen approximate to ideal gases at elevated temperatures, the appropriate form of equation 1.81 can be substituted for each of the bracketed terms. Since the pressure of the gas is one atmosphere in the standard state, the logarithm of the standard pressure is zero and the standard free-energy change is given by

$$-\Delta G^0 = 2RT\ln p_{CO_2}-2RT\ln p_{CO}-RT\ln p_{O_2} \qquad 1.88$$

$$= RT\ln\left[\frac{p^2_{CO_2}}{p^2_{CO}\cdot p_{O_2}}\right] \qquad 1.89$$

Because ΔG^0 has a precise numerical value at any temperature, it follows that the quantity in brackets is dependent only on the temperature and is independent of the total pressure, concentration, etc. In order to determine the equilibrium partial pressure of oxygen at any temperature it is more convenient to re-arrange the terms to obtain the equations:

$$RT\ln p_{O_2} = \Delta G^0+2RT\ln\frac{p_{CO_2}}{p_{CO}} \qquad 1.90$$

or

$$RT\ln p_{O_2} = \Delta G^0-2RT\ln\frac{p_{CO}}{p_{CO_2}} \qquad 1.91$$

The application of this procedure is illustrated by the following example, in which the ratio p_{CO}/p_{CO_2} in equilibrium with pure iron, pure (stoichiometric) ferrous oxide and oxygen gas at 1,027°C (1,300°K) is calculated.

For the reaction:

$$2Fe_{(s)} + O_{2(g)} \rightleftharpoons 2FeO_{(l)} \qquad 1.92$$

the standard free-energy change is [1]

$$\Delta G^0_{(298-1,642°K)} = -124,100 + 29{\cdot}9T \text{ cal mol}^{-1}$$

and at 1,300°K

$$\Delta G^0 = -85,230 \text{ cal mol}^{-1}$$

but

$$\Delta G^0 = RT \ln p_{O_2}$$

therefore

$$p_{O_2} = \text{antilog}_{10}\left(\frac{-85,230}{4{\cdot}575 \times 1,300}\right)$$

$$= 4{\cdot}8 \times 10^{-15} \text{ atmospheres.}$$

For the reaction

$$2CO_{(g)} + O_{2(g)} \rightleftharpoons 2CO_{2(g)} \qquad 1.93$$

the standard free energy of reaction is [1]

$$\Delta G^0_{(298-2,500°K)} = -135,000 + 41{\cdot}50T \text{ cal mol}^{-1}$$

and at 1,300°K

$$\Delta G^0 = -81,050 \text{ cal mole}^{-1}$$

Substituting these values for $RT \ln p_{O_2}$ and ΔG^0 into equation 1.91 gives

$$-85,230 = -81,050 - 9{\cdot}15T \log \frac{p_{CO}}{p_{CO_2}}$$

and

$$\frac{p_{CO}}{p_{CO_2}} = 2{\cdot}34$$

This result can be obtained by a more simple procedure. Because the free energies are additive, the standard free-energy change for equation 1.93 can be subtracted from that for equation 1.92 and the resultant divided by 2 to obtain:

$$Fe + CO_2 \leftrightharpoons FeO + CO \quad \Delta G^0 = 5,500 - 5{\cdot}8T$$

and since for this reaction:

$$\Delta G^0 = RT \ln \frac{p_{CO}}{p_{CO_2}}$$

by substituting the value of ΔG^0 and the temperature we obtain:

$$\frac{p_{CO}}{p_{CO_2}} = 2{\cdot}34$$

as before.

The partial pressure of oxygen in equilibrium with a pure metal and its pure oxide can be calculated directly from the standard free-energy change for the metal–oxygen–metal oxide system. However, it is often necessary to evaluate, for example, the partial pressure of oxygen when the metal and/or the oxide are not pure and hence are not present in the standard states as previously defined, or it may be necessary to determine the partial pressure of oxygen in equilibrium with oxygen dissolved in a metal at lower concentrations than are in equilibrium with the metal oxide. The above calculation can then be applied in reverse and the equilibrium partial pressure of oxygen can be calculated from a knowledge of the ratio p_{CO}/p_{CO_2} in equilibrium with the oxygen in the metal at the temperature under consideration and the standard free-energy change for the gas reaction at that temperature.

1.7 Partial Molar Quantities

The discussion so far has been restricted to consideration of the thermodynamic properties of pure substances, but iron- and steel-making processes are usually concerned with a metal containing varying amounts of elements such as carbon, manganese, silicon, phosphorus and sulphur, and with a liquid solution containing a number of oxides and other compounds. The various components of these solutions are not present in their standard states as previously defined, and their thermodynamic properties in the solution differ from those in the standard states. In order to measure the free energy, entropy, enthalpy, etc., of such a solution in terms of the thermodynamic properties of the components of the solution, it is necessary to introduce some additional thermodynamic factors.

Consider a solution containing n_A moles of component A, n_B moles of B, n_C moles of C, etc. According to equation 1.2 any thermodynamic extensive property can be defined completely in terms of the state variables pressure, temperature and composition and, for example, in terms of equation 1.3 the derivative of the total free energy (G') for $n_A+n_B+n_C\ldots$ moles is completely defined by the relation

$$dG' = \left(\frac{\partial G'}{\partial P}\right)_{Tn_An_Bn_C\ldots}.dP + \left(\frac{\partial G'}{\partial T}\right)_{Pn_An_Bn_C\ldots}.dT + \left(\frac{\partial G'}{\partial n_A}\right)_{PTn_Bn_C\ldots}.dn_A + \left(\frac{\partial G'}{\partial n_B}\right)_{PTn_An_C\ldots}.dn_B + \left(\frac{\partial G'}{\partial n_C}\right)_{PTn_An_B\ldots}.dn_C + \ldots \qquad 1.94$$

The partial derivatives

$$\left(\frac{\partial G'}{\partial n_A}\right)_{PTn_Bn_C\ldots}, \quad \left(\frac{\partial G'}{\partial n_B}\right)_{PTn_An_C\ldots}$$

are called the *partial molar free energies* of components A and B in the solution and are represented by the symbols $\bar{G}_A$, $\bar{G}_B$. At constant temperature and pressure dT and dP are zero, so that

$$dG'_{TP} = \bar{G}_A\,dn_A + \bar{G}_B\,dn_B + \bar{G}_C\,dn_C \qquad 1.95$$

The partial molar free energy of component A is a measure of the rate of change of the total free energy of the system when a very small quantity (dn_A) of A is added at constant temperature and pressure to a large quantity of the solution, such that the concentrations of the other components remain virtually unchanged. In some textbooks this quantity is called the *chemical potential* and is represented by the symbol μ.

It is evident from the definition that the value of the partial molar free energy is dependent only on the composition and not on the amount of the substance. If equation 1.95 is integrated for a system at constant temperature, pressure and composition, we obtain:

$$(n_A+n_B+n_C+\ldots)\,G'_{TP} = n_A\bar{G}_{A_{TPn_Bn_C\ldots}} + n_B\bar{G}_{B_{TPn_An_C\ldots}} + n_C\bar{G}_{C_{TPn_An_B\ldots}} \qquad 1.96$$

That is, the total free energy of a system of constant composition is equal to the sum of the products of the partial molar free energy and the number of moles of each component. Dividing throughout by

$$(n_A+n_B+n_C\ldots)$$

and remembering that the mole fraction N_A of component A is equal to

$$\frac{n_A}{n_A+n_B+n_C\ldots}, \text{ etc,}$$

we obtain

$$G_{TPN} = N_A\bar{G}_{A_{TPN}} + N_B\bar{G}_{B_{TPN}} + N_C\bar{G}_{C_{TPN}} \qquad 1.97$$

where G is the total free energy of one mole of the mixture and is called the *integral molar free energy of the system.*

The discussion has been restricted so far to consideration of partial molar free energies, but any characteristic thermodynamic function of the state of a system whose differential can be written in similar form to equation 1.95 can be expressed as a partial molar property. For example, the partial molar enthalpy and entropy of component A in a solution are related to the partial molar free energy by differentiating the definitional relation, $G=H-TS$, by n_A, whilst holding the temperature, pressure and concentrations of other components constant, to obtain

$$\left(\frac{\partial G}{\partial n_A}\right)_{TPn_Bn_C\cdots} = \left(\frac{\partial H}{\partial n_A}\right)_{TPn_Bn_C} - T\left(\frac{\partial S}{\partial n_A}\right)_{TPn_Bn_C} \qquad 1.98$$

where $\left(\frac{\partial H}{\partial n_A}\right)_{TPn_Bn_C\cdots}$ and $\left(\frac{\partial S}{\partial n_A}\right)_{TPn_Bn_C\cdots}$ are the partial molar enthalpy and partial molar entropy respectively of component A in a solution of fixed composition.

GRAPHICAL METHOD FOR OBTAINING PARTIAL MOLAR QUANTITIES

The values of a partial molar quantity for the component substances of a binary (i.e., two component) mixture can be obtained readily by a graphical method when the corresponding values of the integral molar property are known over the entire composition range. For example, the variation of the integral molar free energy with composition is shown in Figure 1.4*a* for a system which forms a complete range of either solid or liquid solutions at the temperature considered. At any concentration x the partial molar free energies of the substances A and B are given by the intercepts of the tangent to the curve at x with the concentration limits $N_B=0$ and $N_B=1$ respectively.

A thermodynamic proof of this statement can be obtained as follows:

For one mole of a two-component solution at constant temperature and pressure, equation 1.95 can be written as

$$dG = \bar{G}_A dN_A + \bar{G}_B dN_B$$

Multiplying throughout by N_A/dN_B gives

$$N_A\frac{dG}{dN_B} = N_A\bar{G}_A\frac{dN_A}{dN_B} + N_A\bar{G}_B\frac{dN_B}{dN_B}$$

But $N_A+N_B=1$ for a two-component solution and so $dN_A = -dN_B$. Substituting this relation and simplifying, we obtain

$$N_A\frac{dG}{dN_B} = -N_A\bar{G}_A + N_A\bar{G}_B$$

Adding this to equation 1.97 gives

$$G+N_A\frac{dG}{dN_B} = \bar{G}_A(N_A-N_A)+\bar{G}_B(N_A+N_B)$$

or

$$\bar{G}_B = G+N_A\frac{dG}{dN_B}$$

Similarly, by multiplying throughout in the first step by N_B/dN_A we obtain

$$\bar{G}_A = G+N_B\frac{dG}{dN_A}$$

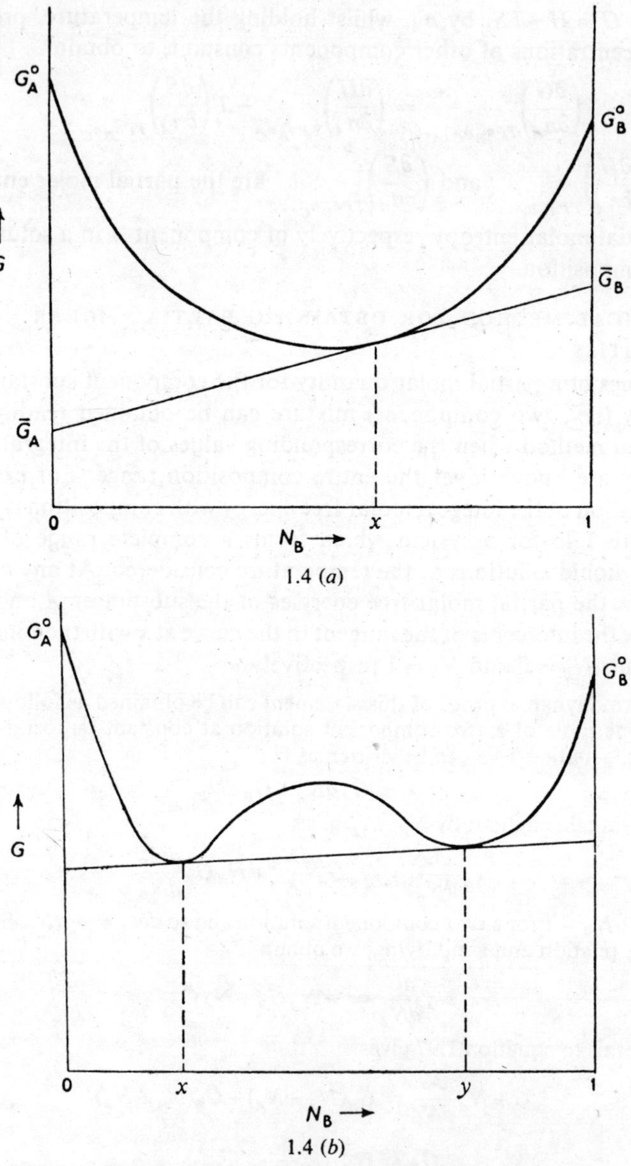

FIG. 1.4. Illustration of the tangent-intercept method for determining partial molar free energies from the curve relating the integral free energy of mixing and the mole fractions in a binary system forming (*a*) a complete range of liquid or solid solutions, and (*b*) partial liquid or solid miscibility.

The slope of the tangent to the G–N curve is equal to dG/dN_A and dG/dN_B. Thus at $N_B = 1$, $\bar{G}_B$ is equal to the value of G at composition x plus the difference between the value of G at x and the value of G where the tangent to the curve at composition x intersects the $N_B = 1$ axis.

The G–N curve for a liquid or solid solution is concave downwards for all compositions, but if the solution breaks down into a mixture of phases the curve exhibits inflexions and becomes concave upwards in the mixed-phase region. Figure 1.4*b* shows the G–N curve for a binary system which forms either liquid or solid solutions at the composition extremities and a mechanical mixture of the two solutions in the central region. The tangent to the curve at the composition x is also the tangent at composition y. Within these limits a mechanical mixture of the two phases has a lower (i.e., more negative) free energy than a simple solution. Hence, in this composition range the mechanical mixture is thermodynamically more stable than a solution. Because the partial molar free energies are given by the extensions of the tangent to the binary limits, it follows that at constant temperature and pressure the partial molar free energies of the component phases in a phase mixture are constant and are independent of the composition. By similar reasoning it can be shown that the partial molar enthalpies and entropies of the component phases of a phase mixture are independent of the composition.

1.8 Fugacity and Activity

It is now convenient to introduce two additional thermodynamic functions which facilitate the application of these equations to a real gas, to any solid or liquid solution, or to mixtures of solutions.

FUGACITY

When dealing with an ideal gas in sections 1.4 and 1.6 use was made of the equation of state, $PV = RT$. No real gas is ideal under all conditions of temperature and pressure; the equation does not hold when the gas deviates from ideal behaviour and, to make the treatment of more general applicability, a quantity called the *fugacity*, $\mathscr{F}$, is so defined that $\mathscr{F}V = RT$ under *all* conditions. Equation 1.66 can then be written for the free-energy change of a real gas which is subjected to a small pressure change at constant temperature in the form:

$$\begin{aligned} dG &= V\,d\mathscr{F} \\ &= \frac{RT}{\mathscr{F}}.\,d\mathscr{F} \\ &= RT\,d\ln\mathscr{F} \end{aligned} \qquad 1.99$$

On integration at constant temperature this equation becomes:

$$G = RT \ln \mathscr{F} + I \qquad 1.100$$

where I is an integration constant whose value is so chosen that $\mathscr{F}$ approaches p as p tends to zero. For an ideal gas $\mathscr{F}$ is equal to p at all pressures and temperatures.

In extraction and refining processes we are usually concerned with gases at low pressures and comparatively high temperatures, where deviations from ideal behaviour of the gases are not usually significant. Under such conditions the replacement of the fugacities by the actual partial pressures of the gases is well within the experimental uncertainty limits, but at high pressures or at low temperatures, approaching the critical temperature and pressure of a gas, the deviation from ideal behaviour is often pronounced and disregard of the fugacities may then give rise to large errors.

The free-energy change accompanying the expansion of a gas at constant temperature from the standard pressure of one atmosphere to any other pressure is equal to the difference between the partial molar and standard free energies of the gas in the two conditions. Thus, integrating equation 1.99 between these pressure limits, the change in the free energy is given by:

$$\Delta \bar{G} = \bar{G} - G^0 = RT \ln \frac{\mathscr{F}}{\mathscr{F}^0} \qquad 1.101$$

ACTIVITY

In the same way as the fugacity is defined as a measure of the behaviour of a real gas relative to the ideal gas behaviour, we can define a quantity, which is called the *activity* and represented by the symbol a, to measure the behaviour of a condensed substance relative to the behaviour of a thermodynamic ideal liquid or solid solution. Solutions are considered in detail in the next Chapter, but an ideal solution can be defined here as one in which, at the temperature considered, the fugacity of the vapour of each substance over the solution is directly proportional to the atomic or mole fraction of that substance over the entire range of compositions of the solution. That is, for a component A of an ideal solution:

$$\mathscr{F}_A = \mathscr{F}_{A\,(P)} N_A \qquad 1.102$$

where $\mathscr{F}_{A(p)}$ is the fugacity of the pure substance at the same temperature and total pressure. This is a statement of *Raoult's law*.

The activity of a substance is defined as the ratio of the fugacity in the actual state to that in the chosen standard state, i.e., for a component A the activity is given by:

$$a_A = \frac{\mathscr{F}_A}{\mathscr{F}_A^0} \tag{1.103}$$

When the standard state for the substance is chosen as the pure solid or liquid, whichever is stable at the temperature considered, the activity tends to unity as the mole or atomic fraction tends to unity, and the activity is unity in the standard state. Substituting equation 1.103 in equation 1.99, for one mole of a component A in a solution since $\mathscr{F}_A^0$ is unity, we have:

$$d\bar{G}_A = RTd\ln a_A \tag{1.104}$$

and integrating from the actual state to the chosen standard state at constant temperature, since a_A^0 is unity,

$$\Delta\bar{G}_A = \bar{G}_A - G_A^0 = RT\ln a_A \tag{1.105}$$

It follows from equations 1.102 and 1.103 that, when the standard state is chosen as the pure component, the activity of a substance in an ideal solution is equal to the atomic or mole-fraction. The equality does not hold for a non-ideal solution, but the extent of the deviation from ideality can be measured in terms of an *activity coefficient* γ, which is equal to the ratio of the activity to the mole fraction of the component, i.e.,

$$\gamma_A = \frac{a_A}{N_A} \tag{1.106}$$

and the activity coefficient is unity for each component of an ideal solution.

1.9 The Gibbs-Duhem Relation

An experimental study is often designed to yield data on the variation with concentration of the activity of one substance in a solution. The concentration dependence of the partial molar free energy of that substance can then be obtained by use of equations 1.104 or 1.105 and, if the study covers a range of both compositions and temperatures, it is possible to assess the partial molar enthalpies and partial molar entropies of that component as functions of concentration within the temperature range studied. It is often useful to apply these data for the one substance so as to obtain the corresponding data for the other substances in the solution, and this information is obtained by application of the Gibbs-Duhem relation.

Consider a binary solution of substances A and B at constant temperature and pressure, for which the activity of component B is known over the entire concentration range. From equation 1.96 the total free energy of n_A+n_B moles of solution is given by:

$$G' = n_A \bar{G}_A + n_B \bar{G}_B \qquad 1.107$$

Differentiating without holding the composition constant we obtain:

$$dG' = (n_A d\bar{G}_A + n_B d\bar{G}_B) + (\bar{G}_A dn_A + \bar{G}_B dn_B)$$

and subtracting equation 1.95 for a two-component solution, we obtain

$$n_A d\bar{G}_A + n_B d\bar{G}_B = 0 \qquad 1.108$$

which is one form of the Gibbs-Duhem relation. Dividing through by (n_A+n_B), for one mole of solution we have:

$$N_A d\bar{G}_A + N_B d\bar{G}_B = 0 \qquad 1.109$$

Similar equations can be written for each of the partial molar thermodynamic properties.

Substituting for $d\bar{G}_A$ and $d\bar{G}_B$ in terms of equation 1.104

$$N_A d\ln a_A + N_B d\ln a_B = 0 \qquad 1.110$$

and integrating between the limits $N_A = 1$ (i.e. pure A) and $N_A = N_A$ (the mole fraction of A at any chosen concentration in the solution, the activity of A at this concentration is given by:

$$\ln a_A = -\int_{N_A=1}^{N_A=N_A} \frac{N_B}{N_A} . d\ln a_B \qquad 1.111$$

which can be integrated graphically by plotting N_B/N_A against $-\ln a_B$. However, the graphical evaluation is difficult for the integrand approaches infinity as N_A tends to zero, whilst $\ln a_B$ approaches minus infinity as N_A tends to unity (Figure 1.5).

A more straightforward evaluation can be performed when the Gibbs-Duhem relation is expressed in terms of activity coefficients. For a binary mixture:

$$N_A + N_B = 1$$

and by differentiation this becomes:

$$dN_A + dN_B = 0$$

Multiplying the first term by $N_A/N_A = 1$ and the second term by $N_B/N_B = 1$ gives:

$$\frac{N_A}{N_A} dN_A + \frac{N_B}{N_B} dN_B = 0$$

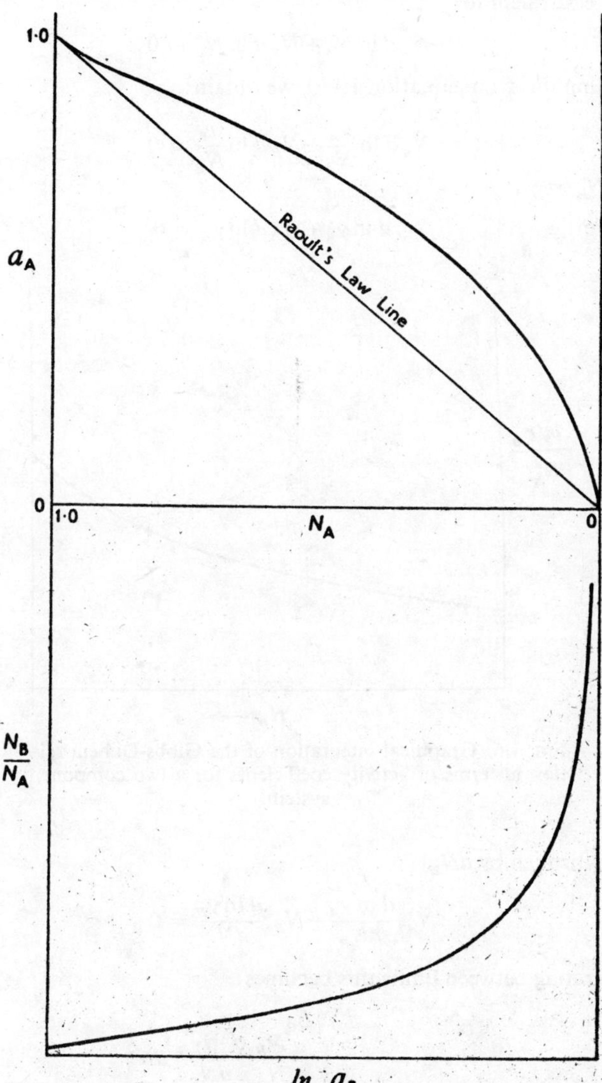

FIG. 1.5. Graphical method for the evaluation of the activities in a binary mixture when the activities of component A are known as a function of the composition.

which is equivalent to:

$$N_A d \ln N_A + N_B d \ln N_B = 0 \quad 1.112$$

Subtracting this from equation 1.110, we obtain:

$$N_A d \ln \frac{a_A}{N_A} + N_B d \ln \frac{a_B}{N_B} = 0$$

but $a_A/N_A = \gamma_A$

Therefore

$$N_A d \ln \gamma_A + N_B d \ln \gamma_B = 0 \quad 1.113$$

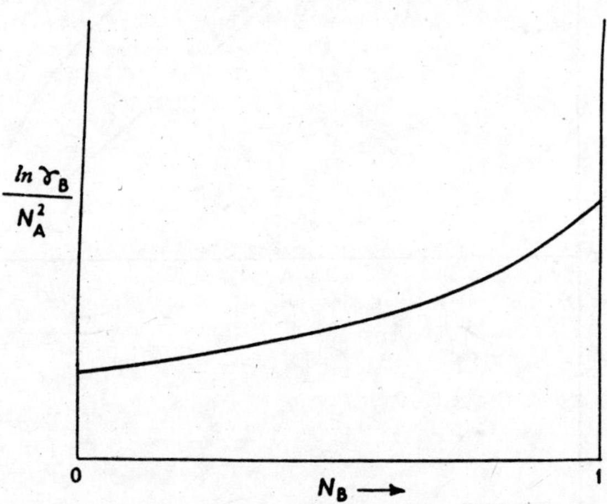

FIG. 1.6. Graphical integration of the Gibbs-Duhem relation in terms of activity coefficients for a two component system.

Dividing through by dN_B

$$N_A . \frac{d \ln \gamma_A}{dN_B} + N_B . \frac{d \ln \gamma_B}{dN_B} = 0$$

and integrating between limits this becomes:

$$\ln \gamma_A = - \int_{N_A=1}^{N_A=N_A} \frac{N_B}{N_A} . \frac{d \ln \gamma_B}{dN_B} . dN_B$$

Wagner [10] has proposed a solution to this equation involving integration by parts to give:

$$\ln \gamma_A = \int_{N_A=1}^{N_A=N_A} \frac{\ln \gamma_B}{N_A^2} . dN_B - \frac{N_B}{N_A} . \ln \gamma_B \quad 1.114$$

and a graphical evaluation is performed by plotting $\ln\gamma_B/N_A^2$ against N_B. In this form the solution presents no difficulty in the vicinity of $N_A = 1$ since γ_B is always finite. Both $\ln\gamma_B$ and N_A approach zero as N_B tends to 1 and evaluation in this region is again dependent on extrapolation to the N_B axis, but the curvature is usually small and the extrapolation introduces no appreciable error (Figure 1.6).

APPLICATION OF THE GIBBS-DUHEM RELATIONS TO TERNARY SYSTEMS

The Gibbs-Duhem relation can be extended to evaluate the partial molar properties of each component in a multi-component system when the corresponding partial molar property is known for one component as a function of composition. Thus for a ternary system containing n_A, n_B and n_C moles of substances A, B and C, holding n_C constant and differentiating with respect to n_B and n_A, equation 1.95 becomes

$$\left(\frac{\partial G'}{\partial n_A\,\partial n_B}\right)_{n_C} = \left(\frac{\partial \bar{G}_A}{\partial n_B}\right)_{n_A n_C} = \left(\frac{\partial \bar{G}_B}{\partial n_A}\right)_{n_B n_C} \qquad 1.115$$

The evaluation of this expression is simplified if the mole ratios are selected as independent variables in place of the number of moles, and for this purpose the following two ratios are selected [11]:

$$x = \frac{n_C}{n_A+n_C}$$

$$y = \frac{n_C}{n_B+n_C}$$

Rearranging the first of these ratios in the form:

$$n_A = \frac{n_C}{x} - n_C$$

and differentiating with respect to x whilst holding n_B and n_C constant gives

$$\left(\frac{\partial n_A}{\partial x}\right)_{n_B n_C} = -\frac{n_C}{x^2} \qquad 1.116$$

From the principles of partial differentiation the following relation is valid:

$$\left(\frac{\partial \bar{G}_B}{\partial x}\right)_{n_B n_C} = \left(\frac{\partial \bar{G}_B}{\partial n_A}\right)_{n_B n_C} \cdot \left(\frac{\partial n_A}{\partial x}\right)_{n_B n_C}$$

or by rearranging terms:

$$\left(\frac{\partial \bar{G}_B}{\partial n_A}\right)_{n_B n_C} = \left(\frac{\partial \bar{G}_B}{\partial x}\right)_{n_B n_C} \Big/ \left(\frac{\partial n_A}{\partial x}\right)_{n_B n_C}$$

The term on the left-hand side of this equation is identical with the third term in equation 1.115, whilst the denominator on the right hand side is equivalent to equation 1.116. Hence:

$$\left(\frac{\partial \bar{G}_B}{\partial n_A}\right)_{n_B n_C} = -\left(\frac{\partial \bar{G}_B}{\partial x}\right)_{n_B n_C} \Big/ \frac{n_C}{x^2}$$

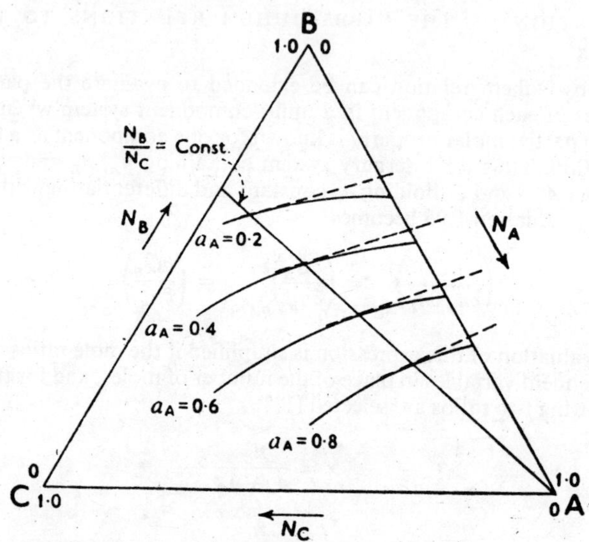

FIG. 1.7. Graphical method for evaluating the activities of components B and C of a three-component system when the activities of component A are known as a function of composition (see text). The curved lines are the iso-activity contours for component A.

By a similar process a relation is obtained in terms of $\bar{G}_A$ and y:

$$\left(\frac{\partial \bar{G}_A}{\partial n_B}\right)_{n_A n_C} = -\left(\frac{\partial \bar{G}_A}{\partial y}\right)_{n_A n_C} \Big/ \frac{n_C}{y^2}$$

Substituting these terms in equation 1.115, rearranging and remembering that constant $n_A n_C$ is equivalent to constant x and constant $n_B n_C$ is equivalent to constant y, we have:

$$\left(\frac{\partial \bar{G}_A}{\partial y}\right)_x = \frac{x^2}{y^2}\left(\frac{\partial \bar{G}_B}{\partial x}\right)_y$$

and substituting for the partial molar free energies from equation 1.104 gives:

$$\left(\frac{\partial \ln a_A}{\partial y}\right)_x = \frac{x^2}{y^2}\left(\frac{\partial \ln a_B}{\partial x}\right)_y \qquad 1.117$$

A number of methods have been suggested for the solution of this relation [10, 12, 13]. Schuhmann [13] has proposed a simple method of evaluation whereby the activity/composition relations for two components B and C can be derived from tangent intercepts to lines of constant activity for the third component A. The relation between activity and mole fraction is expressed by:

$$\left(\frac{\partial \ln a_B}{\partial \ln a_A}\right)_{N_B N_C} = -\left(\frac{\partial N_A}{\partial N_B}\right)_{a_A N_C} \qquad 1.118$$

Integration between the composition limits m and n gives

$$\ln a_{B_{(m)}} = \left[\ln a_{B_{(n)}} - \int_{a_{A(n)}}^{a_{A(m)}} \left(\frac{\partial N_A}{\partial N_B}\right)_{a_A \cdot N_C} . d\ln a_A\right]_{N_B/N_C} \qquad 1.119$$

where m and n are the limits of the integration path along a line of constant molar ratio N_B/N_C. The starting point of the path may be chosen as any convenient point where $\ln a_B$ is known. For example, a_B can be taken as unity at the saturation surface of B if the solubility of both A and C in B is small. The integrand is evaluated by extending tangents to the iso-activity lines for component A at constant N_B/N_C to intercept the A–B side of the ternary system (Figure 1.7). The values of the binary intercepts, corresponding to $\left(\frac{\partial N_A}{\partial N_B}\right)_{a_A \cdot N_A}$ are plotted against $\ln a_A$ and the integrand is equal to the area under the curve between the limits $\ln a_{A_{(m)}}$ and $\ln a_{A_{(n)}}$.

1.10 The Equilibrium Constant

The reaction between a pure metal M and oxygen gas to form the pure oxide MO, i.e.,

$$2M_{(s\,or\,l)} + O_{2(g)} \rightleftharpoons 2MO_{(s\,or\,l)}$$

is considered in section 1.6. Consider now the more general case where the metal and oxide are not pure, but are dissolved in another metal and oxide respectively, and the equilibrium oxygen pressure is less than one atmosphere. At equilibrium the free energy of the reaction is given by:

$$\Delta G = 2\bar{G}_{MO} - 2\bar{G}_M - \bar{G}_{O_2} \qquad 1.120$$

whereas, when all the substances are in their standard states,

$$\Delta G^0 = 2G^0_{MO} - 2G^0_M - G^0_{O_2} \qquad 1.121$$

Subtracting equation 1.121 from 1.120 and remembering that at equilibrium $\Delta G = 0$ we have:

$$-\Delta G^0 = 2(\bar{G}_{MO} - G^0_{MO}) - 2(\bar{G}_M - G^0_M) - (\bar{G}_{O_2} - G^0_{O_2})$$

Substituting from equation 1.105 for the first two bracketed terms and from equation 1.101 for the third term gives:

$$-\Delta G^0 = 2RT \ln a_{\mathrm{MO}} - 2RT \ln a_{\mathrm{M}} - RT \ln \mathscr{F}_{\mathrm{O}_2}$$

or

$$\Delta G^0 = -RT \ln \left[\frac{a^2_{\mathrm{MO}}}{a^2_{\mathrm{M}} . \mathscr{F}_{\mathrm{O}_2}} \right] \quad 1.122$$

Because at any temperature ΔG^0 has a precise numerical value, the term in square brackets must be constant at any temperature and is independent of the concentration, *provided only that the system is in thermodynamic equilibrium*. The quantity in brackets is designated accordingly as the *thermodynamic equilibrium constant*, and is represented by the symbol K. Thus:

$$\Delta G^0 = -RT \ln K \quad 1.123$$

or

$$\Delta G^0 = -4{\cdot}575T \log K \quad 1.124$$

In the general case, for a reaction between v mols of A, w mols of B, etc., to produce x mols of C, y mols of D, etc., i.e.,

$$vA + wB + \ldots \rightleftharpoons xC + yD + \ldots$$

the equilibrium constant is given by:

$$\frac{\Delta G^0}{4{\cdot}575T} = -\log K = -\log \frac{a^x_C . a^y_D \ldots}{a^v_A . a^w_B \ldots} \quad 1.125$$

In the absence of reliable activity data the equilibrium relation is often expressed in terms of concentration units instead of activities, or mixed units are used when the activities of the reactants and products are known in one phase (e.g., the metal) but not in the second phase (e.g., the slag) of a heterogeneous reaction. The replacement of activities by concentrations is only justified thermodynamically for a substance which exhibits ideal behaviour, and if this approximation is applied to a non-ideal solution the 'equilibrium' constant for the reaction is not the true thermodynamic constant but is a function of both the temperature and the composition. This distinction is emphasized in subsequent Chapters by the following terminology:

1. The *true equilibrium constant* K applies only when the equilibrium relation is expressed entirely in terms of the activities of the condensed phases.
2. The '*apparent*' *equilibrium constant* K' is used when one or more of the substances appearing in the equilibrium relation are expressed in terms of the concentration.

It is sometimes found that, within the limits of the sensitivity of a particular experimental investigation, the apparent equilibrium constant is seemingly independent of the composition. This may apply, for example, in a concentration range where the activities are not ideal but vary almost linearly with the concentration, or when the activities of all the reactants and products vary at approximately the same rate with change in the composition. However, it is improbable that these approximations are valid over more than a very limited concentration range and use of the apparent constants at compositions outside the range from which they are evaluated may give rise to serious error.

VARIATION OF THE EQUILIBRIUM CONSTANT WITH TEMPERATURE

From equation 1.125 it is evident that the variation of the equilibrium constant with temperature can be obtained readily if the temperature-dependence of the standard free-energy change is known.

Writing equation 1.66 in terms of the standard free energy and entropy changes and differentiating with respect to temperature at constant pressure we have:

$$\left(\frac{\partial \Delta G^0}{\partial T}\right)_P = -\Delta S^0 \qquad 1.126$$

But by definition

$$\Delta G^0 = \Delta H^0 - T\Delta S^0$$

Substituting equation 1.126 for ΔS^0 gives:

$$\Delta G^0 = \Delta H^0 + T\left(\frac{\partial \Delta G^0}{\partial T}\right)_P \qquad 1.127$$

Dividing through by $1/T^2$ and collecting like terms we obtain:

$$\frac{\Delta G^0}{T^2} - \frac{1}{T}\left(\frac{\partial \Delta G^0}{\partial T}\right)_P = -\frac{\Delta H^0}{T^2}$$

which is equivalent to:

$$-\left(\frac{\partial\left(\frac{\Delta G^0}{T}\right)}{\partial T}\right)_P = \frac{\Delta H^0}{T^2} \qquad 1.128$$

Substituting for ΔG^0 from equation 1.123, this becomes:

$$\left(\frac{\partial \ln K}{\partial T}\right)_P = \frac{\Delta H^0}{RT^2} \qquad 1.129$$

which is known as the *van't Hoff 'isochore'*. The integration is straightforward when ΔH^0 can be expressed as a function of temperature (e.g., in terms of heat capacities as described in 1.3). This equation can be used to estimate the variation of the equilibrium constant with temperature from a single value of ΔH^0. Conversely the value of ΔH^0 at a particular temperature can be obtained from knowledge of the rate of variation with temperature of the equilibrium constant at that temperature.

It is often permissible to approximate and assume that ΔH^0 is independent of temperature over a small range from T_1^0 to T_2^0 K, in which case equation 1.129 can be written as

$$\ln\left(\frac{K_{T_1}}{K_{T_2}}\right) = \frac{\Delta H_T^0}{R}\left(\frac{1}{T_2}-\frac{1}{T_1}\right)$$

or

$$\Delta H_T^0 = \frac{RT_1T_2(\ln K_{T_1} - \ln K_{T_2})}{T_1 - T_2} \qquad 1.130$$

and, for example, the value of K_{T_2} can be estimated from knowledge of of K_{T_1} and ΔH_T^0.

FURTHER CONSIDERATION OF THE OXYGEN POTENTIAL DIAGRAMS

The significance of the oxygen potential diagrams reproduced in Figures 1.2 and 1.3 can be seen more clearly with the activity concept in mind. The diagrams apply only to the specific condition where $a_M = a_{MO} = 1$. If the activity of the oxide is less than unity, whilst the metal is present in the pure form, the oxygen potential is lowered and the oxide is formed from the metal at a lower partial pressure of oxygen than that shown by the diagram. This is illustrated by comparison of the curves for the oxidation of iron to form ferrous oxide. In Figure 1.3 the oxides are present in combination with silica as ferrous silicate with the result that the oxide activities are less than unity and at all temperatures the oxygen potentials are more negative than those in Figure 1.2 for pure ferrous oxide. The converse applies also, for if the activity of the metal is less than unity the oxygen potential is raised and the metal is more difficult to oxidise than when it is present in the pure form. Thus, by dilution, the relative stabilities of the oxides may be altered from those predicted on the diagrams, but the diagrams are still extremely useful for indicating general trends. For example, in liquid steel the activity of iron is approximately unity and for copper to be oxidised in preference to liquid iron under a pure ferrous oxide slag either the activity of copper oxide must be less than unity or the activity of copper must be much greater than unity. However, the activity of copper relative to pure copper as the standard state, cannot exceed unity. Further, under steel-making conditions the ferrous oxide is not pure, but is present in a solution of silica, lime and other oxides, so the oxygen potential for the formation of ferrous oxide is lower than that indicated in Figure 1.2. The activity of copper oxide must, therefore, be very much lower than unity in order to remove copper from iron by oxidation, and in practice it is extremely difficult to effect the removal. The converse applies with respect to the reduction of magnesia or lime in preference to the reduction of iron oxides.

REFERENCES

1. O. KUBASCHEWSKI and E. L. EVANS, *Metallurgical Thermochemistry*, Third Edition, Pergamon, 1958.
2. F. D. ROSSINI, et al., *Selected Values of Chemical Thermodynamic Constants*, National Bureau of Stds., 1952, etc.
3. J. F. ELLIOTT and M. GLEISER, *Thermochemistry for Steel-making*, Vol. 1, Pergamon, 1960.
4. F. D. RICHARDSON and J. H. E. JEFFES, *J. Iron St. Inst.*, 1948, Vol. 160, p. 261.
5. H. T. T. ELLINGHAM, *J. Soc. chem. Ind., Lond.*, 1944, Vol. 63, p. 125.
6. F. D. RICHARDSON and J. H. E. JEFFES, *J. Iron St. Inst.*, 1949, Vol. 161, p. 229; 1949, Vol. 163, p. 397; 1952, Vol. 171, p. 165.
7. F. D. RICHARDSON, J. H. E. JEFFES and G. WITHERS, ibid, 1950, Vol. 166, p. 213.
8. F. D. RICHARDSON, ibid., 1953, Vol. 175, p. 53.
9. L. S. DARKEN and R. W. GURRY, *J. Amer. chem. Soc.*, 1946, Vol. 68, p. 798.
10. C. WAGNER, *Thermodynamics of Alloys*, Addison Wesley, 1952.
11. N. A. GOKCEN, *J. phys. Chem.*, 1960, Vol. 64, p. 401.
12. L. S. DARKEN, *J. Amer. chem. Soc.*, 1950, Vol. 72, p. 2909.
13. R. SCHUHMANN, *Acta Met.*, 1955, Vol. 3, p. 219.

CHAPTER 2

Solutions

In the field of ferrous extraction and refining operations the metallurgist is concerned not with pure materials but with solutions of various elements in iron, or with complex solutions of oxides, sulphates, sulphides and other compounds which form the slags. The thermodynamic treatment of these solutions differs somewhat from that for pure elements and compounds. The more important concepts which are necessary for the treatment of these solutions have already been discussed in Chapter 1 and their application can now be considered.

2.1 Ideal Solutions

An ideal solution has been defined as one which obeys Raoult's law and, in consequence, the activities of the constituents are equal to their respective mole fractions. It can also be shown that an ideal solution undergoes no change in volume or enthalpy on mixing. No known system completely satisfies these requirements over the whole concentration range, although mixtures of two isotopes of one element may approach very closely to ideality, and with some actual solutions the deviations from Raoult's law are less than the experimental error in the determination of the activities. Thus, liquid mixtures of FeO and MnO are treated as ideal for most practical purposes. In the majority of solutions the activities show either a positive or a negative deviation from Raoult's law, but in all cases Raoult's law is approached as a limiting law as the activity approaches unity. These conditions are summarised in Figure 2.1.

From equation 1.105:

$$\bar{G}_A - G_A^0 = RT \ln a_A$$

but for an ideal solution $a_A = N_A$ and, therefore:

$$\bar{G}_A - G_A^0 = RT \ln N_A \tag{2.1}$$

The quantity $\bar{G}_A - G_A^0$ is the change in the free energy of a substance A when it is diluted from the standard state (e.g. pure A) by mixing with other substances to form a solution containing N_A mole-fraction of A. For convenience it is designated G_A^M and called the *partial molar free energy of mixing*, or the *relative partial molar free energy* of A. The corresponding enthalpy and entropy changes on mixing are related to G_A^M through the relation:

$$G_A^M = H_A^M - TS_A^M \tag{2.2}$$

Because the activity in an ideal solution is equal to the mole fraction, the activity is independent of temperature and it follows from equation 1.128 that H_A^M is zero, so that:

$$G_A^M = -TS_A^M$$

and

$$S_A^M = -R \ln N_A \tag{2.3}$$

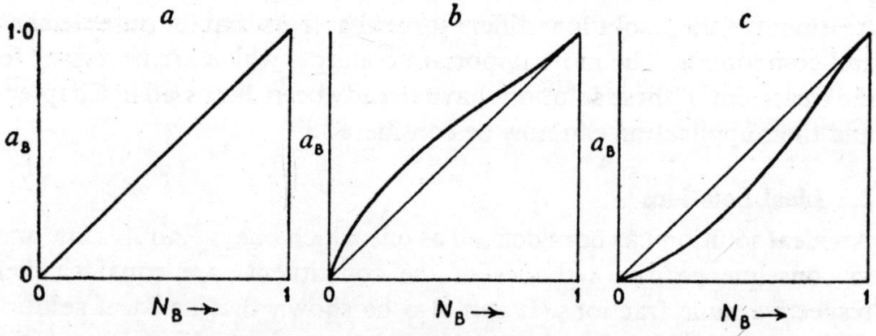

FIG. 2.1. Activity–concentration curves showing (*a*) ideal (Raoult's law) behaviour, (*b*) positive, and (*c*) negative deviations from ideality.

For an ideal binary solution containing N_A and N_B mole fractions of A and B (where $N_A + N_B = 1$), the *integral molar entropy of mixing* (S^M) of the solution is given by:

$$S^M = N_A S_A^M + N_B S_B^M$$

$$= -R(N_A \ln N_A + N_B \ln N_B) \tag{2.4}$$

and the *integral molar free energy of mixing* is given by:

$$G^M = RT(N_A \ln N_A + N_B \ln N_B) \tag{2.5}$$

Converting to common logarithms and substituting for R in equation 2.4 we obtain:

$$S^M = -4{\cdot}575(N_A \log N_A + N_B \log N_B)$$

This equation is evaluated by substituting values of N_A and N_B as follows:

Table 2.1 *Variation of Entropy of Mixing with Composition for an Ideal Solution*

N_B	N_A	$N_A \log N_A$	$N_B \log N_B$	$4{\cdot}575(N_A \log N_A + N_B \log N_B)$
1·0	0·0	0	0	0
0·9	0·1	−0·1	−0·041	−0·475
0·8	0·2	−0·140	−0·077	−0·993
0·6	0·4	−0·159	−0·133	−1·332
0·5	0·5	−0·151	−0·151	−1·378
0·4	0·6	−0·133	−0·159	−1·332
0·2	0·8	−0·077	−0·140	−0·993
0·0	1·0	0	0	0

The values in the end column are symmetrical about the 0·5 mole-fractions and the mixing entropy (S^M), which is equal to the negative of these values, is plotted against the mole-fraction of B in Figure 2.2*a*. The free energy of mixing equals the product of column 5 and the temperature in degrees absolute, and the relation between G^M and the concentration at 1,000°K is shown in Figure 2.2*b*. The curve passes through zero at $N_B = 0$ and at $N_B = 1$, since there is no free energy of mixing with pure A or pure B, but the total free-energy curve has intercepts on these axes of G_A^0 and G_B^0 respectively. In the general case these values are not equal and the total free-energy curve is inclined about the horizontal axis, but the minimum still lies at $N_B = 0{\cdot}5$ (Figure 2.2*c*). The total free energy of an ideal solution is, therefore,:

$$G = [N_A G_A^0 + N_B G_B^0] + RT[N_A \ln N_A + N_B \ln N_B] \qquad 2.6$$

where the first term in brackets is the free energy of a simple mixture of A and B, and the second term is the free energy of formation of an ideal solid solution.

Reference is made in Chapter 1 to the statistical nature of entropy; equation 2.4 and the subsequent relations can be derived also by the application of the Boltzmann relation (equation 1.56) to a solution which conforms to Raoult's law. By this means it can be shown that the entropy of mixing term is, in reality, a measure of the atomic disorder or randomness introduced by mixing the two components to form a solution.

2.2 Actual Solutions

Although ideal solutions are rarely encountered, the concept is a convenient model to use as the starting point for consideration of the changes in the thermodynamic properties accompanying the formation of a real solution of two or more components. In practice it is found that all

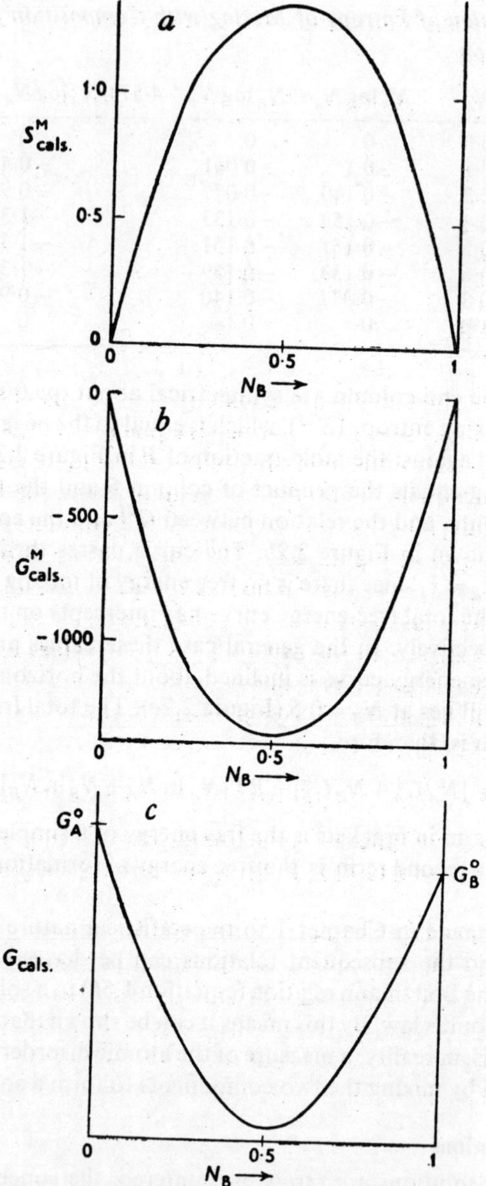

FIG. 2.2: Variation with composition in an ideal liquid or solid solution of (*a*) the integral molar entropy of mixing, (*b*) the integral molar free energy of mixing, and (*c*) the total free energy of the solution.

solutions deviate from Raoultian behaviour; in some cases the deviation is extremely small and may be ignored, but more commonly the deviation is significant and application of the ideal solution equations can give rise to serious errors. A qualitative interpretation of the cause of the deviations from ideality can be obtained from consideration of the forces holding the atoms or molecules together in the solution. Consider the addition of a small number of solute atoms *B* to a very large number of solvent atoms *A*. At high dilution the probability of two *B* atoms occupying adjacent positions is very small, but the probability is increased as the solute concentration is raised. Since, in general, the force holding two solute atoms together is different from the force between two solvent atoms or between a solvent and a solute atom, there is a natural tendency for the atoms to segregate into *A–A* and *B–B* pairs, or into *A–B* pairs. The former is generally associated with a positive heat of mixing whereas a preference for *A–B* groups generally gives rise to a negative heat of mixing. In both cases, however, the atoms show some degree of order and the mixing entropy is decreased relative to that of a purely random mixture of *A* and *B*. For purely random mixing, therefore, the entropy of mixing is a maximum, whilst the heat of mixing is zero, and this statement provides an alternative way of defining an ideal solution.

Since the mixing entropies and enthalpies for a component of a solution are related to the activities of that component through equations 1.105 and 2.2, it is expected that the deviation of the activities from ideal mixing can also be predicted qualitatively from consideration of the bonding energies, and this is generally found to apply in practice. For example, if the solvent and solute atoms tend to form separate groups in preference to a random association the system exhibits a positive deviation from ideal behaviour, and in the extreme case the solution breaks down into a phase mixture of two immiscible liquids or of two solid solutions. This type of behaviour is portrayed, for example, by mixtures of iron and copper or iron and lead. On the other hand, a negative deviation from ideality results from preferential association between solute and solvent atoms, this behaviour being generally associated with a tendency towards compound formation (e.g., Fe–Al and Fe–Ti mixtures).

Inspection of the relevant phase diagram will frequently indicate whether a positive or negative deviation is to be expected. Negative deviations in the activities of the solid substances may be expected if the phase diagram indicates the existence of a stable compound in the solid state and, as the liquid state can be regarded as a more disordered form of the structures present in the solid, a similar but less pronounced negative deviation would be expected in the liquid mixtures. Conversely, positive deviations are probable when the phase diagram shows the existence of eutectic, monotectic or peritectic reactions.

In recent years detailed studies of a number of solid systems which show relatively small departures from ideality have indicated that the deviations may be opposite in sign to that which is predicted from the differences in the bond energies. However, this behaviour is usually interpreted in terms of the properties of the solid state and it is very doubtful whether such considerations apply to liquid solutions.

REGULAR SOLUTIONS

If the solution does not obey Raoult's law the heat of solution cannot be zero. However, to facilitate the treatment of those systems which exhibit only a small deviation from ideality, Hildebrand [1] introduced the concept of a *regular solution* which is defined as one in which the entropy of mixing is the same as that for an ideal solution, but in which the heat of mixing may have any value, positive or negative, up to about one kcal per mole. Thus S^M is given again by equation 2.4 but, as the system deviates from ideality, G^M must be expressed in terms of activities instead of mole fractions. For a binary solution of A and B which exhibits regular behaviour:

$$G^M = RT(N_A \ln a_A + N_B \ln a_B) \quad 2.7$$

but

$$H^M = G^M + TS^M$$

and substituting equations 2.4 and 2.7 into this relation gives:

$$H^M = RT(N_A \ln a_A + N_B \ln a_B) - RT(N_A \ln N_A + N_B \ln N_B)$$

$$= RT\left[N_A \ln\left(\frac{a_A}{N_A}\right) + N_B \ln\left(\frac{a_B}{N_B}\right)\right]$$

But

$$\frac{a_A}{N_A} = \gamma_A$$

Therefore

$$H^M = RT(N_A \ln \gamma_A + N_B \ln \gamma_B) \quad 2.8$$

and

$$H_A^M = RT \ln \gamma_A$$

$$H_B^M = RT \ln \gamma_B \quad 2.9$$

If the deviation from ideality is positive, (i.e., $\gamma_A > 1$) H_A^M is positive and heat is absorbed on mixing and, conversely, for a negative deviation ($\gamma_A < 1$) H_A^M is nègative and heat is evolved. If H_A^M does not vary with temperature then $RT \ln \gamma_A$ must also be independent of temperature so, when γ_A is less than unity, $\ln \gamma_A$ is a negative quantity which becomes less negative and γ_A approaches nearer to unity as the temperature is increased. Likewise if γ_A is greater than one its value decreases towards unity as the temperature rises. In practice H_A^M usually varies with temperature, but in many cases the variation is small and is not sufficient to alter the direction in which γ_A changes with the temperature.

In the absence of experimental data the regular-solution equations are frequently used to obtain some indication of the values of the activities in a homogeneous solution at various temperatures from measurements made at one temperature. The equations have also been used to derive activities in liquid solutions from the liquidus and solidus lines of the appropriate phase diagram, and the calculation by Rey [2] of activities in binary silicate melts is a well known example of this application.

For a regular solution it can be shown that the heat of mixing is related to the composition by the equation:

$$H^M = -bN_A N_B = -b(1-N_B)N_B \qquad 2.10$$

where b is a constant at a fixed temperature for any particular solution. The partial molar heats of mixing are given by:

$$H_A^M = -bN_B^2 \qquad 2.11$$

and

$$H_B^M = -bN_A^2$$

Substituting for H_A^M and H_B^M from equation 2.9

$$RT \ln \gamma_A = -bN_B^2 \qquad 2.12$$

and

$$RT \ln \gamma_B = -bN_A^2$$

These equations are sometimes used to extend measurements made over a limited composition range to a wider range of solutions. From the Gibbs-Duhem relations it can be shown that, when b for component A is constant over the entire range of compositions, the constant for component B has the same numerical value. It is sometimes found that b is a function of the composition but, if this can be expressed for one component by a simple mathematical relation, the value of b for the second component is readily obtained by integration of the Gibbs-Duhem relation.

The concept of regular solutions has proved useful for obtaining approximate values of the thermodynamic properties of metallic systems, although few conform to the regular solution behaviour, but it is extremely doubtful whether the equations are applicable to liquid silicate slags composed of molecules or ions which are complex when compared with the simple atomic structures of metallic solutions (i.e. solutions of two or more metals). The activities which Rey calculated for the silicate melts differ markedly from some of those which have since been determined experimentally, although (*a*) the experimental values obtained by different authors often show wide divergence, and (*b*) the activity-concentration curves, calculated on the basis of regular solution behaviour, often lie within or very close to the scatter bands through all the experimental results. Whilst the application of the regular solution equations to molten slag systems is hence scarcely justified, the calculated activities are in fact as reliable as some of the direct evaluations.

EXCESS THERMODYNAMIC QUANTITIES

Although the departure of a component of an actual solution from ideal behaviour is indicated by the magnitude of the activity coefficient, it is sometimes more convenient to express the degree of non-ideality in terms of the free energy and a quantity called the *excess partial molar free energy* (G_A^E) has been introduced for this purpose by Scatchard [3]. This quantity is defined as the difference between the partial molar free energy of a substance A in an actual solution and the partial molar free energy which it would possess in an ideal solution at the same concentration of A. That is, it is given by:

$$G_A^E = G_A^M - G_A^I \qquad 2.13$$

where G_A^I is the partial molar free energy in the ideal solution. Substituting for G_A^M and G_A^I from equations 1.105 and 2.1 respectively:

$$\begin{aligned} G_A^E &= RT \ln a_A - RT \ln N_A \\ &= RT \ln \gamma_A \qquad 2.14 \end{aligned}$$

The sign of G_A^E is the same as the sign of the deviation of the system from ideality and for an ideal solution G_A^E is zero. The excess partial molar enthalpy H_A^E and entropy S_A^E are related to G_A^E by the normal relation:

$$G_A^E = H_A^E - TS_A^E \qquad 2.15$$

If the solution is regular, then by definition $S_A^M = S_A^I$, and S_A^E is equal to zero. Therefore

$$G_A^E = H_A^E = H_A^M - H_A^I$$

But the enthalpy of mixing for an ideal solution (H_A^I) is also zero, and in a regular solution

$$G_A^E = H_A^M = RT \ln \gamma_A \qquad 2.16$$

For a solution which is neither ideal nor regular:

$$S_A^E = S_A^M - R \ln N_A \tag{2.17}$$

The *integral excess molar quantities* are defined similarly. Thus:

$$G^E = G^M - G^I$$

$$= G^M - RT(N_A \ln N_A + N_B \ln N_B) \tag{2.18}$$

and

$$S^E = S^M - R(N_A \ln N_A + N_B \ln N_B) \tag{2.19}$$

2.3 Dilute Solutions

The foregoing treatment of non-ideal solutions is applicable to a homogeneous solution of any concentration, but the treatment can often be simplified when the concentration of one component (the solvent) is predominant and the other components (the solutes) are present in very small amounts. A low carbon steel fits into this category during the later stages of refining because the solute elements carbon, silicon, manganese, sulphur and phosphorus are present only in small concentrations.

It is found experimentally that at extreme dilution the activity of a solute is directly proportional to the mole fraction:

$$a_A = \kappa N_A \tag{2.20}$$

where κ is a constant for any particular solution. This is a generalised form of *Henry's law* which relates the partial pressure (p_A) of a substance A to the mole fraction:

$$p_A = \kappa N_A \tag{2.21}$$

The relation between the two equations is apparent from the fundamental equations for pressure, fugacity and activity.

Henry's law is a limiting condition which is approached by all solutions at extreme dilution and the constant κ is equal to the slope of the tangent to the activity curve at the origin as shown in Figure 2.3 for both a positive and a negative deviation from ideality. In most cases a positive deviation from the Henry's law relation corresponds to a negative departure from ideality, and vice versa. Further, the deviation from ideal behaviour is negative if κ is less than unity, positive if κ is greater than one, and the system is ideal if κ is equal to unity. Within the concentration range to which Henry's law applies, the activity coefficient is constant and is designated γ^0; this quantity is actually a measure of the slope of the activity curve at the origin and is therefore equal to the constant κ.

Since the activity of the solute is a linear function of the concentration within the concentration range to which Henry's law is applicable, it can readily be demonstrated by means of the Gibbs-Duhem relation that Raoult's law holds for the solvent over the same range of composition.

Sieverts has proposed a useful modification of Henry's law whereby the solubility N_X in a liquid or a solid phase of a gas X in which there are n atoms per molecule is related to the partial pressure of the gas:

$$N_X = \kappa \sqrt[n]{p_{X_n}}$$

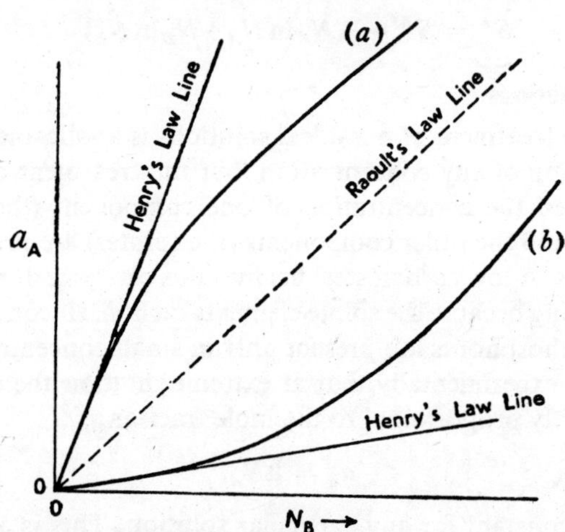

FIG. 2.3. Schematic representation of Henry's law behaviour by solutes showing positive (curve a) and negative (curve b) deviations from ideal (Raoult's law) behaviour.

Most diatomic gases such as hydrogen, nitrogen and oxygen dissolve in metals in the monatomic form, in which case:

$$N_X = \kappa \sqrt[2]{p_{X_n}} \qquad 2.22$$

Sieverts' law holds in the majority of cases for the low concentrations of the gases which are found to dissolve in a metal.

2.4 Alternative Standard States

The activity of a solute in an ideal solution at high dilution is very small when the pure solute is chosen as the standard state and, when the activity shows a marked negative deviation from ideality, at low concentration the value of the activity coefficient with respect to the same reference state can be 0·01 or even lower. To simplify the mathematical manipulation of the activities under these conditions it is frequently more convenient to select a different standard state, such that the activity of the solute is unity at some point within the composition range of interest, in which case it is often permissible to set the activity equal to the concentration for low concentrations of the solute. The use of an alternative standard state may also be more convenient when the pure solute exists in a different form from that in which it is present in the solution at the temperature and pressure considered. For example, pure sulphur exists as a gas and pure titanium is solid at 1,600°C, but both elements form dilute liquid solutions in iron at the same temperature.

A number of alternative standard states are in common use, but the *infinitely dilute solution* is usually selected for consideration of the thermodynamic behaviour of solute elements at low concentration in iron. When the standard state is defined in this way the solute activity approaches the solute concentration as the latter tends to zero.

Atomic Fraction Scale

If the concentrations of the solute are measured in atomic fractions the alternative standard state can be expressed in mathematical form as:

$$a_B = N_{B(\text{limit } N_B \to 0)}$$

or

$$\frac{a_B}{N_B} \to 1 \text{ as } N_B \to 0$$

where a_B and N_B are the activity and atomic fraction of the solute B.

The standard free energy of the solute element, relative to the infinitely dilute standard state, is different from that observed when the pure solute is chosen as the reference condition. The change in the standard free energy accompanying the change in standard state is obtained as follows:

For the reaction:

$$B_{(\text{pure})} \rightleftharpoons B_{(\text{infinitely dilute})}$$

the change in the standard free energy is given by:

$$\begin{aligned} \Delta G^0 &= G^0_{B(\text{dilute})} - G^0_{B(\text{pure})} \\ &= RT \ln a_{B(\text{dilute})} - RT \ln a_{B(\text{pure}} \end{aligned} \qquad 2.23$$

and the standard free energy in the new standard state in relation to the original is given by:

$$G^0_{B(\text{dilute})} = G^0_{B(\text{pure})} + RT \ln \frac{a_{B(\text{dilute})}}{a_{B(\text{pure})}} \qquad 2.24$$

It is stated earlier in this Chapter (see p. 63) that all solutes conform to Henry's law at high dilution; hence the activities in equation 2.24 can be replaced by the mole fractions and the activity coefficients relative to Henry's law and Raoult's law, i.e.

$$\Delta G^0 = RT \ln \frac{\gamma_{B(H)} N_{B(H)}}{\gamma_{B(H)} N_{B(R)}}$$

where the subscripts H and R signify Henry's law and Raoult's law respectively. Extrapolating the Henry's law line to pure B, where $N_H = N_R = 1$ and $\gamma_{B(R)}$ is unity, the intercept of the line on the $N_B = 1$ axis is $\gamma_{B(H)}$

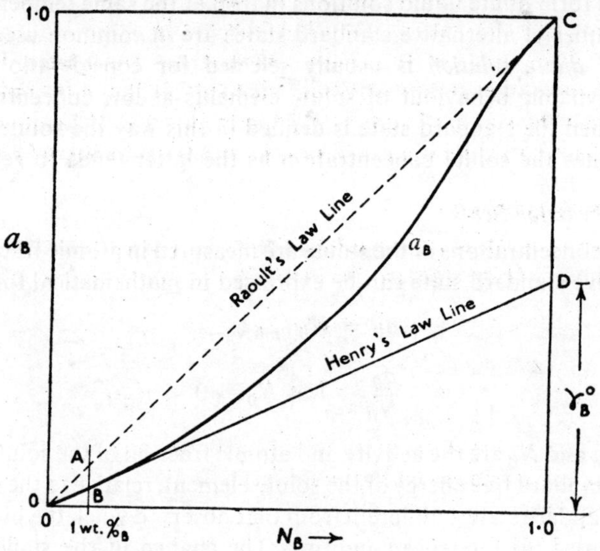

FIG. 2.4. Schematic representation of alternative standard states.

which is equal to γ_B^0, the activity coefficient of the solute at infinite dilution (Figure 2.4). Hence:

$$\Delta G^0 = RT \ln \gamma_B^0$$

$$= 4{\cdot}575T \log \gamma_B^0 \qquad 2.25$$

which is equal to the difference in the standard free energies of the solute with reference to states C and D in Figure 2.4. For a solution which obeys Raoult's law, γ^0 is unity and the change in the standard free energy is zero.

The value of γ^0 for a particular solute cannot be calculated from standard free-energy data and must be determined experimentally. It is usually a function of temperature and should be represented as such, but this information is not always available. However, the temperature dependence can be estimated by use of the following relation when the value of γ_B^0 and the partial molar heat of solution of B at high dilution in iron (H_B^M) are known at the same temperature:

$$\Delta G^0 = H_B^M - \left[\frac{RT\ln\gamma_B^0 - H_B^M}{T}\right].T$$

On substituting the known values of γ_B^0 and H_B^M at temperature T this statement simplifies into an equation of the general type:

$$\Delta G^0 = x - yT$$

where x and y are constants. If the partial molar heat of solution at high dilution has not been determined, a rough assessment of ΔG^0 at other temperatures can be obtained by assuming that the solute conforms to the regular solution laws and by using the relation $H_B^M = RT\ln\gamma_B^0$. This implies that H_B^M is independent of the temperature and that the free-energy change in equation 2.25 is also independent of the temperature. The approximation is often acceptable when γ_B^0 is close to unity, but the error may be large when γ_B^0 differs markedly from unity.

Weight-Per cent Scale

The concentrations of solute elements in iron are usually recorded in units of weight-per cent, and the determination of the solute activities in relation to either the pure solute or the infinitely dilute atom-fraction standard states necessitates first the conversion of the concentrations into atom fractions. This step can be avoided if the standard state is defined such that the activity approaches the weight-per cent concentration at infinite dilution, i.e.,

$$\frac{a_B}{\text{wt.-}\%_B} \rightarrow 1 \text{ as wt.-}\%_B \rightarrow 0$$

This means that the solute activity is unity at one weight-per cent concentration if the solute obeys either Raoult's law (point A) or Henry's law (point B in Figure 2.4) up to this concentration.

The change in the standard free energy accompanying the change to this standard state is equal to the ratio of the activities in the dilute weight-per cent and in the pure element standard states, i.e.

$$B_{\text{(pure)}} \rightleftharpoons B_{\text{(wt.-\% dilute)}}$$

$$\Delta G^0 = G^0_{B\text{(wt.-\% dilute)}} - G^0_{B\text{(pure)}} = RT\ln\frac{a_{B\text{(wt.-\% dilute)}}}{a_{B\text{(pure)}}} \qquad 2.26$$

For a solution which obeys Raoult's law the activity is unity in the two standard states at one weight-per cent B and at pure B respectively, and the standard free-energy change is equal to the atomic fraction ratio of these two concentrations. If the solute is present in the solution at a concentration of X weight-per cent and the atomic weights of the solute and iron are equal to M and 55·85, the atomic fraction of the solute (N_B) is given by the relation

$$N_B = \frac{X/M}{\dfrac{X}{M} + \dfrac{(100-X)}{55{\cdot}85}} \qquad 2.27$$

and this is approximately equal to $0{\cdot}5585/M$ at one weight-per cent concentration. The atom fraction is unity for the pure element and so the change in the standard free energy is given by:

$$\Delta G^0 = RT \ln \frac{0{\cdot}5585}{M}$$

$$= 4{\cdot}575T \log \frac{0{\cdot}5585}{M} \qquad 2.28$$

For example, the atomic weight of manganese is 54·93 and the liquid iron–manganese solution is approximately ideal; the standard free energy of manganese relative to the infinitely dilute weight-per cent solution of manganese in iron is therefore:

$$G^0_{\text{Mn(wt.-\% dilute)}} = G^0_{\text{Mn(pure)}} + 4{\cdot}575T \log \frac{0{\cdot}5585}{54{\cdot}93}$$

$$= G^0_{\text{Mn(pure)}} - 9{\cdot}11T \qquad 2.29$$

When the solute does not conform to Raoult's law, but obeys Henry's law at high dilution, the free-energy change accompanying the change in standard state must include a term which is a measure of the actual deviation from ideal behaviour at high dilution. From Figure 2.4 it is evident that the ratio of the activities for ideal and actual solutions at one weight-per cent (points A and B) is equal to the ratio of the activities at C and D. The latter has already been considered and the accompanying free-energy change is given by equation 2.25. Hence, combining equations 2.25 and 2.28, the general equation for the standard free-energy change accompanying the change of standard state is:

$$\Delta G^0 = G^0_{B\text{(wt.-\% dilute)}} - G^0_{B\text{(pure)}}$$

$$= RT \ln \gamma^0_B + RT \ln \frac{0{\cdot}5585}{M}$$

$$= 4{\cdot}575T \log \gamma^0_B \frac{0{\cdot}5585}{M} \qquad 2.30$$

If the pure solute exists normally as a solid or as a gas, but forms a dilute liquid solution in iron at the temperature considered, the standard free energy of fusion or condensation must be added to equations 2.25 and 2.30 to obtain the overall change in free energy accompanying the change from the pure solid or gaseous element to the chosen alternative standard state. This quantity is readily calculated when either the standard entropy or enthalpy change accompanying the change of phase of the pure solute is known. For example, the free-energy change accompanying the transition from pure solid to pure liquid solute element is zero at the melting point and therefore:

$$S_f = \frac{H_f}{T_f}$$

where the subscript f signifies fusion. At any other temperature the free-energy difference between liquid and solid is given by:

$$G_f = H_f - TS_f \qquad 2.31$$

For example, γ^0_{Ti} in liquid iron is approximately 0·011 at 1,623°C (1,900°K), the atomic weight of titanium is 47·90, and so the free-energy change accompanying the formation of an infinitely dilute weight-per cent solution of titanium in iron in given by:

$$Ti_{(\text{pure liquid})} \rightleftharpoons Ti_{(\text{wt.-\% in Fe})}$$

$$\Delta G^0 = 4{\cdot}575 \times 1900 \log 0{\cdot}011 + 4{\cdot}575 T \log \frac{0{\cdot}5585}{47{\cdot}90}$$

$$= -17{,}020 - 8{\cdot}59T \text{ cal mol}^{-1} \qquad 2.32$$

The melting point of pure titanium is 1,660°C (1,933°K); hence the free energy of fusion must be included. The heat of fusion is 4,500 calories per mole and so, from equation 2.29, the entropy of fusion becomes:

$$S_f = \frac{4{,}500}{1933}$$

$$= 2{\cdot}61 \text{ cal deg}^{-1} \text{ mol}^{-1}$$

Thus for the reaction:

$$Ti_{(\text{pure solid})} \rightleftharpoons Ti_{(\text{pure liquid})}$$

the change in free energy is given by:

$$G_f = +4{,}500 - 2{\cdot}61 T \text{ cal mol}^{-1} \qquad 2.33$$

Adding equations 2.32 and 2.33 we have:

$$Ti_{(\text{pure solid})} \rightleftharpoons Ti_{(\text{wt.-\% in Fe})}$$

$$\Delta G^0 = -12{,}520 - 11{\cdot}2T \text{ cal mol}^{-1} \qquad 2.34$$

The value of γ^0_{Ti} is known only at 1,623°C and the first term in equation 2.34 applies only at this temperature, but values for ΔG^0 at other temperatures could be obtained by assuming that titanium forms a regular solution at low concentrations in liquid iron and setting H^M_{Ti} as a constant at high dilution. However, the value of γ^0_{Ti} quoted indicates that the iron–titanium system is far from ideal and the assumption of regular-solution behaviour is of doubtful validity.

The activities with reference to the alternative standard states can be written into an equilibrium constant in exactly the same manner as that outlined previously when the pure component is chosen as the reference state, and it is also possible to combine activities with respect to different reference states in an equilibrium equation.

For example, the equilibrium constant for the reaction:

$$Ti_{(in\ Fe)} + 2O_{(in\ Fe)} \rightleftharpoons TiO_{2(slag)}$$

is given by:

$$K = \frac{(a_{TiO_2})}{[a_{Ti}][a_O]^2}$$

where round brackets signify a slag component and square brackets indicate a component in solution in the metal. It is convenient to express the activities of oxygen and titanium in iron with respect to the dilute weight-per cent solution, whereas the activities of the titanium oxide are normally given relative to pure TiO_2. The mixed standard states are permissible provided the free-energy equation used to determine K is also expressed in terms of the appropriate standard state for each component of the reaction.

It is frequently convenient to have a measure of the deviation from Henry's law in terms of an alternative standard state and the *activity coefficient* f is defined as the ratio of the activity to the concentration in whatever units are used to define the reference state. Thus, in terms of the dilute weight-per cent standard state, $f_B = a_B/\text{wt.-}\%_B$ and $f_B = 1$ for the concentration range in which the solute obeys Henry's law.

2.5 Activities in Multi-Component Systems

The preceding treatment is applicable only to a dilute binary solution, whereas liquid iron may contain several solutes in dilute solution each of which may be expected to affect the thermodynamic behaviour of the others as a result of the change in the inter-atomic forces created by the addition of the solutes. This interaction frequently causes marked changes in the solute activities and the data for dilute binary solutions do not necessarily apply to more complex systems. It is obvious that the experimental evaluation of the activities of each solute in the presence of all possible combinations and concentrations of other solutes would be an extremely long and tedious procedure. In fact the number of combinations of any two solutes in the solvent iron of direct importance to steelmaking is very large and only a limited number of these have been investigated. Few data are available for dilute solutions containing three or more solutes. However, approximate values of the activities in a multi-component system can be obtained by combination of the thermodynamic data for all the relevant three-component mixtures to which the solvent and the solute considered are common. For example, the activity of a solute X in the presence of solutes Y and Z is obtained by combination of data for the systems Fe–X, Fe–X–Y and Fe–X–Z where in each case iron is the solvent.

Consider first the binary mixture Fe–X. Selecting the infinitely dilute atom-fraction solution as the reference state for X, the activity coefficient of X at any specific concentration is designated f_x^x to denote reference to the binary solution. If a small addition of solute Y is made to this solution, holding constant the concentration of X, the activity coefficient is changed to some new value f_x and the two values of the activity coefficient can be related through a proportionality factor f_x^y. Thus:

$$f_x = f_x^x \times f_x^y \tag{2.35}$$

or

$$\ln f_x = \ln f_x^x + \ln f_x^y \tag{2.36}$$

The factor f_x^y is called the *interaction coefficient* of Y on X. It is a measure of the change in f_x^x caused by the addition of a specific concentration of Y to the solution Fe–X at the *same* concentration of X.

In a similar manner a coefficient f_x^z can be determined for the effect of solute Z on f_x^x. A method is now required to combine the data for these three systems and determine the activity coefficient of X in the quaternary system Fe–X–Y–Z. When the solute Z is added to the solution Fe–X–Y the interaction coefficient f_x^z may change with the concentration of Y and similarly f_x^y may change with the concentration of Z. No general mathematical solution is possible if this condition applies, and each combination of solutes must be investigated experimentally. However, it is frequently found that, within the limits of experimental error, the interaction coefficient f_y^x is independent of the concentration of X and of the concentrations of all other solutes except Y in the dilute solution, in which case the interaction coefficients can be combined by a Taylor-series expansion of f_x. [4] Thus, in the presence of the solutes

X, Y, Z at molar or atomic fractions of Nx, Ny, and Nz the activity coefficient of X is given by:

$$\ln f_x = N_x.\frac{\partial \ln f_x}{\partial N_x} + N_y.\frac{\partial \ln f_x}{\partial N_y} + N_z.\frac{\partial \ln f_x}{\partial N_z} +$$

$$+\tfrac{1}{2} N_x^2.\frac{\partial^2 \ln f_x}{\partial N_x^2} + N_x N_y.\frac{\partial^2 \ln f_x}{\partial N_x \partial N_y} + \ldots \qquad 2.37$$

where the derivatives are taken for the limiting case of zero concentration of all solutes. At very low concentration, N_x^2 and the product of N_x and N_y become vanishingly small and the cross and square terms may be omitted, whilst the function $\partial \ln f_x/\partial N_y$ is equal to $\partial \ln f_x^y$, etc. The equation then simplifies to:

$$\ln f_x = N_x \partial \ln f_x^x + N_y \partial \ln f_x^y + N_z \partial \ln f_z^x \qquad 2.38$$

This approximation is strictly valid only for infinitely dilute solutions, but can usually be applied without introducing a large error to moderately dilute solutions. The partial differential $\partial \ln f_x/\partial N_y$ is usually represented by the symbol ϵ_x^y and it is to be expected that this quantity will be a function of the temperature but, at constant temperature, the value derived for the infinitely dilute solution is constant over the concentration range in which Henry's law is obeyed by the solute X. Within this range the activity coefficient of X is given by

$$\ln f_x = N_x.\epsilon_x^x + N_y.\epsilon_x^y + N_z.\epsilon_x^z + \ldots \qquad 2.39$$

The value of the interaction coefficient is generally obtained from the slope of the graph of $\ln f_x$ against N_Y at constant concentration of X, which is equal to $\ln f_x^y$ at 1 mole per cent of Y if the plot is linear to this concentration.

If the addition of solute Y raises or lowers the value of f_x, one would expect that X would have a corresponding effect on f_y and it has been shown that a simple relationship does exist between the ϵ parameters [4]. Since

$$\partial^2 G/\partial N_x.\partial N_y = \frac{\partial \bar{G}_x}{\partial N_y} = \frac{\partial \bar{G}_y}{\partial N_x}$$

it follows that:

$$\left(\frac{\partial \ln f_x}{\partial N_y} = \frac{\partial \ln f_y}{\partial N_x}\right)_{(N_y \to 0,\, N_z \to 0)}$$

and $$\epsilon_x^y = \epsilon_y^x \qquad 2.40$$

It is often more convenient to express the concentrations in terms of weight-per cent and to use common logarithms; equation 2.37 is then written in the form:

$$\log f_x = \%x\frac{\partial \log f_x}{\partial \%x} + \%y\frac{\partial \log f_x}{\partial \%y} + \%z\frac{\partial \log f_x}{\partial \%z} + \ldots \qquad 2.41$$

The partial derivatives are now represented by the symbols e_x^x, e_x^y, e_x^z etc., and these quantities are again constants over the concentration range in which Henry's law is obeyed. The value of f_x in terms of the weight-per cent reference state is given by:

$$\log f_x = \%x.e_x^x + \%y.e_x^y + \%z.e_x^z + \ldots \qquad 2.42$$

The ϵ and e parameters are related through the following expression:

$$e_x^y = \frac{0{\cdot}2425}{M_y}\epsilon_x^y = \frac{M_x}{M_y}.e_y^x \qquad 2.43$$

where M_x and M_y are the atomic weights of the solutes with iron as the solvent. This relation is strictly valid for infinite dilution, but in normal dilute solutions the error is only significant if the atomic weights of the solutes differ markedly from that of the solvent.

An estimate of the probable effect of one solute on the thermodynamic behaviour of another solute can often be made from consideration of the respective atomic bonding energies. For example, if X interacts more strongly with Y than with the solvent then the activity of X is generally decreased by the addition of Y, so ϵ_x^y is a negative quantity. Conversely, if Y interacts more strongly with the solvent than with X, f_x will tend to increase and ϵ_x^y is positive. Thus, if carbon is the solute considered, in general f_C is lowered by the addition of solutes which tend to form more stable carbides than iron (such as chromium, niobium, vanadium and titanium), whereas less strong carbide formers (such as silicon, sulphur and phosphorus) raise f_C.* This should not be interpreted as a tendency to form chemical compounds in the melt, since all the available evidence points to the contrary. It is more correct to say that in the former case each carbon atom has, on average, a greater number of atoms of the carbide-forming elements surrounding it than the ratio of alloy to iron atoms elsewhere in the solution, whilst in the latter case the silicon and iron atoms tend to associate preferentially, rejecting the carbon atoms. The interaction effects are then seen to arise from the tendency for certain atomic species to form clusters, excluding other species from their

* There are several known exceptions to this general rule and some of these are discussed in Part 4.

immediate vicinity. The atoms forming the clusters are more tightly bound than those in a random atomic solution, whereas the atoms rejected from the clusters are bound less firmly.

The extent of the interaction between two solutes should be influenced by the strength and the number of the various solvent–solute and solute–solute bonds which are formed in the solution. The mean bond energy of any one solute species and its neighbour atoms or molecules is in turn related to the partial molar enthalpy of that species in the solution and it follows that the extent of the interaction should be dependent on the difference between the partial molar enthalpies of the various solutes in the solution. Assuming an ideal random distribution of the atoms, Alcock and Richardson [5] have derived the following relation:

$$RT\epsilon_x^y = H^M_{x(\text{in } x-y)} - H^M_{x(\text{in } x-z)} - H^M_{y(\text{in } y-z)} \qquad 2.44$$

where $H^M_{x(\text{in } x-y)}$ is the relative partial molar enthalpy of x in the solution x–y etc. Rough estimates can be made of the interaction parameter, provided that $H^M_{x(\text{in } x-y)}$ and $H^M_{x(\text{in } x-z)}$ do not differ too markedly, but if this difference is large the assumption of a random atomic distribution is not justified and the estimated values of ϵ may be seriously in error [6].

GRAPHICAL METHOD FOR EVALUATION OF ACTIVITIES

It frequently occurs that $\log f_x$ is not a linear function of wt.-% Y over the concentration range of interest, in which case calculation of the activity coefficients from combination of interaction parameters by means of equations 2.39 or 2.42 may lead to gross error. An approximate method for use in such cases has been suggested by Morris and Buehl [7] and extended by Sherman and Chipman [8]. The principle can be demonstrated by consideration of a system in which the activity coefficient of X is raised by solutes A, B and C and lowered by solutes D, E, F and X. The variation of $\log f_x$ with concentration of a second solute for each of the three-component systems: solvent (Fe)–X–A, solvent–X–B, etc., is shown in Figure 2.5 and the value of $\log f_x$ in the multi-component system is obtained graphically as follows. Consider first the solutes which raise f_x and, starting with the curve which most closely approaches linearity, the value of $\log f_x$ is determined at the actual concentration of the second solute. In Figure 2.5 the line nearest to linear is that for solute A and the value of $\log f_x$ at a concentration a of component A is given by the point l. A point m is now located on the next line of least curvature (curve B) at the same value of $\log f_x$ and the equivalent concentration of B, which causes a change in f_x equal to the change caused by a % of A, is determined. To this is added the actual concentration of solute

B (point *n*). The procedure is repeated by projecting a line horizontally through *n* to intersect the *C* curve at *p*, the actual concentration of *C* is added to the equivalent concentration of *C* and the value of $\log f_x$ in the presence of those solutes which give a positive interaction is given by point *q*. Proceeding in the same way, $\log f_x$ is now determined in the presence of the solutes which give a negative interaction and the value of $\log f_x$ in the solution is finally obtained as the difference between these two quantities. In the limiting case, where all the curves are linear over the appropriate concentration range, the graphical solution is equivalent to the simple addition of the interaction parameters previously described.

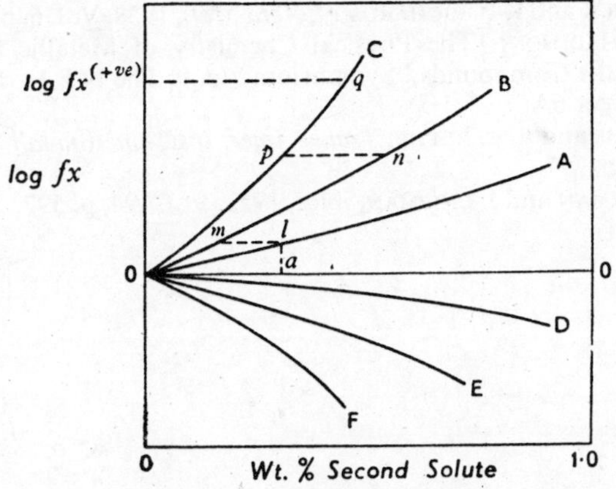

FIG. 2.5. Illustration of the graphical method for evaluating the activity coefficient of one component in a dilute, multicomponent solution.

The methods shown for combining activity coefficients can only be justified at extreme dilution. The error in the evaluation increases as the concentration is raised and the cross and square terms in equation 2·37 become increasingly significant, but the methods can be used with reasonable confidence to determine the activity coefficients of oxygen, carbon, sulphur, phosphorus and other solutes at relatively low concentrations in liquid steel. In fact the graphical method yields values of the activity coefficients which agree remarkably well with experimental values for solute concentrations as high as the range normally found in cast irons [8]. However, it is assumed that the coefficient f_x^y is a function of the concentration of *Y* but is independent of the concentration of *X*. Although this is true within the limits of experimental error for most of the coefficients so far determined, there is no *a priori* reason to assume that it is true in all cases. At high dilution the effect of the variation of the concentration of X on f_x^y should not be large, but extension of this principle to more concentrated solutions may introduce appreciable error.

REFERENCES

1. J. H. HILDEBRAND, *Proc. nat. Acad. Sci., Wash.*, 1927, Vol. 13, p. 267; *J. Amer. chem. Soc.*, 1929, Vol. 51, p. 66.
2. M. REY, *Disc. Faraday Soc.*, 1948, No. 4, p. 257.
3. G. SCATCHARD, *Chem. Rev.*, 1949, Vol. 447, p. 17.
4. C. WAGNER, *Thermodynamics of Alloys*, Addison Wesley, 1952.
5. C. B. ALCOCK and F. D. RICHARDSON, *Acta Met.*, 1958, Vol. 6, p. 385.
6. F. D. RICHARDSON, 'The Physical Chemistry of Metallic Solutions and Intermetallic Compounds,' Symposium No. 9, The N.P.L., London, 1958, Vol. 2, Paper 6A.
7. J. P. MORRIS and R. C. BUEHL, *Trans. Amer. Inst. min. (metall.) Engrs.*, 1950, Vol. 188, p. 317.
8. C. W. SHERMAN and J. CHIPMAN, ibid., 1952, Vol. 194, p. 597.

CHAPTER 3

The Structure and Properties of Slags

3.1 Introduction

The primary function of a refining slag is to serve as a receiver for the unwanted liquid and solid constituents of the metal bath, including the gangue from the iron ores and the elements which are partitioned between the slag and metal. To do this the composition must often be adjusted within narrow limits to secure an optimum removal of the undesirable elements without removing other elements required in the metal. The slag may be required also to fulfil other needs, such as controlling the supply of oxygen from the atmosphere into the metal, or to act as a physical barrier to the passage into the metal of gaseous species such as hydrogen, nitrogen and sulphur. The formation of a liquid slag from solid oxides is endothermic and generally the thermal efficiency of the process is related inversely to the volume of the slag. In processes where heat is supplied from above the metal bath, as in the open-hearth and electric-arc furnaces, the slag acts as a thermal barrier and the rate of rise of the metal temperature is inversely proportional to the thickness of the slag layer.

From thermodynamic considerations it is possible to select the most suitable range of slag compositions which will satisfy the metallurgical requirements with the minimum slag bulk, but the thermodynamic approach gives only the distribution of the elements between the slag and metal in the equilibrium states; it does not predict the rate at which the reactions proceed towards equilibrium. In parts II and III it is demonstrated frequently that an equilibrium condition may not be attained in a reasonable period of time, due to various kinetic factors. The most common rate-controlling factors in the slag for a reaction between the slag and the metal are the diffusion of the reacting species in the slag and the transfer of the species across the slag-metal interface, so the viscosity and surface tension of the slag are also of considerable importance.

The slag must be sufficiently fluid to run freely when it is removed as a liquid from the furnace. Generally, for any composition, the fluidity decreases as the temperature is raised above that of the liquidus and a slag of any composition can be made fluid if the temperature is high enough. However, apart from practical temperature limitations which are imposed by the available fuel and refractory materials, it is desirable economically to operate a process at the lowest temperature at which the chemical reactions can proceed rapidly. Except in certain oxygen-refining treatments, it is rare to find the slag temperature in excess of about 1,700°C, but the melting temperatures of lime, magnesia, alumina and a number of oxides which may be present in liquid slags are considerably greater than this. Thus the slag composition range is frequently a compromise between the physical and the thermodynamic properties and the simplest starting point for the examination of these opposing factors is the consideration of the atomic arrangements which can occur in liquid slags.

3.2 The Structure of Liquid Slags

From the earliest attempts to formulate equations for slag-metal reactions and up to comparatively recent times, it has generally been considered that a liquid slag contains molecular compounds and these are dissociated into the component oxides to an extent which varies with the composition and temperature. The treatment of the subject on these lines by Schenck [1] is a classical example of this approach. However, the results obtained from the measurement of certain physical properties of the melt such as the viscosity, surface tension, electrical conductivity and transport numbers, can only be interpreted satisfactorily if the slag is considered to be a mixture of ions. This has caused a marked change in the type of model proposed to represent the slag structures and the model now accepted envisages a random, disordered mixture of anions and cations.

The direct evidence of the structures in the melt is very limited and this fact reflects the experimental difficulties which are involved in obtaining such data. There is some indirect evidence, such as the low values of the entropies of fusion of the oxides and the comparatively small change in viscosity on melting, which suggests that no major structural changes are involved on changing from the solid to the liquid state, and some information about the structure of a liquid slag can therefore be inferred from examination of the solid structures. Ferrous refining slags invariably contain silica, and silica-rich liquids are characterised by a very weak tendency to crystallise; on rapid cooling a glass is formed which differs from a solid to the extent that the long-range atomic ordering of the crystalline state is absent. A glass is really a supercooled liquid and a structural model of these liquids can be inferred from the examination of glasses by X-ray and other techniques which were developed extensively by Bragg, Warren and others [2–5].

FACTORS AFFECTING THE CO-ORDINATION OF THE IONS

In the solid state the structure of the basic oxides such as CaO, MgO, MnO and FeO is a close-packed array of oxygen ions with the cations accommodated in the intervening interstices. Most of these oxides form structures similar to that of sodium chloride (Figure 3.1), in which each cation is *co-ordinated* with (i.e. surrounded by and in very close proximity to) six oxygen anions and this is called an octahedral co-ordination. Each oxygen ion is co-ordinated similarly with six cations. A few basic oxides exhibit an eight-fold co-ordination but, in contrast, the co-ordination is only four-fold (or tetrahedral) for the cations in the acid oxides

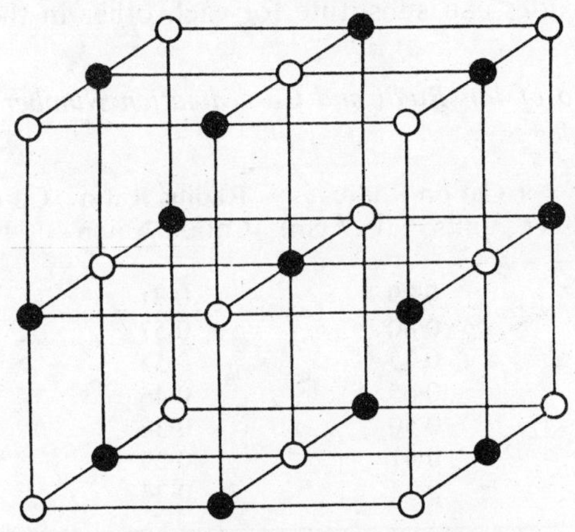

FIG. 3.1. The structure of sodium chloride. Full and open circles represent the centres of the atoms of the two elements.

silica, alumina and phosphorus pentoxide. This is because the co-ordination number of a cation in an oxygen-ion lattice is determined largely by the relative sizes of the cation and the cavities between the oxygen ions. By simple geometry it can be shown that this is related to the ratio of the cation and the anion radii and typical values of these are given in Table 3.1 [6].

If the radius ratio is less than 0·414 it is not possible to place six oxygen ions in contact with one cation and the sodium chloride type of structure is unstable. The tetrahedral co-ordination is usually formed when the radius ratio is in the range 0·225 to 0·414, whilst an eight-fold co-ordination is usually found for radius ratios greater than 0·732.

For a constant electronic change on the cation, the bond energies between the cation and each oxygen ion in the co-ordination shell must increase as the co-ordination number is lowered. That is, a cation in a four-fold co-ordination is more firmly bonded to its oxygen ions and is less readily separated from them than when the co-ordination is six or eight-fold. Further, the electron charge on the cations of the basic oxides listed in the Table is 2, whereas the electron charge on the aluminium, silicon and phosphorus ions is 3, 4, and 5 respectively. Hence, taking into account the differences in co-ordination, it is apparent that the cation-oxygen bonds are considerably stronger in the acid than in the basic oxides.

The bond numbers and bond strengths have an important bearing on the inter-solubilities of the various oxides. Thus the cations of many of the basic oxides can substitute for each other in the anion lattice without changing the structural arrangement and fairly extensive ranges of solid solutions are formed. The cations of the acid oxides can also interchange, but to a very much more limited extent because of the differences in the electron charges, and the replacement of a silicon or phosphorus ion by a basic oxide cation is highly improbable.

Table 3.1 *Ratio of Ion Radii and Co-ordination Number for some Pure Oxides*

Oxide	Cation Radius (K_x units = 10^{-8} cm)	Radius Ratio Cation/Anion	Co-ordination Number
CaO	0·99	0·71	6
MnO	0·80	0·57	6
FeO	0·75	0·54	6
MgO	0·65	0·46	6
Al_2O_3	0·50	0·36	4
SiO_2	0·41	0·29	4
P_2O_5	0·34	0·24	4

In the liquid state the regular arrangement of the ions is destroyed; each cation remains co-ordinated on average with approximately the same number of anions as in the solid state, but the cations are rapidly interchanging their positions and the anions are interchanging likewise. The mobility of a cation in the liquid is related to the energy required to transfer it from one cation site to another, and this in turn is related to the electron charge and co-ordination of the cation. Hence, from the considerations of the solid lattice we should expect to find that the cations in the oxy–acid groups are less mobile than the cations formed by the basic oxides and this general tendency is observed in real liquid solutions.

SILICATE STRUCTURES

The structure of pure silica is now fairly well established and models have been proposed for the structural changes which occur when basic oxides are added to silica-rich melts. In pure solid silica each silicon ion shares one electron charge with each of the four oxygen ions forming the tetrahedron, so that each oxygen ion has one residual negative charge and the tetrahedral group carries four negative charges (i.e. SiO_4^{4-}). Electroneutrality is obtained when each oxygen ion is shared by two tetrahedra (i.e. linked to two Si^{4+} ions) and the cation/oxygen ratio is that for pure silica (SiO_2). The structure can be represented in two dimensions by the pattern in Figure 3.2*a*; Figure 3.2*b* shows the corresponding tetrahedral pattern, where the tetrahedra are linked together to form a three-dimensional pattern of hexagonal cells. The structure of pure liquid silica is probably very similar, although the regularity of the solid lattice is distorted (Figure 3.2*c*), and it can be regarded as a gigantic molecule. This hypothesis is supported by the fact that the viscosity of the liquid is very high for, in general, the viscosity is increased as the dimensions of the molecules are increased.

When a basic oxide is added to this liquid each added oxygen ion enters the network and separates the corners of two tetrahedra, whilst the added cation remains adjacent to the separation and is accommodated within the holes or cells of the structure. With progressive additions of basic oxides the three-dimensional array is steadily broken down until eventually the silica is present as discrete SiO_4^{4-} groups. When this stage is reached further additions of basic oxides cause no change in the structure of the silicate ions. The viscosity of the melt decreases rapidly as the three-dimensional network is disrupted, but the viscosity/composition relation is non-linear [7, 8]. On this evidence it has been suggested [7] that a series of silicate ion complexes is formed, ranging from discrete SiO_4^{4-} units for liquids more basic than the ortho-silicate ($2MO.SiO_2$) composition (i.e. where sufficient oxygen ions are available to satisfy all the charges on the Si^{4+} ions) through two-dimensional chains with N_{SiO_2} in the range 0·33 to 0·50 to two-dimensional rings with 0·50 to 0·67 N_{SiO_2} and so on up to the three-dimensional network of pure silica.

The interpretation of the results of some recent and more sensitive measurements of the viscosity [9], partial molar volumes [10] and electromotive force [11] of liquid slags has required some adjustment to this model and two modifications have been proposed on the basis of these measurements. In one of these [9] it is assumed that a progressive breakdown of the three-dimensional network is continuous up to a concentration of 0·12 mole-fraction of a divalent oxide. When this concentration is exceeded the structure changes abruptly into a mixture of anion complexes with the general formula $(Si_nO_{2n+3})^{6-}$ and further additions of basic oxides cause a reduction in the size of these complexes. Rings, chains and discrete $(SiO_4)^{4-}$ groups are considered to form at the same silica concentrations as in the earlier model. The structural

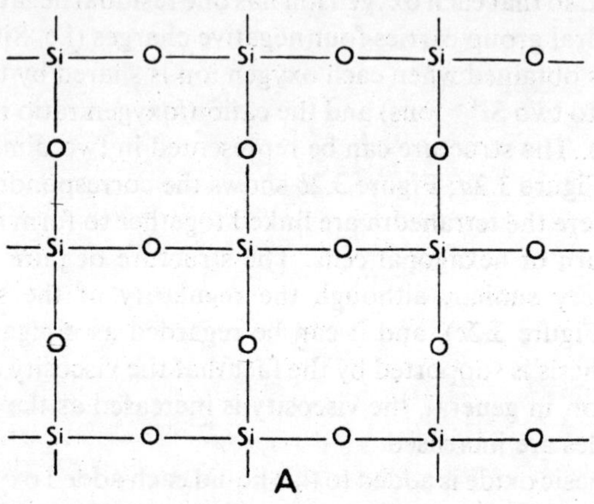

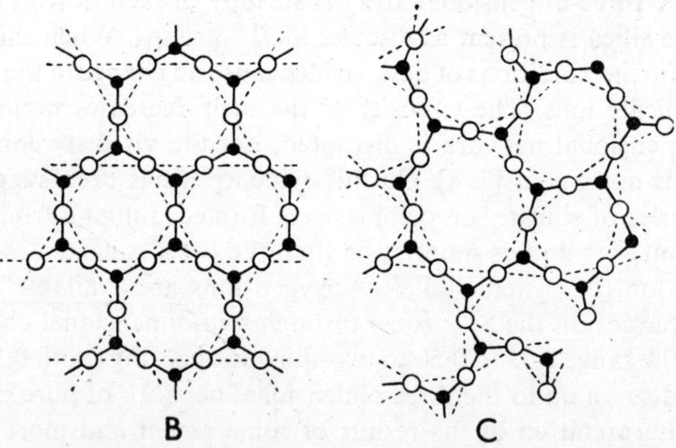

FIG. 3.2. Schematic representation of the structure of pure silica to show (*a*) a simplified two-dimensional pattern which is sometimes used to describe (*b*) the tetrahedral arrangement of crystalline silica. Figure 3*c* represents the structure of silica in the liquid state. Open and solid points are the centres of the atoms of oxygen and silicon respectively. (Figures 3*b* and 3*c* are after Richardson, *Chemistry and Industry*, 1952, p. 50.)

breakdown is summarised in terms of the ratio of oxygen to silicon ions as follows:

Ratio O/Si	*Molecular Formula*	*Structure*
2:1	SiO_2	All corners of tetrahedra shared
5:2	$(MO.2SiO_2)$	One broken link per tetrahedra
3:1	$(MO.SiO_2)$	Two broken links per tetrahedra
7:2	$(3MO.2SiO_2)$	Three broken links per tetrahedra
4:1	$(2MO.SiO_2)$	All links broken. Discrete SiO_4^{4-} tetrahedra

Electro-neutrality considerations require the presence of more than one type of ion complex between each of these fixed ratios and it is probably more correct to consider that each complex can exist over a range of silica concentrations, with a peak in the distribution curve at the silica concentration corresponding to the Si/O ratio in the complex.

The second alternative model is derived with particular reference to the monovalent (alkali) silicates, in which it is assumed that the three-dimensional network is replaced by two immiscible liquids in the concentration range 0·67 to 0·88 N_{SiO_2} [10]. One of these liquids is structurally similar to pure silica and the other, which approximates to the composition $(M_2O.2SiO_2)$, forms a thin film separating islands of the silica-rich structure. At higher basicities the breakdown is considered to be similar to the sequence given for the earlier model.

Either of these alternative models is consistent with the observed changes with composition of the structure-sensitive properties of silicate melts but, from the present point of view, the essential feature is the continuous reduction in the complexity of the silicate units as the basicity is raised up to the orthosilicate composition. The system FeO–SiO_2 forms the only series of binary silicates whose properties are known not to conform to one or other of these models [12, 13], for the viscosity passes through a maximum at a composition corresponding to the orthosilicate ratio. However, in this case it is possible that the change in viscosity which arises from the alteration of the anion linkage is masked by a larger viscosity change caused by the variation with composition in the ratio Fe^{+++}/Fe^{++} in the non-stoichiometric wüstite phase.

At constant composition the viscosity [14, 15], surface tension [16, 17] and electrical conductivity [18, 19] of silica-rich melts all decrease as the temperature is raised and it has been suggested [19] that this could be due to the dispersion of an absorbed layer of silica at the surface of the slag. Increasing the temperature above the liquidus should cause a rapid decrease in the surface concentration and alter the physical properties accordingly. The tendency to form an absorbed layer usually increases with increasing tendency to immiscibility in the liquid state and it is interesting to note that a more normal negative coefficient of electrical conductivity is obtained with the alkali silicates which do not form extensive regions of liquid immiscibility [16]. Alternatively, these changes could arise from a progressive breakdown of the linkages between the tetrahedra as a result of the increase in thermal energy obtained by heating to higher temperatures. It is probable that all the bonds are broken at temperatures close to the boiling point of silica and the silicon and oxygen ions are mixed in a similar manner to the cations and anions in a molten salt. That is, at temperatures close to the boiling point the structure of the liquid silicates approaches the random state of a gas, whereas at temperatures near to the liquidus the structure resembles more closely that of the solid. In the intermediate temperature range there is probably a continuous transition from the one limiting condition to the other.

ALUMINATE AND PHOSPHATE STRUCTURES

Similar considerations apply to mixtures of other network-forming ions and aluminium, boron and phosphorus can each form linked oxygen-anion structures in the solid state. From X-ray and other data there is strong evidence that these oxides can adopt a four-fold co-ordination with oxygen and substitute for silicon in the linked tetrahedral structures, provided that the cation/anion ratio is also adjusted in order to preserve electro-neutrality. It is probable that these trends are followed also in the liquid phase, since the misfit energy arising from the replacement of cations of dissimilar ionic size should be negligible in the liquid state.

The phosphorus ion carries five positive charges and, consequently, when phosphorus replaces silicon in four-fold co-ordination with oxygen ions the electron field strength is increased and the oxygen ions are more firmly tied to the phosphorus than to the silicon. Conversely, the replacement of silicon by aluminium, with only three positive charges, weakens the electron field and the oxygen ions are less firmly tied to the aluminium ions. These changes are reflected in the thermodynamic properties of the melts. For example, it is generally found that the activity of a basic oxide is lowered increasingly by equal additions of alumina, silica and phosphorus pentoxide.

3.3 Liquid Immiscibility

BINARY SILICATE MELTS

Regions of liquid immiscibility are found in many of the binary and multi-component constitutional diagrams of mixtures of the divalent basic oxides with the complex–acid–ion formers, whereas the monovalent basic oxides form a complete range of solutions with these anion groups. When liquid immiscibility exists it is generally, but not invariably, located in that part of the system where extensive anion-linkage is expected, as

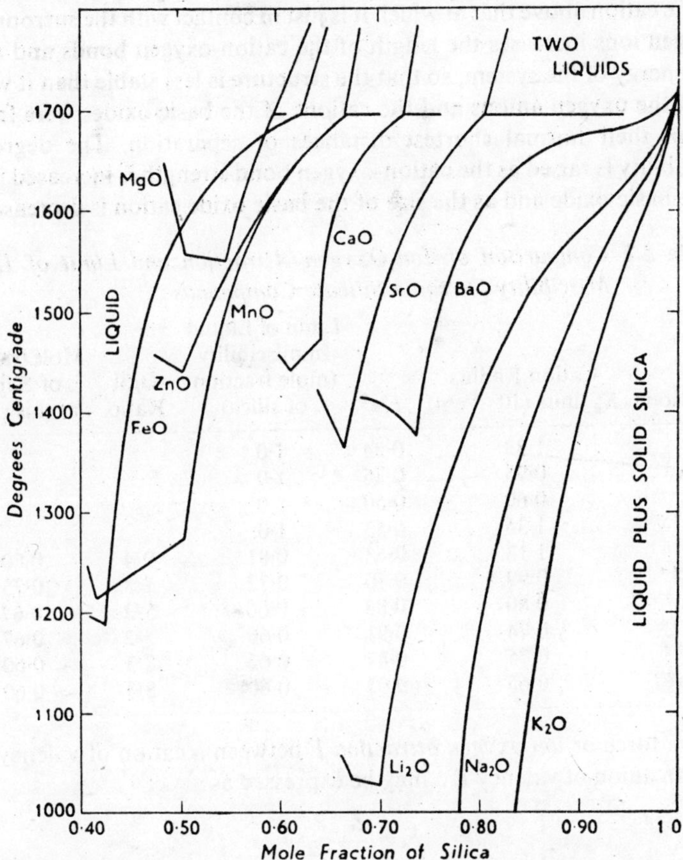

FIG. 3.3. Silica-saturation surfaces for various binary silicate systems.

can be seen from examination of the liquidus-composition curves for a number of the binary silicate systems which are assembled together in Figure 3.3.

The immiscible behaviour has been attributed to the tendency of the basic oxide cation to retain its normal co-ordination with oxygen in preference to the higher co-ordination which it must adopt if it enters a hole or cell in the silicate lattice [20]. In the three-dimensional silica structure each cation is required to enter a hole much larger than itself, in which it is co-ordinated with twelve oxygen ions. The co-ordination decreases as the network is broken down, but only falls to a value of six at the ortho-silicate composition. An increase in the co-ordination of the cation above that at which it is just in contact with the surrounding oxygen ions increases the length of the cation-oxygen bonds and raises the energy of the system, so that the structure is less stable than it would be if the oxygen anions and the cations of the basic oxides were free to adopt their normal shortest distances of separation. The degree of instability is raised as the cation-oxygen bond strength is increased in the pure basic oxide and as the size of the basic oxide cation is decreased.

Table 3.2 *Comparison of Ion-Oxygen Attraction and Limit of Liquid Miscibility for some Silicate Compounds*

Cation	Cation Radius K_x units (10^{-8} cm)	F/e^2	Limit of Liquid Immiscibility (mole fraction of silica)	O/Si Ratio	Mole fraction of Silica at this ratio
K^+	1·33	0·24	1·0		
Na^+	0·95	0·36	1·0		
Li^+	0·60	0·50	1·0		
Ba^{2+}	1·35	0·53	1·0		
Sr^{2+}	1·13	0·63	0·81	9/4	0·80
Ca^{2+}	0·99	0·70	0·72	7/3	0·75
Mn^{2+}	0·80	0·83	0·66	5/2	0·67
Zn^{2+}	0·74	0·91	0·66	5/2	0·67
Fe^{2+}	0·75	0·87	0·62	8/3	0·60
Mg^{2+}	0·65	0·95	0·60	8/3	0·60

The force or *ion-oxygen attraction* F between a cation of valency Z^+ and an anion of valency Z^- may be expressed as:

$$F = \frac{Z^+ Z^- e^2}{r^2} \qquad 3.1$$

where e is the electron charge and r is the mean distance of separation between the centres of the ions. Calculated values of F/e^2 are compared with the limit of liquid immiscibility for a number of binary silicate systems in columns 1 to 4 of Table 3.2. The values of F/e^2 for the pure oxides of aluminium, silicon and phosphorus are respectively 3·2, 4·9 and 6·0.

The ion-oxygen attractions of the alkali-group oxides are evidently of insufficient strength to detach oxygen ions from the silicate complexes and complete miscibility exists at all liquid compositions, although with decreasing basic-cation radius the inflections in the liquidus curves show a tendency towards unmixing. In the divalent cation group this trend is continued with barium oxide, where the liquidus curve indicates that a single-phase liquid is only just stable at high silica concentrations. Immiscible liquids are formed with further increase in the ion-oxygen attraction and the two-liquid region extends to lower silica concentrations as the cation-oxygen bond strength of the basic oxide approaches more closely to the strength of the Si—O bond.

It has been suggested [21] that the limits of miscibility in these systems correspond to the compositions of specific silicate complexes, and the last two columns in Table 3.2 show the simplest ratio of O/Si which corresponds closely to the miscibility limit of each system and the mole fraction of silica at this ratio. It is probable that each of these O/Si ratios corresponds to a peak in the distribution curve of one of the silicate-complex ring or chain structures and, if this is correct, the miscibility

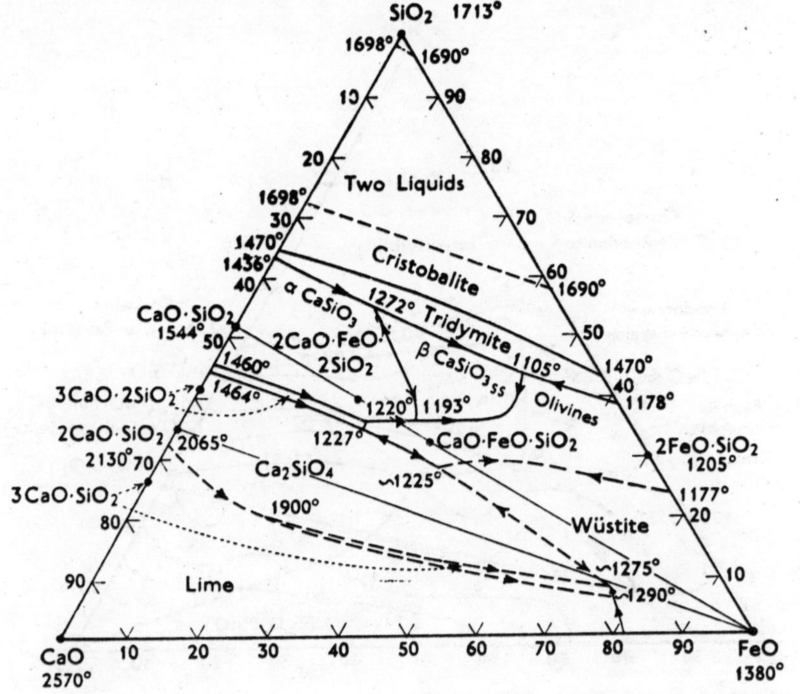

FIG. 3.4. Liquidus surfaces in the ternary system CaO–FeO–SiO_2. (After Bowen, Schairer and Posnjak, *Amer. J. Sci.*, 5th series, 1933, Vol. 26, p. 204.)

limits may be interpreted as evidence that, with increasing silica concentration, the basic oxide cations are accommodated within the silicate lattice until a structure of a specific complexity is predominant. At this stage the change in co-ordination imposed upon the basic cation becomes excessive and a state of lower energy is obtained when the liquid separates into two phases, one of which retains the structure corresponding to the miscibility limit and the other is nearly pure silica. In this respect it seems significant that the miscibility limits for MnO and ZnO correspond closely to the composition for the transition from chain-type to ring-type silicate structures.

The miscibility limit for ferrous oxide is inconsistent with the calculated ion-oxygen attraction force. The value of Z/e^2 which is given in Table 3.2 is calculated for stoichiometric ferrous oxide, but the miscibility limit refers to the solubility of the non-stoichiometric wüstite phase in liquid silica, and it is possible that the miscibility limit for the stoichiometric oxide is at approximately 33 mole-per cent FeO, in accordance with the limits for MnO and ZnO.

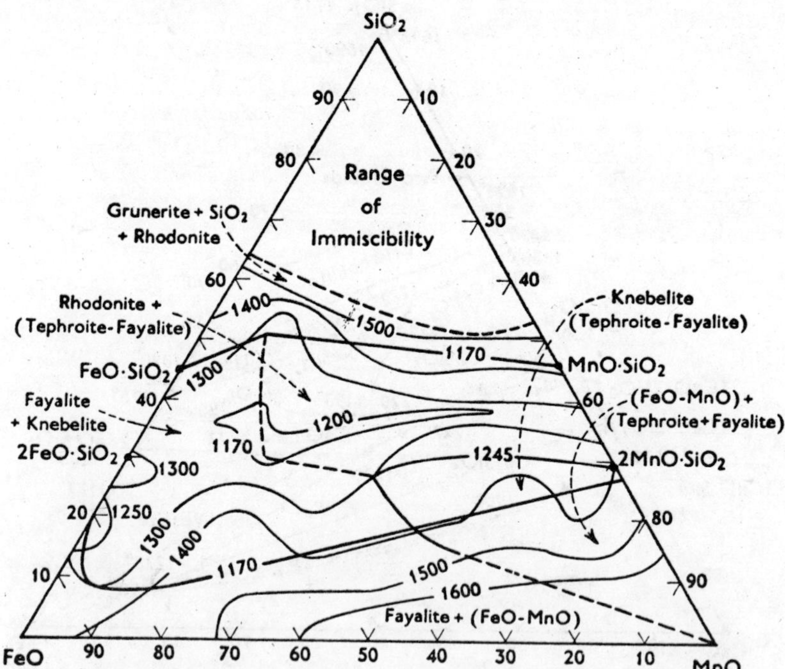

FIG. 3.5. Liquidus surfaces in the ternary system FeO–MnO–SiO_2. (After W. R. Maddocks, *J. Iron St. Inst.*, Carnegie Schol. Mem., 1935, Vol. 24, p. 64.)

TERNARY SILICATE MELTS

When silica is mixed with any two of the basic oxides which form immiscible liquids in the binary silicate systems, a two-liquid region, which joins the miscibility limits of the binary systems, is formed at the silica apex of the ternary system as can be seen from the constitutional

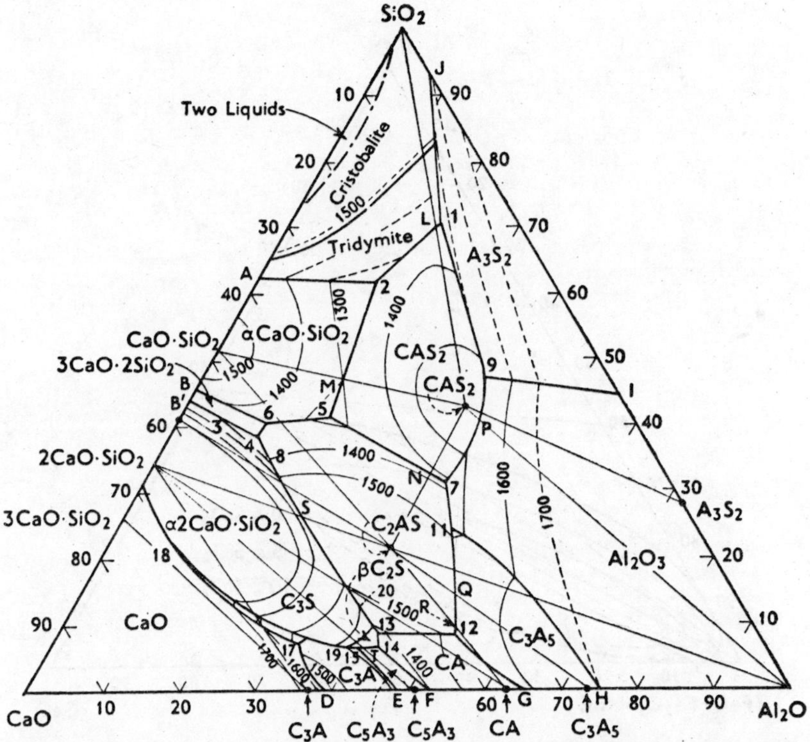

FIG. 3.6. Liquidus surfaces in the ternary system CaO–Al_2O_3–SiO_2. (After Greig, *Amer. J. Sci.*, 5th series, 1927, Vol. 13, p. 41, and Rankin and Wright, *Amer. J. Sci.*, 4th series, 1915, Vol. 39, p. 52.)

diagrams for ternary mixtures of FeO–CaO–SiO_2 and FeO–MnO–SiO_2 which are reproduced in Figures 3.4 and 3.5. Generally, the miscibility limit approximates to a straight line across the ternary system, which suggests that the change in the mean co-ordination of the basic oxide cations is almost a linear function of the ratio of their concentrations in the liquid. When only one of the binary systems exhibits liquid immiscibility the immiscible region extends for only a short distance into the ternary system and this type of behaviour is portrayed, for example, by the system $CaO–Al_2O_3–SiO_2$ (Figure 3.6).

PHOSPHATE MELTS

In phosphate melts the PO_4^{3-} ions exist as separate units at all compositions corresponding to, and more basic than, the ortho-phosphate

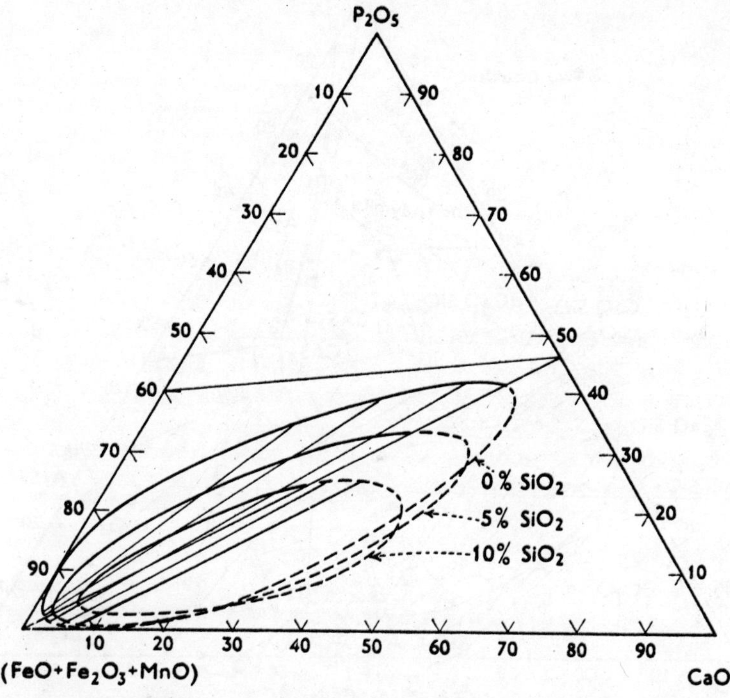

FIG. 3.7. The region of liquid immiscibility in the system $CaO–(FeO+Fe_2O_3+MnO)–P_2O_5$ and showing the effects of various concentrations of silica. (After Oelsen and Maetz, *Mitt. K.-Wilh.-Inst. Eisenforsch.*, 1941, Vol. 23, p. 195.)

compounds (e.g., $3MO.P_2O_5$); at more acid compositions it is probable that a series of linked anion complexes is formed, similar to the behaviour of the silicate melts. Since the tetrahedral P—O bond strength is greater than that of the Si—O bond, it is expected that the liquid miscibility limits of the various P_2O_5–basic oxide systems are located at lower concentrations of the basic oxides than in the corresponding silicate melts. These miscibility limits have not been determined experimentally because of the low boiling point of P_2O_5, but the data which are available indicate that if, in fact, liquid immiscibility could exist with mixtures of P_2O_5 and any of the basic oxides in Table 3.2, the range of existence of the two-liquid region would be very restricted.

In each of the ternary systems FeO–CaO–P_2O_5, MnO–CaO–P_2O_5 and FeO–Na_2O–P_2O_5 a lenticular-shaped region of liquid immiscibility extends from one of the basic oxide apices towards the opposite side of the ternary system. In the first of these systems liquid mixtures which lie on the join between the compositions of the binary compounds 3FeO.P_2O_5 and 3CaO.P_2O_5 are completely miscible, whereas the join FeO–3CaO.P_2O_5 corresponds with the major axis of the two liquid region (Figure 3.7). Similar behaviour is observed in the other two ternary systems. On this evidence it is suggested that the basic oxide cations in the ortho-phosphate liquids can interchange freely their positions, but when FeO or MnO is added the liquid separates into a solution consisting mainly of Ca^{2+} and PO_4^{3-} ions and a second solution which is largely Fe^{2+} (or Mn^{2+}) and oxygen ions. It is probable that when oxygen ions are present in excess of the number required to satisfy all the P^{5+} bonds the basic oxide with the greater ion-oxygen attraction is associated preferentially with the excess of oxygen ions. If this interpretation is correct it is surprising that similar regions of liquid immiscibility are not found in the more basic composition ranges of the ternary silicate melts. The activity data for ferrous oxide in FeO–CaO–SiO_2 melts presented in Chapter 12 can be interpreted as evidence that this system does tend to separate into two liquids at high ferrous oxide contents, but the phase diagram indicates that a complete breakdown of the liquid does not occur.

3.4 The Viscosity of Liquid Slags

The viscosity of a liquid is related directly to the size of the ionic or molecular structural units and therefore it is expected that the viscosity of a liquid slag is raised if the structural complexity is increased, for example, by lowering the basicity. Reference has already been made to the interpretation of viscosity measurements to yield information concerning the structure/composition relation of silicate liquids. The data on slag viscosities are far too extensive to review here in detail, but the salient points can be summarised as follows:

In the systems CaO–SiO_2 [9], FeO–CaO–SiO_2 [13, 22], FeO–MnO–SiO_2 [22] and CaO–MgO–Al_2O_3–SiO_2 [23–28] the viscosity increases rapidly with increasing silica concentration for all compositions more acid than the meta-silicate ratio. Since alumina can replace silica to a limited extent in the network structures, it is expected that the viscosity is raised when alumina is added to an acid melt whilst holding constant the ratio of basic oxides to silica and, in support of this contention, it has been found that silica and alumina are practically equivalent in their effect on the viscosity of a blast furnace slag [23, 25, 26].

Small additions of titanium oxide have a negligible effect on the viscosity of an acid slag. The radius ratio Ti/O is 0·48, which is close to the lower limit for the six-fold co-ordination structures and, from the data on the structure of glasses, it appears that titanium ions can substitute for silicon ions to a very limited extent in the tetrahedra. When this limit is exceeded the titanium enters into a six-fold co-ordination and acts as a flux which lowers the viscosity [22, 29], but the fluxing effect is less pronounced in the presence of alumina [29] and it is probable that the substitution of Al^{3+} for Si^{4+} ions in the tetrahedra allows a more extensive replacement also of Si^{4+} by Ti^{4+} ions. When large amounts of titania are present the slag may become saturated with the oxide and the viscosity is increased by the presence of solid particles which remain suspended in the slag.

For compositions more basic than the ortho-silicate ratio the anions exist as discrete entities and the viscosity is not affected markedly by structural considerations. Further generalisation is then not justified, for it is difficult to distinguish between the viscosity changes brought about by structural alterations and those which result from a change in the liquidus temperature of the slag when other oxides are added to the melt. The small variations in viscosity observed in the more basic melts are due largely, if not entirely, to the latter effect.

The above considerations apply to the bulk properties of the liquid slag, but the structure of the surface layers may differ markedly from that of the bulk liquid. For example the surface tension of liquid basic oxides is lowered progressively by additions of silica [16, 29–31] or of phosphorus pentoxide [31] and it is thought that this is due to the absorption of a thin layer of these anions on to the slag surface. It is estimated that for liquid calcium silicates the absorbed layer is almost pure silica when the bulk silica concentration is greater than 0·60 mole-fraction and when N_{SiO_2} in the bulk liquid is 0·50 the surface concentration is about 0·67 [31]. The anions in the surface layer are joined together into chains, rings or sheet complexes in the same way as those in a bulk slag of similar composition and so, for all compositions less basic than the ortho-silicate, the viscosity at the slag surface is greater than that of the bulk slag. This aspect is very important with regard to the formation of a foaming slag, for the absorption film tends to stabilise the foam [22, 32].

3.5 Departures from Purely-Ionic Bonding

POLARISATION

Many oxides exhibit a mixed type of ionic and covalent (homopolar) bonding in the solid state. Generally the amount of ionic bonding is lowered as the cation size and its co-ordination number in the oxide are decreased and this is illustrated by the estimations of the amount of ionic bond in the following oxides [6].

Oxide	*Per cent Ionic Bond*	*Co-ordination Number of Cation*
CaO	80	6
MgO	75	6
Al_2O_3	64	4
TiO_2	60	4
SiO_2	50	4

More recent work has suggested that these values are too high and that Si—O is only 37 per cent and Mg—O is only 55 per cent ionic [33].

The statement of the bond types in this way tends to be misleading for it gives the impression that two distinct forms of electron bonds are present in the oxides, whereas the deviation from purely ionic binding is due mainly to the *polarisation* of the cation-oxygen bonds. Polarisation is the distortion of the electron field of an ion by the electron fields of other adjacent ions which causes a decrease in the bond length and an increase in the bond strength [34] until, in the limit, an electron is transferred from the cation to the anion and the bond is classed as covalent. The ionic and covalent bonds are thus the limiting conditions of a continuous transition in bonding and the proportions of ionic and covalent linkages in a structure with mixed bonding indicate the position of this structure with respect to the bond transition path.

The polarising power of an anion is proportional to the strength of its electron field and it increases with decreasing radius and increasing charge on the ion. The polarising power is affected also by any electron shell whose energy levels overlap the valence electron shell of the ion. This type of overlap is found in the elements of the *B* sub-groups and the transition elements of the periodic table, with the result that elements such as zinc, manganese and iron are able to exert a greater polarising force on the surrounding anions than would be expected solely in terms of the cation size and charge. The amount of non-ionic bonding in the pure oxides of these elements is probably greater than that for the Mg—O bond.

An ion is not polarised by ions of opposite charge in a symmetrical electron field, for then the electron field around any one ion is uniform in all directions. Thus in mixtures of any one anion with a series of cations the extent of the polarisation of the cation-anion bond is increased as the co-ordination of the ions is lowered and the electron field becomes more asymmetrical. This can be seen by comparing, for example, the proportion of non-ionic bonding in solid lime and silica. Generally, the extent of polarisation is reflected also by a variation in the spacing of the cations from the anions and at room temperature the measured bond length of the Ca—O bond is almost identical with the sum of the ionic radii, whereas the Si—O distance is markedly less than the sum of the radii.

In the liquid state the bonding is very considerably weaker and the ions are able to interchange rapidly their positions, so that for any oxide the cation-oxygen bonds are more ionic when molten than they are in the solid state. However, the existence of complex structures such as are found in liquid silicates implies that the non-ionic bonding may remain fairly extensive in the liquid, and there is fairly strong evidence that a substantial portion of non-ionic bonding is retained also in liquid ferrous oxide (wüstite).

The extent of the ionic polarisation may be modified when cations with differing polarising force are mixed with a common anion. For example, when a divalent basic oxide (MO) is mixed with silica, the polarising force exerted on the oxygen ion in each M^{++}—O^{--}—Si^{4+} linkage by the silicon ion is very much stronger than that applied by the basic oxide cation; as a result the polarisation of the M—O bond is reduced and may be entirely eliminated. Electrical conductivity measurements have been interpreted as evidence that small additions of silica remove completely the small element of covalency in Ca—O and Mg—O bonds [35], whereas the more strongly polarised Fe—O bond is not completely ionic at silica saturation [36, 37]. Anomalous values for the energies of activation for conductance of magnesium, manganese and iron oxides in binary silicate melts have also been ascribed to the influence of polarisation [19].

THE EFFECT OF NON-IONIC BONDING ON THE THERMODYNAMIC PROPERTIES

Heats of Formation

The bond strengths in the solid state of a series of basic oxides with a constant cation/anion ratio increase and approach more closely to the strength of the Si—O bond as the cation size is decreased. Thus as the size of the basic oxide cation is decreased, the cation is less readily accommodated in the interstices between the silica tetrahedra and the stability of the structure is lowered. This can be demonstrated by comparing the heats of formation of the binary silicates which are formed by a series of divalent (MO) oxides (Table 3.3).

Because the number and strength of the Si—O bonds are approximately constant at a constant mole ratio of basic oxide/silica, the heats of formation are related to the strengths of the bonds in the basic oxides. The data in Table 3.3 show that, in general, the heats of formation decrease as the size of the basic oxide cation is decreased and the strength of the M—O bond is raised, but the values for the silicates of manganese, zinc and ferrous oxide are not consistent with this trend. It has been suggested [38] that the asymmetric polarisation of the M—O—Si bonds produces both an increased attraction between the two ion pairs and a negative heat of formation. If a large M^{++} ion is replaced by one which is smaller, or

Table 3.3 *Heats of Formation of Binary Silicate Compounds at 25°C*

Basic Oxide	BaO	SrO	CaO	MnO	ZnO	FeO	MgO
Cation radius	1·35	1·15	0·99	0·80	0·74	0·75	0·65
$-\Delta H_{kcal\,mol^{-1}}$(ortho-silicate)	64·5	50·0	30·2	11·8	7·0	10·3	15·1
$-\Delta H_{kcal\,mol^{-1}}$(meta-silicate)	38·0	31·2	21·5	5·9			8·7

by one whose M—O bonds are more strongly polarised, the attraction is diminished and the heat of formation is less negative or, in the extreme case, is positive and the system tends towards immiscibility. The cation-oxygen bonds of the pure solid oxides of manganese, zinc and iron (FeO) are polarised more extensively than the bonds of the other oxides listed in Table 3.3, and this interpretation would explain their inconsistent behaviour.

Mixing Enthalpies and Entropies

Since the structure of a silicate melt resembles fairly closely the structure of the solid, it is expected that the heats of mixing are negative in the binary silicate liquids and the mixing enthalpies (H^M) show similar trends to those shown by the respective heats of formation. That is, H^M should decrease with decreasing cation size, increasing cation charge and increasing non-ionic bond in the pure basic oxide. Again, in the extreme, H^M is positive and, if the entropy is ideal, the activities deviate positively from Raoultian behaviour. Assuming that liquid binary mixtures of ferrous oxide and silica obey the regular solution equations, it has been calculated [39] that H^M is positive in the miscible liquid region. However, it is improbable that mixtures of a basic oxide with silica possess an ideal entropy of mixing (S^M), for it has been shown that the basic oxide cations which normally enter into a high co-ordination with oxygen are unlikely to interchange positions with the tetrahedral co-ordinated silicon ions, so random mixing is prevented and the mixing entropy is less than ideal. Conversely, the progressive breakdown of the silica network as basic oxides are added increases the randomness of mixing with respect to the pure silica structure and S^M is increased. At constant mole-fraction of silica the increase in S^M is dependent on the valency of the cation in the basic oxide [40], for each divalent cation which is added becomes associated with two silica tetrahedra to form an ion-group, whereas each added monovalent cation is associated with only one tetrahedra. Thus the number of new ionic groups which are formed by each mole of a monovalent basic oxide is equal to twice the number formed by a mole of divalent oxide and the mixing entropy is increased accordingly.

From structural considerations it is expected that the decrease in S^M arising from the restricted mixing will predominate at concentrations up to about 12 mole-per cent of basic oxide, for this appears to be the limit of the stability of the three-dimensional silica network. Thereafter, the increase in S^M due to the structural breakdown becomes predominant and with increasing basic oxide content the excess mixing entropy may change from negative to positive. There are insufficient data to substantiate this hypothesis in the binary silicate systems, but a variation of this type has been suggested [40] for the mixing entropy in $CaO–B_2O_3$ liquid mixtures (Figure 3.8) in which the association between anion groups resembles those in the binary silicate mixtures. In this system the mixing enthalpy is negative at all compositions, and changes linearly from the ortho-borate composition to the limit of liquid miscibility in the borate-rich mixtures.

The above entropy considerations are concerned with the positional entropy of mixing, which is the quantity usually measured by the statistical mechanical approach with equation 1.56. However, the mixing entropy contains a number of terms, including the changes in the positional, (S_p) vibrational (S_v) and magnetic (S_m) entropies and the entropy of mixing is obtained from the relation

$$S^M = S_p + S_v + S_m + \dots \qquad 3.2$$

Often the terms other than S_p are small and can be neglected, but when mixing causes a change in the bonding the other terms may contribute significantly to the value of S^M. For example, the reduction in the amount of the non-ionic bonding when basic oxides are mixed with silica may affect the vibrational entropy and, if the change in bonding is sufficiently pronounced, the change in the vibrational entropy may be greater than the change in the positional entropy. This has been shown to apply to the binary liquids Mg–Bi, Mg–Sb and Ag_2O–B_2O_3 [41] where the mixing entropies are negative, but the positional entropies are always positive. No measurements of the mixing entropy have been made for silicate systems, but some indication of the probable magnitude of the vibrational entropy term can be obtained from a comparison of the entropies of

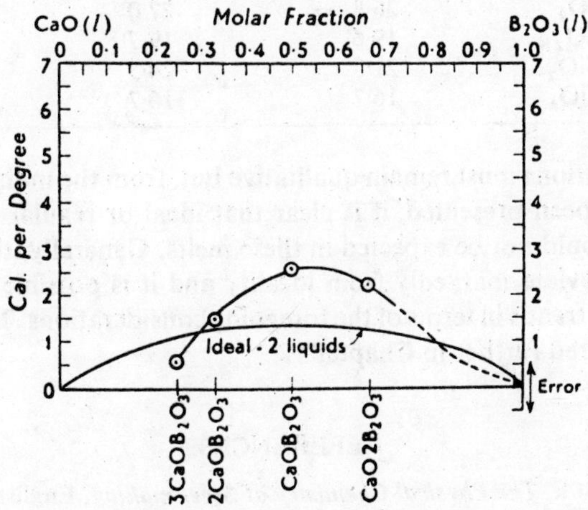

FIG. 3.8. Comparison of ideal and calculated entropies of mixing in calcium borate melts. (After Richardson [40].)

formation of solid silicate compounds. The data in Table 3.4 show that the entropies of most of the ortho- and meta-silicates are approximately equal to the sum of the entropies of the constituent oxides. A difference of the order of one cal deg^{-1} mol^{-1} can be discounted as within the accuracies of the measured values, but the entropy differences for the iron, manganese and zinc silicates are larger than this and are probably significant. Since the randomness of mixing is essentially constant in the ortho-silicates and also in the meta-silicates, the large ΔS values for these compounds probably originate in the mixing entropy terms other than the positional entropy.

Until such time as the necessary data are obtained from experimental studies, consideration of the mixing enthalpies and entropies of liquid

Table 3.4 *Entropies of Formation of Silicate Compounds at 298°K (cal deg⁻¹ mol⁻¹)*

Compound	Entropy Sum of Constituent Oxides	Measured Entropy	Difference (ΔS)
$2BaO.SiO_2$	43·5	43·8	+0·3
$2CaO.SiO_2$	30·3	29·2	−1·1
$2FeO.SiO_2$	34·7	38·1	+3·4
$2ZnO.SiO_2$	31·4	33·4	+2·0
$2MgO.SiO_2$	22·7	23·3	+0·5
$BaO.SiO_2$	26·8	27·0	+0·2
$CaO.SiO_2$	19·6	19·7	+0·1
$MnO.SiO_2$	21·3	24·5	+3·2
$MgO.SiO_2$	16·2	16·7	+0·5

silicate solutions must remain qualitative but, from the indirect evidence which has been presented, it is clear that ideal or regular solution behaviour should not be expected in these melts. Generally, the measured activities deviate markedly from ideality and it is possible to interpret some of the trends in terms of the foregoing considerations. These aspects are considered further in Chapter 12.

REFERENCES

1. H. SCHENCK, *The Physical Chemistry of Steel-making*, English Translation by B.I.S.R.A., London, 1945.
2. W. L. BRAGG, *The Structure of Silicates*, Oxford Univ. Press, 1932.
3. B. E. WARREN and A. G. PINCUS, *J. Amer. ceram. Soc.*, 1940, Vol. 23, p. 301.
4. J. BISCOE, A. PINCUS, C. S. SMITH and B. E. WARREN, ibid., 1941, Vol. 24, p. 116.
5. W. H. ZACHARAISEN, *J. Amer. chem. Soc.*, 1932, Vol. 54, p. 3841.
6. L. PAULING, *The Nature of the Chemical Bond*, second edition, New York, Cornell Univ. Press, 1948.
7. K. ENDELL and J. HELLBRUGGE, *Naturwissenschaften*, 1942, Vol. 30, p. 421.
8. LO CHING CHANG and G. DERGE, *Trans. Amer. Inst. min. (metall.) Engrs.*, 1947, Vol. 172, p. 90.
9. J. O'M BOCKRIS and D. C. LOWE, *Proc. roy. Soc.*, Series A, 1954, Vol. 226, p. 423.
10. J. O'M BOCKRIS, J. W. TOMLINSON and J. WHITE, *Trans. Faraday Soc.*, 1956, Vol. 52, p. 299.
11. O. E. ESIN, L. G. GARRILOV and B. M. LEPINSKIKH, *Dokl. obsch. Sobr. Ak. Nauk. S.S.S.R.*, 1953, Vol. 88, p. 714.
12. P. RÖNTGEN, H. WINTERHAGER and R. KAMMEL, *Z. Erzbergb. Metallhüttenw.*, 1956, Vol. 9, p. 201.
13. P. KOZAKEVITCH, *Rev. Met.*, 1949, Vol. 46, p. 505.

14. P. HERASYMENKO, *Disc. Faraday Soc.*, 1948, No. 4, p. 324.
15. H. TOWERS and J. KAY, *Trans. Brit. ceram. Soc.*, 1950, Vol. 49, p. 341.
16. T. B. KING, *J. Soc. Glass Tech.*, 1951, Vol. 35, p. 241; *The Physical Chemistry of Melts*, Inst. Min. Met., London, 1953, p. 35.
17. R. E. BONI and G. DERGE, *Trans. Amer. Inst. min. (metall.) Engrs.*, 1956, Vol. 206, p. 53.
18. K. MORI and Y. MATSUSHITA, *Tetsu to Hagane*, 1952, Vol. 38, p. 283; 1952. Vol. 38, p. 365.
19. J. W. TOMLINSON, *The Physical Chemistry of Melts*, Inst. Min. Met., London, 1953, p. 22.
20. B. E. WARREN and A. G. PINCUS, *J. Amer. ceram. Soc.*, 1940, Vol. 23, p. 301.
21. R. HAY, P. T. CARTER and S. K. KABI, *Trans. Soc. Glass Tech.*, 1956, Vol. 40, p. 429.
22. P. KOZAKEVITCH, *Rev. Met.*, 1949, Vol. 46, p. 572.
23. B. G. BALDWIN, *J. Iron St. Inst.*, 1957, Vol. 186, p. 388.
24. E. F. OSBORN, R. C. DEVRIES, K. H. GEE and H. M. KRAMER, *Trans. Amer. Inst. min. (metall.) Engrs.*, 1954, Vol. 200, p. 33.
25. J. S. MACHIN, T. B. YEE and D. L. HANNA, *J. Amer. ceram. Soc.*, 1954, Vol. 51, p. 569.
26. Y. KAWAI, *Sci. Rep. Res. Inst. Tohoku Univ.*, Series A4, 1952, p. 615.
27. N. L. ZHILO, A. V. RUDNEVA, G. A. SOKOLOV and L. M. TSYLEV, *Izv. Ak. Nauk Otdelenie Tekn.*, 1957, (2), p. 27.
28. E. E. HOFFMANN, *Stahl. u. Eisen, Düsseldorf*, 1959, Vol. 79, p. 846.
29. G. BEHRENDT and T. KOOTZ, ibid., 1949, Vol. 69, p. 399.
30. S. M. MIKIASHVILI, A. M. SAMARIN and L. M. TSYLEV, *Izv. Ak. Nauk Otdelenie Tekn.*, 1957, (4), p. 54.
31. S. I. POPLL and O. A. ESIN, *Zh. Neorganisk Khimii*, 1957, (2), p. 632.
32. C. F. COOPER and J. A. KITCHENER, *J. Iron St. Inst.*, 1959, Vol. 193, p. 48.
33. A. F. WELLS, *Structural Inorganic Chemistry*, second edition, London, Clarendon Press, 1950.
34. O. K. RICE, *Electronic Structure and Chemical Binding*, McGraw Hill, 1940.
35. J. INOUYE, J. W. TOMLINSON and J. CHIPMAN, *Trans. Faraday Soc.*, 1953, Vol. 49, p. 796.
36. M. T. SIMNAD, G. DERGE and J. GEORGE, *Trans. Amer. Inst. min. (metall.) Engrs*, 1954, Vol. 200, p. 1386.
37. M. T. SIMNAD, L. YANG and G. DERGE, ibid., 1956, Vol. 206, p. 690.
38. J. WHITE, 'Ceramics, A Symposium', *Brit. Ceram. Soc.*, 1953, p. 94.
39. M. REY, *The Physical Chemistry of Melts*, Inst. Min. Met., London, 1953, p. 63.
40. F. D. RICHARDSON, ibid., p. 75.
41. O. KUBASCHEWSKI and E. LL. EVANS, *Metallurgical Thermochemistry*, third edition, London, Pergamon Press, 1958, p. 52.

14. [illegible], 1948, No. 4, p. 729.
15. [illegible], 1950, Vol. 36, p. 341.
16. [illegible], 1951, Vol. 35, p. 761. The Structure of Metals, Inst. Min. Met., London, 1953, p. 45.
17. [illegible] and G. [illegible], Trans. Amer. Inst. Min. (Metall.) Engrs., 1946, Vol. 20, p. 35.
18. [illegible] and [illegible], Phil. Mag., 1932, Vol. 18, p. 263; 1933, Vol. 15, p. 163.
19. [illegible], The Physical Chemistry of Metals, Inst. Min. Met., London, 1953, p. 22.
20. [illegible] and [illegible], J. Amer. Ceram. Soc., 1940, Vol. 23, p. 101.
21. [illegible], Trans. Soc. Glass Tech., 1936, Vol. 20, p. 443.
22. [illegible], 1949, Vol. 46, p. 527.
23. [illegible], 1957, Vol. 180, p. 98.
24. [illegible], Trans. Amer. Inst. Min. Metall. Engrs., 1954, Vol. 200, p. 1.
25. [illegible], J. Amer. Ceram. Soc., 1952, Vol. 35, p. 303.
26. [illegible], Sci. Rep. Res. Inst. Tohoku Univ., Series A4, 1952, p. 645.
27. [illegible], 1957 (2), p. 21.
28. [illegible], 1948, Vol. 73, p. 845.
29. [illegible], 1949, Vol. 44, p. 299.
30. [illegible], 1952, (4), p. 84.
31. [illegible], 1953 (2), p. 663.
32. [illegible], 1953, Vol. 19, p. 645.
33. A. F. WELLS, Structural Inorganic Chemistry, second edition, London, Clarendon Press, 1950.
34. [illegible], McGraw-Hill, 1949.
35. [illegible], 1951, Vol. 47, p. 591.
36. [illegible], 1948, Vol. 175, p. 118.
37. [illegible], 1955, Vol. 23, p. 990.
38. [illegible], 1954, p. 94.
39. [illegible], London, 1953.
40. [illegible], ibid. p. 27.
41. [illegible], London, 1957.

CHAPTER 4

Ionic Theory and Reactions in Slags

Theoretical analyses of the chemical reactions in which a liquid slag participates have in the past been based on the assumption that the components of the slag are present in the melt in molecular form, but more recently the appreciation of the ionic nature of these melts has led to a number of attempts to formulate the slag–metal reactions in ionic rather than molecular concentration units. On thermodynamic grounds the choice of either of these units of concentration is immaterial; although a melt is known to be ionic, the equilibrium constant appertaining to a reaction therein can be stated in molar quantities without implying that such molecules actually exist in the liquid, and the choice of units is usually decided by the convenience of the mathematical manipulation of the data. For example, it is sometimes found that when the concentrations are expressed as ion fractions the activities conform more closely to the ideal solution law than when molar units are used.

When the melt contains only mixtures of oxides such as lime, magnesia, manganese oxide and ferrous oxide (which can be regarded as completely dissociated into basic cations and oxygen anions), the ionic approach is less complicated than when the complex anion-forming oxides are also present. Consequently the application of the ionic theory is here restricted initially to consideration of reactions involving only basic oxides and the effect of the presence of silica and other acid oxides is considered later.

4.1 Application to Simple Ionic Melts

In one of the earliest attempts to apply an ionic theory to slag–metal reactions it is assumed that mixing of the cations and the anions is completely random [1] and the ionic fraction of any species is defined as the number of ions of that component divided by the total number of ions present. Equilibrium 'constants' have been calculated by assuming ideal solution behaviour and inserting the ion fractions in the appropriate equilibrium equations, but it is found necessary to introduce empirical constants to make these fit experimental data [2].

If a solution behaves ideally according to Raoult's law, the components are required to mix randomly, but completely random mixing cannot be expected in a mixture of particles carrying positive and negative charges, for ions of similar charge repel, whereas ions of opposite charge are attracted towards each other. A cation may be replaced by another cation of similar charge without disturbing the electron balance, but replacement of a cation by an anion is opposed by the repulsive forces from the anions surrounding the cation site. Thus cations may mix randomly with each other but not with anions, and vice versa, and it follows that the entropy of mixing of the cations in an ideal solution should be determined in terms of the ionic fraction of each of the cations in relation to the *total number of cations* present and not to the total number of ions in the solution. Similarly the ionic fraction of each of the anions should be determined relative to the total number of anions present. That is, for a solution containing A^{2+}, B^{2+}, C^{2-} and D^{2-} ions the ionic fractions of the A^{2+} and C^{2-} ions are

$$N_{A^{2+}} = \frac{n_{A^{2+}}}{n_{A^{2+}}+n_{B^{2+}}} \text{ and } N_{C^{2-}} = \frac{n_{C^{2-}}}{n_{C^{2-}}+n_{D^{2-}}} \qquad 4.1$$

where N_X is the ionic fraction of X and n_X is the number of X ions in the solution.

The entropy of mixing is calculated as follows [3, 4]. For a reaction

$$A^+.C^- + B^+.D^- \rightleftharpoons A^+.D^- + B^+.C^-$$

the number of ions of each species are given by:

$$n_{A^+} = n_{A^+C^-} + n_{A^+D^-}$$

$$n_{B^+} = n_{B^+D^-} + n_{B^+C^-}$$

$$n_{C^-} = n_{A^+C^-} + n_{B^+C^-}$$

$$n_{D^-} = n_{B^+D^-} + n_{A^+D^-}$$

where n_{A^+}, n_{B^+}, n_{C^-} and n_{D^-} are obtained directly by chemical analysis of the mixture. If cations mix only with cations and mixing is ideal, the entropy of mixing for the cations can be calculated using equation 2.4:

$$S_+^M = -k\left[n_{A^+}.\ln\left(\frac{n_{A^+}}{n_{A^+}+n_{B^+}}\right) + n_{B^+}.\ln\left(\frac{n_{B^+}}{n_{A^+}+n_{B^+}}\right)\right] \qquad 4.2$$

Similarly the entropy of mixing the anions only is given by:

$$S_-^M = -k\left[n_{C^-}.\ln\left(\frac{n_{C^-}}{n_{C^-}+n_{D^-}}\right) + n_{D^-}.\ln\left(\frac{n_{D^-}}{n_{C^-}+n_{D^-}}\right)\right] \qquad 4.3$$

The total entropy of mixing is the sum of these quantities and, if the solution is ideal (i.e. the heat of mixing is zero), the free energy of mixing is given by:

$$G^M = -TS^M = -T(S_+^M + S_-^M) \tag{4.4}$$

This treatment can be applied to such reactions as:

$$\text{FeS}_{(\text{metal})} + \text{CaO}_{(\text{slag})} \rightleftharpoons \text{CaS}_{(\text{slag})} + \text{FeO}_{(\text{metal})}$$

which is equivalent to:

$$\text{S}_{(\text{metal})} + \text{O}^{2-}{}_{(\text{slag})} \rightleftharpoons \text{S}^{2-}{}_{(\text{slag})} + \text{O}_{(\text{metal})}$$

where the mixing involves anions only, and it applies similarly to mixing of cations such as:

$$\text{Fe}_{(\text{metal})} + \text{MnO}_{(\text{slag})} \rightleftharpoons \text{Mn}_{(\text{metal})} + \text{FeO}_{(\text{slag})}$$

In the last case, where only one species of anion is involved, the entropy of mixing the anions is zero since the reaction involves substitution of one cation for another in a constant anion lattice in which the number of anions, and hence the number of cation sites, are unaltered. Equation 4.4 then takes the form:

$$G^M = kT\left[n_{\text{Fe}^{2+}}.\ln\left(\frac{n_{\text{Fe}^{2+}}}{n_{\text{Fe}^{2+}}+n_{\text{Mn}^{2+}}}\right) + n_{\text{Mn}^{2+}}.\ln\left(\frac{n_{\text{Mn}^{2+}}}{n_{\text{Fe}^{2+}}+n_{\text{Mn}^{2+}}}\right)\right] \tag{4.5}$$

and, from equation 2.1 the partial molar free energy of the ferrous ions is

$$G^M_{\text{Fe}^{2+}} = kT\ln\left(\frac{n_{\text{Fe}^{2+}}}{n_{\text{Fe}^{2+}}+n_{\text{Mn}^{2+}}}\right) \tag{4.6}$$

$$= RT\ln N_{\text{Fe}^{2+}}$$

But, since $n_{\text{Fe}^{2+}} = n_{\text{FeO}}$ and $n_{\text{Mn}^{2+}} = n_{\text{MnO}}$, therefore

$$\left(\frac{n_{\text{Fe}^{2+}}}{n_{\text{Fe}^{2+}}+n_{\text{Mn}^{2+}}}\right) = \left(\frac{n_{\text{FeO}}}{n_{\text{FeO}}+n_{\text{MnO}}}\right) = N_{\text{FeO}}$$

and it follows that:

$$N_{\text{Fe}^{2+}} = N_{\text{FeO}} = a_{\text{FeO}} \tag{4.7}$$

Likewise

$$N_{\text{Mn}^{2+}} = N_{\text{MnO}} = a_{\text{MnO}} \tag{4.8}$$

Where mixing involves both cations and anions as in the reaction:

$$\text{FeS}_{(\text{slag})} + \text{MnO}_{(\text{slag})} \rightleftharpoons \text{MnO}_{(\text{slag})} + \text{FeS}_{(\text{slag})}$$

G^M is obtained from the sum of the entropies of mixing for the cations and for the anions. The partial molar free energy of mixing for the ferrous ions ($G^M_{Fe^{2+}}$) is again given by equation 4.6, and the partial molar free energy of mixing for the oxygen anions is given by:

$$G^M_{O^{2-}} = kT \ln\left(\frac{n_{O^{2-}}}{n_{O^{2-}} n_{S^{2-}}}\right) \quad 4.9$$

By adding equations 4.6 and 4.9 and replacing by mole fractions, since $R = N_0 k$ where R is the gas constant, N_0 is Avogadro's number and $N_0 n_A = N_A$, etc., for one mole of solution we obtain:

$$G^M_{FeO} = RT \ln N_{Fe^{2+}} . N_{O^{2-}} \quad 4.10$$

and hence: $$a_{FeO} = N_{Fe^{2+}} . N_{O^{2-}}$$

Likewise $$a_{MnO} = N_{Mn^{2+}} . N_{O^{2-}}$$

and $$a_{FeS} = N_{Fe^{2+}} . N_{S^{2-}}$$

$$a_{MnS} = N_{Mn^{2+}} . N_{S^{2-}}$$

EFFECT OF CHANGE OF VALENCY

In the foregoing examples a cation or an anion is replaced by an ion of like sign and charge, but further restrictions are imposed on purely random mixing when ions of different valency are involved in the reaction. For example, if divalent cations are substituted for monovalent cations in a mixture of oxides or sulphides, two monovalent cations must be displaced by each divalent cation added in order to preserve electroneutrality. The number of anions, and hence the number of cation sites, remains unaltered; hence one vacant cation site must be created for each divalent cation introduced into the mixture. Thus if sodium and manganese oxides are mixed, the total number of cation sites equals:

$$n_{Na^+} + n_{Mn^{2+}} + n_{C\square}$$

where $n_{C\square}$ is the number of vacant cation sites, which is equal to $n_{Mn^{2+}}$. Similarly, the insertion of a trivalent cation will create two cation vacancies and so on. The entropy of mixing for the cations is proportional to the number of different ways of arranging the reacting species on the available cation sites; hence the vacant sites must be taken into account in estimating the entropy value [5, 6]. From equation 1.56, $S = k \ln \omega$ where ω is the number of different arrangements possible and, because the number of cations is equal to the number of cation sites, if we do not distinguish between the individual ions of the same species, the number of ways of arranging the cations in the pure monovalent oxide (ω_1) is unity. In the pure divalent oxide there is one vacant site for each cation present and:

$$\omega_2 = \frac{(\text{number of cation sites})!}{(\text{number of cations})!\,(\text{number of vacancies})!} \qquad 4.11$$

$$= \frac{(2n_{Mn^{2+}})!}{(n_{Mn^{2+}})!\,(n_{C\square})!} = \frac{(2n_{Mn^{2+}})!}{(n_{Mn^{2+}})!\,(n_{Mn^{2+}})!} \qquad 4.12$$

where the sign (!) denotes a permutation.

When the two oxides are mixed the number of cation sites is given by:

$$((n_{Na^+})+(2n_{Mn^{2+}}))$$

and hence

$$\omega_{(\text{mixture})} = \frac{(n_{Na^+}+2n_{Mn^{2+}})!}{(n_{Na^+})!\,(n_{Mn^{2+}})!\,(n_{Mn^{2+}})!} \qquad 4.13$$

The entropy of mixing is the difference between the entropy of the mixture and that of the pure oxides and is given by:

$$S_+^M = k\ln\omega_{(\text{mixture})} - k\ln(\omega_1+\omega_2)$$

$$= k\ln\left(\frac{\omega_{(\text{mixture})}}{(\omega_1+\omega_2)}\right) = k\ln\left(\frac{(n_{Na^+}+2n_{Mn^{2+}})!}{(n_{Na^+})!\,.\,(2n_{Mn^{2+}})!}\right) \qquad 4.14$$

When X is a large number, greater than 10, a permutation can be expanded by Sterling's approximation, which states that $\ln X! = X\ln X - X$. Hence

$$S_+^M = k\left[n_{Na^+}.\ln\left(\frac{n_{Na^+}+2n_{Mn^{2+}}}{n_{Na^+}}\right)+2n_{Mn^{2+}}.\ln\left(\frac{n_{Na^+}+2n_{Mn^{2+}}}{2n_{Mn^{2+}}}\right)\right]$$

$$= -k\left[n_{Na^+}.\ln\left(\frac{n_{Na^+}}{n_{Na^+}+2n_{Mn^{2+}}}\right)+2n_{Mn^{2+}}.\ln\left(\frac{2n_{Mn^{2+}}}{n_{Na^+}+2n_{Mn^{2+}}}\right)\right] \qquad 4.15$$

But

$$n_{Na^+} = 2n_{Na_2O}$$

and

$$n_{Mn^{2+}} = n_{MnO}$$

Substituting these relations in the last equation and remembering that, for one mole of solution, $R = N_0 k$

$$S_+^M = -k\left[2n_{Na_2O}\ln\left(\frac{2n_{Na_2O}}{2n_{Na_2O}+2n_{MnO}}\right)+2n_{MnO}\ln\left(\frac{2n_{MnO}}{2n_{Na_2O}+2n_{MnO}}\right)\right]$$

$$= -R[N_{Na_2O}\ln(N_{Na_2O})^2+N_{MnO}\ln(N_{MnO})^2] \qquad 4.16$$

Thus even if the mixture of ions is ideal the *activities are not proportional to the mole fractions*, but to the square of these values and the activity of each oxide is not a linear function of the ionic concentration.

For a mixture of divalent and trivalent oxides such as FeO and Fe_2O_3 we may assume that each mole of ferrous oxide contributes one ferrous cation and one oxygen ion, whereas a mole of ferric oxide contributes two cations, a cation vacancy and three oxygen ions to the mixture. Proceeding in an analogous manner to that just described [7] we obtain:

$$S_{+}^{M} = -k\left[n_{Fe^{2+}}.\ln\left(\frac{n_{Fe^{2+}}}{n_{Fe^{2+}}+\frac{3}{2}n_{Fe^{3+}}}\right)+\tfrac{3}{2}n_{Fe^{3+}}.\ln\left(\frac{\frac{3}{2}n_{Fe^{3+}}}{n_{Fe^{2+}}+\frac{3}{2}n_{Fe^{3+}}}\right)\right] \quad 4.17$$

Replacing the number of ions by the mole fraction gives:

$$S_{+}^{M} = -R\left[N_{FeO}\ln\left(\frac{N_{FeO}}{N_{FeO}+3N_{Fe_2O_3}}\right)+N_{Fe_2O_3}\ln\left(\frac{3N_{Fe_2O_3}}{N_{FeO}+3N_{Fe_2O_3}}\right)^{3}\right] \quad 4.18$$

and, if the solution is ideal,

$$G_{FeO}^{M} = RT\ln\left(\frac{N_{FeO}}{N_{FeO}+3N_{Fe_2O_3}}\right) \quad 4.19$$

whence

$$a_{FeO} = \frac{N_{FeO}}{N_{FeO}+3N_{Fe_2O_3}} \quad 4.20$$

The terms in round brackets in equations 4.15 and 4.17 are called the *electrically equivalent ion fractions* [6, 8].

In the general case, for a mixture of ions A^{+}, B^{2+}, C^{3+}, D^{4+}, the equivalent ion fraction N' of B^{2+} is given by the relation:

$$N'_{B^{2+}} = \frac{2n_{B^{2+}}}{n_{A^{+}}+2n_{B^{2+}}+3n_{C^{3+}}+4n_{D^{4+}}} \quad 4.21$$

If the anions also exhibit different valencies a similar treatment is applied to the anion fractions; alternatively it can be assumed that the anion lattice is unchanged, but the number of cation sites is reduced when divalent anions are replaced by monovalent anions and the number increased if divalent anions are replaced by trivalent ones. A number of other models are also possible, and the relations derived for the activities depend on the model assumed.

4.2 Melts Containing Complex Anions

Reactions involving the oxides of silicon, phosphorus and aluminium are less amenable to ionic treatment because each of these oxides can form complex, stable anion groups of varying cation/oxygen ratio such as SiO_4^{4-}, $Si_2O_7^{6-}$, SiO_3^{2-}, PO_4^{3-} and AlO_3^{3-}. Furthermore, as discussed in

Chapter 3, the complexes may exist as single units or linked together to form larger complexes, depending on the ratio of these cations to oxygen anions. The cation sites occupied by the highly-charged silicon, phosphorus and aluminium ions are not available for occupation by cations of lower charge such as Fe^{2+}, Ca^{2+} and Mn^{2+}, and completely random or ideal mixing of the cations would not therefore be expected. A further complication arises in deciding which complexes exist over any given range of composition, for the actual species present cannot be determined experimentally. This decision can only be made with certainty in basic melts where sufficient oxygen ions are present to satisfy all the cation bonds of the anion complex, and most of the ionic treatments in use are restricted to basic steel-making practice for this reason.

ACID AND BASIC MELTS

The division between acid and basic melts is usually taken as the composition at which just sufficient oxygen ions are donated by the cations to satisfy all the oxygen-ion bonds of the acidic polymers. Thus, a mixture of calcium oxide and silica would be classed as neutral at the orthosilicate composition, where two moles of lime are present for each mole of silica to give two Ca^{2+} cations and one SiO_4^{4-} anion. All the oxygen bonds of the silicon ion are satisfied and an increase in the lime concentration results in the presence of free oxygen ions. Conversely, if the lime content is reduced, some of the SiO_4^{4-} ions become unsaturated and degenerate into $Si^2O_7^{6-}$ or similar groupings. Similarly, since one molecule of the oxide of a monovalent element forms two cations and one oxygen ion, the neutral melt here contains cations and oxy-acid silica groups in the ratio of 4/1. Each molecule of P_2O_5 or Al_2O_3 requires three oxygen ions for the formation of the stable, saturated PO_4^{3-} and AlO_3^{3-} groupings respectively and neutrality obtains at a higher cation/oxygen ratio than with silica. In the light of the foregoing considerations a basic melt can be defined as one containing saturated acid polymers and free oxygen ions, whereas an acid melt contains unsaturated anion groups.

When any of the complex anions are contained in the melt the calculation of the oxygen ion concentration must be modified to take into account the number of oxygen ions bound-up in the complexes. Thus in a basic melt containing CaO, MnO, FeO, SiO_2, P_2O_5 and Al_2O_3 the number of 'free' oxygen ions is obtained from a relation such as:

$$n_{O^{2-}} = n_{CaO} + n_{MnO} + n_{FeO} - 2n_{SiO_2} - 3n_{P_2O_5} - 3n_{Al_2O_3} \qquad 4.22$$

because each mole of silica is associated with two O^{2-} ions to form SiO_4^{4-} and so on.

Alumina is amphoteric; in acid or weakly basic melts it may ionise in accordance with the following reaction:

$$Al_2O_3 = 2Al^{3+} + 3O^{2-}$$

and under these conditions alumina acts as a base, whereas in more strongly basic melts it behaves as an acidic oxide. Ferric oxide appears to behave similarly.

The phosphorus content of neutral or acid slags is small, the alumina content of steel-making slags is usually low and so, according to equation 4.22, the oxygen ion concentration effectively falls to zero when the number of oxygen ions supplied by the basic oxides is approximately twice the number of silicon ions present in the melt, or when the ratio of oxygen to silicon is 4/1. This implies that more acid melts would be non-oxidising but in the following chapters it is shown that this is not true and it appears highly probable that oxy-acid polymers of lower O/Si ratio than 4/1 exist even at the neutral composition. For example, a mixture of the two ortho-silicates 2 $FeO.SiO_2$ and 2 $CaO.SiO_2$ would normally be classed as a neutral melt in which one would expect to find only Fe^{2+}, Ca^{2+} and SiO_4^{4-} ions. However, in view of the measured activities of ferrous oxide in these mixtures, it seems reasonable to suggest that some of the oxygen ions are detached from the silicate groups to give O^{2-} and $Si_2O_7^{6-}$ or other silicate ions in addition to those listed above. There is some evidence to support the contention that unattached oxygen ions are also present in more acid melts. The extent to which the silicate anions are dissociated cannot be determined from the thermodynamic approach and the calculation of the oxygen ion fraction in acid slags must be based on an arbitrary assumption of the anion species present in the melt. It is primarily for this reason that no comprehensive ionic treatment of acid steel-making reactions has yet been postulated.

THE ACTIVITY COEFFICIENT

Because the presence of oxy-acid polymers precludes the possibility of completely random mixing of the cations, even in melts of high basicity, the equilibrium constant for an ionic reaction should include the activity coefficients of the ions present and thermodynamically, in such cases, there appears to be no advantage in the ionic approach. However, by consideration of a cyclic process for a reaction such as:

$$A^-_{(slag)} + C_{(metal)} \rightleftharpoons C^-_{(slag)} + A_{(metal)}$$

Flood et al. [8, 9] have derived the following relation for a reaction which involves ions of one sign only, but which occurs in the presence of ions of opposite sign:

$$\ln K' = \Sigma^{i} N' \ln K_i' + \Sigma^{i}(\gamma) \quad 4.23$$

The first summation represents all the possible binary interchange reactions between the anion under consideration and a system of i species of cations, and the term $\Sigma^{i}(\gamma)$ represents the activity coefficients in all the binary equilibria. In the general case this last term tends to cancel out and, to a first approximation, the equilibrium constant can be obtained by evaluation of the first term.

To illustrate the application of this approach, the transfer of phosphorus from metal to slag may be considered [8]. The basic equation is:

$$2P_{(metal)} + 5O_{(metal)} + 3(O^{--})_{(slag)} \rightleftharpoons 2(PO_4^{3-})_{(slag)} \quad 4.24$$

If the slag contains Fe^{2+}, Mn^{2+}, Mg^{2+} and Ca^{2+} cations the expanded form of equation 4.23 is:

$$\ln K' = N'_{Fe^{2+}} . \ln K'_{Fe} + N'_{Mn^{2+}} . \ln K'_{Mn} + N'_{Mg^{2+}} . \ln K'_{Mg} + N'_{Ca^{2+}} . \ln K'_{Ca} \quad 4.25$$

where N' is the electrically equivalent ion fraction, K'_{Fe} is the equilibrium constant for the reaction:

$$2P_{(metal)} + 5O_{(metal)} + 3FeO_{(slag)} \rightleftharpoons 3FeO . P_2O_{5(slag)} \quad 4.26$$

K'_{Mn} is the constant for the reaction:

$$2P_{(metal)} + 5O_{(metal)} + 3MnO_{(slag)} \rightleftharpoons 3MnO . P_2O_{5(slag)} \quad 4.27$$

and so on. The constant for the complex system is readily calculated if the equilibrium constants or the standard free energies are known for the simple reactions. It will be noticed that the presence of other anions in the slag is not considered in deriving the constant.

This treatment is exact only when the ionic solutions are ideal, but in practice the departure from ideal mixing is often pronounced and the error in the calculated constant is then dependent on the magnitude of the term $\Sigma^{i}(\gamma)$. However, complex phosphate ions are present in the reaction considered and the slag may also contain silicate and aluminate ions, so large deviations from ideal mixing would be expected; nevertheless the calculated values of K' agree closely with those obtained experimentally [9]. This is not necessarily true for all systems, but it is always worth testing an approximate equation to ascertain whether or not the experimental data can be fitted adequately. If a satisfactory fit is obtained it is again important to realise that the approximate equation may not apply outside the composition range over which it has been tested because, for any reaction, a truly universal constant can only be obtained by inclusion of all the relevant activity coefficients.

REFERENCES

1. P. HERASYMENKO, *Trans. Faraday Soc.*, 1938, Vol. 34, p. 124.
2. P. HERASYMENKO and G. SPEIGHT, *J. Iron St. Inst.*, 1950, Vol. 166, p. 169; 1950, Vol. 166, p. 269.
3. M. TEMKIN, *Acta Physicochimica URSS*, 1945, Vol. 20, p. 411.
4. A. SAMARIN, M. TEMKIN and L. SCHWARTZMANN, ibid., 1945, Vol. 20, p. 421.
5. J. WHITE, 'Ceramics, A Symposium', *Brit. Ceram. Soc.*, 1953, p. 94.
6. H. FLOOD and A. MUAN, *Acta chem. Scand.*, 1950, Vol. 4, p. 359.
7. C. BODSWORTH and I. M. DAVIDSON, *Physical Chemistry of Process Metallurgy*, 1961, Interscience, p. 233.
8. H. FLOOD and K. GRJOTHEIM, *J. Iron St. Inst.*, 1952, Vol. 171, p. 64.
9. H. FLOOD, T. FORLAND and K. GRJOTHEIM, *The Physical Chemistry of Melts*, Inst. of Min. Met., London, 1953, p. 46.

CHAPTER 5

Kinetics of Reactions

A detailed review of the kinetic theory would occupy considerable space and is unnecessary here since several good textbooks are available which are devoted exclusively to this topic; the reader is advised to consult texts such as those by Mellor [1], Hinshelwood [2], Glasstone et al. [3] and Eyring [4] if a more extensive treatment is required.

It is well recognised that during refining certain reactions deviate from the conditions of thermodynamic equilibrium, but there are as yet few quantitative measurements of the reaction rates. However, from consideration of the formal basis of kinetic theory, it is often possible to decide which factors may affect the rate of a reaction under a particular set of conditions and how changing the conditions may alter the rate.

5.1 Reaction Rate Theory

The foundation of the theory of reaction rates in chemical systems dates back to the second half of the last century when Guldberg and Waage concluded from experimental observations, that 'the rate at which a substance reacts is proportional to its *active mass* in the Law of Mass Action relation for the reaction'. Thus, for a reaction between u moles of component A, v moles of B, etc.

$$uA+vB+wC \rightleftharpoons xD+yE+zF+\ldots \qquad 5.1$$

the rate of the reaction in the forward direction (i.e., from left to right in equation 5.1) is given by:

$$\kappa_{(f)} = [a_A]^u . [a_B]^v . [a_C]^w \qquad 5.2$$

and the rate of the reaction in the reverse direction is given by:

$$\kappa_{(r)} = [a_D]^x . [a_E]^y . [a_F]^z \qquad 5.3$$

where $\kappa_{(f)}$ and $\kappa_{(r)}$ are called the *velocity* or *rate constants*. The net reaction rate is the difference between these two quantities, that is:

$$\kappa = \kappa_{(f)} - \kappa_{(r)} \qquad 5.4$$

When equilibrium is established the forward and reverse reactions proceed at the same rates and the net rate is zero, hence

$$\frac{[a_A]^u \cdot [a_B]^v \cdot [a_C]^w}{[a_D]^x \cdot [a_E]^y \cdot [a_F]^z} = \frac{\kappa_{(\mathrm{f})}}{\kappa_{(\mathrm{r})}} = K \qquad 5.5$$

where K is the equilibrium constant for the reaction.

From experimental observations Arrhenius found that for many reactions the plot of the logarithm of the rate constant against the reciprocal of the temperature in degrees absolute is a straight line, i.e.,

$$\ln \kappa = \ln \mathrm{A} - \frac{Q}{RT} \qquad 5.6$$

where A is a constant, equal to the intercept on the $\ln \kappa$ axis at $1/T = 0$, and Q/R is the slope of the plot; Q is a constant and R is the gas constant. In its more usual form the Arrhenius relation is written as:

$$\kappa = \mathrm{A}e^{-(Q/RT)}. \qquad 5.7$$

ENERGY OF ACTIVATION

The free-energy change accompanying a reaction is negative and the free energy of the system decreases to a minimum at the equilibrium condition. However, most reactions proceed at a finite rate and it is evident that the initial and final states of the reaction are separated by some energy barrier, otherwise the reaction would occur spontaneously as soon as the reactants are brought into contact. An additional amount of energy is required to enable the reactants to surmount this barrier to form the products and, from statistical mechanical considerations, it can be shown that the constant Q in equation 5.7 is related to the fraction of the atoms or molecules of the reactants which at any instant possess sufficient energy to surmount the barrier. The constant Q is called accordingly the *energy of activation* for the process.

The energy barrier is readily visualised from the following mechanical analogy. A rectangular box is in meta-stable equilibrium when it is stood on one end with the largest dimension vertical. If a small amount of energy is applied to the box to tilt the axis slightly away from the vertical it will return to the meta-stable position when the energy is released but, if the energy supplied is increased in small increments, a condition is eventually attained at which the box becomes unstable and falls on to one of the broad faces with the largest dimension horizontal. The latter condition is a state of minimum energy and the critical amount of energy which is required to displace the box from the meta-stable into the stable position can be regarded as the activation energy for the process.

In a similar manner the atoms or molecules of the reactants in a chemical reaction can be regarded as residing in an energy trough and any small change which causes a displacement from the initial condition is accompanied by an increase in energy (Figure 5.1). The reactants are also located in an energy trough and the difference between the minima in the two troughs is equal to the free energy of the reaction. For the reactants to move out of their energy trough and change into products

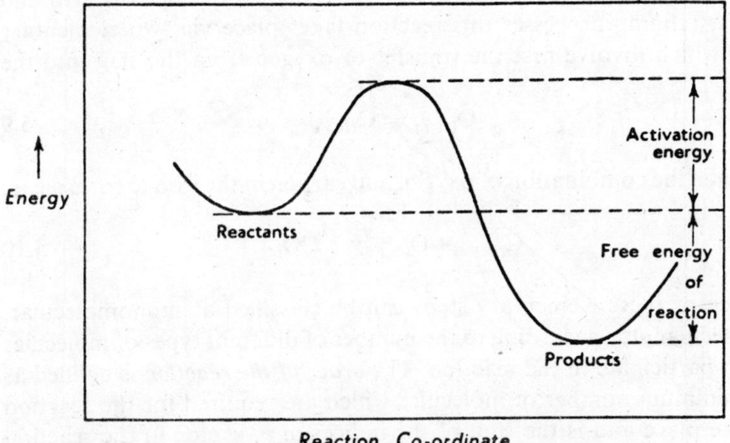

FIG. 5.1. Schematic representation of the activation energy for a reaction.

additional energy must be supplied to surmount the energy barrier, and this is the activation energy for the process.

Two theories are in common use to describe the kinetics of a reaction. The *collision theory* [1, 2] assumes that the rate is determined by collisions between the reacting atoms or molecules. The constant A in equation 5.7 is called the frequency factor and is considered as a measure of the frequency with which an atom or molecule is in a favourable position to take part in the reaction. The theory of *activated complexes* by Eyring and others [3, 4] assumes that the first step in the reaction is the formation of an activated complex from the reactants, which then decomposes rapidly to form the products in the second step. The overall reaction rate is determined by the speed of the first step and this can be treated as a simple chemical reaction. Both approaches are described fully in the references given and these can be consulted for further details.

5.2 The Order of a Reaction

Many chemical reactions are more complex than appears from a cursory inspection, for the chemical equation representing the overall reaction is frequently a summation of a number of elementary steps. For example, the oxidation of carbon in liquid iron by oxygen dissolved in the slag is usually represented by:

$$C_{(\text{in Fe})} + O_{(\text{slag})} \rightleftharpoons CO_{(g)} \tag{5.8}$$

However, there is good reason to believe that in the open-hearth and similar refining processes this reaction takes place via two elementary steps which involve first the transfer of oxygen from the slag into the metal:

$$O_{(\text{slag})} \rightleftharpoons O_{(\text{in Fe}} \tag{5.9}$$

and then the combination of oxygen and carbon in the iron to form a gas:

$$C_{(\text{in Fe})} + O_{(\text{in Fe})} \rightleftharpoons CO_{(g)} \tag{5.10}$$

Each of these elementary steps can be classified as monomolecular, bimolecular etc., according to the number of different types of molecules which participate in the reaction. The *order of the reaction* is defined as the minimum number of molecules which are required for the reaction to take place and is the sum of the indices u, v, w etc., in the reaction parameter $\kappa[a_A]^u . [a_B]^v . [a_C]^w$. Thus reaction 5.9 is first order in either direction, whilst reaction 5.10 is second order in the forward direction and first order in the reverse.

The overall reaction cannot strictly be classified in this way when it involves a number of elementary steps, nor can the elementary steps be classified for conditions close to equilibrium if the forward and reverse reactions are of different orders, for the net order is then fractional. Overall reactions are sometimes classified empirically as the sum of the powers of the reactants, but it is often found that the order is fractional and has no real meaning in terms of molecularity. If the reaction parameter is expressed in terms of concentration units, instead of the activities of the reactants, the order of the reaction is not constant but varies with composition for non-ideal solutions. However, in many complex reactions it is found that the rate of one of the elementary steps is so much slower than the rate of the other steps that this one elementary step controls the overall reaction rate. If this step is first order then the overall reaction is first order, and so on.

For a first order reaction in which the initial concentration x of the reactant changes to $(x-y)$ at time t, the reaction rate is:

$$\frac{dy}{dt} = \kappa(x-y) \qquad 5.11$$

On integration this gives:

$$-\ln(x-y) = \kappa t + C \qquad 5.12$$

where C is an integration constant; but $y = 0$ at $t = 0$ and hence $-\ln x = C$. Therefore the rate constant is:

$$\kappa = \frac{1}{t}\ln\left(\frac{x}{x-y}\right) \qquad 5.13$$

and the time taken to reach the half-way stage in the reaction is:

$$t = \frac{\ln 2}{\kappa} \qquad 5.14$$

Thus for a first order reaction the time taken to reach any chosen point along the reaction co-ordinate is independent of the concentration of the reactant.

Proceeding in similar manner for a second order reaction in which the initial concentrations x and b of the reactants change to $(x-y)$ and $(b-y)$ at time t, the reaction rate is:

$$\frac{dy}{dt} = \kappa(x-y)(b-y) \qquad 5.15$$

and

$$\kappa = \frac{1}{t(x-b)}.\ln\left[\frac{b(x-y)}{x(b-y)}\right] \qquad 5.16$$

For the simple case where x is equal to b equation 5.15 is written as

$$\frac{dy}{dt} = \kappa(x-y)^2$$

which, on integration, becomes

$$\kappa = \frac{1}{xt}\left(\frac{y}{x-y}\right) \qquad 5.17$$

and the half-reaction time is:

$$t = \frac{1}{\kappa x} \qquad 5.18$$

which shows that the reaction time is dependent on the concentration terms for a second order reaction.

It is usual to determine the order of an elementary reaction by substituting the experimental data in the integrated rate equations for these and higher-order reactions and deciding which gives the best fit. Fortunately, reactions of the third and higher orders are very rare in liquid melts and it is normally possible to decide between first and second order by determining whether or not the reaction time is concentration dependent.

5.3 Reactions Near to Equilibrium

At the start of a reaction, when the concentration of the products is either zero or very low, the effect of the reverse reaction on the net reaction rate can reasonably be disregarded, but as equilibrium is approached the rates of the forward and reverse reactions become similar and both rates must be considered. This is important in relation to iron and steel refining because, except at the start of refining, it is probable that most of the refining reactions are at, or very close to, equilibrium and the rate of approach to the equilibrium is determined by the difference between the forward and back reaction rates. For a first order reaction of the type:

$$X \rightleftharpoons B$$

these rates are $r_{(f)} = \kappa_{(f)} x$

and $r_{(r)} = \kappa_{(r)} b$

where x and b are the concentrations of X and B respectively, and the net reaction rate is given by:

$$r = r_{(f)} - r_{(r)} = \kappa_{(f)} x - \kappa_{(r)} b \qquad 5.19$$

At any stage in the reaction there is an equilibrium concentration of B which is equal to the product of the concentration x and the equilibrium constant K. Designating this concentration of B as b_e, the net reaction rate is:

$$r = \kappa_{(r)}(b_e - b) \qquad 5.20$$

Designating similarly the concentration of X in equilibrium with b as x_e, then:

$$r = \kappa_{(f)}(x - x_e) \qquad 5.21$$

and it follows that:

$$\frac{\kappa_{(f)}}{\kappa_{(r)}} = \frac{(b_e - b)}{(x - x_e)} \qquad 5.22$$

Proceeding in the same manner for a reaction which is close to equilibrium and which involves two slow steps with similar rates we have:

$$X \rightleftharpoons B \tag{5.23}$$

and

$$B \rightleftharpoons C \tag{5.24}$$

The rates for the two reactions can be expressed as:

$$r_{XB} = \kappa_{XB(r)}(b'_e - b) \tag{5.25}$$

and

$$r_{BC} = \kappa_{BC(f)}(b - b''_e) \tag{5.26}$$

where b'_e and b''_e are the concentrations of B which would be in equilibrium with the instantaneous concentrations of X and C respectively. If the concentrations remain essentially constant a steady state is reached at which the two steps proceed at the same rate (i.e., $r_{XB} = r_{XC}$) and

$$\frac{(b'_e - b)}{(b - b''_e)} = \frac{\kappa_{BC(f)}}{\kappa_{XB(r)}} \tag{5.27}$$

In many refining reactions the concentration of one reactant is very much smaller than those of the others and, in comparison, the concentrations of the other reactants remain almost constant. When this condition applies the reaction rate can often be expressed in terms of the relations for a first order reaction. It is probable that this approximation can be applied, for example, to the oxidation of carbon during the refining of steel, since at any instant the oxygen content of the metal is very small in relation to the amount of oxygen which is required to combine with all the carbon. Assuming that the reaction involves two slow steps as expressed in equations 5.9 and 5.10, the components b, b'_e and b''_e can be identified respectively with the actual amount of oxygen dissolved in the iron and the amounts of oxygen which would be in equilibrium with the slag and with the carbon in the iron [5].

5.4 Diffusion

When a reaction occurs locally between two elements dissolved in the same liquid phase, the concentrations of those elements are lowered at the site of the reaction and a concentration difference is established with the adjacent liquid. For the reaction to continue at that site the concentrations of the reactants must be replenished continuously and this can be achieved either by diffusion or by induced mixing of the liquid. The argument can be extended to a heterogeneous reaction which involves the transfer of some species from one phase and across an interface to react with an element dissolved in a second phase. For example, the oxidation of silicon in iron by oxygen ions in the slag to produce silica in the slag can be broken down into four separate stages which occur simultaneously:

1. The continuous supply of silicon from within the bulk metal to replenish the concentration of silicon in the metal at the interface.
2. A similar continuous replenishment of oxygen from the bulk slag to the vicinity of the interface.
3. The transfer of one or more species across the interface. (This in turn may involve several elementary steps.)
4. The continuous removal of silica from the interface and into the bulk slag.

Steps 1, 2 and 4 involve the transport of some species to or from the site of the reaction and each of these may be diffusion controlled. By analogy with the preceding discussion of reaction kinetics it is evident that the overall reaction rate may be controlled by any one of these steps, depending on which has the slowest rate under the particular set of conditions.

Very high reaction rates are obtained in the Bessemer converter processes, which suggests that in general the various refining reactions are inherently very fast. A characteristic feature of the Bessemer processes is the violent agitation of the metal by the passage of a gas from the bottom of the vessel, and it is notable that the reaction rates are very considerably slower in the refining processes where turbulence is absent or is less pronounced. Thus, in general, it appears that transfer across the interface is not the rate-limiting step and the refining reactions are mainly, if not entirely, diffusion-controlled. However, the comparatively few measurements which have been made in the temperature range of interest for the diffusivities of solutes in liquid iron and liquid slags show that the rate of migration of an atom or ion in a stagnant liquid is not more than about one centimetre per hour. If the transport of the reactants and products were entirely dependent on diffusion, refining would not be a practical proposition in any of the present-day steel-making furnaces, but even in the absence of induced turbulence the atoms or ions in the metal and slag layers are always mixed to some extent by convective flow. With the model upon which most of the present treatments are based, it is assumed that regions of localised concentration difference are stretched out into very thin layers by turbulence or by convective flow and diffusion is required to equalise the concentration difference between the layers. The thickness of the layers is reduced as the turbulence of the liquid is increased.

FICK'S LAWS OF DIFFUSION

The formal basis of diffusion theory is stated by *Fick's Laws* [6]. For a two-component, single-phase solution at constant temperature and pressure the first of these laws states that 'the quantity of diffusing substance, or the flux (J), passing in unit time through unit area of a plane at right angles to the direction of diffusion is proportional to the concentration of the diffusing substance':

$$J = -D\frac{\partial c}{\partial x} \qquad 5.28$$

where c is the concentration, x is the distance along the diffusion path and D is the diffusion coefficient which is usually measured in units of $cm^2\ sec^{-1}$.

Fick's second law is an expression for the rate of accumulation of the diffusing substance in a given volume element:

$$\frac{\partial c}{\partial t} = \frac{\partial}{\partial x}\left(D\frac{\partial c}{\partial x}\right) \qquad 5.29$$

and, in the special case where D is constant, this simplifies to:

$$\frac{\partial c}{\partial t} = D\left(\frac{\partial^2 c}{\partial x^2}\right) \qquad 5.30$$

Generally it is found that D is not constant but is a function of the concentration and with an ideal solution there is some evidence that the variation is related to the energies of the different types of atomic or ionic bonds which are broken by the diffusion of one of the species in the solution [7]. For a non-ideal solution the direction of diffusion is not entirely random but takes place preferentially in a direction which lowers the free energy of the system. Evidence has been obtained of diffusion down a free-energy gradient but against the concentration gradient [8], and this suggests that the diffusivity of a substance is dependent on the rate of change of the excess free energy of the substance with distance in the solution, rather than on the rate of change of concentration as implied by Fick's law. However, it is expected that generally the free energy and concentration gradients change in the same direction, particularly when the concentration difference is fairly large.

The diffusivity of a substance is related exponentially to the temperature by an equation of the form:

$$D = D_0 e^{-E/(RT)} \qquad 5.31$$

where D_0 is a constant. This relation is similar to the Arrhenius rate equation and by analogy the quantity E is called the activation energy for diffusion. In the solid state this is the energy which is required to displace the adjacent atoms and allow the diffusing atom to leave its co-ordination shell. With the close-packed crystalline structures the activation energies are large and the diffusivities are markedly dependent on the temperature, but in the liquid state the atoms or ions are more loosely packed and the energy required to displace an atom from its co-ordination shell is very much smaller than in the solid state. Consequently, the activation energies are small for diffusion in a liquid relative to those in a solid, and the diffusivities do not vary markedly over a fairly wide temperature range.

DIFFUSION IN HETEROGENEOUS REACTIONS

The solution of Fick's equations for any particular diffusion problem requires a knowledge of the boundary conditions and a number of solutions for particular boundary conditions in homogeneous systems are available in the literature [9–11]. Since most refining reactions are heterogeneous, it is more appropriate to examine solutions for Fick's equations which are applicable to the transport of the reacting species to or from the interface at the phase boundary. The mathematical treatment of these conditions has developed from consideration of problems of heat flow, for which the equations are analogous to those for diffusion.

The transfer of heat from a circulating gas to a solid surface is described satisfactorily by assuming that a uniform gas of constant temperature is separated from the solid by a static or dead film of gas and that the rate of heat transfer to the solid is dependent on thermal flow across this static film. In a similar manner, it is assumed that a homogeneous liquid of uniform composition is separated from the phase boundary interface by a static film of liquid. The composition of the bulk liquid is maintained constant by turbulent and convective flow, whilst the rate of supply of the reactants to or from the interface is controlled by diffusion across the static film.

If the boundary conditions and the film thickness remain constant the rate of transport across the static film can be calculated from Fick's first law, but generally the concentration gradient across the film cannot be determined directly. Figure 5.2 shows schematically a concentration profile which varies from the average composition (c_0) of the diffusing substance in the bulk liquid to the concentration (c) at the interface. The effective film thickness (Δx) is defined as the intersection of the tangent to the concentration profile at the interface with the extrapolated line for the concentration in the bulk liquid [12, 13]. That is:

$$\Delta x = \frac{c_0 - c}{\left(\frac{\partial c}{\partial x}\right)_{x=0}} \qquad 5.32$$

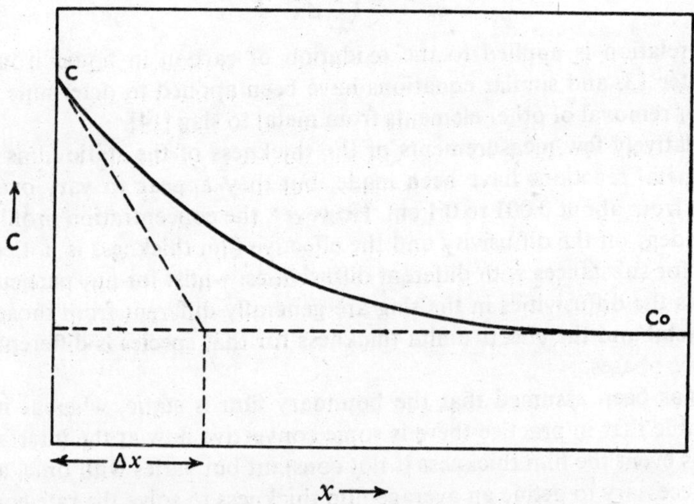

FIG. 5.2. Variation of concentration with distance from a phase boundary.

Substituting Fick's first law in this equation gives the relation:

$$J = D\left(\frac{c - c_0}{\Delta x}\right) \qquad 5.33$$

and, if transport across the interface is diffusion-controlled, the rate of transport is given by:

$$-\frac{\partial c}{\partial t} = \left[D'\left(\frac{c' - c_0'}{\Delta x'}\right)\right] = \left[D''\left(\frac{c'' - c_0''}{\Delta x''}\right)\right] \qquad 5.34$$

where the first term in brackets is the rate of transport across the static film in one phase and the second term in brackets is the corresponding rate of transport in the second phase on the opposite side of the interface. It is frequently found that the concentration of the diffusing substance is very much smaller in one phase than in the other and the diffusion rate is then determined by the rate across the boundary layer in the phase with the smaller concentration of the substance [13].

If the concentration of the diffusing substance in the metal at the interface which would be in equilibrium with the instantaneous concentration of the substance in the bulk slag is designated c_{SE}, then from equation 5.20 the rate of transport is:

$$\frac{\partial c}{\partial t} = D\left(\frac{c_{SE} - c_0}{\Delta x}\right) \quad 5.35$$

This relation is applied to the oxidation of carbon in liquid iron in Chapter 13, and similar equations have been applied to determine the rate of removal of other elements from metal to slag [14].

Relatively few measurements of the thickness of the static films for slag-metal reactions have been made, but they appear to vary over a range from about 0·001 to 0·1 cm. However, the concentration profile is dependent on the diffusivity and the effective film thickness is not constant for substances with different diffusivities, whilst for any particular species the diffusivities in the slag are generally different from those in the metal and the effective film thickness for that species is different in the two phases.

It has been assumed that the boundary film is static, whereas it is probable that in practice there is some convective flow at the interface. In this event the film thickness is not constant but varies with time, and it is necessary to define an average film thickness to solve the rate equations. Since the extent of convective flow in the boundary layer may be expected to vary from one experiment to another and the conditions in the refining practices may differ from those which are imposed in experimental studies, it is evident that the experimentally evaluated rate constants for a diffusion-controlled reaction may be vastly different from those which are actually obtained during refining.

In practice a number of transport-controlled reactions may occur simultaneously and, because the rate of transport across the boundary layer is different for each of the reacting species, it is possible that one or more of the reactions transiently overshoots the final equilibrium state. In addition, the concentrations of the reactants and products are changed during the reaction and this may cause a change in the rate-controlling mechanism as equilibrium is approached. Some examples of these features have been examined theoretically [13], but so far there have been very few attempts to obtain experimental confirmation.

REFERENCES

1. J. W. MELLOR, *Chemical Statistics and Dynamics*, Longmans, 1904.
2. C. N. HINSHELWOOD, *Kinetics of Chemical Change*, Oxford Univ. Press 1942.
3. H. EYRING, *Chem. Rev.*, 1935, Vol. 17, p. 65.
4. S. GLASSTONE, K. J. LAIDLER and H. EYRING, *The Theory of Rate Processes*, McGraw Hill, 1941.
5. L. S. DARKEN and R. W. GURRY, *The Physical Chemistry of Metals*, McGraw Hill, 1953.
6. A. FICK, *Pogg. Ann.*, 1855, Vol. 94, p. 59.
7. A. D. LE CLAIRE, *Acta Met.*, 1953, Vol. 1, p. 438.
8. L. S. DARKEN, *Trans. Amer. Inst. min. (metall.) Engrs.*, 1948, Vol. 175, p. 184.
9. R. M. BARRER, *Diffusion in and Through Solids*, Cambridge Univ. Press, 1941.
10. W. JOST, *Diffusion in Solids, Liquids, and Gases*, New York, Academic Press, 1952.
11. J. CRANK, *The Mathematics of Diffusion*, Clarendon Press, Oxford, 1957.
12. K. F. GORDON and T. K. SHERWOOD, *Chem. Engng. Prog.*, Symposium No. 10, 1954, p. 15.
13. C. WAGNER, *The Physical Chemistry of Steel-making*, New York, John Wiley, 1958, p. 237.
14. L. S. DARKEN, *Basic Open-Hearth Steel-making*, second edition, A.I.M.E., 1951, p. 592.

PART II

THE PRODUCTION OF IRON THE BLAST-FURNACE PROCESS

CHAPTER 6

Gas Reactions in the Blast Furnace

6.1 Introduction

The essential functions of the blast furnace are two-fold. Primarily, the oxygen combined with the iron in the ore must be removed and this is accomplished by chemical reaction between the iron oxides and carbon in the form of coke to produce carbon monoxide, carbon dioxide and metallic iron. Secondly, the process must separate the metal so produced from the remaining non-metallic or *gangue* content of the ore and from the ash residue of the coke. This is effected by melting the charge and allowing differences in density to cause a separation into a layer of slag containing most of the unwanted non-metallic components which float on top of the liquid metal.

On economic grounds the process is required to operate at the lowest possible temperature, commensurate with a high production rate and, as the reduction of the iron oxides is practically complete before the charge is fully molten, the minimum operating temperature is determined by that required to melt the slag and the metal. Pure iron melts at 1,537°C, but this melting point is lowered progressively by additions of most of the common solute elements. The metal from the blast furnace is saturated with carbon and also contains varying amounts of other elements, such as manganese, silicon, phosphorus and sulphur, which are reduced from the ore at the oxygen potentials prevailing in the furnace. The presence of these elements in solution lowers the melting point of the iron to a temperature in the region of 1,200°C.

The final slag consists principally of lime, magnesia, silica and alumina, but no mixture of these oxides is completely liquid below about 1,200°C and the liquid composition range is still relatively small at 1,400°C. Clearly then the fusion temperature of the slag sets the lower limit of temperature at which the process can be operated, but the actual temperature must be somewhat greater than this minimum in order to ensure that the slag is sufficiently fluid to run freely from the furnace. In normal practice the slag is at 1,400 to 1,500°C, whilst the average temperature of the metal is about 50°C lower, owing to thermal losses through the hearth.

It follows that the burden must be adjusted to form a slag whose composition lies within the range which ensures the formation of a liquid of low viscosity at this temperature. At the same time the slag composition should ideally be adjusted so as to present the lowest attainable activities of the elements which are not wanted in the iron, whilst the activities in the slag should be high for those elements which it is desirable to recover in the iron.

It is extremely rare that the gangue content of an ore, together with the ash from the coke, conforms to all these requirements and two or more ores with suitable gangue compositions are blended to give a 'self-fluxing' burden; alternatively a flux is added to the charge to adjust the slag composition. In practice the working composition is usually a compromise. The composition range which gives suitable fluidity at the operating temperature is not very great and the selection of the optimum composition to control the concentration of the solute elements in the metal must be made within this range. In recent years intensive studies of the liquidus surfaces of the relevant multi-component phase diagrams and investigations of the thermodynamic relationships of the reactions occurring in the blast furnace have facilitated marked advances in this aspect of slag control. These investigations have also disclosed that equilibrium distribution of certain elements between slag and metal is not attained even at the high temperatures prevailing in the hearth and examination of the factors controlling the kinetics of these reactions is now recognised as being of comparable importance to the determination of the equilibrium distribution.

THE BLAST FURNACE

The blast furnace can be divided structurally into three sections or zones (Figure 6.1) and it is convenient to consider the reactions which occur in each of these. The highest temperatures are attained in the *hearth* immediately in front of the tuyeres where the oxygen in the air blast reacts with the coke to give local temperatures of about 1,900°C. Outside the combustion zone the temperature falls steeply and is about 1,600°C on the same horizontal plane but closer to the centre line of the furnace. Immediately preceding the tapping operation the pool of molten metal and slag extends upwards from the bottom of the hearth almost to the tuyere zone and at intermediate times the liquid level varies between these limits. All the ore is molten by the time it descends to the hearth, but a column of coke extends to below the upper surface of the slag and possibly to the bottom of the furnace.

The inverted, truncated-cone section over the hearth is called the *bosh* and above this is the *stack*. The temperature falls fairly uniformly with increasing height above the tuyeres to about 1,100°C at the top of the bosh and to about 800°C mid-way up the stack. At this height the temperature falls rather rapidly to between 550 and 600°C and then continues to fall at a slower rate, reaching 200 to 250°C at the top of the furnace.

The hot gas from the combustion zone travels up the furnace countercurrent to the descending burden, and a large part of the sensible heat in the gas is transferred to the charge. The high stack is necessary to obtain

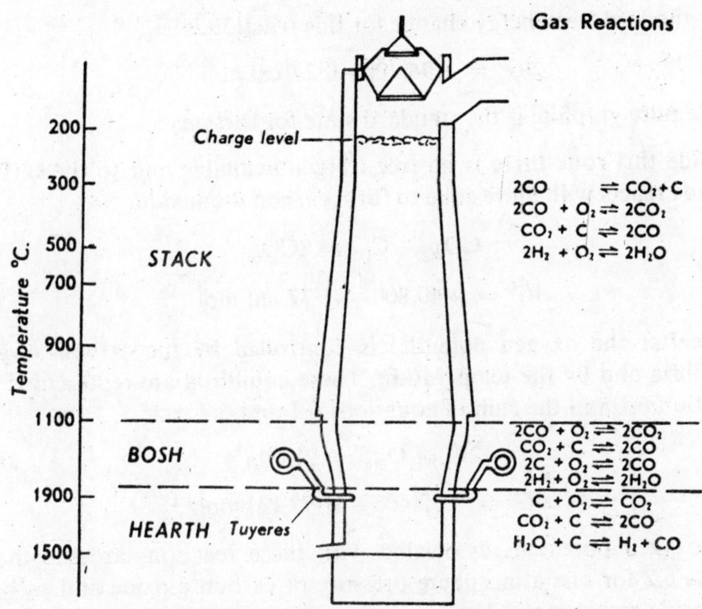

FIG. 6.1. Schematic representation of the temperature distribution and the gas reactions in the blast furnace.

the maximum thermal recovery from the gas and, ideally, the height of the furnace should be such that the top gas is almost at atmospheric temperature. However, as the height is increased the load within the furnace which the charge materials have to support is raised proportionately and there is increasing risk that part of the charge may be crushed to powder. This fine material becomes packed in between the larger particles, decreasing the voids through which the gas flows and thus obstructing the passage of the gas. The maximum overall height of the furnace has remained fairly constant at about 100 feet for the last 20 years, but the exit gas temperature has been lowered by up to 100°C through closer control of both the burden preparation and the charge distribution in the furnace in order to obtain better gas–solid contact and hence improved thermal transfer.

6.2 Gas-Solid Reactions

The oxygen in the air blast is entirely consumed by reaction with the coke charge within a small radius of the tuyeres:

$$C_{(s)} + O_{2(g)} \rightleftharpoons CO_{2(g)} \qquad 6.1$$

The standard free-energy change for this reaction is:

$$\Delta G^0 = -94{,}200 - 0{\cdot}2T \text{ cal mol}^{-1}$$

where pure graphite is the standard state for carbon

Outside this zone there is no free oxygen available and so the carbon dioxide reacts with more coke to form carbon monoxide:

$$CO_{2(g)} + C_{(s)} \rightleftharpoons 2CO_{(g)} \qquad 6.2$$

$$\Delta G^0 = +40{,}800 - 41{\cdot}7T \text{ cal mol}^{-1}$$

Thereafter the oxygen potential is controlled by the carbon-oxygen equilibria and by the temperature. These equilibria are represented by equation 6.1 and the sum of equations 6.1 and 6.2, viz.,

$$2C_{(s)} + O_{2(g)} \rightleftharpoons 2CO_{(g)} \qquad 6.3$$

$$\Delta G^0 = -53{,}400 - 41{\cdot}9T \text{ cal mol}^{-1}$$

The oxygen potentials associated with these reactions are shown in Figure 6.2 for one atmosphere pressure of carbon dioxide and carbon monoxide respectively. The two lines intersect at 700°C and at all temperatures above the intersection the CO line is at a lower oxygen potential than the CO_2 line. Thus, above 700°C carbon monoxide is more stable and reaction 6.2 goes to the right, whilst at lower temperatures carbon dioxide is more stable and the direction of reaction 6.2 is reversed.

The ratio of p_{CO}/p_{CO_2} as a function of temperature can be obtained from equation 6.2, i.e.,

$$\Delta G^0 = -RT \ln p^2_{CO}/p_{CO_2} \cdot a_C$$

and, since graphite is the standard state for carbon in equation 6.2, this simplifies to:

$$\frac{\Delta G^0}{4{\cdot}575T} = -\log \frac{p^2_{CO}}{p_{CO_2}} \qquad 6.4$$

The ratios thus obtained are as follows:

Temperature °C	1,900	1,460	1,180	980	820	670	530	410	300
CO/CO_2	$10^5/1$	$10^4/1$	$10^3/1$	$10^2/1$	10/1	1/1	1/10	$1/10^2$	$1/10^3$

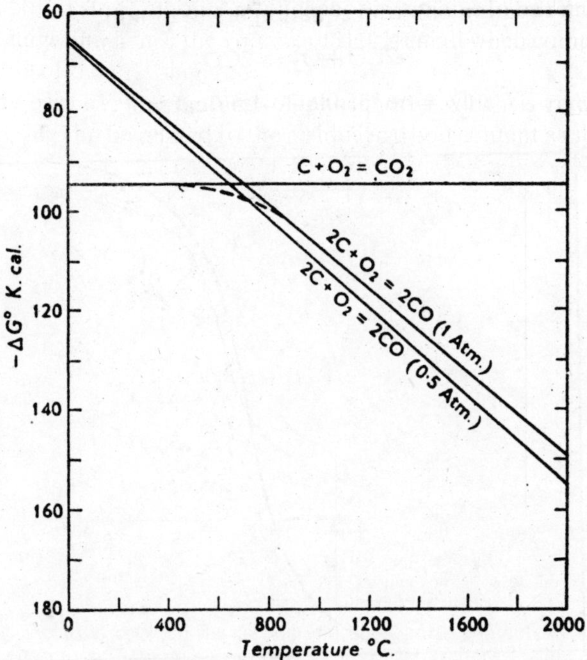

FIG. 6.2. Oxygen potential diagram for the formation from oxygen and solid carbon of carbon monoxide and carbon dioxide.

and the variation of the carbon monoxide concentration with the temperature is shown in Figure 6.3.

At temperatures in excess of 1,000°C the concentration of carbon dioxide is negligible and the total pressure of the carbonaceous gases is approximately equal to the partial pressure of the carbon monoxide, whilst below 400°C the partial pressure of carbon monoxide is negligible and the reverse applies. Between these temperature limits the partial pressures of both gases are significant and, if $(p_{CO}+p_{CO_2})$ is equal to one atmosphere pressure, the partial pressure of each gas is less than atmospheric. Since the solid lines in Figure 6.2 represent the oxygen potentials in equilibrium with either p_{CO} or p_{CO_2} at one atmosphere pressure, it follows that at temperatures between 400 and 1,000°C the oxygen potentials in equilibrium with the mixed gas are not represented by the solid lines but by a curve which joins the two straight lines. This is shown by the broken line in Figure 6.2, the location of which is determined by subtracting equation 6.2 from equation 6.1 to obtain:

$$2CO + O_2 \rightleftharpoons 2CO_2$$

$$\Delta G^0 = -135{,}000 + 41{\cdot}5T \text{ cal mol}^{-1}$$

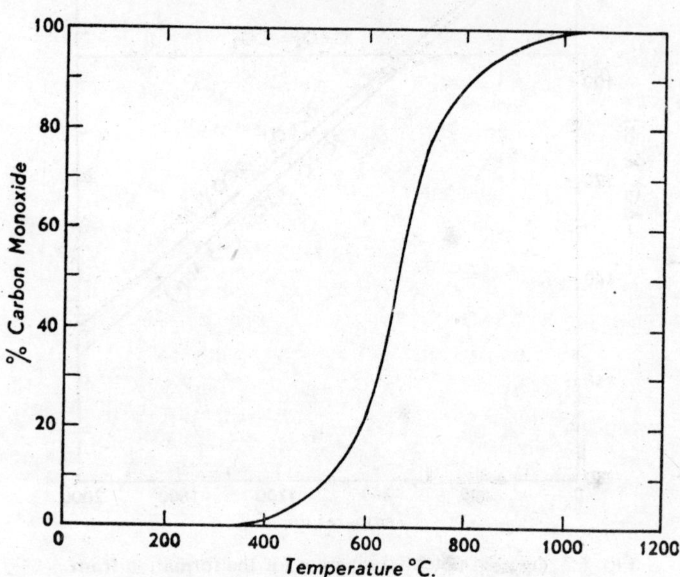

FIG. 6.3. Variation with temperature of the concentration of carbon monoxide in equilibrium with carbon dioxide and solid carbon at a total pressure of one atmosphere.

which can be rearranged in similar manner to equation 6.4 to give:

$$RT \ln p_{O_2} = \Delta G^0 - 2RT \ln p_{CO}/p_{CO_2} \qquad 6.5$$

The values of the ratio p_{CO}/p_{CO_2} at various temperatures determined from equation 6.4 are substituted in equation 6.5 to obtain the corresponding partial pressures of oxygen.

VARIATIONS IN GAS CONCENTRATIONS WITHIN THE STACK

The actual conditions within the blast furnace are somewhat more complex than the simple equilibria which have been described. The gas pressure is usually close to atmospheric at the top of the stack and is between 1·3 and 1·5 atmospheres at the tuyere level, but the carbonaceous gases are diluted with large amounts of nitrogen and smaller quantities of hydrogen and steam from the air blast. In consequence ($p_{CO}+p_{CO_2}$) is less than one atmosphere, and this affects the equilibrium ratio p_{CO}/p_{CO_2} at a given oxygen potential and temperature. The effect of this can be judged from the position of the line in Figure 6.2 for the reaction:

$$2C_{(g)}+O_{2(g)} \rightleftharpoons 2CO_{(g)}(0{\cdot}5 \text{ atmosphere})$$

which corresponds more closely to the partial pressure of carbon monoxide in the hearth and bosh of the blast furnace. The CO and CO_2 lines now intersect at a lower temperature and the equilibrium p_{CO}/p_{CO_2} ratios in the blast furnace are slightly lower than those previously tabulated.

From consideration of the temperature distribution within the furnace and the equilibrium gas ratios, it is evident that below the base of the stack and outside the combustion zone the atmosphere is almost entirely free from carbon dioxide, but during its passage up the stack the gas is cooled rapidly, the carbon dioxide concentration is increased and, if equilibrium is maintained, the exit gas contains only a negligible amount of carbon monoxide. However, the total transit time of the gas from the tuyere level to the top of the furnace is only about 10 to 12 seconds, during which time the gas temperature falls by about 1,700°C, and there is little opportunity for equilibrium to be maintained.

Owing to the heterogeneous nature of the gas distribution within the furnace it is difficult to obtain a relation between the analysed gas composition and the sampling position in the stack which has any significance, but the following values, quoted by Joseph [1] for American practice, give some indication of the way in which the gas composition varies from bottom to top of the furnace.

	Volume Per cent					
	CO_2	CO	H_2	H_2O	N_2	CO/CO_2
Bosh gas	0	35·0	1·0	0	64·0	
800°C level	7·8	33·4	—	—	58·0	4·3
Top gas	13·6	27·6	0·4	0·4	58·0	2·03

The wide range in gas composition on a horizontal plane is well illustrated by the following figures from Russian practice [2]:

Sampling position	CO/CO_2	*Gas Temperature*
Just below stock level, centre line	1·3	290°C
do. do. periphery	5·0	580°C

The concentration of CO_2 in the exit gas varied over the range 6·4 to 18·7 per cent, owing partly to channelling and partly to variations in the reducibility of the ore. In British and American practice the mean top gas CO/CO_2 ratio is generally between 1·6 and 2·2.

The conditions existing within the furnace are made clearer by following the change in the composition of a small volume of gas from the time of entering the furnace in the air blast to the time when it is discharged from the furnace top [3, 4]; the relevant data are presented in Figure 6.4. The shaded band represents the range of oxygen potentials which are

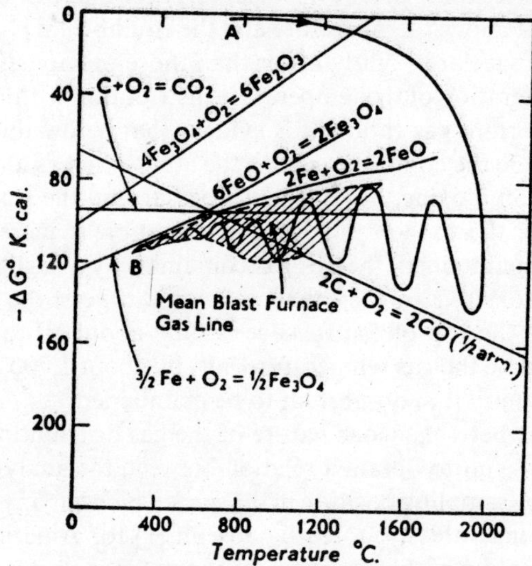

FIG. 6.4. Changes in the composition of a small volume of gas during its ascent through the blast furnace. (After Richardson and Jeffes [4].)

found by taking simultaneous gas samples and temperature measurements at various levels in the furnace stack and the broken line within the band gives the mean gas composition.

The air blast enters the furnace at a preheat temperature of 500 to 800°C and the oxygen potential is atmospheric ($p_{O_2} = 0{\cdot}21$ atmospheres). These conditions set the starting point (A) of the compositional changes which are marked on the diagram by the heavy line. Immediately in front of the tuyeres the oxygen in the blast reacts with the coke, the temperature rises rapidly to about 1,900°C and the oxygen potential is lowered towards the C–CO line, which is the lowest attainable potential under the prescribed conditions. Before this limit is reached the gas will probably come into contact and react with unreduced iron oxides; this reaction raises the oxygen potential of the gas towards the appropriate iron-oxygen line. It is highly improbable that an oxide of iron other than FeO is present at this level in the furnace; hence the upper limit is fixed by the Fe–FeO line. Alternate contact with ore and coke particles causes the gas composition to cycle between these limits as the gas rises up the furnace and its temperature falls by thermal transfer to the burden. The magnitude of the composition fluctuations diminish progressively as the two limiting lines converge with decreasing temperature, and below 1,100°C they lie within the limits of the band of measured gas compositions. Since there is insufficient time at the lower temperatures for the attainment of equilibrium, the band is situated at oxygen potentials which are lower (more negative) than those for the CO–CO_2 equilibrium. With further decrease in temperature the speed of the reactions diminishes rapidly and the limits of the band converge towards the point (*B*) where the gas is discharged from the furnace at about 250°C.

6.3 Factors affecting Reducibility

EFFECT OF TEMPERATURE

From the above considerations it is apparent that any oxide whose oxygen potential lies above the minimum potential attained in the blast furnace can be reduced to the elemental form and, as the oxygen potential lines for the three oxides of iron lie above or within the gas composition band at all temperatures, theoretically the oxides could be reduced by the gas even at room temperature. However, the reaction times increase exponentially with decreasing temperature and the rate of reduction only becomes significant at temperatures above about 400°C.

The reactions occurring and the associated heats of reaction may be written as follows:

$$3Fe_2O_3 + CO \rightleftharpoons 2Fe_3O_4 + CO_2; \ \Delta H^0_{298°K} = -12{\cdot}6 \text{ kcal mol}^{-1} \quad 6.6$$

$$Fe_3O_4 + CO \rightleftharpoons 3FeO + CO_2; \ \Delta H^0_{298°K} = +9{\cdot}6 \text{ kcal mol}^{-1} \quad 6.7$$

$$Fe_3O_4 + 4CO \rightleftharpoons 3Fe + 4CO_2; \ \Delta H^0_{298°K} = -3{\cdot}7 \text{ kcal mol}^{-1} \quad 6.8$$

$$FeO + CO \rightleftharpoons Fe + CO_2; \ \Delta H^0_{298°K} = -4{\cdot}4 \text{ kcal mol}^{-1} \quad 6.9$$

and the carbon dioxide produced in each of these reactions can react with the carbon in the coke to produce further quantities of carbon monoxide according to reaction 6.2.

These reactions represent consecutive steps in the reduction of the higher oxides of iron to metallic iron. Ferric oxide is first reduced to magnetite according to equation 6.6 and magnetite is then reduced either directly to metallic iron (reaction 6.8), or by a two-stage process to ferrous oxide and then to metallic iron (reactions 6.7 and 6.9). The single-stage reduction of magnetite occurs at temperatures below 570°C and the two-stage process occurs above this temperature. The reason for this is readily seen from an examination of the constitutional diagram for the

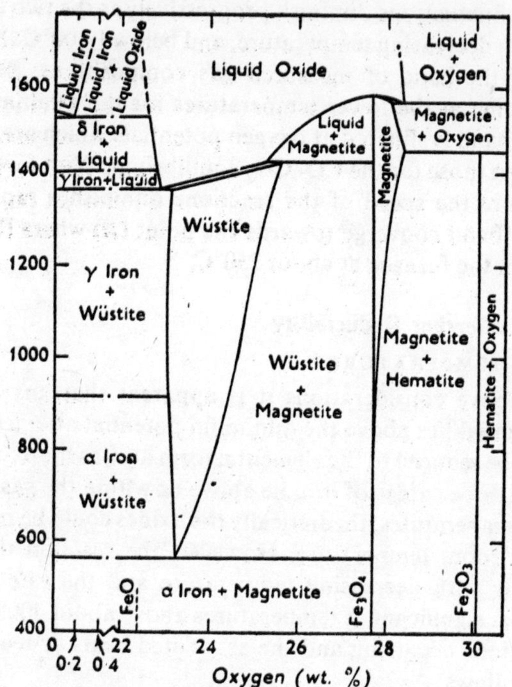

FIG. 6.5. The binary system iron–oxygen at a total pressure of one atmosphere. (After Darken and Gurry [5].)

iron-oxygen binary system (Figure 6.5). This diagram shows that the non-stoichiometric FeO phase (wüstite) is unstable below 570°C and decomposes into a mixture of metallic iron and Fe_3O_4. Thus, reading from right to left across the phase diagram at constant temperature, below 570°C the phase sequence is Fe_2O_3–Fe_3O_4–Fe, whereas above 570°C the sequence is Fe_2O_3–Fe_3O_4–FeO–Fe.

The heats of reaction for each step amount to only a few kcal per mole, for the endothermic reduction of the oxides is almost exactly balanced by the heat liberated in the oxidation of carbon monoxide to carbon dioxide. The overall heat of reaction for the reduction of one mole of ferric oxide to metallic iron is the same at 570°C for either sequence of reactions.

EFFECT OF DIFFUSION OF OXYGEN

The speed of the gaseous reduction is controlled by the rate at which oxygen atoms in the ore can combine with the carbon monoxide. This largely depends on how readily the oxygen atoms are brought into contact with carbon monoxide and also on the rapidity with which the carbon dioxide produced in the reaction is replaced by carbon monoxide in order to continue the reaction. Once the oxygen atoms are removed from the surface layer of an ore particle, the rate of reduction of the particle is controlled primarily by the rate at which oxygen atoms from the centre can reach the gas-solid interface to react with the gas mixture. It has been shown [6] that oxygen ions can diffuse readily through ferric oxide when an oxide layer is formed on metallic iron, but the metal ion is the more mobile species in magnetite and wüstite. This behaviour should apply also to the reduction of the oxide and it is expected that the first stage of reduction from Fe_2O_3 to Fe_3O_4 proceeds more rapidly than the subsequent steps. In the first stage, the removal of the oxygen ions from the surface leaves vacant lattice sites through which oxygen ions can migrate from the unreduced core, whereas the removal of oxygen in the later stages of reduction forms a dense metallic layer which opposes oxygen-gas contact. Thus the reducibility of a rich magnetite ore is improved by a prior oxidising roast to convert the oxide to Fe_2O_3 for, during the subsequent reduction to Fe_3O_4. the removal of the oxygen ions makes the ore porous and the further reduction to metallic iron is enhanced by gas-solid contact within the pores.

EFFECT OF VOLUME CHANGE

The one-step reduction of Fe_3O_4 to metallic iron is accompanied by a large decrease in volume. Before the onset of fusion the ore has negligible plasticity and this shrinkage causes the ore particles to become deeply fissured, which increases the surface area per unit volume and facilitates reduction. In contrast the reduction of Fe_3O_4 to FeO in the two-step process is accompanied by a negligible change in volume; most of the change occurs during the second step of the reduction to metallic iron. Fissures do not form until the reduction is well advanced and the reaction rate tends to be lower than for the one-step reduction process, even though the higher temperature range favours the kinetics of the two-step process.

EFFECT OF MOISTURE IN THE AIR-BLAST

The air blast at the tuyeres normally contains a small amount of moisture, and this reacts with carbon to produce hydrogen and carbon monoxide according to the reaction:

$$H_2O + C \rightleftharpoons H_2 + CO \qquad 6.10$$

$$\Delta G^0 = +32{,}200 - 34{\cdot}05T \text{ cal mol}^{-1}$$

The carbon monoxide is then available to react with the oxygen in the ore and effect reduction by a sequence of reactions similar to those given in equations 6.6 to 6.9.

Up to about 800°C the rate of reduction by hydrogen is more rapid than by carbon monoxide [7, 8, 9], whereas at equilibrium the extent of the reduction is less with hydrogen than with carbon monoxide. This difference suggests that the step which is rate controlling is the diffusion of the reducing gas through the pores in the ore particles, for the diffusivity of hydrogen is about four times as great as that of carbon monoxide. At higher temperatures the time for complete reduction is less with carbon monoxide than with hydrogen, but this may be due to a change in the mechanism of the carbon reduction. It has been suggested [10] that at high temperatures carbon monoxide dissociates at the gas-solid interface and carbon atoms diffuse through the compact metal layer in the same manner as in the gas carburising process. The carbon atoms then react with the oxides in the core of the particle and re-form carbon monoxide, which builds up a pressure sufficient to crack the reduced outer shell and expose the oxide core to the reducing gas.

EFFECT OF POROSITY AND PARTICLE SIZE IN THE ORE

Laboratory experiments have confirmed that, other conditions remaining constant, the reducibility of an ore increases as the porosity is raised [12,14] and as the particle size is decreased [14–16], but the shape is important only with particles smaller than about 0·01 inch [17]. The time required for the complete reduction of an ore particle is dependent on the longest path which an oxygen atom must travel to reach a gas-solid interface, and so the maximum particle size which can be reduced completely before the ore reaches the hearth level is determined largely by the chemical composition and the porosity of the ore. Reduction may also be hindered by the thermal gradient through the ore lump but, with the size range which is normally charged to the furnace, thermal diffusion is sufficiently rapid to prevent the formation of any appreciable temperature gradients [18].

EFFECT OF SLAG FORMATION

Gaseous reduction is impeded at higher temperatures by the onset of slag formation. The first-formed slag is very viscous and coats the ore particles with a relatively impermeable layer, which prevents the access of reducing gas to the surface of the ore. The temperature at which this occurs is determined by the composition of the charge, but low melting-point, ferrous alumino-silicate compounds (melting commences at about 1,100°C) may be formed in the furnace or may be present in the ore. In general, gaseous reduction has practically ceased by the time the ore has been heated to between 1,100° and 1,200°C (i.e., at the base of the stack), and from this level onwards the remaining iron oxides are reduced by reaction with the solid coke in the charge.

EFFECT OF SOLID CARBON

Theoretically, solid carbon is a potential reducing agent at all temperatures for iron oxides in the blast furnace, but the speed of a reaction between two solids is much slower than between a solid and a gas. Since the coke cannot penetrate the pores in the ore particles, the oxygen atoms are removed from the ore at the points of contact of the ore and coke and thereafter the reaction is dependent on the supply of oxygen atoms from the interior of the particle to the points of contact. Hence the rate of solid reduction by coke is negligible compared with gaseous reduction in the upper part of the furnace and becomes significant only above about 1,000°C when the gaseous reactions are impeded by slag formation. However, elemental carbon can make a significant contribution to the extent of the reduction in the lower temperature range, for it was shown earlier that the stability of carbon monoxide decreases and the carbon dioxide content of the mixed gas increases rapidly as the temperature falls below 700°C. Part of the conversion of carbon monoxide into carbon dioxide required by this change in the equilibrium gas ratio is obtained by reaction with the oxides in the ore, the remainder being supplied by driving reaction 6.2 to the left, so that carbon monoxide breaks down into carbon dioxide and graphitic carbon. This carbon is deposited in a finely divided form upon the charge materials and some is accommodated in the pores of the ore particles, where it is carried down the stack with the descending burden. Because the ore is associated intimately with the deposited carbon, reduction by means of this carbon attains a significant rate at somewhat lower temperatures than those required for the reduction by coke.

The deposited carbon also plays a secondary role in the gaseous reduction reactions. The carbon dioxide produced by a reaction within the pores must be replaced by carbon monoxide before the reaction can proceed further and the carbon in the pores can react in situ with the carbon dioxide to replenish continuously the gas.

The two mechanisms of reduction by gaseous and solid carbon are conventionally known as *indirect* and *direct* reduction respectively.

EFFECT OF GANGUE COMPONENTS

The aspects of reducibility have so far been restricted to consideration of pure iron oxides, but in many ores the oxides are present as compounds with the gangue materials, such as silicates (nFeO.SiO_2), aluminates (nFeO.Al_2O_3), chromates (FeO.Cr_2O_4), carbonates ($FeCO_3$), hydrates ($Fe(OH)_2$) and more complex mixtures of these. This is particularly true of the calcareous Lincolnshire and the silicious Northants ores which are mined extensively in the Scunthorpe and Corby districts of Great Britain; the presence of such compounds in the ore markedly affects the rate of reduction.

Iron carbonates and hydrates decompose at relatively low temperatures (400 to 600°C) with the evolution of carbon dioxide and/or steam:

$$FeCO_3 \rightleftharpoons FeO + CO_2;\ \Delta H^0_{298^\circ K} = +21{\cdot}5 \text{ kcal mol}^{-1} \qquad 6.11$$

$$Fe(OH)_2 \rightleftharpoons FeO + H_2O;\ \Delta H_{298^\circ K} = +10 \text{ to } +20 \text{ kcal mol}^{-1} \qquad 6.12$$

The effects of the gas evolution are threefold:

1. The reactions are endothermic, so the temperature is lowered at the level in the stack where decomposition is occurring and the thermal load is increased. Unless there is a surplus of stack heat available, additional coke must be burnt to provide the extra heat.
2. The carbon dioxide and/or water vapour produced by the decomposition decrease the reducing potential of the gas. The dilution of the furnace gas occurs at the surface and within the pores of the ore particles; reduction is hindered by the necessity for the gas produced in this way to diffuse out of the pores and away from the surface of the ore.
3. Dissociation of the compounds causes a large decrease in volume and leads to the formation of deep fissures within the ore, thus increasing the surface area exposed for gaseous reduction, but in

extreme cases the ore is shattered and falls to a powder which obstructs the free flow of the gas up the furnace stack.

The magnitude of these effects is directly related to the carbonate and hydrate content of the ore. For example, gaseous reduction is restricted in proportion to the change in the CO/CO_2 ratio and the temperature drop in the decomposition zone, but it is enhanced at higher temperatures by the increased surface area for reaction.

Limestone ($CaCO_3$) is frequently added to the burden as a flux for slag formation. The dissociation of this carbonate is also endothermic:

$$CaCO_3 \rightleftharpoons CaO + CO_2; \Delta H^0_{298°K} = +41{\cdot}9 \text{ kcal mol}^{-1} \qquad 6.13$$

and lowers the CO/CO_2 ratio but the overall effect on the reduction is even less desirable than the decomposition of iron carbonates, for the reaction occurs in the 700 to 900°C range where the higher oxides of iron have been reduced to FeO and a high reduction potential is required for the final step to metallic iron. If reduction is hindered at this stage there is not much further opportunity for the gas reactions to occur before they are brought to an end by the start of liquid slag formation, and the proportion of the total reduction which must be performed subsequently by solid carbon is increased. An increase in the amount of direct reduction means that more coke must be charged per ton of ore to complete the reduction of the oxides and hence the carbon efficiency is decreased. There is thus a strong case for a prior calcination treatment of the limestone or, if carbonate fluxes are included in the charge, the particle size should be small to ensure that decomposition occurs in the lower part of the temperature range. In this connection it is interesting to note that the rate of decomposition is more rapid for a porous dolomite flux, $(Ca.Mg)CO_3$, than for a dense limestone [19, 20].

If the iron oxides are present in the ore as silicates, aluminates, chromates or similar compounds, the activities of the iron oxides are less than unity and at any temperature the equilibrium oxygen potentials are lower (more negative) than those for the corresponding pure oxides. This is illustrated in Chapter 1 by comparison of the relevant lines in Figures 1.2 and 1.3. Since at low temperatures the mean blast furnace gas line is almost coincident with the $Fe–Fe_3O_4$ line, it is evident that, when the iron oxides are present as compounds of this type, the minimum temperature for reduction is raised and the time available for gaseous reduction is restricted with respect to the time available for the reduction of the pure oxides. If the physical factors affecting reducibility are unchanged the proportion of the reduction which must be performed by solid carbon in the higher temperature zones is increased.

THE USE OF SINTER IN THE BURDEN

The significance of the effect of gangue components on the reducibility is important with respect to the production and smelting of sinter. Ore fines form an undesirable constitutent of the furnace burden for, by filling the voids between the larger lumps of ore, they restrict the upward flow of gas and cause non-uniform rates of reduction in any plane of cross-section. In extreme cases unreduced ore may extend downwards almost to the slag level. This cause of irregular performance was early recognised and the fines were screened-out from the burden, but as the heaps of ore fines accumulated some means was sought of converting them into a usable form and the sintering process, which had been developed by the non-ferrous industries, was adapted for the purpose.

The essential feature of sintering is the rapid heating of the ore to a temperature at which a small amount of liquid is formed [21, 22]. The liquid is whipped into a froth by a blast of air and the fines are cemented into a porous aggregate. Considerable difficulty has been encountered, however, in producing a suitable sinter from lean ores for, if only small amounts of liquid are formed, the sinter has a low crushing strength and readily reverts to fines during handling and charging to the furnace, but when sufficient bond is formed to give adequate strength the particles are often coated with slag and the reducibility is poor [23]. Even with ore fines from high grade ores it has frequently been found that the reducibility of the sinter is poor, relative to the reducibility of the same ore in lump form, and sinter came to be regarded as an undesirable material in the blast furnace. In numerous papers published during the period 1920–1940 it is stated that sinter should not constitute more than some 20 to 30 per cent of the charge weight if the furnace is to operate smoothly.

During the last decade rapid advances have been made in the technology of sinter production, to the extent that the attitude of the blast furnace operator has changed completely with regard to the use of sinter in the charge. A number of plants are now operating with more than 75 per cent of sinter and, following the pioneer work at the Appleby Frodingham Branch of the United Steel Companies, Ltd., there is a growing opinion in favour of crushing all the ore to fines and operating with an all-sinter charge [24–29]. The increase in the proportion of sinter in the charge has been accompanied by an increase in the furnace throughput, with lower coke rates per ton of iron than when untreated ores are fed directly to the furnace.

This remarkable change has been brought about largely by a clearer appreciation of the factors affecting reducibility. In the early days of sinter production the sintering temperature was often too high, with the result that the iron oxides formed compounds with the gangue materials and the reducibility of the sinter was poorer than that of the original iron ore. Today the carbon and moisture additions to the sinter strand are closely controlled to regulate the transit speed of the combustion zone and allow the formation of just sufficient liquid bond to provide the necessary strength requirements without coating the fines with a glazed layer of slag. Lime and magnesia additions can be made to silicious and aluminous ores to displace the iron oxides from their compounds during sintering, and so raise the activity of the oxide and improve the reducibility with respect to the untreated ore [30, 31, 32]. The requisite flux additions can be incorporated in the sinter, giving better mixing with the gangue constituents and earlier slag formation in the furnace; this results in more uniform burden descent and smoother working of the furnace. In addition all the carbonates and hydrates in the ore and in the flux are decomposed by sintering, so reducing the thermal load on the blast furnace. This means, particularly in the case of lean ore practice, that (i) less coke is required in the furnace charge and (ii) as the fuels used for sintering are coke breeze or fines which are unsuitable for use in the blast furnace, the demand for high quality coking coals is reduced.

One further and very important advantage arising from the use of sinter is a decrease in the sulphur content of the charge. The strongly oxidising conditions which prevail during the sintering operation cause about 70 per cent of the sulphur in the ore to be removed as sulphur dioxide [21, 24, 33, 34], but this is usually only a small part of the total decrease in the amount of sulphur present, as about 95 per cent of the sulphur in the furnace burden comes from the coke. Since the use of sinter decreases the amount of coke required for smelting, there is a very significant decrease in the sulphur load, so that an iron of the same sulphur content can be made with a slag of either lower basicity or smaller volume, thus reducing the quantity of fluxes needed in the burden and the heat required for melting the slag. This leads in turn to a further decrease in the coke rate and a reduction in the amount of sulphur which enters the furnace. For example, a change from 45 per cent to 100 per cent of sinter in the burden at Appleby Frodingham decreased the sulphur load by 13·66 lb per ton of iron, of which 87 per cent arose through reduced coke consumption and 13 per cent by the removal of sulphur from the ore during sintering [28].

The benefits arising from an all-sinter charge appear to be impressive and one is tempted to ask why the practice has not been adopted more widely. It is important to realise, however, that the actual improvement in furnace performance which is obtained is largely dependent on the physical and chemical nature of the ore used. For instance the haematite ores of the Great Lakes district of the U.S.A., which are relatively easily reduced, have negligible contents of carbonate and hydroxide and with such ores the improvement in furnace performance is unlikely to justify the heavy capital cost of crushing and sintering equipment. As these high-grade deposits become exhausted and are replaced by mechanically-concentrated taconite ores, the present trend favours briquetting or pelletising the fines in preference to sintering.

Marked improvements both in furnace output and in the coke rate (i.e. the weight of coke consumed per ton of iron produced) have been realised in Britain when using a lean ore practice with sintered Lincolnshire and Northants ores. Sintering is used also extensively with the rich Swedish magnetite ores to convert the oxide to Fe_2O_3 and improve the reducibility. Close control of the sintering operation is less important with ores of the last mentioned type, for the amount of low-melting-point bond present in the ore is often negligible and to a large extent the fines are bonded together by the actual fritting of the ore particles.

6.4 The Effects of Direct and Indirect Reduction on Thermal Efficiency

The reduction of iron oxides by solid and gaseous reactions can be summarised in the general form:

$$yC + Fe_xO_y \rightleftharpoons xFe + yCO \quad \text{(direct reduction)} \qquad 6.14$$

$$yCO + Fe_xO_y \rightleftharpoons xFe + yCO_2 \quad \text{(indirect reduction)} \qquad 6.15$$

In the first of these reactions carbon is oxidised to carbon monoxide, but the further oxidation to carbon dioxide is prevented by the CO/CO_2 equilibrium at the high temperatures where the direct reduction reactions are predominant. The heat liberated by the oxidation of carbon to form carbon monoxide is only about 27 kcal/mole of carbon, whereas some 94 kcal/mole are released by the oxidation of carbon to form carbon dioxide; i.e. almost three quarters of the available heat is obtained during the oxidation of CO to CO_2. The removal of oxygen atoms from each of the oxides of iron is accompanied by a marked absorption of heat, and it follows that direct reduction is strongly endothermic, but indirect reduction liberates the maximum available heat from the carbon and the overall change in the heat content during the reaction is comparatively small. Whilst the gaseous reaction to form carbon dioxide and the subsequent reaction of this carbon dioxide with coke or with deposited carbon is stoichiometrically identical with direct reduction, the resulting effects on furnace performance and efficiency in the two cases are very different.

The carbon which is burned to supply the hearth heat generates a large quantity of carbon monoxide which is immediately available for indirect reduction, whereas direct reduction requires the supply of additional coke at hearth level in excess of that required to meet the thermal needs of melting the charge. This carbon is partly needed for the direct reduction reaction and partly to compensate for the increased thermal load in the high-temperature zone arising from the endothermic reduction step [35]. Indirect reduction together with a reaction between carbon dioxide and carbon causes an overall thermal loss similar to that in direct reduction, but the heat is lost from the upper part of the furnace where the thermal balance is usually less critical.

The available heat in the stack is dependent on the composition of the burden. With lean ores containing a large proportion of gangue materials the thermal load is high and the provision of the hearth heat required to melt the slag and metal results in the production of a large volume of gas which carries heat up the stack. In this case there is ample stack heat available to satisfy the needs of the indirect reduction process and, for maximum thermal efficiency, all the oxide should be reduced by the gas to minimise the heat requirements at hearth level. Conversely, with a rich ore practice the thermal load is low and insufficient heat may be available in the stack for complete gaseous reduction; hence part of the oxide must be reduced by the direct reaction, but the maximum amount of gaseous reaction should still be the aim so as to obtain a high thermal efficiency [1, 36]. For any given burden the optimum amount of indirect reduction is set by the lowest CO/CO_2 ratio which can be obtained, for an increase in this ratio above the practical minimum often is indicative of an increase in direct reduction due to poor gas-solid contact. Thus with a lean ore burden the ore must be more readily reducible for the same top gas ratio to be realised as with a rich ore burden, where the thermal burden is low and hence the gas volume is also low.

With improved burden preparation the carbon efficiency has improved markedly during the present century, but there is still a considerable margin for further improvement. Using rich Swedish ores, coke rates of 13 cwt per ton and a top gas CO/CO_2 ratio of 1·4 have been realised [37], but in British and American practice the corresponding values are rarely less than 16 cwt and 1.6. However, the top gas ratio is not a good criterion for comparison of different practices, for the CO_2 content may vary over a wide range, owing to the decomposition of carbonates. Some of the carbon dioxide originating from this source and particularly that from the high-temperature breakdown of calcium carbonate may react with the coke, but the rest passes unchanged up the stack and lowers the top gas ratio. Any hydrogen which is present in the gas at low temperatures has the reverse effect, for it reacts with the carbon dioxide and raises the CO/CO_2 ratio. A number of attempts have been made to correct the top gas CO/CO_2 ratios for these various effects, notably by Killian [38], but the problem is complicated by the difficulty in obtaining representative gas and burden samples at different levels in the stack.

6.5 The Effect of the Temperature and Composition of the Hot Blast

The thermal energy which must be made available in the hearth is dependent on the heat required to melt the slag and metal and on the amount of oxide which has to be reduced at this level by the solid carbon. Part of this energy is supplied as sensible heat in the hot blast and, within certain limits, the amount of coke which is burnt in the hearth varies inversely with the blast temperature. If this temperature is raised, less coke is required to supply the same amount of hearth heat, a smaller volume of blast is required and less sensible heat is carried upwards from the hearth by the gas. This effect displaces downwards the thermal levels within the furnace, so that the fusion zone moves nearer to the hearth and becomes narrower. The zone in which the initial slag is formed is also restricted, causing the charge to become sticky, and in consequence the burden tends to hang in the bosh and descend in an intermittent manner. This irregular descent can increase the amount of iron oxide which enters the bosh and raise the amount of direct reduction. In general, it is more difficult to operate a high-blast temperature with lean ores which form a large volume of slag than it is with rich ores.

HIGH HUMIDITY PRACTICE

The reaction between the moisture in the hot blast and the coke in the combustion zone is endothermic. Thus the effect of an increase in the moisture content is the opposite to an increase in blast temperature, for it cools the hearth, decreases the gas-solid temperature difference, and widens the fusion zone. It is estimated that one per cent of steam, or 4·6 grains of moisture per cu. ft of blast, is equivalent to an increase in blast temperature of 74°C [39, 40], and this is the basis of the *high humidity practice* which has been developed in Russia and is now being applied fairly extensively in other countries [39–46]. Sufficient steam is introduced in the blast just to compensate for an increase in the blast temperature so that the hearth heat remains constant, whilst the hydrogen generated in the hearth is available for gaseous reduction, liberating additional stack heat and increasing the speed of indirect reduction by virtue of the greater diffusivity of hydrogen than of carbon monoxide. It is claimed that increased production and decreased coke rates per ton of iron are obtained, together with lower top-gas temperatures and more uniform burden descent. These improvements are possible, however, only when the hot-blast temperature is limited by the smooth working of the furnace and not by the hot-blast stove capacity.

OXYGEN ENRICHMENT

Part of the hearth heat is consumed in raising the blast temperature to that of the combustion zone. With dry air 79 per cent of the blast is inert nitrogen, which serves no useful purpose in the hearth but which is responsible for about half of the sensible heat in the stack gas. If the oxygen content of the blast is increased less coke is burnt to supply the same amount of hearth heat and the overall effect is similar to that of an increase in the blast temperature. However, when the oxygen content is raised above approximately 30 per cent the decrease in the gas volume is accompanied by the development of a steep temperature gradient in the bosh and the charge is only at a dull red-heat at the bottom of the stack. The amount of indirect reduction is then negligible and both the top gas CO/CO_2 ratio and the coke rate increase rapidly [47–50]. Smaller additions of oxygen may increase the rate of production and decrease the coke rate when there is a deficiency of hearth heat, but there is no advantage in using oxygen if the same results can be obtained by increasing the blast temperature, and oxygen enrichment is only justified when the blast temperature is limited by inadequate stove capacity.

OXYGEN–STEAM ADDITIONS

Since steam additions have an effect opposite to that of oxygen enrichment on the available hearth heat, the two can be added together to the blast to maintain a constant hearth temperature whilst increasing the rate of production and lowering the coke rate. An increase of 4 per cent of oxygen requires about 15 grains per cu. ft of moisture to maintain the thermal balance [40, 51, 52]. The main advantage of oxygen–steam additions to the blast lies in the greater degree of control which can be exercised over the hearth temperatures, as compared with operation with a normal air blast. Even with the most careful burden preparation the amount and composition of the slag-forming materials which enter the hearth vary slightly from time to time. The amount of coke in the burden can be varied to meet the thermal needs of the hearth but the effect is not obtained in the hearth until the burden has descended the full height of the furnace, which may take between 8 and 15 hours, depending on the geometry of the furnace and the rate of driving. On the other hand, an adjustment can be made to the oxygen or steam supplies and the hearth heat is adjusted almost immediately. Similar control can be exercised by adjustment of the hot-blast temperature and moisture content, provided that the necessary heat is available from the stoves.

Recently attention has been directed to the enrichment of the blast with oil or hydrocarbon gases. The results obtained suggest that more marked improvements can be obtained in the smoothness of the furnace operation and in both the production and coke rates than with oxygen–steam additions [53].

HIGH-TOP PRESSURE OPERATION

Considerable interest in recent years has been focused on the possibility of increasing the output of the blast furnace by raising the top gas pressure. The throughput can be raised by stepping-up the rate of blowing to increase the rate of combustion of carbon in the hearth, but the tendency for chanelling of the gas increases in proportion to the linear velocity of the gas and the coke consumption per ton of iron produced is increased disproportionately. Also the exit gas carries with it small particles of dust, the sizes of which are dependent on the gas velocity, and these particles must be removed before the gas is fed to the blowing engines, hot-blast stoves or other combustion units. As the gas velocity is raised the amount of dust carried by the gas is increased rapidly and a more extensive gas cleaning plant is required. However, the pressure drop in the furnace decreases if the pressure of the gas at the top of the furnace is raised whilst keeping the volume velocity of the gas constant. That is, the blowing pressure increases less rapidly than the top pressure and the linear velocity of the gas is lowered as the top pressure is raised [54]. The immediate result is a more uniform gas flow which diminishes the amount of channelling and increases both the extent and the time of gas-solid contact, so that the proportion of indirect reduction is increased and the coke rate is lowered. Apart from this the increased pressure has negligible effect on the reduction mechanism, for the rate of reduction is dependent mainly on fissure diffusion and this is affected more by the mass rate of flow than by the pressure of the gas [55, 56].

An increase in the total pressure of a gas reaction tends to drive the reaction in the direction which decreases the number of gas molecules. This would be expected to affect those reactions in the furnace which involve a change in the number of gas molecules, but the change arising from a pressure rise of less than one atmosphere is in fact negligible [55].

Experience of the effects of high-top pressure operation on European and American practice is contradictory but recent reports, mainly from

Russia, indicate that marked improvements have been obtained. With an increase in top pressure to between 1·3 and 1·8 atmospheres the productivity is raised by from 4 to 9 per cent, the coke rate is decreased by 2 to 4 per cent and the dust loss is approximately halved [40, 41, 45, 57–60]. A small increase in the blast temperature is also possible without interference with the burden descent; this was predicted some time ago [54] from the fact that the density of gas is increased by raising the pressure or by lowering the temperature.

The improvement in performance resulting from pressure operation depends largely on the composition and preparation of the burden. It has been stated that the results are better with sinter or a sized burden [60, 61], but the reduced tendency to channelling would be expected to cause a more significant improvement when working with an unsized burden [10].

6.6 Carbon Deposition

It was shown earlier that the CO_2 content of the furnace gas is negligible above 1,100°C, but increases steadily at temperatures below this level, and that carbon dioxide is formed by the dissociation of carbon monoxide with the deposition of solid carbon. According to Figure 6.3 the equilibrium ratio CO/CO_2 changes rapidly in the temperature range 500 to 700°C and, as the gas already departs markedly from equilibrium in this temperature range, the driving force for carbon deposition increases rapidly as the temperature falls. The reaction occurs most rapidly in the blast furnace at 500 to 550°C [55, 62], but the actual gas composition remains well below the equilibrium value, owing partly to lack of time for the attainment of equilibrium, but mainly because the reaction does not proceed readily unless a suitable catalyst is also present. The specific nature of the catalyst is uncertain and it has been variously described as metallic iron [63], the Fe_3O_4–Fe interface [62] and iron carbide [63–65], but the available evidence indicates that the reaction is catalysed by an iron carbide. In the absence of a catalyst the CO/CO_2 ratio may remain greatly in excess of the equilibrium value without the deposition of carbon.

Small amounts of nitrogen or sulphur compounds in the gas, such as cyanogen, ammonia, hydrogen sulphide and carbon disulphide, inhibit the catalytic action, whilst hydrogen and steam are activators for the reaction [63, 66, 67]. (The action of hydrogen is important in relation to the controlled high humidity practice.) The inhibiting action of sulphur is opposed by zinc oxide and by alkali compounds [66], the concentrations of which are usually negligible in the burden, but these compounds are volatilised at the high temperatures which prevail in the hearth and bosh and are re-deposited at lower temperatures in the stack. Hence the actual content of the compounds in the furnace is cumulative and, after a short time on blast, the concentrations may be sufficient to prevent the inhibiting effect of any traces of sulphur in the gas.

It is difficult to assess the total amount of carbon which is deposited in the furnace, and the estimates vary from about 9 per cent of the carbon gasified with normal lump-ore burdens [68] to 13 per cent with fine ores [69], which present a greater area of catalytic surface for the reaction, although a recent report suggests that these estimates are excessive [70]. An appreciable part of the carbon is deposited within the pores of the ore particles, causing them to swell and open up deep fissures which increase the area of gas-solid contact, but excessive deposition, particularly if the ore is highly friable, causes the lumps to fall to a powder. Some of the carbon is also deposited on any active iron spots in the corresponding temperature zones of the fireclay refractory lining and gives rise to the slow disintegration of the lining.

Whilst the last effect is obviously deleterious, the overall effect of carbon deposition within the charge is debatable. The principle features have been considered previously and may be summarised as follows:

1. The reaction decreases the top gas CO/CO_2 ratio, but it is exothermic and the top gas temperature is increased, so that the chemico-thermal efficiency is improved at the expense of purely thermal efficiency.
2. The deposited carbon is carried downwards with the ore and ultimately promotes the direct reduction reaction.
3. Owing to the fine state of division and intimate contact with the ore, the deposited carbon reacts more readily than the coke with the oxides and it probably contributes markedly to the process of reduction at temperatures as low as 800°C.

Thus the net effect of moderate amounts of carbon deposition is to re-circulate the carbon in the gas, with the result that the time available for reaction is increased and the thermo-chemical balance is improved, but excessive deposition which results in disintegration of the ore and irregular gas flow is definitely detrimental. The critical level beyond which carbon deposition is undesirable depends on the physical nature of the burden and the particular practice.

The deposition of volatilised zinc oxide and alkalis in the charge gives rise to fissuring and crumbling of the ore in a similar manner to the deposited carbon, but the amounts of these compounds circulating in the furnace are rarely sufficient to interfere in this way with the smooth descent of the ore. The cumulative build-up of the compounds in the refractory lining is more serious, for this can give rise to disintegration of the brick lining. The compounds may also form a very viscous liquid on the surface of the refractory lining to which lumps of ore and coke adhere, and an accretion, or scaffold, is gradually built outwards from the walls which restricts the free descent of the burden.

REFERENCES

1. T. L. JOSEPH, *Trans. Amer. Inst. min. (metall.) Engrs.*, 1951, Vol. 189, p. 37.
2. I. Z. KOZLOVICH, *Stal'*, 1939, Nos. 4, 5, 6, I.S.I. trans. No. 19.
3. C. GOODEVE, *J. Iron St. Inst.*, 1951, Vol. 166, p. 379.
4. F. D. RICHARDSON and J. H. E. JEFFES, ibid., 1949, Vol. 163, p. 397.
5. L. S. DARKEN and R. W. GURRY, *J. Amer. chem. Soc.*, 1946, Vol. 68, p. 799.
6. M. H. DAVIES, M. T. SIMNAD and C. E. BIRCHENALL, *J. Metals, N.Y.*, 1951, Vol. 3, p. 142.
7. M. TENEBAUM and T. L. JOSEPH, *Trans. Amer. Inst. min. (metall.) Engrs.*, 1940, Vol. 140, p. 106.
8. J. O. EDSTRÖM, *J. Iron St. Inst.*, 1953, Vol. 175, p. 289.
9. A. M. KUZNETSOV, *Zh. Fiz. Khimii*, 1953, Vol. 27, p. 1807.
10. J. TAYLOR, *Iron & Steel, Lond.*, 1953, April, p. 11.
11. E. DIEPSHLAG, M. ZILLGEN and H. POTTER, *Stahl. u. Eisen, Düsseldorf*, 1935, Vol. 52, p. 1154.
12. T. L. JOSEPH, *Trans. Amer. Inst. min. (metall.) Engrs.*, 1936, Vol. 120, p. 72.
13. H. L. SAUNDERS and H. J. TRESS, *J. Iron St. Inst.*, 1945, Vol. 152, p. 291.
14. A. E. EL MEHAIRY, ibid., 1955, Vol. 179, p. 219.
15. M. C. UDY and C. H. LONG, *Trans. Amer. Inst. min. (metall.) Engrs.*, 1943, Vol. 154, p. 162.
16. J. R. LEWIS, ibid., 1947, Vol. 172, p. 27.
17. S. Y. M. EZZ and R. WILD, *J. Iron St. Inst.*, 1959, Vol. 194, p. 211.
18. H. L. SAUNDERS and R. WILD, ibid., 1950, Vol. 165, p. 198.
19. H. L. SAUNDERS and H. J. TRESS, ibid., 1946, Vol. 154, p. 67.
20. T. L. JOSEPH, H. M. BEATTY and G. BITSIANES, *Trans. Amer. Inst. min. (metall.) Engrs.*, 1943, Vol. 1954, p. 148.
21. E. M. MCBRIAR, W. JOHNSON, K. W. ANDREWS and W. DAVIES, *J. Iron St. Inst.*, 1954, Vol. 177, p. 316.
22. R. D. BURLINGAME, G. BITSIANES and T. L. JOSEPH, *Trans. Amer. Inst. min. (metall.) Engrs.*, 1956, Vol. 206, p. 853.
23. R. WILD, *J. Iron St. Inst.*, 1953, Vol. 174, p. 131.
24. G. D. ELLIOTT, J. A. BOND and T. E. MITCHELL, ibid., 1953, Vol. 175, p. 241.
25. G. D. ELLIOTT, ibid., 1955, Vol. 181, p. 75.
26. R. D. SUNDQVIST, *J. Metals, N.Y.*, 1955, Vol. 7, p. 737.
27. A. LEDGARD, D. RIST and P. K. GLEDHILL, *Iron Coal Tr. Rev.*, 1957, Vol. 175, p. 249.
28. H. S. AYRES, ibid., 1957, Vol. 175, p. 601.
29. I. I. BORNATSKII, *Stal'*, 1957, (5), p. 475.
30. B. G. BALDWIN, *J. Iron St. Inst.*, 1954, Vol. 177, p. 312.
31. B. G. BALDWIN, ibid., 1955, Vol. 179, p. 142.

32. H. B. WENDEBORN, *J. Iron St. Inst.*, 1953, Vol. 175, p. 280.
33. V. GIEDROYC, ibid., 1955, Vol. 180, p. 129.
34. B. G. BALDWIN and L. F. BURGESS, *Iron Coal Tr. Rev.*, 1957, Vol. 175, p. 425.
35. B. G. BALDWIN, *J. Iron St. Inst.*, 1955, Vol. 179, p. 30.
36. J. F. PETERS, *Trans. Amer. Inst. min. (metall.) Engrs.*, 1954, Vol. 200, p. 466.
37. M. TIGERSCHIÖLD, *Stahl. u. Eisen, Düsseldorf*, 1950, Vol. 70, p. 397.
38. S. T. KILLIAN, *Trans. Amer. Inst. min. (metall.) Engrs.*, 1951, Vol. 192, p. 742.
39. T. P.COLCLOUGH, *J. Iron St. Inst.*, 1958, Vol. 189, p. 113.
40. M. I. KOROBOVA and N. I. KOROBOV, *Bull. Acad. Sci. URSS*, Classe Sci. Tech., 1946, p. 567.
41. A. A. MARKIN and M. YAOSTROSKHIN, *Izv. Ak. Nauk Otdelenie Tekn.*, 1956, (11), p. 58.
42. B. J. WILSON, *J. Metals, N.Y.*, 1958, Vol. 10, p. 268.
43. D. REJC and M. GORUP, *Rud. Met. Zbornik*, 1958, (4), p. 429.
44. V. YA MILLER and S. A. EL'KIN, *Stal'*, 1958, (3), p. 193.
45. A. N. POCHWISNEW, *J. Iron St. Inst.*, 1958, Vol. 190, p. 337.
46. A. GRIEVE, ibid., 1960, Vol. 194, p. 37.
47. R. J. ROCCA and M. B. BEVER, *Proc. Amer. Inst. Min. (Metall.) Engrs. Blast Furn. Comm.*, 1947, Vol. 6, p. 67.
48. W. GUNZ, *Gas Producers and Blast Furnaces*, New York, John Wiley, 1950, p. 260.
49. J. H. STASSBURGER, *Steel*, 1956, Vol. 138, No. 24, p. 122.
50. V. A. KHROMOV, *Kislorod*, 1958, Vol. 4, p. 11.
51. J. H. STASSBURGER, *Rev. Tech. Luxemb.*, 1958 (I.S.I. Symposium 1958)
52. T. E. DANCY, A. T. SADLER and H. N. LANDER, *Proc. Amer. Inst. Min. (Metall.) Engrs. Blast Furn. Comm.*, 1958, Vol. 17, p. 16.
53. See for example papers presented for discussion at the Autumn meeting of the Iron and Steel Institute, London, Nov. 1961.
54. T. L. JOSEPH, *Blast Furn.*, 1945, Vol. 33, p. 669.
55. L. F. MAREK, A. BOGROW and G. W. KING, *Trans. Amer. Inst. min. (metall.) Engrs.*, 1947, Vol. 172, p. 46.
56. E. IWANCIW, *Archiv. Gornictiva Hutnic.*, 1953, Vol. 1, p. 131.
57. B. N. ZHEREBIN, G. E. KOZARNOVSKII, N. V. KREPYSHEV and V. N. MINKIN, *Stal'*, 1955, (12), p. 1063.
58. P. F. SERGEEV and N. I. VASIL'CHENKO, ibid., 1956, (11), p. 963.
59. E. G. TETEREVYATNIKOV and V. N. ANDRONOV, ibid., 1957, (3), p. 200.
60. I. A. KUZ'MIN, *Metallurg.*, 1957, (7), p. 3.
61. M. A. STEFANOVICH, *Stal'*, 1955, (5), p. 398.
62. J. TAYLOR, *J. Iron St. Inst.*, 1956, Vol. 184, p. 1.
63. S. KLEMENTASKI, ibid., 1952, Vol. 171, p. 176.
64. W. R. DAVIS, R. T. SLAWSON and G. R. RIGBY, *Trans. Brit. ceram. Soc.*, 1957, Vol. 56, p. 67.
65. W. R. DAVIS and G. R. RIGBY, ibid., 1954, Vol. 53, p. 511.
66. T. F. BERRY, R. N. AMES and R. B. SNOW, *J. Amer. ceram. Soc.*, 1956, Vol. 39, p. 308.
67. W. A. BONE, H. L. SAUNDERS and H. J. TRESS, *J. Iron St. Inst.*, 1938, Vol. 137, p. 85.
68. H. L. SAUNDERS, G. B. BUTLER and J. H. TWEEDY, ibid., 1949, Vol. 163, p. 173.
69. E. NOWAK, *Stahl. u. Eisen, Düsseldorf*, 1950, Vol. 70, p. 829.
70. H. SCHENCK and H. BARTH, ibid., 1960, Vol. 80, p. 1681.

CHAPTER 7

The Partitioning of Solute Elements between the Iron and the Slag in the Blast Furnace

7.1 Introduction

The end product of the blast furnace is not pure iron, but a metal which is nearly saturated with carbon. It contains also variable amounts of manganese, silicon, phosphorus, sulphur and other elements which are reduced from compounds in the burden at the prevailing oxygen potentials and the concentrations of which can be controlled to some extent by adjustment of the slag composition. The thermodynamic equilibrium distribution of the elements between the slag and the metal is not attained in the blast furnace process, but this is not surprising in view of the almost stagnant nature of the molten bath. The slag and metal are in contact across a plane interface which is disturbed only by the droplets of liquid metal passing through the slag layer; the temperature of the metal decreases with increasing depth below the interface, with the result that circulation of the metal by thermal convection currents is very restricted.

When a reaction involves the transfer of an element across the interface, a layer on one side of the interface tends to become saturated, while the corresponding layer on the other side of the interface is depleted of that element. Further transfer necessitates the replenishment of the element in the depleted layer from the bulk material and/or the removal of the element from the saturated layer and into the bulk material on the other side of the interface. These conditions are examined in Chapter 5.5 where it is shown that in the absence of agitation this replacement is dependent on the rates of diffusion of the species in the slag and metal layers, but even at the temperatures found in the hearth of the blast furnace the rate of transport by diffusion is comparatively slow. The diffusion coefficients of solutes in liquid carbon-saturated iron are usually in the range 10^{-4} to 10^{-5} cm^2 sec^{-1}, whilst the corresponding rates in the liquid slag are lower by one or two orders of magnitude and vary over a wide range, for the temperature of the slag is usually fairly close to the liquidus and the viscosity is then markedly dependent on the temperature. The equilibrium distribution is approached more closely if the period of contact of the slag and metal is increased, but even under the most favourable laboratory conditions a true equilibrium can be obtained only after the elapse of several hours at the normal blast-furnace temperatures.

The kinetics for certain of the reactions are not changed significantly if the metal is agitated mechanically and this implies that these reactions are not diffusion-controlled, but that some other factor such as the activation energy for the reaction is unfavourable. Clearly, therefore, the composition of the metal obtained from the blast furnace is influenced markedly by the kinetics of the various reactions, but in order to indicate which of the reactions may be rate-controlling, it is preferable first of all to examine the factors affecting the equilibrium distributions.

7.2 The Transfer of Elements into the Iron above the Hearth

MINIMUM TEMPERATURES FOR REDUCTION OF OXIDES

It was shown in the previous chapter that an element may be reduced from its oxide in the blast furnace when the temperature is reached at which the oxygen potential for the formation of that oxide is greater than that in equilibrium with the furnace atmosphere, which in turn is determined by the oxygen potential in equilibrium with carbon monoxide or with the mean blast furnace gas line, whichever is the lower. The minimum temperatures at which the pure oxides are reduced can be obtained by superimposing the gas lines from [illegible]ure 6.4 upon Figure 1.1; for the more common oxides the minimum te[illegible]atures are as follows:

Oxide	*Temperature* °C
P_2O_5	485
Cr_2O_3	1,150
MnO	1,380
SiO_2	1,490
TiO_2	1,520
Al_2O_3	2,010

These temperatures refer to a furnace which is operating at normal gas pressures for, if the top pressure is raised, or the blast is enriched with oxygen, the partial pressure of the carbon monoxide is raised and the minimum temperatures for reduction are increased accordingly.

The above data show that, with the exception of phosphorus, the oxides are not reduced to the metallic form until a temperature is reached corresponding to that in the lower bosh or in the tuyere zone. However, the values refer to the reduction of the pure oxide to form the pure metal at unit activity, whereas in the blast furnace the reduced elements form dilute solutions in the iron and their activities are considerably below unity. This factor displaces upwards the oxygen potential lines for the oxides and the temperatures are lowered at which they intersect the blast furnace gas line. Conversely, the activities of the oxides are less than unity and the minimum temperatures for reduction are raised when the oxides are present as silicates, aluminates or other compounds.

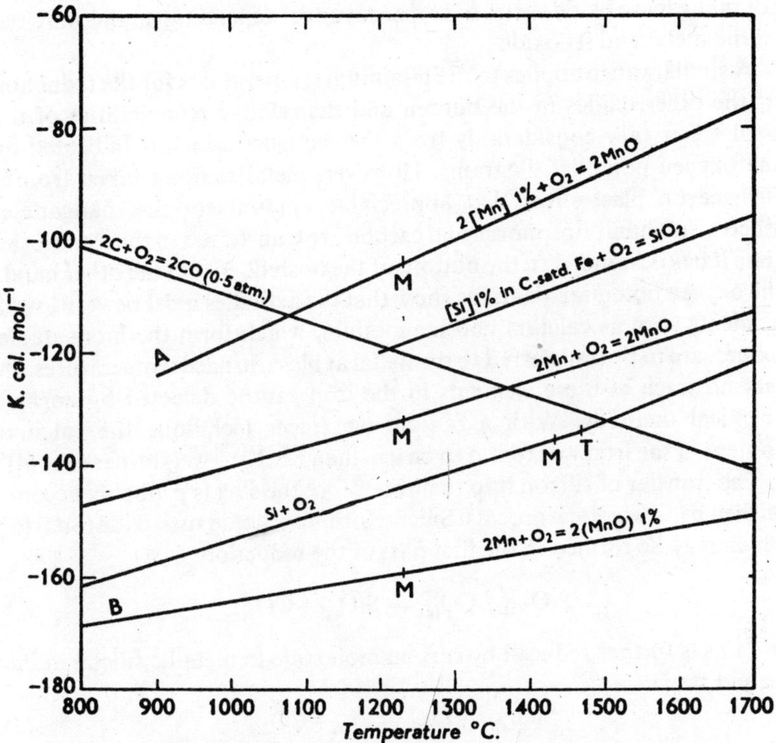

FIG. 7.1. The effect of dilution of a metal or its oxide on the oxygen potentials for reduction. M and T are melting and transformation points. (After Richardson and Jeffes [1].)

The effect of a change in the activity of the element or its oxide on the minimum temperature for reduction has been demonstrated graphically by Richardson and Jeffes [1], and some of their data are summarised in Figure 7.1. Assuming that manganese forms an ideal solution in liquid iron and that γ_{Mn} is not affected by other elements dissolved in the iron then, theoretically, pure manganese oxide can be reduced to form a 1 weight-per cent solution of manganese in liquid iron at 1,080°C (line A,

Figure 7.1), but the rate of solution of manganese in the iron is very slow when the iron is solid. The impure iron in the blast furnace starts to melt at about 1,150°C and the manganese then enters into solution more rapidly; this places a practical lower limit upon the reduction temperature. Similarly, if the activity of the manganese oxide is 0·1, the minimum temperature for the reduction of the oxide to form manganese metal is about 1,800°C (line B). Thus the minimum reduction temperature for manganese oxide varies over a wide range, depending on the activities of the metal and its oxide.

A similar range applies to the minimum temperatures for the reduction of the other oxides in the burden and the relative reducibilities of the oxides can vary considerably from the sequence which is indicated by the oxygen potential diagrams. However, metal samples taken from a furnace on blast show that appreciable amounts of the manganese, silicon, sulphur, phosphorus and carbon are transferred to the iron by the time it has descended to the bottom of the bosh [2, 3]. On the other hand, the oxygen potential diagrams show that the activities must be very low if elements such as calcium and magnesium, which form the more stable oxides, are to be transferred to the metal at blast furnace temperatures. In fact no trace of these elements in the iron can be detected by normal chemical methods. With a radioactive tracer technique the calcium content of the iron was found to be less than 6×10^{-5} weight-per cent [4].

The transfer of silicon into the iron above the slag is probably accomplished by a two-stage process. Silicon forms a volatile sub-oxide (SiO) [5] which may be formed in the first part of the reduction:

$$SiO_{2(s)} + CO_{(g)} \rightleftharpoons SiO_{(g)} + CO_{2(g)} \qquad 7.1$$

and this is further reduced by carbon monoxide to metallic silicon in the second stage:

$$SiO_{(g)} + CO_{(g)} \rightleftharpoons Si_{(s)} + CO_{2(g)} \qquad 7.2$$

Whilst the overall free-energy change for the reduction of silica via silicon monoxide is identical with that for the direct reduction to silicon, the reaction rate is more rapid when the gaseous oxide is produced in the intermediate step, for the reaction products are then removed by the gas and a surface of unreduced silica is continuously exposed to the reducing gas. The silicon monoxide is in intimate contact with the carbon monoxide in the atmosphere and is readily reduced to metallic silicon or to silicon in iron.

From the data for the minimum temperatures for reduction it is evident that the phosphates are readily reduced in the blast furnace.

Phosphorus boils at 280°C; hence the product of the reduction is a gas which is absorbed readily by the metallic iron and, as the top gas temperature is below the boiling point of phosphorus, any traces of the element remaining in the gas at the top of the stack are condensed and re-cycled in the burden. Thus practically all the phosphorus in the burden is transferred ultimately to the iron.

Most of the sulphur enters the blast furnace in the coke and is burnt to form gaseous oxides of sulphur. Some of this sulphur combines with lime to form calcium sulphide at high temperatures, most of the remainder is transferred to the iron and only a very small amount leaves the furnace in the exit gas.

In general, when an element is present as a gas the rate of transfer into the iron is greater and the transfer becomes significant at a lower temperature than when the element is in the solid form. The reason for this is that the gas penetrates into the interstices of the ore particles and the element starts to transfer as soon as metallic iron is present. As much as 1 weight-per cent of carbon is dissolved from the atmosphere before the iron starts to melt [2], but when the iron is molten the transfer rate is increased by the mixing of the liquid iron with the coke and the carbon content may reach 3 per cent before the iron passes below the slag surface.

With both gaseous and solid reduction products the amount of each element which is transferred to the iron at any temperature level within the furnace is restricted by the solubility of the element in the iron at that temperature. Thus only small amounts of antimony, zinc or lead can dissolve in the iron without the formation of a second phase. Consequently the activity of these elements increases rapidly with the concentration and the conditions become increasingly less favourable for the transfer reaction.

RE-OXIDATION IN THE TUYERE ZONE

As the droplets of liquid iron fall through the upper regions of the hearth some of them pass through the combustion annulus of the tuyere zone and the dissolved elements are partly re-oxidised in the strongly oxidising atmosphere, but below the annulus and above the surface of the slag the atmosphere is again reducing, so that any oxides of iron which form in the oxidising zone are rapidly reduced. Some of the other elements are also reduced back into the iron, but it is probable that an appreciable amount remains as oxides which separate from the iron into the slag, and large variations in composition can occur at the tuyere level. This is illustrated by the following analyses of metal samples taken from a blast furnace [2].

Element	*Sample Position*		
	1·5 ft from tuyere nozzle weight-per cent	Centre line of furnace weight-per cent	Tap metal weight-per cent
Si	0·014	0·62	0·75
Mn	nil	1·50	2·95
P	0·033	0·123	0·115

The composition of the metal droplets may change appreciably as they descend through the slag layer, but the composition of the tap metal is determined by the reactions across the slag–metal interface.

7.3 The Formation of the Slag

The first liquid slag which forms in the upper part of the bosh is rich in iron and consists principally of ferrous oxide, silica and alumina, but the iron oxide is rapidly reduced as the slag flows downwards through the column of coke. Analysis of the tap slag usually gives the concentration of iron in the range 0·2 to 1·0 per cent, but it is probable that this represents mainly particles of metallic iron which are mechanically entrapped in the slag and which are not differentiated from ferrous oxide by the normal assay methods. Simultaneously with the reduction of the iron oxides from the slag, the lime in the charge slowly dissolves and in the lower bosh the slag consists principally of lime, silica and alumina. Generally the lime dissolves more readily if the flux is mixed intimately with the ore, as with a sinter burden, and the start of slag formation may then occur at lower temperatures in the lower part of the stack.

As the slag drains downwards into the hearth it absorbs the alumina and silica which are left in the ash from the combustion of the coke, and the basicity (i.e., the ratio $N_{CaO}/N_{SiO_2}+N_{Al_2O_3}$) of the slag in the hearth is lowered relative to that of the bosh slag. This can have an important bearing on the smoothness of the descent of the burden through the bosh, for the slag which is formed in the bosh prior to the solution of the coke ash may become over-saturated if the charge is selected to form a hearth slag which is almost saturated with lime. The lime which is in excess of the solubility limit remains as solid particles and raises the viscosity of the slag; in extreme cases the slag becomes pasty and the burden descends in an irregular manner. A high basicity can be obtained in the hearth slag without interference with the burden descent by lowering the lime content in the burden and injecting the additional lime in the air blast. Suggestions have been made from time to time for performing this operation [6, 7], but the technique does not appear to have found favour.

The changes in the metal composition which occur in the hearth of the furnace are to a large extent determined by the composition of the hearth slag, but the volume of slag is increasing continuously between the times when the slag is flushed, Thus any variation with time in the composition of the burden is reflected by a corresponding change in the analysis of the bulk slag. The ability to produce an iron to a constant composition is greater, therefore, the more homogeneous is the burden, and adequate burden preparation by blending, sintering, or other means is very important.

7.4 Slag–Metal Reactions

THE CARBON CONTENT OF THE IRON

It is probable that the iron is almost completely saturated with carbon before it descends through the slag layer into the bath of molten metal. The column of coke extends through the slag to the interface with the metal and possibly extends to the bottom of the furnace, so carbon can continuously transfer into the iron to reach carbon saturation right up to the time when the metal is tapped from the furnace.

Carbon is insoluble in the slag and there is no transfer of this element across the slag–metal interface, but the carbon in solution in the iron affects the other interface reactions through the control of the oxygen content of the metal. At 1,500°C the partial pressure of oxygen in equilibrium with graphite (i.e. carbon-saturated iron) and carbon monoxide is about 10^{-16} atmospheres. If the oxygen content is greater than this the excess of oxygen can combine with the carbon to form carbon monoxide which is insoluble in both the slag and the metal and is continuously removed into the furnace atmosphere. It is extremely difficult to nucleate a bubble of gas within a homogeneous liquid solution, in practice the product $[a_C].[a_O]$ is invariably greater than the equilibrium value, and this has been suggested as a probable cause of the failure to reach equilibrium in the other hearth reactions.

The solubility of carbon in pure liquid iron in equilibrium with graphite is 4·3 weight-per cent at the iron-graphite eutectic temperature (1,153°C) and increases to about 5·5 per cent at 1,600°C [8]. At any temperature the atomic fraction of carbon which is soluble in pure iron is given by the relation [9]:

$$\log N_C = -\frac{560}{T} - 0{\cdot}375 \qquad 7.3$$

The effect of various solutes on the solubility of carbon in iron has been studied in a number of investigations and recently the available data have been critically assessed [10, 11]. The difference (Δ%C) between the solubility of carbon in the ternary system Fe–C–X (where X is a solute at low concentration) at a fixed concentration of X and the solubility in a pure iron–carbon alloy is approximately independent of the temperature [12, 13] and, at low concentrations of X, the ratio $\Delta\%C/\%X$ is also independent of the concentration of X. The data are summarised in Table 7.1 in terms of weight and molar concentration units and are represented diagrammatically in Figure 7.2.

Table 7.1

	Temp °C	$\frac{\Delta\%C}{\%X}$ [10, 11]	$\frac{\partial \ln N_C}{\partial N_X}$ [11]	$\frac{\partial \ln N_C}{\partial N_X}$ [10]
Aluminium	1600	−0·25	−2·7	−2·7
Chromium	1600	+0·09	+1·3	+0·97
Cobalt	1600	−0·03	−0·4	
	1550			−0·50
Copper	1600	−0·20	−4·2	−4·1
Manganese	1600	+0·04	+0·5	
	1500			+0·5
Nickel	1600	−0·07	−0·5	
	1490			−0·99
Niobium	1600	+0·12		+4·1
Phosphorus	1600	−0·35	−4·6	−4·6
Silicon	1600	−0·30	−3·7	−4·0
Sulphur	1600	−0·40	−5·2	−4·0
Vanadium	1600	+0·13	+2·3	
	1560			+2·2
Titanium	1600	+0·17	+3·3	

The agreement between the two sets of data is reasonably good, but for most elements the two evaluations were obtained from the same experimental data and the differences reflect the disagreement in the interpretations of the results of the various investigations. The effects of these elements on the activity coefficient of carbon in the saturated solutions relative to the infinitely dilute standard state are obtained by inserting the differential quantities into equation 13.41; the values thus obtained are tabulated in Appendix 1.

The solubility of carbon in liquid iron is raised and the activity is lowered by manganese, chromium, vanadium and titanium increasingly in that order. These elements all form carbides which are more stable

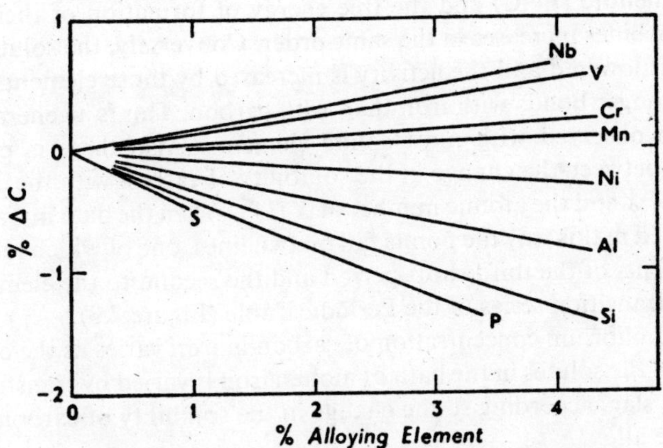

FIG. 7.2. The effect of alloying elements on the solubility of carbon in liquid iron. (After Ohtani and Gokcen [11].)

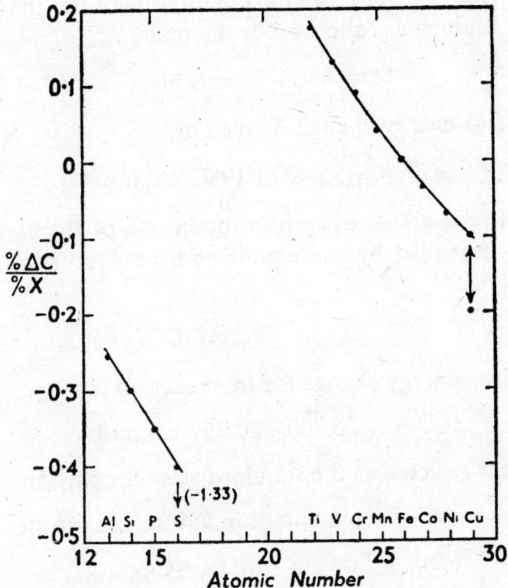

FIG. 7.3. Variation of $\Delta\%C/\%X$ with the atomic number of the element X for carbon-saturated solutions at 1,600°C.

than cementite (Fe_3C) and the free energy of formation of their more stable carbides increases in the same order. Conversely, the solubility of carbon is lowered and the activity is increased by those elements which form stronger bonds with iron than with carbon. This is a general effect which is observed with most solute elements. Also there is periodic relation between the change in the solubility of carbon with the concentration of X and the atomic number of X [11]. When the data in Table 7.1 are plotted in this way the points fall on two lines, one of which relates to the elements of the third short period and the second to the elements of the first transition series of the Periodic Table (Figure 7.3).

The equilibrium concentration of carbon in iron varies as the concentration of the solutes in the bath of molten iron is varied by transfer to or from the slag, according to the change in the solubility of carbon in the iron-base alloy.

MANGANESE DISTRIBUTION

The Equilibrium Constant

The equilibrium distribution of manganese between the slag and the metal can be obtained as follows. For the reaction:

$$Mn_{(l)} + \tfrac{1}{2}O_{2(g)} \rightleftharpoons MnO_{(s)} \qquad 7.4$$

The standard free-energy change is given by:

$$\Delta G_4^0 = -95{,}400 + 19{\cdot}7T \text{ cal mol}^{-1}$$

but the partial pressure of oxygen in the hearth of the blast furnace is, theoretically, controlled by the equilibrium between solid carbon and carbon monoxide:

$$C_{(graphite)} + \tfrac{1}{2}O_{2(g)} \rightleftharpoons CO_{(g)} \qquad 7.5$$

The standard free-energy change for this reaction is:

$$\Delta G_5^0 = -26{,}700 - 20{\cdot}95T \text{ cal mol}^{-1}$$

where graphite is selected as the standard state for carbon.

Subtracting equation 7.5 from equation 7.4 we obtain the relation:

$$Mn_{(l)} + CO_{(g)} \rightleftharpoons MnO_{(s)} + C_{(graphite)} \qquad 7.6$$

$$\Delta G_6^0 = -68{,}700 + 40{\cdot}65T \text{ cal mol}^{-1}$$

and the equilibrium constant for the reaction is given by:

$$K_{Mn} = \frac{(a_{MnO})[a_C]}{[a_{Mn}]p_{CO}} \qquad 7.7$$

where round and square brackets refer to components of the slag and metal respectively. Since $\Delta G^0 = -RT\ln K$, it follows that:

$$\log K_{Mn} = \frac{15{,}020}{T} - 8{\cdot}884 \qquad 7.8$$

The constant K varies inversely with the temperature and changes from 2·40 at 1,350°C to 0·30 at 1,525°C. In carbon-saturated iron the activity of carbon is unity and the partial pressure of carbon monoxide in the hearth is essentially constant. Hence, from equation 7.7, the ratio $(a_{MnO})/[a_{Mn}]$ also varies inversely with the temperature and, if the activities of manganese and its oxide are not markedly temperature dependent, the extent of the manganese recovery in the metal is increased as the temperature is raised.

It is necessary to convert the activities into concentration units to determine the extent of the recovery at any given temperature and bath composition, and for this purpose it is better to express the equilibrium constant in terms of the molar concentrations and the activity coefficients:

$$K_{Mn} = \frac{(N_{MnO})(\gamma_{MnO})}{[N_{Mn}][\gamma_{Mn}]\cdot p_{CO}} \qquad 7.9$$

The activity coefficient of the manganese dissolved in the metal can be determined from the tabulations in Appendix 1, but for carbon-saturated iron the value lies in the range 0·75 to 0·85 depending on the amounts of the other solutes which are present in the iron.

Manganese Oxide Activities

The activity coefficients of manganese oxide have been measured at 1,650°C for a limited number of compositions in the ternary system CaO–SiO_2–Al_2O_3 and values at the binary edge CaO–SiO_2 have been estimated from more extensive data for the ternary system MnO–CaO–SiO_2 [14]. The iso-activity coefficient lines are shown in Figure 7.4, which indicates that as the alumina concentration is raised from zero (γ_{MnO}) is at first raised and then lowered at low and moderate silica contents, but is raised progressively in the more acid melts. However, the evidence is limited and is not strictly applicable to iron-making as the temperature of the slag in the blast furnace is between 100 and 250°C lower than the experimental temperature and some magnesia is usually present in the slag.

Approximate values of (γ_{MnO}) for conditions more applicable to blast furnace practice can be obtained by the substitution of data for the observed manganese distribution [15, 16, 17] into equation 7.9. These literature references do not give complete metal analyses and so for the calculations $[\gamma_{Mn}]$ is arbitrarily taken as 0·80 and p_{CO} is assumed to be one atmosphere pressure. From the data for silicon distribution, which is discussed later, it is clear that many of the melts listed in the cited references had not attained thermodynamic equilibrium and it is necessary to exclude from consideration all melts which differ markedly from the equilibrium state. For this reason the calculated values of (γ_{MnO}) are regarded as minimum values.

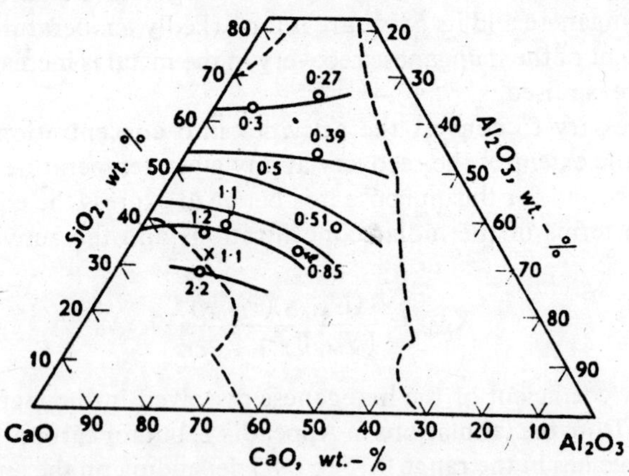

FIG. 7.4. Activity coefficients of manganous oxide in the quaternary system $MnO–CaO–Al_2O_3–SiO_2$. Solid curves represent iso-activity-coefficient contours and numbers against open circles are experimental values of the coefficient, which apply for manganese concentrations up to 0·7 mole fraction. (After Abraham, Davies, and Richardson [14].)

The calculated variation of (γ_{MnO}) with the lime–silica ratio is shown in Figure 7.5 for the range of slag compositions 28–48 per cent CaO, 0–12 per cent MgO, 26–56 per cent SiO_2, 8–35 per cent Al_2O_3 and 0·5–7 per cent MnO. At 1,540°C the points are best fitted by two lines which intersect at about $CaO/SiO_2 = 1·0$. The data available at 1,450°C are restricted to lime–silica ratios of less than unity, but within this range the points lie slightly above the line drawn for 1,540°C. This suggests that (γ_{MnO}) decreases as the temperature is raised but, within the composition range for which data are available at the two temperatures, the activity shows a negative deviation from ideality. The temperature dependence thus has opposite sign to that which is normally expected from thermodynamic reasoning, but it is interesting to note that similar behaviour has recently been found for the activity of ferrous oxide in silicate melts [18, 19]. With lime–silica ratios greater than unity, (γ_{MnO}) increases rapidly with increasing basicity and the deviation from ideality is positive at lime–silica ratios greater than about 1·3. For blast furnace slags containing less than approximately 45 weight-per cent of silica, the calculated values of

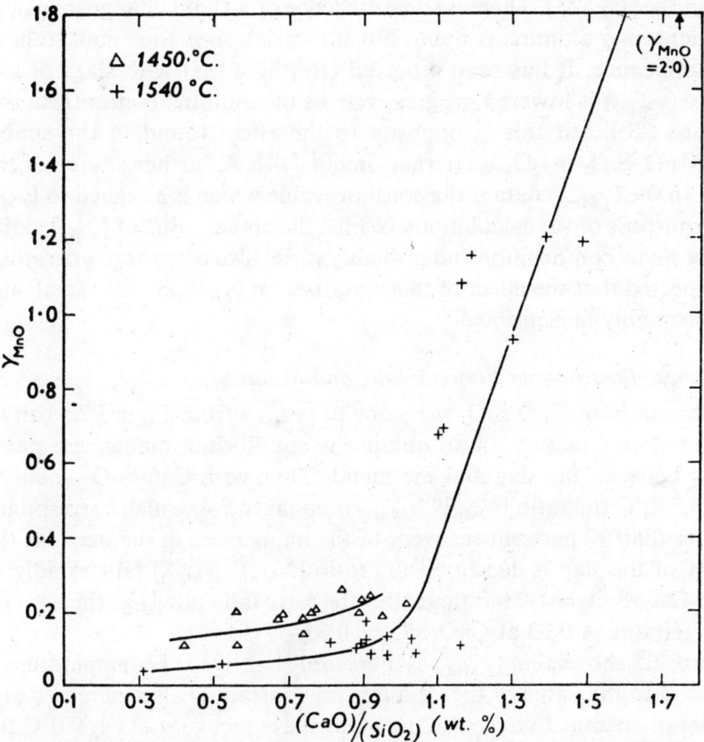

FIG. 7.5. Variation of the activity coefficient of manganese oxide with the lime/silica ratio for blast furnace slags. The points are determined from observed manganese distribution data.

(γ_{MnO}) are consistently lower than those obtained experimentally for the CaO–MnO–SiO_2–Al_2O_3 quaternary melts and, in view of the suggested temperature dependence of (γ_{MnO}), the difference is increased if the calculated values are extrapolated to 1,600°C, but it must be remembered that the calculated points are regarded as minimum values. The agreement is more satisfactory for melts of higher basicity.

There is some scatter of the calculated points which is considered to arise mainly from slight deviations of the melts from the equilibrium distributions. A consistent trend in (γ_{MnO}) could not be detected with variation of the magnesia content, the points for magnesia-containing slags being distributed randomly about the points for magnesia-free melts, and it has been reported elsewhere that the manganese distribution is apparently unaffected by the presence of up to 12 weight-per cent of MgO in the slag [17]. There is some evidence of a slight change in (γ_{MnO}) with increasing alumina content, but the variation is too small to have any significance. It has been reported elsewhere that with slags of low basicity (γ_{MnO}) is lowered progressively as the alumina concentration is increased [20], but this is opposite to the effect found in the simple $CaO–MnO–SiO_2–Al_2O_3$ quaternary melts [14]. A further source of the scatter in the (γ_{MnO}) data is the constant value which is assigned to $[\gamma_{Mn}]$ for the purpose of the calculations. Whilst the actual value of $[\gamma_{Mn}]$ varies with the metal composition and probably varies also with the temperature, it is expected that the effect of this variation on (γ_{MnO}) is very small and can reasonably be neglected.

Manganese Partitioning Between Slag and Metal

At any ratio of CaO/SiO_2 the value of (γ_{MnO}) from Figure 7.5 can be substituted in equation 7.9 to obtain the equilibrium manganese partitioning between the slag and the metal. Thus with CaO/SiO_2 equal to 1·5 at 1,540°C the ratio $[N_{Mn}]/(N_{MnO})$ is equal to 7·4, which corresponds to better than 90 per cent recovery of the manganese in the iron. As the basicity of the slag is decreased the ratio $[N_{Mn}]/(N_{MnO})$ falls rapidly to 0·71 at $CaO/SiO_2 = 1{\cdot}0$ but thereafter the ratio falls slowly as the basicity is lowered and is 0·33 at $CaO/SiO_2 = 0{\cdot}33$.

Figure 7.5 shows that (γ_{MnO}) is increased slightly as the temperature is lowered, but the value of the equilibrium constant is also increased and the overall effect is to decrease the manganese recovery. At 1,450°C the ratio $[N_{Mn}]/(N_{MnO})$ is 0·50 and 0·27 at CaO/SiO_2 ratios of 1·0 and 0·33 respectively. Thus the higher the temperature and the basicity of the slag the better is the recovery of the manganese in the metal.

The manganese recovery which is found in practice is smaller than that which would be obtained if the equilibrium was reached and, under the most basic slags which are normally encountered in the production of pig iron, the recovery rarely exceeds 60 per cent. For a slag with the ratio CaO/SiO_2 equal to 1·2 at 1,505°C the ratio $[N_{Mn}]/(N_{MnO})$ is 0·85 [21], whereas the ratio at equilibrium is about 2·7. It could be argued that this difference arises because the gas pressure in the hearth of the blast furnace is greater than one atmosphere, but the partial pressure of carbon monoxide is not greater than unity and is probably about 0·7 atmospheres.

Thus any correction to the calculated distribution ratios to allow for the actual pressure of carbon monoxide in the blast furnace would increase the theoretical manganese recovery and cause a still greater discrepancy compared with the recovery which is obtained in practice. It seems reasonable to conclude, therefore, that the manganese equilibrium is not attained in the blast furnace.

SILICON DISTRIBUTION

The distribution of silicon between the slag and the metal can be evaluated from a similar treatment to that adopted to study the manganese distribution.

For the reaction:

$$SiO_{2(s)} + 2C_{(graphite)} \rightleftharpoons Si_{(l)} + 2CO_{(g)} \qquad 7.10$$

The standard free-energy change is [22]:

$$\Delta G^0_{10} = 146{,}000 - 81{\cdot}2T \text{ cal mol}^{-1}$$

For carbon-saturated iron and $p_{CO} = 1$ atmosphere the equilibrium constant simplifies to:

$$\log K_{Si} = \log \frac{[a_{Si}]}{(a_{SiO_2})} = \frac{-31{,}900}{T} + 17{\cdot}75 \qquad 7.11$$

The value of K_{Si} increases with the temperature and changes from 0·048 at 1,400°C to 5.13 at 1,600°C.

The activities of silica in the ternary system CaO–SiO_2–Al_2O_3 are known for wide ranges of composition and temperature [22, 23, 24] and the variation of (a_{SiO_2}) with slag composition and temperature is shown in Figure 7.6 for alumina contents of 0·06 and 0·125 mole fractions, corresponding to about 10 and 20 weight-per cent respectively. In the more acid slags the activity of silica is lowered as either the alumina concentration is raised or the temperature is lowered, but the reverse trends are found at more basic compositions. On a weight-per cent basis the effect of magnesia on (a_{SiO_2}) is approximately equivalent to the effect of a corresponding amount of lime, but on a mole fraction basis (a_{SiO_2}) is decreased when lime is replaced by magnesia [22].

The activities of silica in the ternary system have been obtained [23, 24] by the substitution of data for the equilibrium distribution of silicon into equation 7.11 in similar way to that previously shown for estimating (a_{MnO}), and the reverse procedure can be adopted to determine the equilibrium partitioning of silicon for a slag of known composition. The activity coefficient of silicon in carbon-saturated iron is known for a wide range of temperatures and silicon contents [25], and the effect of other solutes on [γ_{Si}] can be determined from the data in Appendix 1.

The silicon distribution has been calculated in this way for pure Fe–C–Si alloys in equilibrium with slags whose compositions are represented by the lines in Figure 7.6, and the results obtained are presented in Figure 7.7, which shows that silicon recovery in the metal increases rapidly with rising temperature and with increasing silica content in the

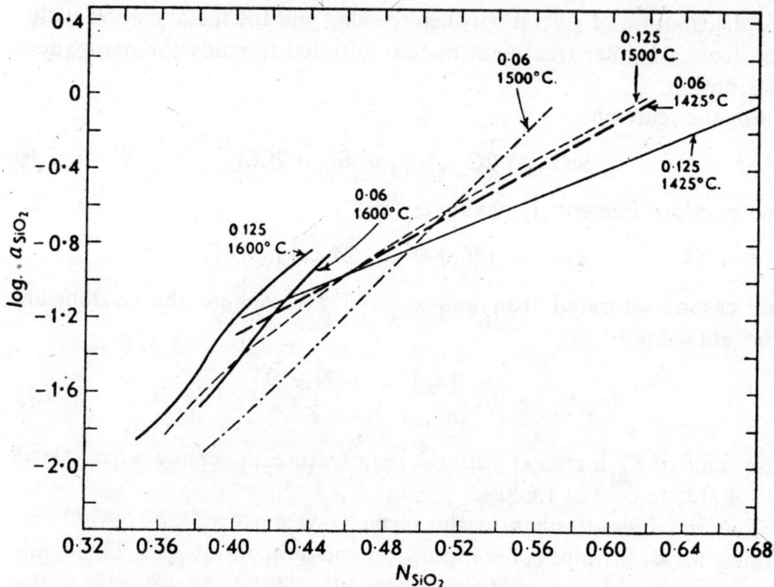

FIG. 7.6. Relation between the activity and the mole fraction of silica in blast furnace slags at 1,425 and 1,500° C (after Fulton and Chipman [23]) and at 1,600°C (after Langenberg and Chipman [24]). The solid and broken lines show the activity-composition relations at constant alumina concentrations of 0·125 and 0·06 mole fraction respectively.

slag. With 0·44 mole fraction of silica in the slag at 1,600°C the equilibrium silicon content in the metal is about 23 weight-per cent and at higher silica contents the slag is in equilibrium with silicon carbide (SiC) [24]. Even at 1,425°C, which is towards the lower limit of the temperatures normally encountered in the hearth of the blast furnace, the equilibrium silicon content is several per cent.

The silicon contents which are usually present in the metal are considerably lower than the equilibrium amounts and the difference is increased if allowance is made for the actual partial pressure of carbon monoxide in the blast furnace. In addition to carbon and silicon, the blast furnace metal contains other solutes which affect the value of $[\gamma_{Si}]$ and modify the silicon equilibrium distribution, but the change is negligible in relation to the difference between the calculated and the actual concentrations. Thus it is evident that the silicon equilibrium is not attained in the blast furnace. Under the most favourable laboratory conditions, in which the metal was violently agitated to increase the area of the slag-metal interface, only a small amount of silicon was transferred

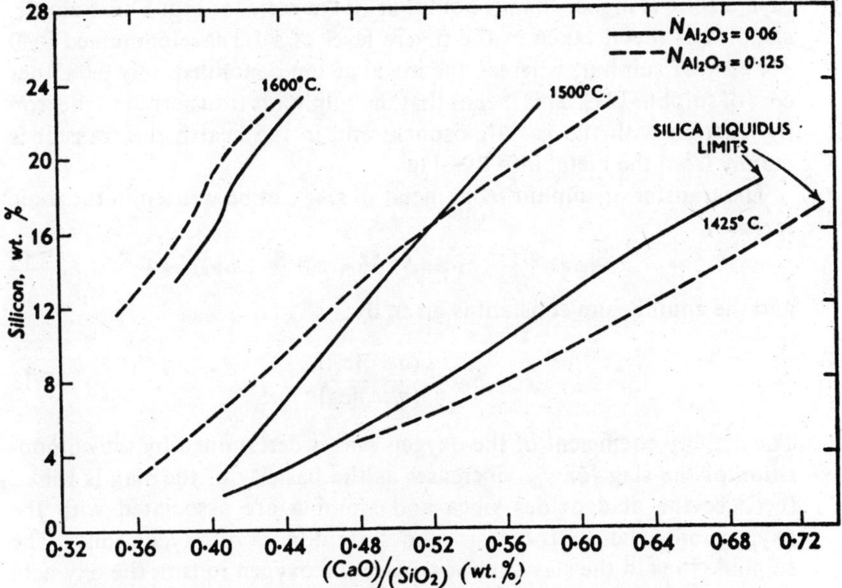

FIG. 7.7. Relation between the silicon content of carbon-saturated iron and the lime/silica ratio of the slag for equilibrium conditions at various temperatures.

to the iron in 8 hours at 1,425 or 1,500°C and equilibrium had not been achieved after 4 hours at 1,700°C [25, 26]. The rate of silicon transfer increases rapidly as the temperature is raised and, from the rate of change with the temperature, the heat of activation has been estimated as 130 ± 30 kcal mole^{-1} [26]. This value is unusually high for a reaction at elevated temperatures; for most of the iron- and steel-making reactions the heat of activation is in the range from 5 to 30 kcal mole^{-1}.

One factor which is clearly established from the equilibrium considerations is the pronounced effect of the composition of the slag on the manganese and silicon contents of the metal. With increasing basicity

(a_{SiO_2}) is lowered but (a_{MnO}) is increased, and the driving force for the transfer of these elements from slag to metal is related directly to their activities in the slag phase. Hence, although the equilibrium distributions are not attained in the blast furnace, for a constant silicon and manganese input in the burden the silicon content of the metal is lowered and the manganese content is raised as the slag basicity is increased.

DISTRIBUTION OF SULPHUR

The mean sulphur content of the metal droplets falling through the tuyere zone is higher than the sulphur in the metal at tap. For example, samples of metal taken at the tuyere level of a furnace contained 0·40 per cent of sulphur, whereas the metal at tap contained only 0·085 per cent of sulphur [27]. This means that the sulphur is transferred to the iron by reaction with the gas atmosphere and in the hearth the transfer is mainly from the metal into the slag.

The transfer of sulphur from metal to slag can be written in the ionic form:

$$S_{(in\,Fe)} + O^{--}{}_{(slag)} \rightleftharpoons O_{(in\,Fe)} + S^{--}{}_{(slag)} \qquad 7.12$$

and the equilibrium constant is given by:

$$K_S = \frac{(a_{S^{--}})[a_O]}{(a_{O^{--}})[a_S]} \qquad 7.13$$

The activity coefficient of the oxygen ions is determined by the composition of the slag for γ_O increases as the basicity of the slag is raised (because the acid oxides silica and alumina are associated with the oxygen ions whereas the basic oxides are donors of oxygen ions). The sulphide ions in the slag cannot replace the oxygen ions in the oxy-acid groups and so $\gamma_{S^{--}}$ is high in acid slags and is decreased as the basicity is raised.

The concentration of oxygen ions which are not associated with the acid oxides is low in acid melts and only becomes significant when all the positive bonds of the acid ions are saturated (see Chapter 4.2). For example, in calcium silicate melts the proportion of oxygen ions which can exchange positions with sulphur atoms in the metal is very low at high silica contents and increases rapidly only when the ratio $(N_{CaO})/(N_{SiO_2})$ is greater than 2·0. Thus, in slags which consist principally of silica and the oxides of the divalent elements, $(\gamma_{O^{--}})$ is low and $(\gamma_{S^{--}})$ is high when the mole fraction of the basic oxides is less than 0·66. Within this range, however, $(\gamma_{S^{--}})$ changes rapidly with change in basicity and in calcium silicate melts it decreases by a factor of 5, from 50 to 10, as the mole fraction of lime is raised from 0·51 to 0·61 [28]. In basic melts, at constant mole fraction of lime, ($\gamma_{S^{--}}$) is raised slightly as silica is replaced by alumina. Under these conditions, where aluminium functions as an anion, the oxygen ions are attached less firmly to the aluminium ions than to the silicon ions and are more readily available for exchange with the sulphur ions.

The activity coefficient of sulphur in liquid iron has been determined for a wide range of metal compositions, relative to the dilute solution standard states, and for blast furnace metal in the normal range of compositions the value of [f_S] varies between 4·5 and 6·5 [29]. The activity of the oxygen in the iron is less than 0·0001.

When equation 7.13 is examined in the light of these remarks it is evident that the sulphur distribution between the slag and the metal should increase rapidly as the basicity is raised and this has been confirmed experimentally. With carbon-saturated iron and blast-furnace type slags the equilibrium ratio (%S)/([%S] varies from about 50 near silica saturation to 350 adjacent to the lime saturation surface [30]. The melts from which these data have been obtained contained more silicon than is normally present in blast furnace metal, but the distribution ratios can be adjusted through the change in [f_S] with weight-per cent of silicon to obtain values which are comparable with those observed in practice. The distribution ratios are almost halved when this correction is made, but they are still from 2 to 5 times as great as the distributions which are actually obtained at comparable basicities in the blast furnace. The effect of changing the temperature on the distribution is uncertain, but it is improbable that a change of, say, 100°C could account for differences of this order of magnitude. The obvious conclusion is that the equilibrium partitioning of sulphur between the slag and the metal is not attained in the blast furnace and this has been confirmed by an experiment in which slag and metal samples were taken simultaneously from a blast furnace and re-melted in a graphite crucible under an atmosphere of carbon monoxide. After 2 hours the ratio (%S)/[%S] had increased by 60 per cent at 1,400°C and by almost 400 per cent at 1,600°C [31].

THE PROBABLE CAUSE OF THE NON-ATTAINMENT OF EQUILIBRIUM

The sulphur transfer reaction has been investigated extensively in recent years and, whilst the mechanism is still uncertain, the factors affecting the rate of the reaction are now known. These investigations have also clarified the causes of the non-equilibrium distributions of manganese and silicon.

The interfacial tension is very low between iron rich in sulphur and a blast-furnace slag, which causes the metal droplets entering the slag from the tuyere zone to remain dispersed as a mechanical suspension in the slag. As the sulphur transfers from the metal to the slag the surface tension of the metal increases and the droplets coalesce into larger masses which can then descend through the slag and into the bath of molten metal [32, 33]. The rate of transfer of sulphur to the slag from the dispersed metal droplets should be very rapid, but it is probable that sulphur transfer continues after the coalesced particles have entered the bath of metal and the deviation from the equilibrium distribution could then arise from kinetic factors operating in the hearth. These factors include the diffusion of sulphur through the metal to reach the interface, the transfer of sulphur across a static film at the interface, and the diffusion of sulphur into the slag and away from the interface. It was noted earlier that the bath is not turbulent and, by observing the movement of radioactive tracers, it has been shown that the slag is generally quiescent [34], so that diffusion is not assisted markedly by convection.

If the reaction rate is controlled only by the diffusion of sulphur, then equilibrium should be attained more rapidly when the bath is stirred and the interfacial area is increased by some form of mechanical agitation. However, in laboratory tests the rate of sulphur removal from the metal is affected only slightly by variation of the rate of stirring the melt [26, 35, 36], and this implies that the rate is not determined by the diffusion of either sulphur atoms in the metal or sulphur ions in the slag.

Since the rate at which sulphur is transferred across the interface also appears to be fairly rapid, it seems that the sulphur distribution is controlled by a reaction other than that expressed in equation 7.12. However, the reaction can be written as two processes which proceed simultaneously. In one of these the sulphur atom gains two electrons as it crosses the interface into the slag:

$$S_{(\text{in Fe})}+2e^- \rightleftharpoons S^{--}{}_{(\text{slag})} \qquad 7.14$$

whilst the second is an oxidising reaction in which electrons are donated to the slag. A number of such reactions are possible [37] including the following:

$$\text{Fe} \rightleftharpoons \text{Fe}^{2+}{}_{(\text{slag})}+2e^- \qquad 7.15$$

$$\text{Mn}_{(\text{in Fe})} \rightleftharpoons \text{Mn}^{2+}{}_{(\text{slag})}+2e^- \qquad 7.16$$

$$\text{C}_{(\text{in Fe})}+\text{O}^{--}{}_{(\text{slag})} \rightleftharpoons \text{CO}_{(g)}+2e^- \qquad 7.17$$

$$\text{Si}_{(\text{in Fe})}+4\text{O}^{--}{}_{(\text{slag})} \rightleftharpoons \text{SiO}_4^{4-}+4e^- \qquad 7.18$$

The equilibrium iron content of a blast-furnace slag is very low and any oxide produced by reaction 7.15 is rapidly reduced again by reactions

7.17 and 7.18. The carbon–oxygen reaction (7.17) is notoriously sluggish (see Chapter 13.3) and it has frequently been suggested that the sulphur equilibrium is not reached because the oxygen potential in the hearth is in excess of the equilibrium level, owing to the slow approach to the equilibrium for reaction 7.17. If this is true then the rate of the carbon–oxygen reaction should be enhanced by bubbling carbon monoxide gas through the bath and the sulphur distribution should come to equilibrium more rapidly, but actually the sulphur distribution is virtually unaffected by this treatment [26]. In fact the amount of carbon monoxide evolved from the bath is less than the amount of sulphur transferred to the slag and a sulphur balance is only attained when the amount of sulphur transferred is equated to the number of chemical equivalents of iron, silicon, manganese, etc., which are transferred simultaneously to the slag, and carbon which is evolved as carbon monoxide [37, 38].

Further experiments in which oxidising and reducing agents have been added to the melt have confirmed the critical dependence of sulphur removal from the metal on the oxygen potential of the bath. Additions of silicon, manganese or aluminium to combine with the oxygen ions released by sulphur transfer, cause an immediate, temporary increase in the reaction rate [39–41], whereas additions of the oxides of these elements cause a temporary reversion of sulphur to the metal [15, 16, 41].

When the silicon content of the metal is raised to the concentration in equilibrium with the slag, the rate at which sulphur is removed from the metal is increased very considerably and is approximately the same as that calculated by assuming a diffusion-controlled mechanism. It thus appears that the slow reduction of silica is the rate-determining step in the process.

Various theories have been put forward from time to time to explain the low rate of reduction of silica, but it seems most probable that the sluggishness of the reaction is due to the difficulty in breaking the Si–O bonds in the slag [26]. This is supported by the high heat of activation for the process, which has been referred to, and by the fact that the rate of diffusion of silicon into the metal from the interface is about twelve times as great as the rate of arrival of silicon ions in the slag at the interface [41].

Relation Between the Manganese and Sulphur Reactions

The non-equilibrium state of the manganese distribution can now be re-examined from this viewpoint. The equilibrium between manganese oxide in the slag and manganese in the metal is related directly to the oxygen potential of the system and it is evident that the slow reduction of silica opposes the manganese reduction reaction in a similar manner to its effect on sulphur removal. If it is assumed that both the manganese and the sulphur reactions proceed rapidly with respect to the rate of the silicon reaction, then at any instant the distribution of manganese and sulphur between slag and metal should be determined by the instantaneous oxygen potential of the system. Whether or not this assumption is valid can be ascertained by re-arranging equation 7.7, together with the corresponding ionic form of equation 7.13, to obtain the relation:

$$\frac{(a_O)}{[a_O]} = \frac{1}{K_S}\cdot\frac{(a_S)}{[a_S]} = \frac{1}{K_{Mn}}\cdot\frac{(a_{Mn})}{[a_{Mn}]} \quad 7.19$$

or

$$\frac{(a_S)}{[a_S]} = \frac{K_S}{K_{Mn}}\cdot\frac{(a_{Mn})}{[a_{Mn}]} \quad 7.20$$

and comparing with this the manganese and sulphur distributions between slag and metal samples taken simultaneously from the furnace. If the activities of manganese and sulphur in the slag and metal do not change markedly with small changes in composition then, to a first approximation, equation 7.20 can be expressed in terms of concentration units as:

$$\frac{(\%S)}{[\%S]} = K'_{S.Mn}\cdot\frac{(\%Mn)}{[\%Mn]} \quad 7.21$$

where $K'_{S.Mn}$ is the 'apparent' equilibrium constant for the reaction. In a number of investigations it has been shown that $K'_{S.Mn}$ is approximately constant over a limited composition range [15, 16, 21, 42] and from this observation it is evident that the manganese and sulphur reactions do rapidly attain a state of metastable equilibrium in the blast furnace.

If the oxygen potential is determined by the silicon distribution the latter can be related to the sulphur distribution by the equation:

$$\frac{(\%SiO_2)}{[\%Si]} = K'_{S.Si}\cdot\frac{(\%S)^2}{[\%S]^2} \quad 7.22$$

and to the manganese distribution by the equation:

$$\frac{(\%SiO_2)}{[\%Si]} = K'_{Si.Mn}\cdot\frac{(\%MnO)^2}{[\%Mn]^2} \quad 7.23$$

and it is found that both $K'_{S.Si}$ [43] and $K'_{Si.Mn}$ [20, 44] are constant over limited composition ranges. The significance of these relations is the implicit assumption that the partitioning of the other elements between the slag and the metal is at all times in equilibrium with the instantaneous silicon distribution and, when the silicon distribution is known, the partitioning of the other elements is readily determined. These, and other relations which are derived later, are summarised in Figure 7.8. It is important to realise, however, that equations 7.21 to 7.23 are empirical and the published 'constants' should not be applied to metal or slag compositions outside the range from which they have been evaluated. Universal constants can be derived only if the relations are expressed in terms of activities instead of concentration units.

FACTORS INFLUENCING THE SILICON AND SULPHUR CONTENTS OF THE METAL

Equation 7.22 shows that at constant temperature the silicon distribution is related directly to the sulphur distribution. This is evident from the earlier equilibrium considerations of the two reactions, where it is shown that both the silicon and the sulphur contents of the metal are lowered as the basicity of the slag is raised. The relation applies only at constant temperature, however, for the equilibrium in reaction 7.13 is displaced to the right as the temperature is raised and more sulphur should transfer to the slag. The extent of the silica reduction is also raised as the temperature is increased, hence more oxygen atoms become available to oppose the removal of sulphur, and this last effect can equal or exceed the change in K_S due to the increase in temperature, with the result that the amount of sulphur transferred to the slag may actually decrease as the temperature is raised.

In addition to these factors the kinetics of the two reactions are affected by changes in the viscosity and the surface tension as the slag basicity and the temperature are varied. Since the stable equilibrium is not established, the extent to which the reactions proceed is influenced also by the physical characteristics of the furnace and, in particular, by the area of the slag-metal interface per unit volume of slag and metal. The interface area is determined by the geometry of the furnace, while the volume of slag and metal are increasing continuously between the flushings. Hence the prediction of the composition of the tap metal from equations such as 7.22 or 7.23 becomes extremely hazardous.

THE DISTRIBUTION OF OTHER ELEMENTS

Experimental studies of the partitioning of chromium and titanium between slag and metal under conditions appertaining to the blast furnace have shown that the extent to which these elements are reduced is also related to the oxygen potential associated with the silica reaction, and the equilibrium distribution is not attained for either element.

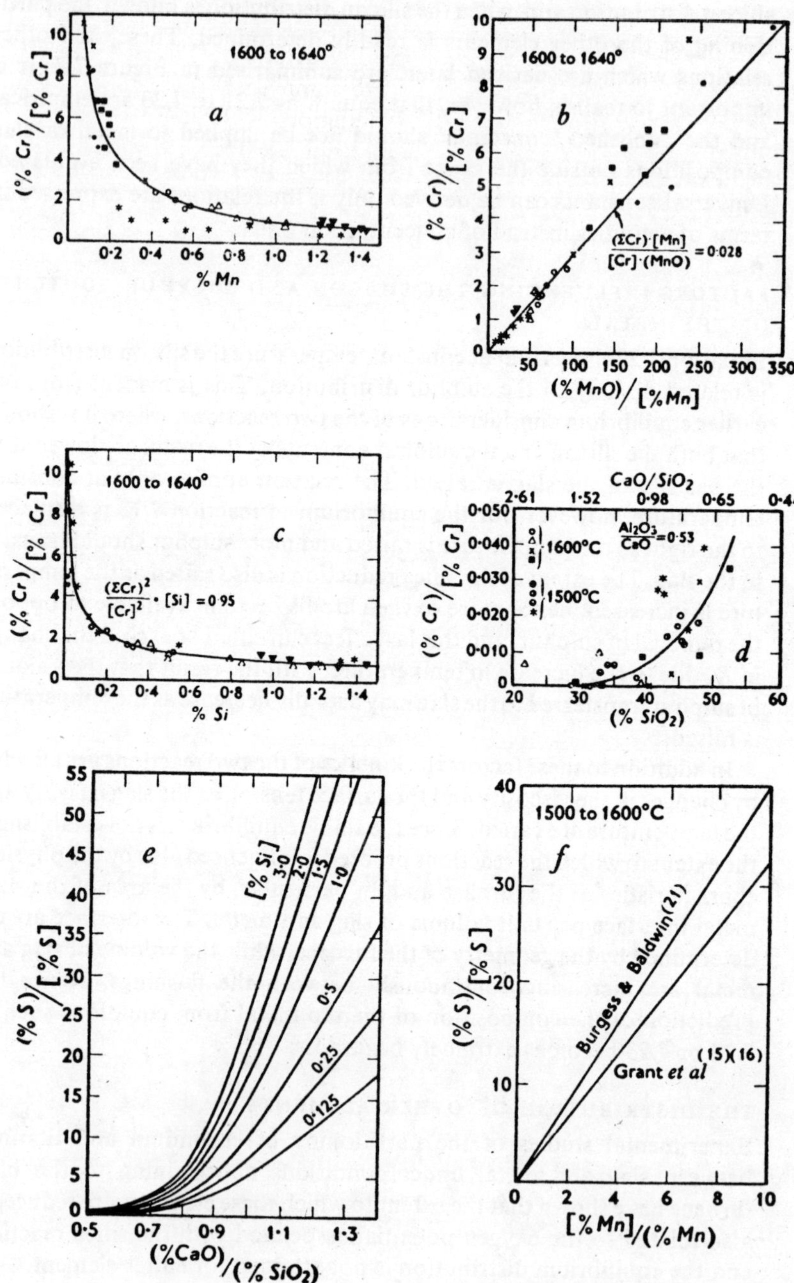
1600 to 1640°
a
(% Cr)/[% Cr]
% Mn
1600 to 1640°
b
(% Cr)/[% Cr]
$\frac{(\Sigma Cr)\cdot[Mn]}{[Cr]\cdot(MnO)} = 0{\cdot}028$
(% MnO)/[% Mn]
1600 to 1640°
c
(% Cr)/[% Cr]
$\frac{(\Sigma Cr)^2}{[Cr]^2}\cdot[Si] = 0{\cdot}95$
% Si
CaO/SiO₂
$\frac{Al_2O_3}{CaO} = 0{\cdot}53$
1600°C
1500°C
d
(% Cr)/[% Cr]
(% SiO₂)
e
[% Si]
3·0
2·0
1·5
1·0
0·5
0·25
0·125
(% S)/[% S]
(%CaO)/(% SiO₂)
1500 to 1600°C
f
Burgess & Baldwin(21)
(15)(16)
Grant et al
(% S)/[% S]
[% Mn]/(% Mn)

Chromium

The oxygen potential diagrams show that chromium oxides should be reduced in the blast furnace in the same temperature range as that required for the reduction of the oxides of manganese and silicon. In practice some chromium is absorbed by the iron before it reaches the hearth, but the chromium content in the metal at tap is determined by the reactions in the hearth.

Under an atmosphere of pure carbon monoxide, chromium is present in the slag as chromous oxide (CrO) [45–47] and the reduction of this oxide by solid carbon is given by the equation:

$$CrO_{(slag)} + C_{(graphite)} \rightleftharpoons Cr_{(l)} + CO_{(g)} \qquad 7.24$$

The equilibrium constant can be expressed as:

$$\frac{(a_O)}{[a_O]} = \frac{1}{K_{Cr}} \frac{(a_{Cr})}{[a_{Cr}]} \qquad 7.25$$

and, if the activities of chromium and its oxide do not change markedly with composition, this can be related also to the silicon distribution:

$$\frac{(\%Cr)^2}{[\%Cr]^2} = K'_{Si.Cr} \frac{(\%SiO_2)}{[\%Si]} \qquad 7.26$$

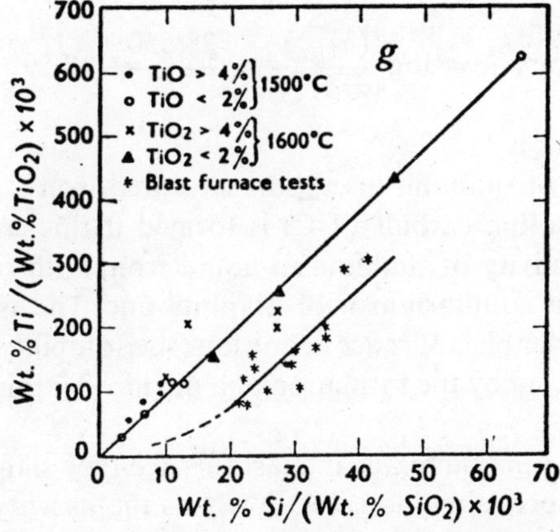

FIG. 7.8. Relations obtained from laboratory and production data between the partitioning of various elements in the slag and metal. ((*a*, *b*, *c*) after Körber and Oelsen [45], (*d*) after McCoy and Philbrook [48], (*e*) after Oelsen and Maetz [43], (*g*) after Delve, Meyer and Lander [51].)

Within a limited composition range $K'_{Si.Cr}$ is approximately constant [45, 48]. The dominating effect of the silica reaction is confirmed by the observation that the rate of reduction of chromous oxide is 20 times as fast as the rate of silica reduction at 1,600°C and 60 times as fast at 1,500°C [48].

Chromous oxide is a very weak acid, hence its activity should decrease as the basicity is raised. However, it is probable that the activity of silica falls more rapidly with increase of basicity so, from equation 7.25, the chromium recovery by the metal should increase as the basicity of the slag is raised. If the silicon distribution is at equilibrium practically all the chromium is transferred into the metal when the ratio CaO/SiO_2 is greater than 1·4, but at CaO/SiO_2 equal to 0·60 the ratio (%Cr)/[%Cr] is about 0·04 [48].

Titanium

According to the oxygen potential diagrams the temperature required for the reduction of titanium dioxide (TiO_2) by solid carbon is somewhat greater than that required for the reduction of silica and it is expected that transfer of titanium into the metal occurs mainly by a reaction in the hearth. This reaction can be written as:

$$Ti_{(l)} + 2CO_{(g)} \rightleftharpoons TiO_{2(s)} + 2C_{(graphite)} \qquad 7.27$$

$$\Delta G^0_{27} = -165{,}000 + 83{\cdot}68T \text{ cal mol}^{-1}$$

For carbon-saturated iron the equilibrium constant is:

$$\log K_{Ti} = \log \frac{(a_{TiO_2})}{[a_{Ti}]p^2_{CO}} = \frac{28{,}550}{T} - 18{\cdot}29 \qquad 7.28$$

and at 1,400°C K_{Ti} is 112.

The solubility of titanium in carbon-saturated iron is less than 0·30 per cent and titanium carbide (TiC) is formed if this concentration is exceeded. The activity of titanium in liquid iron, with respect to pure solid titanium, in equilibrium with graphite and TiC is $7{\cdot}0 \times 10^{-5}$ at 1,500°C [49]. Under blast furnace conditions the solubility of titanium in iron is also restricted by the formation of a nitride (TiN) in the nitrogen-rich atmosphere.

The activity of titanium oxide in blast furnace type slags has not been measured, but it probably increases slowly as the basicity is raised. On comparing equation 7.28 with the equilibrium constants given before it is expected that at equilibrium the titanium recovery by the metal is very low, relative to the recovery of the other elements which have been considered, and that the extent of the recovery increases as the basicity is raised. In practice the extent of the titanium reaction is limited by the high oxygen potential caused by the low silica reduction rate and a linear relation obtains between the distributions of titanium and silicon [50, 51].

Titaniferrous iron ores are often regarded as undesirable materials for the blast furnace on account of the increased tendency for the burden to hang in the lower stack and in the bosh. At high basicity the viscosity and the melting point of the slag increase rapidly as the TiO_2 content is raised, but in acid-rich slags the effect is negligible [52]. At high basicities or at high temperatures titanium oxides are reduced to a greater extent and more titanium carbide and titanium nitride are formed. The melting points of the carbide and the nitride are 3,150°C and 2,950°C respectively, so they remain as solids which increase the viscosity of the slag and which settle out on the floor of the furnace hearth and build up accretions to restrict the hearth capacity. Thus titaniferrous ores can only be smelted satisfactorily if the temperature is low and the slag is acid-rich. The TiO_2 content of a basic slag should not be greater than about 2 per cent [53].

Phosphorus

Phosphorus is present in most iron ores as mineral phosphates which are readily reduced in the blast furnace and the phosphorus is transferred into the iron. It is highly probable that the element is absorbed completely by the droplets of liquid iron before they pass into the slag layer and the partitioning of phosphorus between the slag and the metal in the hearth can be evaluated from data for the reaction

$$2P_{(in\,Fe)} + 5O_{(in\,Fe)} \rightleftharpoons P_2O_{5(slag)} \qquad 7.29$$

(see Chapter 15.1). However, it is evident that the equilibrium in this reaction lies far to the left, for the activity of oxygen is raised to the fifth power in the equilibrium expression. For a partial pressure of carbon monoxide equal to one atmosphere, the equilibrium oxygen activity in carbon-saturated iron is about 5×10^{-5} at 1,500°C and it follows that the activity of the P_2O_5 in the slag is vanishingly small, the phosphorus content of the slag is negligible, and practically all the phosphorus remains dissolved in the metal.

7.5 Selection of the Slag Composition

ACID AND BASIC IRONS

The metal from the blast furnace is traditionally classified as acid or basic according to the sulphur and phosphorus contents in the iron. Neither of these elements can be removed during steel-making when an acid slag is used and the concentration of both elements is usually required to be less than 0·05 weight-per cent. The silicon content of an acid iron is often greater than 1·0 weight per-cent and may be as high as 2·5 per cent if the iron is to be refined in the acid Bessemer converter, whereas basic irons contain only about 0·5 weight-per cent of silicon. The sulphur content of a basic iron should be low, but is less critical than for an acid iron and the phosphorus may be as high as 2·0 weight-per cent when the iron is destined for the basic Bessemer converter. It is possible to alter the silicon and sulphur contents of the hot metal while it is held in the ladle after tapping from the furnace, but these treatments increase the metal cost and, wherever possible, it is preferable to control the composition of the iron to be tapped from the furnace to within the limits of the metal specification. From the preceding considerations it is clear that the analysis of the iron can be changed by making adjustments to the burden, to the composition of the slag and to the hearth temperature. These three factors are to some extent inter-related, for the burden controls the slag composition, whilst the latter is also limited by the requirement that the slag must be free-running at the hearth temperature.

EFFECT OF COMPOSITION ON THE VISCOSITY OF THE SLAG

Although the equilibrium distributions of the elements are not obtained in the blast furnace, it is still true that the silicon and sulphur contents of the iron are decreased as the basicity of the slag is raised at constant temperature, but there is obviously a limit to the basicity which can be usefully employed. At any temperature the viscosity of the liquid slag is markedly dependent on the basicity. The structure of liquid slags is discussed in Chapter 3, where it is shown that the viscosity decreases progressively with increasing concentration of the basic oxides up to the point where all the Si–O–Si bonds are broken. This point corresponds to the ratio $MO/SiO_2 = 2{\cdot}0$, but the liquidus surface for basic blast furnace slags is located at lower ratios than this and, before the minimum value is reached, the viscosity rises again because of the presence of the undissolved solids in the slag.

In comparison with the effect of a change in basicity, the presence of other oxides causes only minor changes in the viscosity. Alumina increases the viscosity in acid melts, but the reverse is true at higher basicities [54], the change being due probably to the fact that aluminium ions can substitute for the silicon ions in an anion network but not when the network is broken down. On a weight-per cent basis, magnesia is slightly more effective than lime in lowering the viscosity [55]. Generally, at constant composition, the viscosity increases with increasing temperature above the liquidus, but the rate of change of the viscosity with the temperature is greater for the more acid slags [56], and it is possible for an added oxide to raise or lower the viscosity solely as a result of a change in the liquidus temperature.

By trial and error methods it was established some years ago that a fluid slag is obtained at a low temperature when the ratio $(CaO + MgO)/SiO_2$ is about 1·4 and the alumina content is in the range from 15 to 18 per cent. If the alumina content is outside this range, either the temperature must be raised or the basicity must be lowered to retain the free-running characteristics of the slag [57]. This conclusion was modified when the liquidus

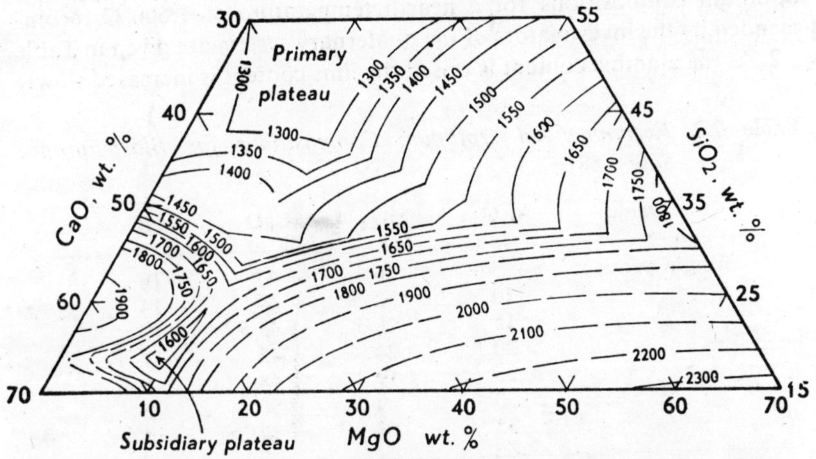

FIG. 7.9. Liquidus surfaces in the system CaO–MgO–Al_2O_3–SiO_2 at 15 weight-per cent of alumina. The optimum basic blast furnace slag compositions lie on the small triangle at the lower left of the diagram. (After Osborn, DeVries, Gee and Kramer [58]).

surfaces were determined for the quaternary system CaO–MgO–Al_2O_3–SiO_2, these four oxides constituting the major portion of normal blast-furnace slags. The liquidus data have been presented [58] as a series of pseudo-ternary sections across the system at constant concentrations of alumina and Figure 7.9, which is taken from the original paper, shows the liquidus isothermals at 15 per cent of Al_2O_3. Adjacent to the CaO–SiO_2 side of the section is a region of low-melting point liquids in the form of a plateau with a gentle slope towards the silica apex. All the compositions within the limits of the plateau are entirely liquid at normal blast furnace temperatures, but the plateau is bounded by liquidus surfaces which rise very steeply. With increasing alumina content the plateau is displaced towards the CaO–MgO edge of the system, and when the alumina is greater than 20 per cent a shallow valley extends from the plateau towards the basic oxide edge, which gradually broadens into a second smaller plateau as the alumina content is further increased.

OPTIMUM COMPOSITIONS FOR BASIC SLAGS

The slag compositions which are suitable for the production of an iron low in both sulphur and silicon lie on the plateau and close to the liquidus wall on the CaO–MgO side. In practice the composition of the slag must be sufficiently remote from this liquidus wall to ensure that the normal fluctuations in the composition lie within the limits of the plateau. The optimum compositions for a hearth temperature of 1,500°C recommended by the investigators of the quaternary system are given in Table 2. As the alumina content is raised the lime content is increased slowly

Table 7.2 *Recommended Optimum Compositions for Blast-Furnace Slags* (58)

Oxide	Al_2O_3	SiO_2	CaO	MgO
Weight-per cent	5	36	43	16
	10	32	44	14
	15	28·5	44·5	12
	20	24	45	11
	25	19	48	8
	30	9	56	5
	35	7	54	4

at first and more rapidly when the alumina exceeds 20 per cent, where the optimum composition moves from the large to the subsidiary plateau.

It is interesting to note that the ratio $(CaO+MgO)/(Al_2O_3+SiO_2)$ lies between 1·3 and 1·4 for each of the tabulated compositions. However, if the magnesia content is raised or lowered, the composition in the high alumina range is displaced from the subsidiary plateau and the basicity must then be lowered to bring the composition on to the main plateau.

A blast-furnace slag consists largely of the four oxides, hitherto considered, but there are also present small amounts of other oxides which could change the temperatures of the liquidus surfaces and alter the optimum compositions from those given in Table 7.2. The effect of titania in this respect has already been mentioned, but it is found that the liquidus temperatures are not changed significantly by the range of concentrations of sulphur, manganese and iron oxides which are normally found in the slag [56].

At constant CaO/SiO_2 ratio the activity of the silica is lowered [26] and the activity of the slag sulphur is raised [28] as the alumina content is increased, but (a_{SiO_2}) changes more rapidly than (a_S) and so, from equation 7.22, the ratio (%S)/[%S] should increase as the alumina content of the slag is raised. Thus there is apparently an advantage in using iron ores with a moderately high-alumina content, but this can raise the amount of sulphur remaining in the metal if the formation of a high-alumina slag increases the thermal load on the furnace, for more coke is consumed to provide the additional heat and the total sulphur in the burden is raised. Increasing the alumina content may also introduce operating problems due to the formation of a viscous bosh slag. In this respect the use of a sinter burden is advantageous to promote the formation of a calcareous slag and lower the viscosity of the bosh slag. When only low-grade silicious ores are available it is often advantageous to maintain a low-slag basicity (i.e., 0·8 to 1·0), together with a low hearth temperature to minimise silica reduction, and to desulphurise the metal in the ladle. Acid-burdening practices of this type have been employed for example with the silicious Northamptonshire [57] and Lorraine [59] ores. On the other hand, if a surplus of hearth heat is available, or if additional hearth heat can be supplied by increasing the temperature or oxygen content of the hot blast and without increasing the coke rate, then this heat can be usefully employed to raise the alumina content of the slag and lower the sulphur in the metal.

7.6 Treatment of the Metal after Tapping from the Furnace

In this chapter conditions have been referred to which may make it impossible or uneconomical to produce an iron from the blast furnace with a composition which is suitable for use in the subsequent steel-making process. Further, the times at which hot metal is required for refining are not readily adjusted to coincide with the times at which the blast furnace is tapped, particularly when the iron is required at frequent intervals as in the Bessemer converter. The *hot-metal mixer* is then a convenient intermediary in which to store the iron until it is required.

The mixer is often a refractory-lined vessel of large capacity with some form of ancillary heating, which serves to smooth-out variations from tap-to-tap in the composition and temperature of the metal and from which the hot metal can be drawn off as required. However, when an off-grade iron is unavoidably produced at the smelter the metallurgical load on the steel plant can be eased by performing some preliminary refining in the mixer. The silicon content can be lowered by feeding oxides or oxygen gas to the bath, but the slag tends to form a froth and a large furnace volume is required to accommodate it. Alternatively, the silicon, sulphur and phosphorus contents can be lowered if the mixer is lined with a basic refractory and a lime-rich, oxide slag is formed. The mixer is then employed as a modified open-hearth furnace and the process is generally called an *active mixer practice*.

In recent years a number of ladle treatments have been developed in which the metal composition can be changed more rapidly and sometimes more extensively than in the active mixer. The treatments are applied to the metal in the transfer ladle after separation of most of the blast-furnace slag and the treated metal is then stored until required in a non-active mixer. A detailed discussion of these treatments would occupy too much space and is not necessary, since the principles upon which they are based are discussed in some detail in Part 3, hence only a brief résumé is given, together with suitable references from which more detailed accounts can be obtained.

Sulphur can be removed by forming a soda slag in the ladle [60] or by blowing through the metal a jet of nitrogen containing a suspension of finely divided lime [61–63], calcium carbide [61–64] or calcium cyanamide [62, 63, 65]. If the lime remains as a solid then (a_{CaO}) and (a_{CaS}) can be taken as unity. It can be shown by a thermodynamic calculation that, when the metal is at 1,300°C and the partial pressure of the carbon monoxide is one atmosphere, the sulphur content of carbon-saturated iron which is in equilibrium with solid lime is about 0·001 per cent. For a metal containing 1·8 per cent of phosphorus, 3·3 per cent of carbon and smaller amounts of silicon and manganese the equilibrium sulphur content is lowered further to about $2{\cdot}5 \times 10^{-5}$ per cent [61]. The equilibrium is not established rapidly because the reaction takes place between a liquid and a solid, but when the solid is finely divided the surface area per unit volume is high and the diffusion distances are small. In practice the sulphur content of the metal can be lowered to about 0·010 per cent with a powder blow of only a few minutes duration.

Most of the silicon and some of the manganese, vanadium, chromium and carbon can be oxidised-out by blowing oxygen gas at high pressure either through or on to the metal [66–69], but a dense fume is formed and an elaborate gas cleaning plant is necessary to deal with the emission.

A variety of methods have also been proposed for preliminary refining of the iron in ancillary vessels other than the transfer ladle. For example, sulphur is rapidly removed by treating molten iron with powdered lime in the Kalling rotary process [70–72]. The iron, together with lime equivalent to about 10 per cent by weight of the metal charged, and fine coke are charged into a cylindrical container which is sealed to maintain a reducing atmosphere and is rotated at 30 to 40 r.p.m. If the oxidation of silicon is prevented, (a_{CaO}) and (a_{CaS}) are approximately unity and the equilibrium sulphur content of the metal is identical with that which obtains when lime powder is injected into the ladle. In practice a sulphur content of 0·10 per cent in the hot metal can be lowered to about 0·020 per cent after 10 minutes and to 0·005 per cent after a further 20 minutes [71].

A suggestion has been made for pouring the molten metal as a fine spray from the top of a tower and allowing the droplets of metal to fall through a blast of oxygen. In pilot plant trials about half of the carbon and most of the silicon and manganese were oxidised and the exothermic reactions caused an appreciable increase in the metal temperature [73].

A similar spray-refining process has been developed by the British Iron and Steel Research Association and a throughput of 2·5 tons per minute has been attained in pilot plant trials. A stream of molten metal falling through a spray device resembling an inverted gas ring is disseminated into a fine spray by jets of pure oxygen, oxygen-enriched air or oxygen–steam mixtures. In addition to the removal of silicon, significant amounts of phosphorus can be removed by injecting powdered lime with the oxidising gas; this is shown by the following specimen analyses:

	Metal Composition (weight-per cent)					*Temperature °C*
	C	Mn	Si	S	P	
Before treatment	3·32	0·79	1·26	0·054	1·32	1,240
After treatment	2·00	0·03	0·01	0·053	1·00	1,500

The extent of the pre-refining is controlled by regulating the rate of flow of the oxidising gas.

Brymbo Steel Works Ltd. have recently introduced a pre-refining treatment in which the carbon, silicon, phosphorus and sulphur contents of the hot metal are lowered by use of an oxygen–lime powder jet in an oil-fired, circular-hearth furnace [74, 75]. The partly refined metal is then fed directly to the arc furnaces for further refining.

REFERENCES

1. F. D. RICHARDSON and J. H. E. JEFFES, *J, Iron St. Inst.*, 1949, Vol. 163, p. 397.
2. S. P. KINNEY, *U.S. Bureau of Mines Bulletin*, No. 397, 1926.
3. Anon., *Iron Coal Tr. Rev.*, 1954, Vol. 168, p. 1265.

4. W. Q. PHILBROOK, K. M. GOLDMAN and M. H. HELZEL, *Trans. Amer. Inst. min. (metall.) Engrs.*, 1950, Vol. 188, p. 361.
5. N. C. TOMBS and A. J. E. WELCH, *J. Iron St. Inst.*, 1952, Vol. 172, p. 69.
6. G. C. HOGBERG, *Trans. Amer. Inst. min. (metall.) Engrs.*, 1943, Vol. 164, p. 96.
7. G. STEUDEL, *Blast Furn.*, 1950, Vol. 28, p. 775.
8. J. CHIPMAN et al., *Trans. Amer. Soc. Metals*, 1952, Vol. 44, p. 1215.
9. E. T. TURKDOGAN and L. E. LEAKE, *J. Iron St. Inst.*, 1955, Vol. 179, p. 39.
10. T. FUWA and J. CHIPMAN, *Trans. Amer. Inst. min. (metall.) Engrs.*, 1959, Vol. 215, p. 708.
11. M. OHTANI and N. A. GOKCEN, ibid., 1960, Vol. 216, p. 533.
12. E. T. TURKDOGAN, R. A. HANCOCK, S. I. HERLITZ and J. DENTON, *J. Iron St. Inst.*, 1956, Vol. 183, p. 69.
13. K. SANBONGI and M. OHTANI, *Sci. Reps. Res. Inst. Tohoku Univ.*, 1957, Series A9, p. 147.
14. K. P. ABRAHAM, M. W. DAVIES and F. D. RICHARDSON, *J. Iron St. Inst.*, 1960, Vol. 196, p. 82.
15. N. J. GRANT, H. KALLING and J. CHIPMAN, *Trans. Amer. Inst. min. (metall.) Engrs.*, 1951, Vol. 191, p. 666.
16. N. J. GRANT, J. W. DOWDING and R. J. MURPHEY, ibid., 1953, Vol. 197, p. 1451.
17. J. E. STUKEL and J. COCUBINSKY, ibid., 1954, Vol. 200, p. 353.
18. C. BODSWORTH, *J. Iron St. Inst.*, 1959, Vol. 193, p. 13.
19. I. M. DAVIDSON and C. BODSWORTH, ibid., 1960, Vol. 195, p. 163.
20. E. T. TURKDOGAN, ibid., 1956, Vol. 182, p. 74.
21. A. J. BURGESS and B. G. BALDWIN, ibid., 1957, Vol. 186, p. 227.
22. A. R. KAY and J. TAYLOR, *Trans. Faraday Soc.*, 1960, Vol. 59, p. 1372.
23. J. C. FULTON and J. CHIPMAN, *Trans. Amer Inst. min. (metall.) Engrs.*, 1954, Vol. 200, p. 1136.
24. F. C. LANGENBERG and J. CHIPMAN, *Trans. Amer. Inst. min. (metall.) Engrs.*, 1959, Vol. 215, p. 958.
25. J. CHIPMAN, J. C. FULTON, N. A. GOKCEN and G. R. GASKEY, *Acta Met.*, 1954, Vol. 2, p. 439.
26. J. C. FULTON and J. CHIPMAN, *Trans. Amer. Inst. min. (metall.) Engrs.*, 1959, Vol. 215, p. 888.
27. D. JOYCE, *J. Iron St. Inst.*, 1948, Vol. 159, p. 291.
28. J. CHIPMAN, *Physical Chemistry of Process Metallurgy*, Interscience, 1961, Vol. 7, p. 27.
29. C. W. SHERMAN and J. CHIPMAN, *Trans. Amer. Inst. min. (metall.) Engrs.*, 1952, Vol. 194, p. 597.
30. G. G. HATCH and J. CHIPMAN, ibid., 1949, Vol. 185, p. 274.
31. E. W. FILER and L. S. DARKEN, ibid., 1952, Vol. 194, p. 253.
32. P. KOZAKEVITCH, *Centre Doc. Sid. Circ. Inform. Tech.*, 1954, (2), p. 354; *The Physical Chemistry of Steel-making*, New York, John Wiley, 1958, p. 89.
33. P. KOZAKEVITCH, G. URBAIN and M. SAGE, *Compte Rend.* 1954, Vol. 239, p. 166.
34. A. KOHN, *Proc. First UNESCO Internat. Conf.*, 1957, Paris, 1958, p. 302.
35. J. C. FULTON, N. J. GRANT and J. CHIPMAN, *Trans. Amer. Inst. min. (metall.) Engrs.*, 1953, Vol. 197, p. 185.

36. C. E. A. SHANAHAN, *J. Iron St. Inst.*, 1955, Vol. 180, p. 140.
37. S. RAMACHANDRAN, T. B. KING and N. J. GRANT, *Trans. Amer. Inst. min. (metall.) Engrs.*, 1956, Vol. 206, p. 1549.
38. T. B. KING and S. RAMACHANDRAN, *The Physical Chemistry of Steel-making*, New York, John Wiley, 1958, p. 125.
39. N. J. GRANT, O. TROILI and J. CHIPMAN, *Trans. Amer. Inst. min. (metall.) Engrs.*, 1951, Vol. 191, p. 672.
40. K. M. GOLDMAN, G. DERGE and W. O. PHILBROOKE, ibid., 1954, Vol. 200, p. 534.
41. L. D. KIRKBRIDE and G. DERGE, ibid., 1960, Vol. 218, p. 87.
42. J. P. MORRIS, ibid., 1952, Vol. 194, p. 939.
43. W. OELSEN and H. MAETZ, *Arch. Eisenhüttenw.*, 1942–3, Vol. 16, p. 283.
44. H. WENTRUP, H. MAETZ and P. HELLER, ibid., 1949, Vol. 20, p. 139.
45. F. KÖRBER and W. OELSON, *Mitt. K.-Wilh. Inst. Eisenforsch.*, 1935, Vol. 17, p. 231.
46. T. NIINOMI, Y. MATSUSHITA and K. KANAMORI, *Tetsu to Hagane*, 1955, Vol. 41, p. 1075.
47. C. W. MCCOY, *Proc. Amer. Inst. Min. (Metall.) Engrs. Blast Furn. Comm.*, 1957, Vol. 16, p. 50.
48. C. W. MCCOY and W. O. PHILBROOK, *Physical Chemistry of Steel-making*, New York, John Wiley, 1958, p. 93; *Trans. Amer. Inst. min. (metall.) Engrs.*, 1958, Vol. 212, p. 226.
49. F. D. DELVE, ibid., 1958, Vol. 212, p. 183.
50. E. FAUST, *Arch. Eisenhüttenw.*, 1939, Vol. 2, p. 361.
51. F. D. DELVE, H. W. MEYER and H. N. LANDER, *Physical Chemistry of Process Metallurgy*, 1961, John Wiley, p. 1111.
52. I. P. SEMIK, *Bull. Acad. Sci. URSS*, 1941, Vol. 4, p. 55.
53. E. WENDEL, *Stahl u. Eisen, Düsseldorf*, 1958, Vol. 78, p. 1603.
54. G. BEHRENDT and T. KOOTZ, ibid., 1949, Vol. 69, p. 399.
55. E. E. HOFMAN, ibid., 1959, Vol. 79, p. 846.
56. B. G. BALDWIN, *J. Iron St. Inst.*, 1957, Vol. 186, p. 388.
57. T. P. COLCLOUGH, ibid., 1946, Vol. 154, p. 547.
58. E. F. OSBORN, R. C. DEVRIES, K. H. GEE and H. M. KRAMER, *Trans. Amer. Inst. min. (metall.) Engrs.*, 1954, Vol. 200, p. 33.
59. C. G. THIBAUT, *Centre Doc. Sider. Circ.*, 1958, Vol. 5, p. 357.
60. Anon., *Iron Coal Tr. Rev.*, 1952, Vol. 164, p. 531.
61. B. TRENTINI, L. WAHL and M. ALLARD, *Trans. Amer. Inst. min. (metall.) Engrs.*, 1957, Vol. 209, p. 1133; *J. Iron St. Inst.*, 1956, Vol. 183, p. 124.
62. D. E. PARSONS and S. L. GERTSMAN, *Amer. Foundrym.*, 1955, Vol. 27, p. 60.
63. J. N. HORNACK and E. J. WITTENBERG, *J. Metals, N.Y.*, 1956, Vol. 8, p. 425.
64. R. B. COATES and J. V. HARDING, *B.C.I.R.A. Journal*, 1960, Vol. 8, p. 89.
65. G. G. HATCH, G. E. WAGNER and M. TROTTIER, *Proc. Amer. Inst. Min. (metall.) Engrs., Blast Furn. Comm.*, 1956, Vol. 15, p. 120.
66. P. S. LEROY, *J. Metals, N.Y.*, 1953, Vol. 5, p. 396.
67. J. SITTARD, *Stahl u. Eisen, Düsseldorf*, 1956, Vol. 76, p. 1554.
68. P. K. GLEDHILL, P. J. CARNALL and K. H. SARGENT, *J. Iron St. Inst.*, 1957, Vol. 186, p. 198.
69. W. WOLF, W. ESCHE, H. WYSOCKI and O. STEINHAUER, *Stahl. u. Eisen, Düsseldorf*, 1958, Vol. 78, p. 1100.

70. B. KALLING, C. DANIELSSON and O. DRAGGE, *Trans. Amer. Inst. min. (metall.) Engrs.*, 1951, Vol. 191, p. 732.
71. S. FORNANDER, ibid., 1951, Vol. 191, p. 739.
72. B. KALLING and S. EKETORP, *Blast Furn.*, 1957, Vol. 45, p. 494.
73. J. M. GAINES and D. C. HILTY, *J. Metals, N.Y.*, 1958, Vol. 10, p. 452.
74. Anon., *J. Iron St. Inst.*, 1959, Vol. 192, p. 117.
75. E. DAVIES, ibid., 1961, Vol. 197, p. 271.

PART III

STEEL-MAKING

CHAPTER 8

The Converter Process

8.1 Introduction

The processes by which blast-furnace metal and ferrous scrap are converted into steel are conventionally named after the furnace in which the refining is performed and the nature of the slag (i.e., acid, basic or neutral) which is employed to absorb the products of oxidation. The acidity or basicity of the slag is in turn decided by the type of refractory lining with which it is in contact. When Bessemer patented the converter process in 1856 the only refractories in use which were capable of withstanding the temperature of molten steel were of a siliceous character and the process which he developed is now known as the acid Bessemer. The basic process followed in 1877 with the development of the magnesian limestone or dolomite lining by Thomas and Gilchrist.

The primary difference between the two methods of steel-making is that phosphorus and sulphur can only be removed effectively from the metal under a cover of basic slag. In the acid process there is often a slight increase in the concentration of these elements, for the number of sulphur and phosphorus atoms in the metal remains more or less constant whilst most of the carbon, silicon and manganese, together with some of the iron, are removed by oxidation. The choice between the two processes rests mainly on the type of ore available in any locality. The acid process was preferred in America where ores of fairly low phosphorus content are widely distributed, whereas the phosphoric ores of Europe are suited only to the basic process.

The advent of the open-hearth furnace caused a rapid decline in the importance of the Bessemer process and by the 1920s only a small percentage of the world steel production was made in converters. However, the high-phosphorus ores of France and West Germany are more suited to the basic Bessemer process than to the open-hearth and, mainly as a result of experiments conducted in these countries, a marked improvement has been effected in the quality of steel made in the converter, to the extent that other steel producing areas are showing renewed interest in the process. A number of new converter plants have recently been commissioned and others are under construction and it may be stated that a considerably greater proportion of the low- and medium-carbon grades of steel will be produced by this means in the foreseeable future. In view of this trend a detailed consideration is given here of the principles of the process, commencing with the traditional or orthodox practice before considering more recent developments.

The characteristic feature of the Bessemer process is the passage of a blast of air through the bath of liquid metal and slag, via tuyeres situated in the bottom of a pear-shaped vessel. This feature gives rise to the present tendency to refer to the practice as the *bottom-blown, pneumatic process.* The gas emerging from the nose of the converter is rich in carbon monoxide which is formed by the oxidation of the carbon in the metal and burns at the nose with the appearance of a bright flame, from which the progress of the blow is assessed. The passage of the blast through the bath creates violent turbulence which brings the reactants into rapid contact and the various reactions proceed at a high speed; for example, a 50-ton heat of blast-furnace metal can be refined to a low-carbon steel in less than twenty minutes. This high-speed reaction provides one of the major difficulties of the process, for there is insufficient time for chemical analysis of the metal to be performed during the blow, and extremely close control of the composition of the charge metal, the temperature and the duration of the blow, is essential to produce steel conforming to the limits of a given specification. The steel-maker has been greatly assisted in this difficulty by the development of a number of instruments which indicate continuously the progress of the blow and remove the element of chance from the determination of the end of the refining.

The oxidising reactions which occur in the converter are exothermic and, with a suitably balanced iron composition, the temperature of the metal increases continuously from that of the molten iron charge (1,250 to 1,400°C) to between 1,600 and 1,650°C. Although no external supply of heat is required, at the end of the blow even low-carbon steel is sufficiently superheated to permit pouring via a ladle into moulds or castings.

8.2 The Acid Bessemer Process

Although today the acid Bessemer process is used for the production of only a very small part of the world's annual steel output, consideration in some detail of the chemical reactions involved is worthwhile for two reasons. Firstly, the limitations to the extent to which various elements can be removed under an acid slag in the converter apply also to the other acid steel-making processes, and in later chapters it will be necessary only to point out any major differences which may arise. Secondly, the principles of the converter process are more readily understood because the reactions involved are fewer in number than in the basic process.

HEAT REQUIREMENTS

The converter is charged with molten iron at a temperature of about 1,350°C and heat must be supplied during refining to raise the metal temperature and ensure that it is sufficiently fluid for pouring when the reactions are completed. In addition thermal energy is required to meet the following needs:

(*a*) To raise the temperature of the air blast to that of the metal.
(*b*) To melt the solids which are added to form the slag.
(*c*) To compensate for thermal losses by conduction and convection through the converter lining and by radiation from the upper surface of the slag layer.

This thermal energy is obtained from the exothermic reactions between the elements dissolved in the molten metal and the oxygen in the air blast. The principal impurities in the iron are carbon, silicon and manganese, and at liquid-steel temperatures the heats of formation of the oxides from a 1 weight-per cent solution of these elements in the iron are as follow:—

$$Si_{(1\,\%\,in\,Fe)}+O_{2(g)} \rightleftharpoons SiO_{2(s)};\quad \Delta H_{Si} = -188 \text{ kcal mol}^{-1} \qquad 8.1$$

$$Mn_{(1\,\%\,in\,Fe)}+\tfrac{1}{2}O_{2(g)} \rightleftharpoons MnO_{(s)};\quad \Delta H_{Mn} = -95{\cdot}4 \text{ kcal mol}^{-1} \qquad 8.2$$

$$C_{(1\,\%\,in\,Fe)}+\tfrac{1}{2}O_{2(g)} \rightleftharpoons CO_{(g)};\quad \Delta H_{CO} = -31{\cdot}8 \text{ kcal mol}^{-1} \qquad 8.3$$

$$xFe_{(l)}+\tfrac{1}{2}O_{2(g)} \rightleftharpoons Fe_xO_{(l)};\quad \Delta H_{Fe} = -83{\cdot}7 \text{ kcal mol}^{-1} \qquad 8.4$$

where $Fe_xO_{(l)}$ is the composition of the liquid iron oxide in equilibrium with pure liquid iron.

Some carbon is oxidised to carbon dioxide in the bath but, within the range of temperature and oxygen potentials to which the metal is exposed, carbon monoxide is more stable and the greater portion of the carbon is oxidised according to equation 8.3.

The iron charge contains 3·5 to 5·0 weight-per cent of carbon (i.e., 14–20 atom-per cent) but the heat generated when this is oxidised to carbon monoxide is only about 5 per cent of the total thermal requirements. The carbon monoxide is carried out of the converter with the nitrogen remaining from the air blast and burns in the air at the converter mouth, but the heat generated here is lost to the atmosphere and only a small fraction (ΔH_{CO}) of the total heat of oxidation of the carbon is available for heating the metal. The oxidation of iron into the slag is obviously undesirable because the metal yield would be lowered; hence the additional heat is supplied by the oxidation of silicon and manganese. In British practice it is found that 1·80 to 2·20 per cent of silicon and 0·5 to 1·0 per cent of manganese in the iron are necessary to meet the heat requirements [1] but, with faster blowing rates, American converters operate successfully with 1·5 to 1·7 per cent of silicon and a similar manganese content [2]. If the silicon is lowered further the manganese must be increased to maintain the heat balance, but the respective concentrations by weight do not vary directly in proportion to the molar heats of reaction. As the atomic weight of manganese is almost double that of silicon, the number of moles of oxide formed by the oxidation of 1 weight-per cent of silicon is almost twice the number that are formed by the same weight of manganese. On the other hand the volume of air required to oxidise an atom of silicon is twice the amount needed to combine with one manganese atom and, therefore, approximately the same volume of air is required to oxidise 1 weight-per cent of each element and similar amounts of energy are lost as sensible heat in the blast. Thus if the silicon is lowered to 1 weight-per cent the manganese must be increased to greater than 3 per cent in order to restore the thermal balance. This composition is typical of Swedish practice.

THE PROGRESS OF THE BLOW

Oxygen from the air blast reacts with the metal immediately in front of the tuyeres and is entirely consumed within a few inches of the tuyere nozzles. According to the oxygen potential diagrams in Chapter 1 the oxygen should combine with the silicon, manganese and carbon in preference to iron but, even in the initial metal charge, the number of iron atoms is several times as great as the total number of atoms of the other species, and most of the oxygen is consumed initially by the formation of iron oxides. It is probable that Fe_3O_4 is formed by the highly oxidising conditions in front of the tuyeres, but this oxide is unstable at high temperatures in the presence of an excess of liquid iron and it rapidly dissociates, the oxygen atoms entering into solution in the liquid iron. Thus the higher oxides have only a transient existence and the oxygen atoms are rapidly distributed through the metal, where they come into contact with the atoms of the solute elements. A typical plot of the change of metal composition as a function of the blowing time is shown in Figure 8.1. Manganese and silicon begin to oxidise almost from the start of the blow and the concentrations of both elements are reduced rapidly to a very low level. Meanwhile the carbon content remains relatively constant and falls rapidly only when the above reactions are practically complete.

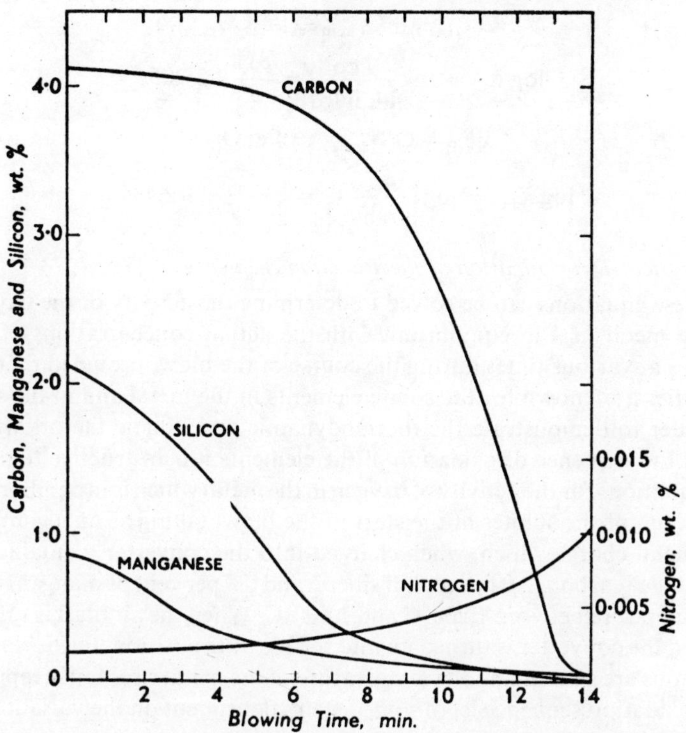

FIG. 8.1. Change of metal composition with time in the acid Bessemer converter process.

EQUILIBRIUM CONDITIONS

The available thermodynamic data do not permit a rigorous treatment of the reactions from the start of refining, but the data are adequate to explain the reaction sequence. When the dilute weight-per cent solution is taken as the standard state for the various solutes, the numerical values of the equilibrium constants are as follow*:

$$Si_{(\%\,in\,Fe)} + 2O_{(\%\,in\,Fe)} \rightleftharpoons SiO_{2(s)} \qquad 8.5$$

$$\log K_{Si} = \log\frac{(a_{SiO_2})}{[a_{Si}][a_O]^2} = \frac{29{,}700}{T} - 11{\cdot}24$$

$$Mn_{(\%\,in\,Fe)} + O_{(\%\,in\,Fe)} \rightleftharpoons MnO_{(l)} \qquad 8.6$$

$$\log K_{Mn} = \log\frac{(a_{MnO})}{[a_{Mn}][a_O]} = \frac{12{,}440}{T} - 5{\cdot}33$$

* The free-energy data used in this and subsequent chapters are considered in detail and references to the original sources are given in Chapter 13.

$$C_{(\%\,\text{in Fe})} + O_{(\%\,\text{in Fe})} \rightleftharpoons CO_{(g)} \qquad 8.7$$

$$\log K_C = \log \frac{p_{CO}}{[a_C][a_O]} = \frac{811}{T} + 2{\cdot}205$$

$$xF_{(l)} + O_{(\%\,\text{in Fe})} \rightleftharpoons Fe_xO_{(l)} \qquad 8.8$$

$$\log K_{Fe} = \log \frac{(a_{FeO})}{[a_{Fe}][a_O]} = -\frac{5{,}762}{T} + 2{\cdot}435$$

Theoretical Determination of the Reaction Sequence

These equations can be solved to determine the activity of the oxygen in the metal $[a_O]$ in equilibrium with the actual concentrations of the solutes at various times during the course of the blow, provided that the activities are known for the solute elements in the metal and in the slag. In order to demonstrate the thermodynamic and kinetic factors which affect the sequence of oxidation of the elements it is instructive to solve the equations for the activity of oxygen in the metal which is in equilibrium with each of the solutes at the start of the blow, taking as an example a hot-metal charge which, when charged into the converter, contains 4·0 per cent of carbon, 2·0 per cent of silicon and 0·8 per cent of manganese at a temperature between 1,250°C and 1,400°C. When the air blast is turned on and the converter is turned up into the blowing position, the oxidation reactions are started and the temperature of the metal rises fairly rapidly. The oxidation sequence is, to some extent, dependent on the variation of the metal temperature with time during refining and this is demonstrated by solving the equations for 1,300° and 1,600°C.

Towards the end of the blow, when the metal is almost pure iron, the activities of the solutes in the metal with respect to the dilute solution standard state can be evaluated from the data in Appendix 1. However, these data are strictly only valid for very dilute solutions and the error in the estimated values may be large if they are used to assess the activities at an earlier stage in the blow, when the concentration of carbon is still relatively high. A more reliable assessment of the activities in the carbon-rich solutions in relation to the same standard state is obtained from the data for graphite-saturated solutions.

The activity coefficient of carbon $[f_C^C]$ in pure iron–carbon alloys is known as a function of the temperature for all carbon contents up to graphite-saturation [3] and the effects of manganese and silicon on $[f_C]$ are known for graphite-saturated solutions. The interaction coefficients for these elements can be reasonably applied to a metal of the present composition and, using these data, $[f_C]$ is 6·9 at 1,600°C. It is necessary to extrapolate the data for $[f_C^C]$ outside the temperature range for which they have been determined in order to obtain the activity coefficient in the iron at 1,300°C, but the error is probably not great and the estimated value of $[f_C]$ becomes 7·4. The activity coefficient of silicon in graphite-saturated iron has been determined as a function of temperature and silicon concentration [4], but it is again necessary to perform a small extrapolation from the experimental temperature range to determine $[f_{Si}]$ at 1,300°C. The effect of manganese on $[f_{Si}]$ in these solutions has not been determined, but at high dilution the interaction coefficient is approximately zero and this is arbitrarily assumed to hold also for the present solution. In this manner $[f_{Si}]$ is estimated as 9·5 at 1,300°C and 9·3 at 1,600°C. It is similarly assumed that $[f_{Mn}]$ is independent of the silicon concentration and depends only on the carbon and manganese contents and the temperature. At 1,530°C the activity coefficient of manganese in carbon-saturated, iron–carbon–manganese melts is about 0·95 [5]. The

Table 8.1

	Reaction			
	$Si–SiO_2$	Mn–MnO	C–CO	Fe–FeO
$K_{(1,300°C)}$	$4{\cdot}4\times10^{7}$	374	526	16·8
$[a_O]_{(1,300°C)}$	$3{\cdot}5\times10^{-5}$	$2{\cdot}8\times10^{-3}$	$6{\cdot}4\times10^{-5}$	0·074
$K_{(1,600°C)}$	$4{\cdot}2\times10^{4}$	20·4	435	4·4
$[a_O]_{(1,600°C)}$	$1{\cdot}1\times10^{-3}$	$5{\cdot}2\times10^{-2}$	$8{\cdot}3\times10^{-5}$	0·12

effect of temperature is unknown but is probably small and can be neglected. It should be noted that the activity coefficients for each element are also dependent on the concentrations of oxygen, phosphorus, sulphur and any other elements which are present in the iron, but these elements are usually present in only small amounts and the correction of the coefficients for their presence would not affect the interpretation.

The activities of the manganese oxide and the silica in the slag and the partial pressure of carbon monoxide gas over the metal are also required, but for the time being it is assumed that these activities are unity and that the partial pressure of carbon monoxide is one atmosphere. Using these values the equations have been solved and the data so obtained for the two temperatures considered are summarised in Table 8.1.

The equilibrium value of $[a_O]$ for each of the above reactions increases as the temperature is raised, but the rate of change with temperature is more rapid for the markedly exothermic oxidation of silicon than for the other reactions and, in comparison, $[a_O]$ in equilibrium with the carbon is almost independent of the temperature. In fact the order of the change in $[a_O]$ with the temperature for the reactions shown in Table 8.1 is identical with the order in which the elements are placed in terms of the heats of reaction.

Considering first the data for 1,300°C, the value of $[a_O]$ in equilibrium with the actual concentrations of the elements decreases in the order Fe > Mn > C > Si. One of the foremost concepts of thermodynamics is that the equilibrium condition applies to the system as a whole and different component parts of a system cannot be at different equilibrium states. Thus in the present case the actual activity of the oxygen in the metal is determined by the element which is in equilibrium with the lowest value of $[a_O]$ and, if equilibrium is maintained, that element will oxidise first. The other elements cannot oxidise until, as a result of the diminished concentration of that particular element, $[a_O]$ has increased to the level at which it is in equilibrium with a second element. The second element can now start to oxidise and the reactions continue until such time as $[a_O]$ has risen to the value in equilibrium with a third element, and so on.

According to this argument, at 1,300°C silicon should start to oxidise as soon as the value of $[a_O]$ exceeds $3{\cdot}5 \times 10^{-5}$. As the silicon content falls $[a_O]$ is increased and, if the temperature remains constant, carbon should start to oxidise when $[a_O]$ is $6{\cdot}4 \times 10^{-5}$. This is not strictly correct because silicon raises the activity coefficient of carbon, so the value of $[a_C]$ is lowered as the silicon concentration falls and the oxygen potential at which the carbon starts to oxidise is somewhat greater than the value shown in the Table. However, the values of $[a_O]$ in equilibrium with the carbon and the silicon lie close together and the difference is less than the uncertainty in the thermodynamic data from which $[a_O]$ is evaluated. Thus a small change in the data or a similar change in the composition and temperature of the metal could reverse the sequence and, at equilibrium, the carbon should then start to oxidise before the silicon. If the equilibrium conditions continue to hold, the condition is next reached where manganese starts to oxidise and iron is the last element in the sequence to transfer into the slag.

Comparison of the Predicted and the Actual Reaction Sequence

Figure 8.1 shows that the actual sequence in which the elements are oxidised is silicon first, followed by manganese, and carbon is removed only when these elements are almost completely eliminated from the iron. Further, the slag which starts to form at the beginning of the blow contains a large proportion of iron oxides whereas, at equilibrium, iron should be the last element to oxidise.

There are two reasons for the apparent deviation from the equilibrium conditions. The first of these is the erroneous assumption of unit activities for MnO and SiO_2 in the slag. The initial ferrous oxide-rich slag contains only about 30 per cent of silica. Therefore the value of (a_{SiO_2}) is markedly less than unity and is about 0·52 [6], which lowers $[a_O]$ in equilibrium with the silicon in the metal to $2{\cdot}5 \times 10^{-5}$ at 1,300°C. Similarly the value of (a_{FeO}) is less than unity and is about 0·55 [6, 7] in the initial slag. As the silicon is oxidised the slag rapidly approaches saturation with silica and (a_{SiO_2}) approaches unity, but the concentration of silicon in the metal has decreased correspondingly and the value of $[a_O]$ in equilibrium with the residual silicon continues to rise. When the manganese starts to oxidise rapidly the slag is almost saturated with silica, so the value of (a_{MnO}) is about 0·02 [8] and, if the temperature of the metal remained constant as the silicon is oxidised, the value of $[a_O]$ in equilibrium with the initial manganese concentration concentration is lowered to about $5{\cdot}6 \times 10^{-5}$. Actually the metal temperature begins to rise as soon as the silicon starts to oxidise, but $[a_O]$ in equilibrium with the silicon changes more rapidly with the temperature than that in equilibrium with the manganese, and it may be safely concluded that the start of the manganese oxidation is very close on the time scale to the start of the silicon oxidation.

It is assumed that p_{CO} is equal to one atmosphere in the calculations, but a large volume of nitrogen is passing continuously through the metal and it might be concluded that p_{CO} is less than unity. However, there is evidence from the basic converter practice that p_{CO} over the metal is approximately atmospheric and the values listed in Table 8.1 are reasonable for $[a_O]$ in equilibrium with the carbon.

Kinetic Considerations

The second cause of the departure from the equilibrium conditions is concerned with the kinetics of the reactions. Kinetic factors may so operate that the start of the first reaction, which is possible thermodynamically, is delayed or the reaction is retarded to the extent that oxygen is supplied to the metal at a greater rate than that at which it is consumed by the reaction, and the oxygen content of the metal increases until the level is reached at which the second element starts to oxidise. If the second reaction is favoured by the kinetics, this second element may be oxidised partly or entirely whilst the first reaction has still barely started.

The strongly turbulent conditions which are created by the passage of the blast through the bath ensure that the reactants are brought rapidly into contact, but the slag is usually very viscous and does not take a direct part in the reactions. The oxides are formed in the metal and there is no transfer of the elements across the slag-metal interface; hence the reaction rates are not dependent on transfer across a diffusion-gradient layer at the interface. It may therefore be expected that the reaction rates are rapid, and this expectation is supported by comparison of the refining time in the converter process with the much longer refining times in the open-hearth and arc-furnace processes.

The factors controlling the kinetics of the removal of the carbon from the metal are somewhat different from those affecting the removal of the other elements. The carbon combines with the oxygen supplied by the air blast and is removed as a gas, but it is not easy to nucleate a bubble of gas in a homogeneous liquid. Instead the carbon and oxygen atoms diffuse into the bubbles of carbon monoxide as they rise through the bath until either the partial pressure of carbon monoxide in the bubble is in equilibrium with the carbon and oxygen remaining in the metal, or the bubble escapes from the metal into the atmosphere. It is probable that the equilibrium partial pressure of carbon monoxide is approached very closely in each bubble of nitrogen and that the actual oxygen activity of the metal in equilibrium with the carbon is only slightly greater than that which is in equilibrium with the carbon and carbon monoxide at one atmosphere pressure. It is certain that this equilibrium is approached more closely in the Bessemer process than in the conventional hearth-refining processes, but any deviation from the equilibrium condition in the direction indicated means that the activity of the oxygen in the metal at the start of the carbon oxidation is greater than the value given in Table 8.1, and this small change is sufficient to tilt the balance in favour of the oxidation of the silicon and manganese before the carbon.

Figure 8.1 shows that the carbon oxidation is rapid only when the silicon and manganese concentrations are less than about 0·5 per cent, but Table 8·1 shows that K_C is almost independent of the temperature whereas K_{Si} is markedly temperature dependent and, therefore, if the charge metal is hot or the temperature rises too rapidly, the equilibrium can be displaced in favour of the carbon reaction. Conversely, if the temperature rise is too slow the metal becomes pasty, for the melting temperature of the metal increases as the impurities are oxidised out. The temperature rise is regulated by making suitable additions to the bath and it is stated [9] that in one practice the temperature of a 20-ton heat is lowered 10°C by the addition of 16·5 lb of steel scrap, which absorbs heat as it melts, or it is raised a similar amount by 2·2 lb of 45 per cent ferro-silicon which generates heat by the exothermic oxidation of the silicon.

Metal Composition at the end of the Blow

Assuming that equilibrium is reached, the concentration of each element in the metal at the end of the blow can be determined from the equilibrium constants and the appropriate values of the activity coefficients in this temperature range. The concentrations of silicon and manganese remaining in the metal are very low and it is probable that $[f_{Mn}]$ and $[f_{Si}]$ are very close to unity, so that the residual concentrations of these elements can be calculated from the oxygen activity in equilibrium with the carbon content at the end of the blow and the appropriate values for $K_{Si} \cdot (\gamma_{SiO_2})$ and $K_{Mn} \cdot (\gamma_{MnO})$. Since the final slag is saturated with silica, (γ_{SiO_2}) is approximately 1·8, whilst (γ_{MnO}) is about 0·5. Using these values and those for the equilibrium constants at 1,600°C from Table 8.1, the residual manganese content should be in the range 0·03 to 0·05 weight-per cent, depending on the mole fraction of manganese oxide in the slag, and the residual silicon content should be less than 0·02 weight-per cent. These values agree closely with the concentrations which are found in practice, and it appears that equilibrium within the metal phase is approached very closely at the end of the blow.

CONTROL OF THE BLOW

During the first 4 to 6 minutes of the blow, whilst only silicon and manganese are being oxidised, a short, reddish-white flame appears at the converter mouth, giving way to a white flame as the carbon starts to oxidise. The flame increases in intensity until, after about 12 minutes on blast it drops quite suddenly, reappears for a short while and is then replaced by a brown fume.

Provided that the composition of the iron charged to the converter remains constant, the time at which the carbon flame appears provides a good indication of the temperature of the bath and steps can then be taken to control the final temperature. The two points at which the flame drops correspond closely to carbon contents in the metal of 0·12 and 0·05 per cent respectively [10] and visual observation of the flame thus shows the end of the carbon reaction and indicates when the converter should be turned down to halt the refining. However, the flame does not entirely disappear and considerable skill is needed to detect the change in its appearance. Further blowing beyond this change point results only in metal loss, since iron alone is now being oxidised, but if the converter is turned down before the end-point is reached the carbon content of the metal is uncertain and it is probable that the metal will not satisfy the specification required. Recently instruments have been devised to assist in the control of the temperature and the determination of the end-point; these are described briefly in a later section (see p. 229).

THE CONVERTER SLAG

The slag which forms during the first minute or so of the blow consists of an almost pure mixture of ferrous oxide and silica in the ratio of 2 parts to 1, and is entirely molten at about 1,200°C. As the silicon is oxidised the slag composition changes rapidly towards saturation with silica, whilst the iron oxide content falls by dilution. In the temperature range 1,300–1,600°C the slag is saturated when the silica content is about 50 per cent and, if the ratio of Si/Mn in the charge is greater than 2/1, the slag is saturated before the start of the carbon blow. The slag then becomes very viscous and is termed a *dry*, or *cinder*, *slag*. At this stage the composition is usually in the following range:

	SiO_2	FeO	MnO	Al_2O_3
Weight-per cent	56/62	16/22	14/18	2/4

Because of the high viscosity practically no reaction occurs between the slag and metal, and the slag serves only as a receiver for the products of the oxidation processes, taking no direct part in the refining reactions other than fixing the activities of the oxidation products. In consequence the slag composition remains virtually unchanged during the carbon blow, but if the metal is over-blown the rapid generation of iron oxides fluxes the silica and the slag again becomes fluid.

If the Mn/Si ratio in the charge metal is greater than 1/2 the slag may remain fluid throughout the blow for the following reasons:

1. The solubility of silica in the slag increases very slightly with increasing MnO content, but the effect of this increase is of relatively minor importance.
2. The ratio of MnO/SiO_2 entering the slag from the metal increases in proportion to the ratio of Mn/Si in the metal. Silica is also absorbed slowly into the slag from the converter lining, but the saturation limit may not be reached before the end of the blow.

Further, manganese oxide decreases the surface tension of a silica-rich melt [11] and thereby increases the fluidity of an unsaturated slag. The oxygen content of the metal rises rapidly at the end of the carbon blow and is about 0·10 per cent when the converter is turned down under a dry slag, but a fluid slag reacts with the metal when the blast is terminated and the final oxygen content of the metal is lower. Thus with an initial Mn/Si ratio of 0·6 the oxygen content falls to about 0·08 per cent after the blast is turned off [10].

DE-OXIDATION

Before the metal is poured, its oxygen content is adjusted to the level required for the particular class of ingot which is to be produced (i.e. killed, semi-killed or rimming), and the concentrations of silicon, manganese, carbon and other elements are increased by suitable additions to bring the composition within the limits of the specification.

When the converter is turned down the activity of the oxygen in the metal is very close to equilibrium with the residual silicon, manganese and carbon concentrations. Hence, when the concentration of any one of these elements is raised, its activity in the metal is also raised and the oxygen concentration must be decreased to lower $[a_O]$ and restore the equilibrium. The reactions involved are identical with those occurring during the oxidation of the solutes and the residual oxygen concentration in equilibrium with any chosen concentration of the deoxidising elements can be calculated in a similar manner. Conversely, it is possible to calculate the concentration of a deoxidising element which is necessary to obtain a specific concentration of oxygen in the metal and the procedure is outlined in Chapter 13.5.

Before the deoxidisers are added the oxygen activity in the metal is often slightly greater than that in equilibrium with the activity of the ferrous oxide in the slag but, when they are added and $[a_O]$ is lowered, there is a driving force for the transfer of oxygen atoms from the slag into the metal. With a dry-slag practice the rate of oxygen transfer into the metal is negligible, the slag becomes depleted in oxygen at the interface, but the high viscosity retards the replenishment of oxygen in the depleted layer from the bulk slag, and very little of the deoxidising addition is lost by the reaction with the slag.

When the slag is fluid the rate of oxygen transfer is more rapid and appreciable amounts of oxygen are transferred across the interface whilst the slag and metal are in contact. The extent of the transfer is largely determined by the following factors:

1. The FeO content of the slag may vary over a fairly wide range, depending on the carbon content at which the converter is turned down.
2. At constant silica concentration (a_{FeO}) increases with decreasing basic oxide content [7, 12].
3. For an acid slag of constant composition both (a_{FeO}) and the fluidity of the slag are increased as the temperature is raised.

Because these conditions are not exactly reproducible from cast to cast, the recovery of the deoxidising elements is less certain than in the dry slag practice.

The transfer of oxygen from the slag can be prevented by completely removing the slag from the metal before the deoxidisers are added and this can be accomplished fairly readily with a dry slag. The separation of a fluid slag is more difficult, for it inevitably results in some metal being poured away with the slag, and it is safer to decant most of the slag and thicken the remainder with additions of silica sand prior to deoxidation. Alternatively, in the production of medium-carbon steels the blow can be 'caught' at the required carbon content and before the rapid generation of iron oxides, when both the oxygen content in the metal and the slag FeO content are very low. The amount of deoxidation required is correspondingly less, but there is increased probability that the carbon content of the metal will not meet the specification.

The situation is more complex in the production of rimming steels, where an oxygen content in the metal of about 0·10 per cent is required to give a good rimming action. With a dry slag the oxygen at the normal end-point corresponds closely with this level, but it is too low with a fluid slag. In the latter case it is possible to decant most of the slag at the normal end-point and then to continue the blow for a few seconds to increase the slag FeO content [10], but it is difficult by this method to obtain consistency in the final oxygen concentration and it is obviously preferable to charge metal with a high Si/Mn ratio for the production of rimming heats.

8.3 The Basic Bessemer Process

COMPOSITION OF THE CHARGE

The acid converter practice is very limited in the range of metal compositions which can be refined since neither phosphorus nor sulphur can be removed from the metal. In contact with oxygen, phosphorus in the metal is oxidised readily to form phosphorus pentoxide (P_2O_5), but this compound is unstable in the presence of carbon, silicon, manganese or other more powerful oxide-forming elements. The stability is raised if the activity of the P_2O_5 is lowered, but it is insoluble in the metal, its activity is high in a silica-rich slag and is lowered markedly only in a very basic slag. This means that, in contrast to the acid process, the slag is required to take an active part in the basic refining reactions and it must therefore be fluid if reasonable reaction speeds are to be attained. Thus the optimum conditions obtain for the removal of phosphorus when the slag in contact with the metal contains only sufficient acidic oxides to give a fluid melt at the operating temperature.

The removal of sulphur from the metal is also improved as the basicity of the slag is increased but, in contrast to the elimination of phosphorus, sulphur is removed to any great extent only under a reducing atmosphere. The oxygen potential in the Bessemer process is thus unfavourable for the sulphur transfer reaction, but with suitable control of the conditions as much as 80 per cent of the sulphur can be removed from the metal under a basic slag.

With the development of the basic converter lining and the possibility of operating under a basic slag cover the range of iron compositions suitable for refining in the pneumatic process was vastly extended but, in order to obtain the necessary slag basicity, the concentration of acid oxides produced by oxidation from the metal must be low. This factor restricts the composition of the charge metal to low silicon contents. If high silicon irons, similar to those used in the acid practice, are refined in the basic converter, the slag volume required to achieve the necessary basicity is very large. The converter capacity must then be increased to accommodate the larger volume, more thermal energy is needed to fuse the slag and the metallic yield decreases, as the amount of iron lost by oxidation varies in relation to the total slag volume. The silicon content of the hot metal can be lowered by pre-refining the iron in an active mixer or in the ladle; alternatively the slag volume can be reduced by pouring-off most of the slag after the silicon has oxidised, but prior to the start of the carbon blow, and forming a new slag by further lime additions.

The manganese content of the charge metal can be the same as, or greater than, that in the acid process, for the oxidation product is a basic oxide which is not therefore detrimental. In fact, manganese oxide lowers the surface tension of the slag and slightly increases the solubility limit, so that more lime can be incorporated in a fluid slag at any given temperature.

The deficiency in heat generated owing to the lower silicon content is compensated by the oxidation of phosphorus, which is strongly exothermic:

$$2P_{(\%\,in\,Fe)} + 2\tfrac{1}{2}O_{2(g)} \rightleftharpoons P_2O_{5(gas)};\ \Delta H_P = -290 \text{ kcal mol}^{-1} \qquad 8.9$$

With 1·0–1·5 per cent of phosphorus in the metal the heat available from the oxidation of the impurities is similar to that in the acid practice, but additional heat is required in basic practice to melt the lime-rich slag and this is obtained by increasing the phosphorus to between 1·5 and 2·5 per cent.

A typical basic iron composition is

	Carbon	*Silicon*	*Sulphur*	*Phosphorus*	*Manganese*
Weight-per cent	3·5	0·5	0·05	2·0	1·0

Lime is charged to form the basic slag before the converter is turned up to start the blow, the quantity varying according to the total amount of acid oxides which will be formed and is usually in the range from 3 to 5 cwt. of lime per ton of hot metal.

THE SEQUENCE OF THE REACTIONS

The change in the composition of the metal as a function of the blowing time is shown in Figure 8.2. As in the acid process, silicon and manganese

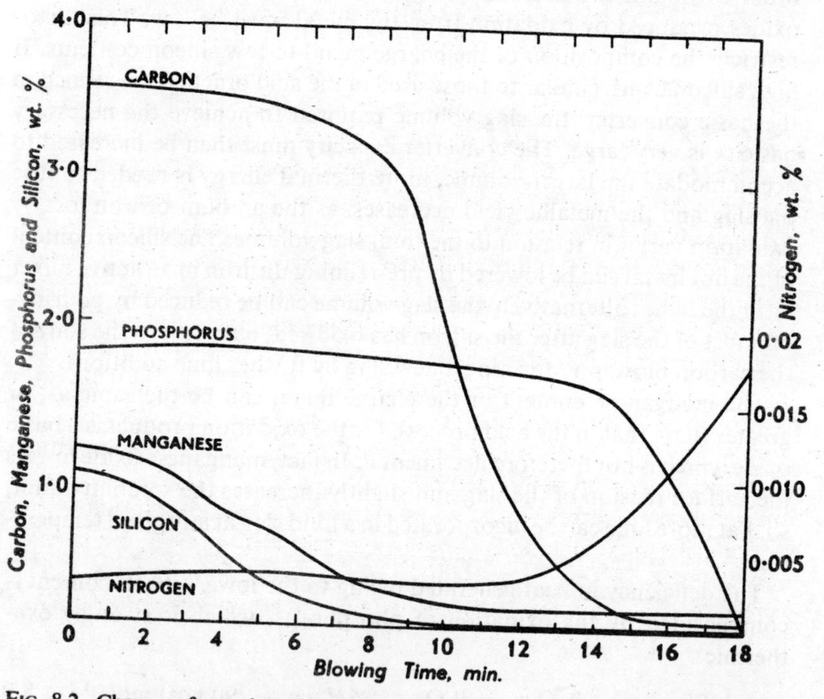

FIG. 8.2. Change of metal composition with time in the basic Bessemer converter process.

start to oxidise immediately after the converter is turned up, but the silicon is removed more rapidly and to a lower level, whilst the manganese is oxidised less rapidly and to a smaller extent, the residual content being usually in the range from 0·05 to 0·15 per cent depending on the initial manganese concentration and the tapping temperature. The carbon flame appears after about 4 minutes and drops at about 15 minutes, but at this stage the phosphorus content is only a little below the initial level and does not fall rapidly until the carbon is almost completely oxidised. The period during which the phosphorus is removed is called the *after-blow* and is characterised by the appearance of a brown fume with only a very short reddish flame at the converter mouth. The duration of the after-blow is usually between 2 and 2½ minutes, depending on the phosphorus content of the charge metal.

Thermodynamic Treatment of the Reactions

The thermodynamic reasoning which has been adopted to describe the sequence of oxidation of the elements in the acid process can be used to determine the equilibrium reaction sequence in the basic process. The values of the equilibrium constants for the reactions which are listed in Table 8.1 apply also to the basic process at the same temperatures, but the equilibrium oxygen activities are changed because the activities of the elements in the slag and metal phases are different This is demonstrated by calculating the value of $[a_O]$ in equilibrium with each element at 1,300°C for a metal of the composition which is given above.

Using the data in Chapter 13 and in Appendix I, the estimated values for $[f_{Si}]$, $[f_{Mn}]$ and $[f_P]$ in the metal are 7·2, 0·95 and 4·3 respectively. The activity coefficient of carbon in a pure iron–carbon alloy at 1,300°C is again determined by an extrapolation of the higher temperature data and, combining this result with the interaction coefficients for the effect of solute atoms on $[f_C]$ at 1,600°C, the value of 6·6 is calculated for $[f_C]$. The effect of oxygen and other trace elements has not been taken into account in the evaluation of the activity coefficients but, as indicated earlier, this omission should not give rise to serious errors in the interpretation of the reaction sequence.

The slag which is formed initially is essentially a ferrous silicate in which the activity of the silica is about 0·5 and, if 0·01 mole fraction of manganese oxide is dissolved in this slag, (a_{MnO}) is about 0·05. The activity of the phosphorus pentoxide in this slag is unknown, but the basicity is low and $(\gamma_{P_2O_5})$ is high; hence a value of unity is assumed initially. The partial pressure of the carbon monoxide is again taken to be one atmosphere.

The remaining quantity which is required is the equilibrium constant for the oxidation of phosphorus. For the reaction:

$$2P_{(\%\text{ in Fe})} + 5O_{(\%\text{ in Fe})} \rightleftharpoons P_2O_{5(l)}$$

the equilibrium constant is given by:

$$K_P = \frac{(a_{P_2O_5})}{[a_P]^2\,[a_O]^5};\ \log K_P = \frac{35{,}880}{T} - 30{\cdot}31 \qquad 8.10$$

and at 1,300°C, K_P is $3{\cdot}2 \times 10^{-8}$.

With these data the activity of the oxygen in the metal can be calculated for equilibrium with each of the elements at their initial concentrations and at a temperature of 1,300°C. The values obtained are as follow:

Reaction	$Si–SiO_2$	$Mn–MnO$	$C–CO$	$P–P_2O_5$
$[a_O]_{(1,300°C)}$	$5{\cdot}6 \times 10^{-5}$	$1{\cdot}4 \times 10^{-4}$	$8{\cdot}2 \times 10^{-5}$	$0{\cdot}09$

According to this calculation the silicon, manganese and carbon should start to oxidise more or less simultaneously, but the phosphorus cannot be oxidised from the metal unless $(a_{P_2O_5})$ in the slag is very much lower than unity. The actual sequence of oxidations in Figure 8.2 is: manganese and silicon first, followed by the carbon and finally phosphorus. Carbon is the only element which is misplaced from the equilibrium sequence and this anomaly can be attributed again to the kinetic factors which require the metal to be super-saturated with oxygen and carbon before the start of the carbon reaction.

Lime, which is introduced with the charge, floats on the surface of the metal and dissolves slowly in the initial slag. As the temperature of the bath is increased by the oxidation of the solutes the rate of solution of the lime is also raised, but it is not completely dissolved until towards the end of the carbon blow. Consequently, the composition of the slag is changing rapidly during the removal of the carbon and is approaching, or may actually exceed, the lime saturation surface by the stage at which the carbon flame drops. This change in the composition of the slag has a marked effect on the equilibrium partitioning of the elements between the slag and the metal, as can be seen by comparison of the activity coefficients of the oxides in the slag in the acid and basic processes at the end of the carbon reaction. In the acid slag (γ_{SiO_2}) is high and (γ_{MnO}) is very low, but in the lime-saturated slag (γ_{SiO_2}) is less than 0·01 whereas (γ_{MnO}) is about 1.5. Thus at the same metal temperature and carbon content the residual silicon should be lower and the residual manganese should be higher in the basic process than in the acid practice. There may be a temporary reversion of manganese to the metal if the basicity increases rapidly while the carbon content is still high, for (γ_{MnO}) is raised while $[a_O]$ in the metal remains more or less constant, but $[a_O]$ is increased as the carbon is removed and the manganese in the metal is partly re-oxidised in the later stages of the carbon blow. These trends are commonly observed in practice.

PHOSPHORUS REMOVAL

During the after-blow the oxygen activity in the metal is determined by the phosphorus concentration. Since P_2O_5 is unstable and insoluble in molten iron the oxidation of the phosphorus occurs mainly at the slag-metal interface, where the oxidation product can be absorbed into the

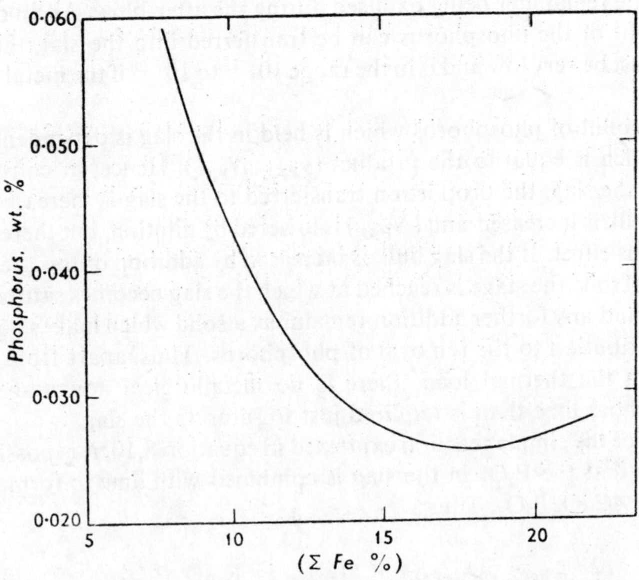

FIG. 8.3. Relation between the residual phosphorus content of the metal and the total iron content of the slag. (After Weber, Kootz and Obst [13].)

slag, and therefore it is expected that the phosphorus in the metal approaches equilibrium with the iron oxides in the slag. Figure 8.3 shows a typical relation which has been obtained in practice using a lime-saturated slag [13]; when the slag is not saturated the position of the curve varies with the amount of lime in the charge. The results of numerous investigations show that equilibrium is not fully attained as (a_{FeO}) in the slag is always greater than the value in equilibrium with the metal, and this is interpreted as evidence that the rate at which iron oxides are formed and enter the slag is greater than the rate at which $[a_O]$ changes with the residual phosphorus content of the metal. However, there is strong evidence to suggest that at any stage during the after-blow the manganese, silicon and phosphorus reactions are at equilibrium with respect to each other and, therefore, the residual manganese and silicon contents are determined by the oxygen activity in equilibrium with the residual phosphorus.

The activity coefficient of the phosphorus pentoxide is lowered rapidly as the slag approaches saturation with lime, and about 10 per cent of the phosphorus is transferred to the slag by the time the carbon flame drops, most of the remainder being oxidised during the after-blow. As much as 98 per cent of the phosphorus can be transferred into the slag; hence $(\gamma_{P_2O_5})$ must be very low and is in the range 10^{-15} to 10^{-20} if the metal is at 1,600°C.

The amount of phosphorus which is held in the slag is dependent on $(a_{P_2O_5})$ which is equal to the product $(\gamma_{P_2O_5} . N_{P_2O_5})$. Hence, at constant $(\gamma_{P_2O_5})$ in the slag, the proportion transferred to the slag is increased as the slag bulk is increased and $(N_{P_2O_5})$ is lowered by dilution, but there is a limit to this effect. If the slag bulk is increased by addition of too great a quantity of lime the stage is reached at which the slag becomes saturated with lime and any further addition remains as a solid which makes negligible contribution to the removal of phosphorus. Thus, apart from the increase in the thermal load, there is no metallurgical advantage in charging more lime than is required just to saturate the slag.

Instead of the simple reaction expressed in equation 8.10, it is possible to assume that the P_2O_5 in the slag is combined with lime to form the compound $4CaO.P_2O_5$. Thus:

$$2P_{(\text{in Fe})} + 5O_{(\text{in Fe})} + 4CaO_{(\text{slag})} \rightleftharpoons 4CaO.P_2O_5 \qquad 8.11$$

If the slag is saturated with CaO and with $4CaO.P_2O_5$ the activities of both of these oxides are unity, so that at equilibrium $[a_P]$ in the metal is dependent only on $[a_O]$ and we have:

$$K'_P = \frac{1}{[\%P]^2 [\%O]^5} \qquad 8.12$$

where K' signifies that the 'apparent' reaction constant is not the true thermodynamic constant, but is dependent on the variation of $[a_P]$ and $[a_O]$ with the concentration of these elements. In two separate investigations it has been shown that K'_P is almost independent of composition but is markedly temperature dependent [14, 15], (Figure 8.4) and that the residual phosphorus in the metal is increased as the temperature is raised. This increase is shown by the following figures [15]:

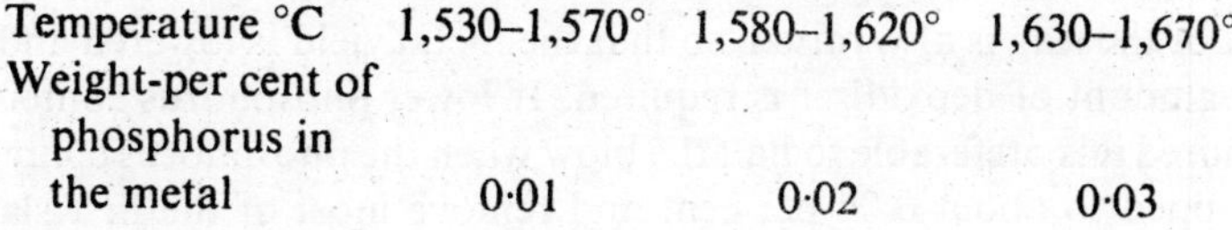

Temperature °C	1,530–1,570°	1,580–1,620°	1,630–1,670°
Weight-per cent of phosphorus in the metal	0·01	0·02	0·03

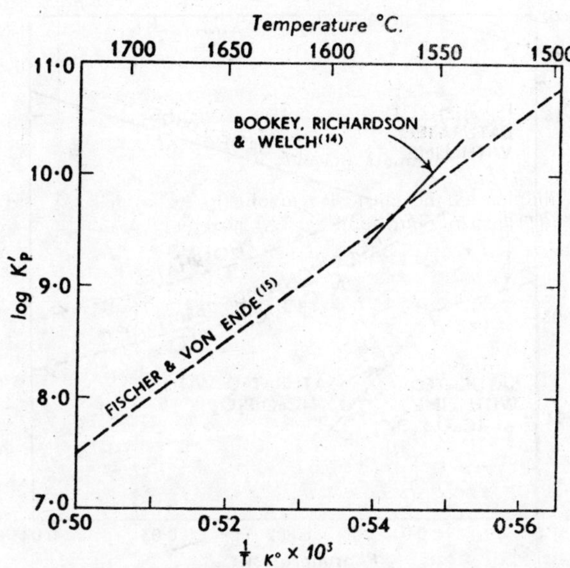

FIG. 8.4. Variation with temperature of the apparent equilibrium constant for phosphorus oxidation in slags saturated with lime and calcium phosphate.

A similar change in the phosphorus content with temperature should apply to slags which are not saturated, for the equilibrium constant for reaction 8.10 decreases to $7{\cdot}1 \times 10^{-12}$ at 1,600°C. The residual phosphorus in the metal is lower when the slag is saturated with phosphate only than when it is saturated with lime or with lime and the phosphate (Figure 8.5) [16].

If the blow is halted when the phosphorus in the metal is about 0·04 per cent the composition of the slag is usually within the following range:

	FeO	CaO	MgO	MnO	P_2O_5	SiO_2
Weight-per cent	13–20	51–56	0–4	2–5	16–22	5–8

and the slag is saturated with 4 $CaO.P_2O_5$ when the P_2O_5 content is greater than approximately 18 per cent [17]. If the blow is continued beyond this point the oxygen in the metal increases rapidly and the FeO content of the slag is also raised, so that the metal yield is lowered and a greater amount of deoxidiser is required. If lower phosphorus contents are required it is preferable to halt the blow when the phosphorus content has dropped to about 0·04 per cent and remove most of the first slag,

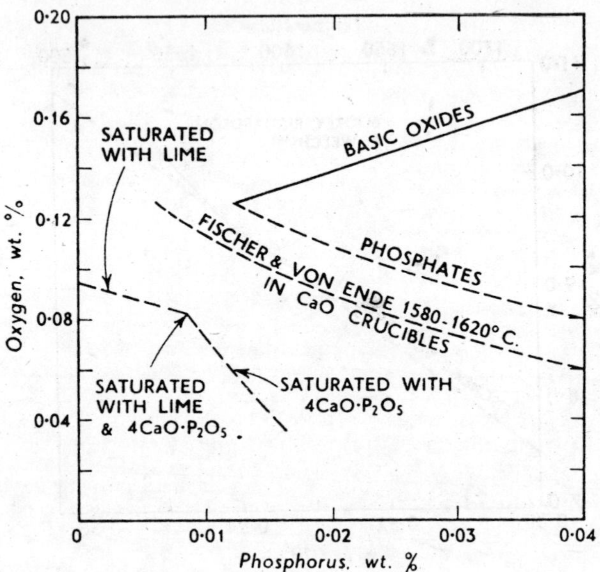

FIG. 8.5. Relation between the concentrations of phosphorus and oxygen in the metal for equilibrium at 1,550° C to 1,700° C with saturated and under-saturated slags. (After Tromel and Fritze [16].)

after which lime is added, together with a small quantity of flux to hasten the formation of a new slag, and the blow is continued. Most of the phosphorus is removed with the first slag and the residual phosphorus is lower when the second slag is saturated with phosphate than is the case in the one-slag practice, but the total heat time is raised and the production rate is lowered correspondingly.

SULPHUR REMOVAL

Some sulphur is transferred from the metal to the slag during the course of the blow, but the Bessemer slag is strongly oxidising and, although the high temperature allows the formation of a slag which is more basic than that which can be produced in the blast furnace, the ratio of sulphur in the slag to sulphur in the metal, (%S)/[%S], is rarely greater than 10·0. Thus, if the weight of slag is one-tenth of the weight of the metal, not more than 50 per cent of the sulphur is transferred to the slag. Actually the amount transferred is greater than this towards the end of the carbon blow, when the slag approaches saturation with lime, but reversion of sulphur into the metal occurs as the phosphorus is oxidised and $[a_O]$ in the metal is increased, the extent of the reversion increasing rapidly towards the end of the after-blow, when the oxygen in the metal increases rapidly (Figure 8.5). A lower residual sulphur content is obtained with the two-slag process when the first slag is removed whilst the oxygen content of the metal is still fairly low.

FINISHING THE HEAT

When the air blast is turned off at the end of the after-blow the metal is rich in oxygen and this content must be lowered to the desired level with deoxidisers before the metal is cast, but the slag is fluid and reacts with the deoxidants across the slag–metal interface. Also, in the one-slag process, the phosphorus in the slag is only just stable and can readily revert to the metal if the oxygen potential of the slag–metal system is lowered. Therefore, the slag is removed as completely as possible and lime is added before the deoxidisers to enrich the remaining slag; alternatively the deoxidants are added to the metal in the transfer ladle when there is only a thin covering of slag. It is, however, difficult to estimate the oxygen content of the metal at the end of the blow, and the amount and composition of the slag which remains on the metal when the deoxidants are added vary from blow to blow. In consequence the extent of the deoxidation and the recovery of the deoxidisers in the metal can be predicted with less certainty than when the slag is stiff, as in the acid process, or when there is time for a rapid slag and metal analysis, as in the open-hearth process.

8.4 Nitrogen in Bessemer Steels

With a normal air blast the air entering the molten metal through the tuyeres contains 79 per cent by volume of nitrogen. On contact with the metal the oxygen reacts to form oxides and, except during the carbon reaction, the gas passing through the metal is almost pure nitrogen. During the carbon reaction the carbon is removed as a gas and the partial pressure of nitrogen is decreased in proportion to the rate of carbon monoxide evolution, but it rises again if the blow is continued after the carbon is oxidised and during the after-blow in the basic process.

The molecules of nitrogen in the gas dissociate and the atoms formed can dissolve in the metal until the solubility limit of nitrogen is reached at the particular temperature and partial pressure of nitrogen. The solubility is given by the Sievert's equation:

$$\%N = \kappa\sqrt{p_{N_2}} \qquad 8.13$$

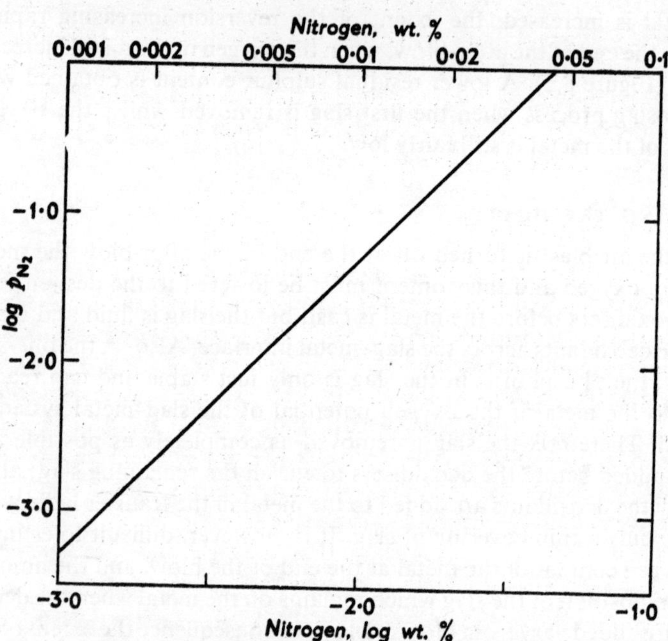

FIG. 8.6. Relation between the solubility of nitrogen in pure liquid iron and the partial pressure of nitrogen gas at 1,600°C.

where κ is a constant which is equal to 0·043 at 1,600°C for pure iron in equilibrium with nitrogen at a pressure of one atmosphere (Figure 8.6), but it increases by approximately 13 per cent for each 100°C increase in temperature [18] and the solubility is markedly temperature dependent. Even at these moderately low concentrations nitrogen in iron does not conform strictly to Sievert's law and the solubility shows a positive deviation from that calculated by equation 8.13 [19], but the deviation is small and can be neglected for the pressure range with which we are concerned. The solubility of nitrogen is also dependent on the composition of the metal, for elements such as carbon, phosphorus, silicon and sulphur in solution in the iron raise $[f_N]$ and decrease the solubility, whereas manganese has the opposite effect. Thus at the end of the carbon reaction the high phosphorus content of the metal in a basic heat should result in a lower nitrogen content than that in an acid heat at the same temperature.

The variation in the nitrogen content of the metal in the course of the basic process is shown in Figure 8.2. During the silicon and manganese oxidation period the gas passing through the bath is almost pure nitrogen, but the low metal temperature and the low activity of the nitrogen in the presence of the carbon and phosphorus in the metal hold the nitrogen solubility at a low level. Towards the end of this period the metal temperature is rising steadily and continues to rise during the carbon blow. As the carbon is oxidised the solubility of the nitrogen increases slowly and may in fact decrease slightly, for the partial pressure of nitrogen is considerably lower than unity owing to dilution with large volumes of carbon monoxide, but the boil becomes less vigorous when the carbon content is about one per cent. Thereafter the carbon monoxide content of the gas decreases rapidly, the partial pressure of the nitrogen increases correspondingly and the concentration of nitrogen in the metal rises rapidly during the final seconds of the carbon oxidation. During the after-blow the partial pressure of nitrogen in the gas is approximately unity and the solubility of nitrogen is raised both by the rise in the temperature of the metal and by the decrease in $[f_N]$ as phosphorus is eliminated from the metal. Thus the driving force for nitrogen solution is increased and at the end of the after-blow the nitrogen content of the metal is 0·015 to 0·025 weight-per cent, the lower limit being approached more closely as the duration of the after-blow is shortened and as the superheating of the metal is diminished.

The nitrogen content in the acid process follows a similar sequence but, compared with the basic process, the solubility of nitrogen is greater at the start of the carbon blow because the metal temperature is higher and the phosphorus content is very much smaller. At the end of the carbon blow the nitrogen content of the metal is again slightly higher and is rising rapidly, but the converter is turned down immediately the carbon flame drops and the final nitrogen content of acid Bessemer steels is only about 0·010 weight-per cent (Figure 8.1).

At the end of the blow in both the acid and basic practices the metal is approaching pure iron in composition and the partial pressure of nitrogen in the gas is close to one atmosphere. Whilst the solubility of nitrogen in the steel should be close to the value for pure iron at the same temperature, the actual values are very considerably lower. This may be due to the fact that the bubbles of gas are not in contact with the metal for a sufficient length of time to allow the equilibrium partitioning, but it has also been suggested that the solution of nitrogen is retarded by the presence of a film of iron oxide surrounding each bubble of nitrogen [20]. It is commonly found that the nitrogen content is greater with a cold heat, in which the silicon is oxidised early, than in a hot blow where silicon oxidation is retarded, and in the latter case the silicon dissolved in the metal would be expected to prevent the formation of oxide films on the bubbles.

If the blow is continued after the drop in the carbon flame the nitrogen content of the metal increases with time and approaches closer to the saturation limit. In both acid and basic practices the nitrogen increases by about 0·005 weight-per cent for each minute of extra blowing time after the carbon drop, so that over-blowing for only a few seconds causes a significant change in the nitrogen content. Thus, for example, the two-slag process increases the total time of the after-blow and the lower sulphur and phosphorus contents are obtained at the expense of a higher nitrogen concentration.

Nitrogen in the finished steels is undesirable, particularly in the low carbon grades of steel for which the orthodox Bessemer process is best suited for, on cooling to room temperature, the solubility of nitrogen in α-iron is exceeded. The solubility of nitrogen in pure iron decreases with temperature as follows [21]:

Temperature °C	400	300	200	100
Nitrogen weight-per cent	0·025	0·008	0·0009	0·0005

On cooling below about 300°C the Bessemer iron becomes supersaturated with nitrogen, which can give rise to the formation of a nitride precipitate with consequent embrittlement and loss of ductility on ageing. In comparison, the nitrogen content of open-hearth and electric-arc furnace steels is only 0·003–0·005 per cent and, whilst the solubility limit is exceeded again at low temperature, the degree of supersaturation is usually not sufficient to cause trouble in service.

The loss of ductility on ageing was one of the primary causes for the change from Bessemer steel-making to the open-hearth process in the early years of this century. Although the significance of the nitrogen content was recognised at that time [22], it is only during recent years that marked progress has been made in the control of the gas content of the metal, but it is now possible to produce steels in the modified converter processes with nitrogen contents on a par with, and even lower than, those produced in the open-hearth process.

CONTROL OF THE NITROGEN CONTENT

The nitrogen content of the Bessemer metal can be lowered by one or more of the following methods:

1. Maintaining the operating temperature at the minimum practicable level, consistent with the avoidance of excessive metal loss through the formation of skulls (i.e. metal which has solidified within the converter vessel). This also favours phosphorus removal.
2. Decreasing the transit time of the gas through the bath, thus lowering the gas–metal contact time.
3. Lowering the partial pressure of nitrogen in the gas passing through the metal.
4. Reducing the duration of the blow, thereby decreasing the total volume of air blown.

The contact time can be reduced within limits by the use of a wider and shallower metal bath. The ferrostatic pressure of the molten metal is then decreased, thus lowering slightly the mean pressure of the gas passing through the bath and contributing to the reduction in the dissolved nitrogen content. With a 25-ton charge and a new lining the bath depth is usually about 30 inches and this decreases to about 25 inches as the lining is eroded, but the converter can be re-designed to give an initial metal depth of 24 inches, falling to about 20 inches before re-lining. If the depth is decreased further the loss of metal through 'slopping' or splashing is increased, some oxygen is left in the blast, the thermal loss to the lining is excessive and the heat reserve in the normal charge is inadequate to reach the casting temperature [23].

Additions of iron oxide to the slag during the carbon blow absorb heat and lower the bath temperature, whilst the oxygen made available for the oxidation reactions by the iron oxides lowers the quantity of air required. Consequently the total blowing time can be decreased by 20 per cent or more. Thus ore is added towards the end of the carbon blow in the German HPN process and mill scale (Fe_3O_4) is added when the carbon flame appears in the Corby LPN practice [23]. These methods, together with close control of the finishing temperature, are capable of reducing the nitrogen content of the metal by as much as 60 per cent.

Somewhat similar conditions are obtained by the addition of limestone to the slag during the carbon blow [24, 25]. The endothermic decomposition and melting of the lime cools the bath, while part of the carbon dioxide evolved from the limestone reacts with the iron to form FeO and carbon monoxide and decreases the amount of oxygen which must be supplied in the air blast.

The final nitrogen content is critically dependent on the time of adding the cooling agents. If the additions are made too early in the carbon blow the drop in temperature checks the rate of decarburisation, whereas additions made at the carbon-flame drop have insufficient time to react with the bath before the rapid increase in nitrogen content occurs in the after-blow. The optimum time appears to be when the carbon blow is 70–80 per cent complete [25], depending on the nature of the additions.

8.5 The Use of Oxygen in the Converter

The nitrogen content of the metal remains fairly low until towards the end of the carbon blow, when the decreasing vigour of the boil, coupled with the high temperatures then prevailing, gives conditions which are ideal for nitrogen absorption. If at this stage the nitrogen content of the blast is reduced the rate of nitrogen absorption by the metal will also decrease. Although the solubility limit of nitrogen is not reached, some indication of the effect of lowering the partial pressure of the nitrogen can be obtained from equation 8.13 and Figure 8.6. Thus, with nitrogen partial pressures of 0·5 and 0·05 atmospheres the nitrogen solubility in pure iron at 1,600°C is 0·030 and 0·009 weight-per cent respectively.

One method of lowering the nitrogen partial pressure is to increase the oxygen concentration in the blast, so reducing the total volume of blast required to oxidise the phosphorus during the after-blow or, with the same rate of blowing, curtailing the duration of the after-blow. However, the extent to which the oxygen content of the air can be enriched is limited, for the amount of heat liberated by oxidation of the iron immediately in front of the tuyeres increases in proportion to the degree of oxygen enrichment, which causes more rapid erosion and a shorter useful life of the bottom plug. The rate of refractory wear becomes excessive with greater than 40 per cent of oxygen in the blast.

The increase in heat available from the enriched blast is due primarily to the reduction in the amount of sensible heat transferred to the nitrogen in the gas, so that the metal temperature increases proportionately with the oxygen content of the blast. This increases the solubility of nitrogen and tends to counteract the reduction in its total volume, but the metal temperature can be held at the normal level by utilising the additional heat available from the blast to extend the range of irons suitable for the converter process to compositions with lower phosphorus contents. Alternatively, it is possible to melt increased quantities of steel scrap, which is a less costly raw material than blast furnace metal, and its use partly offsets the increased operating costs resulting from the use of oxygen. Thus, if scrap equal to 5 per cent of the charge weight can be melted with a normal air blow, the quantity can be increased to 15 per cent with 30 per cent of oxygen in the blast. By the addition of iron oxides or limestone, just prior to the commencement of the oxygen-enriched blast, the final nitrogen content in the metal is lowered to about 0·004 per cent [26–31] which is comparable with open-hearth steels. In addition, oxygen enrichment promotes the removal of sulphur and phosphorus, for the slag temperature is higher and the iron oxide content is increased. Thus the lime dissolves more rapidly, a higher basicity is obtained earlier in the heat and, in consequence, the final phosphorus and sulphur contents are lowered to about 0·025 per cent [27, 29].

The short plug life obtained with an oxygen-enriched blast is serious, for the refractory cost per ton of steel is greater and the process is halted more frequently to allow replacement of the bottom section, resulting in a loss of output. Nevertheless the production of low nitrogen steels by the Bessemer process demands the use of the maximum practical oxygen enrichment.

The operating conditions are markedly improved if a gaseous compound containing oxygen is added to the blast which decomposes or dissociates endothermically at the operating temperature, but does not contaminate the metal. The most obvious low-cost additives are steam and carbon dioxide for, in contact with the molten metal, these gases dissociate into hydrogen and oxygen, or carbon monoxide and oxygen, with considerable heat absorption. With a blast containing O_2/CO_2 or O_2/H_2O in the ratio of 1/1 by volume, approximately the same amount of heat is generated in front of the tuyeres as with a normal air blast, so that the plug life is normal and no additional coolants are required [26, 31–38]. If, however, steam is used, it must be superheated to avoid condensation of water in the tuyeres and hydration of the dolomite refractory.

As with straight oxygen enrichment, the enriched blast is substituted towards the end of the carbon blow, but the nitrogen content of the blast is only that contained in the oxygen, and when commercial purity oxygen is used the final nitrogen content is about 0·002 weight-per cent. A higher finishing temperature can be used without risk of nitrogen pick-up, the carbon flame is more intense, particularly with O_2/CO_2 mixtures, and the end-point is more clearly defined. The increased flame intensity is due to the combustion of larger amounts of carbon monoxide or of hydrogen at the converter mouth and loss of metal by skull formation is lowered in consequence of the increased temperature above the slag.

Hydrogen is soluble in iron and steel and the use of hydrogen–steam mixtures to reduce the nitrogen content could give rise to a loss of ductility and other types of defect resulting from a high hydrogen content.

Fortunately, this effect is not usually serious here, for the Bessemer is used largely for the production of low-carbon grades of steel which are subsequently fabricated into thin-gauge products such as sheet and strip. The diffusivity of hydrogen in solid iron is high, relative to that of the other solute elements; hence hydrogen can diffuse out of the metal during the hot-working operations to reach a value in equilibrium with the atmosphere and it is usually found that the ultimate hydrogen content of the fabricated metal is only slightly greater than that of metal produced with the conventional air blast. The residual hydrogen content has greater significance if the converter is used for the production of either steels of low alloy content or plain carbon steel products of heavier cross section. In these cases the hydrogen content can be lowered by following the O_2/H_2O blow by a short after-blow with a hydrogen-free gas.

THE LD PROCESS

An entirely different technique for refining with gaseous oxygen in the converter has been developed in Austria during the past decade. This is known as the LD (Linz-Donawitz or Linz-Düsenverfahren) process. The refining is carried out in a pear-shaped vessel which is similar to the normal converter but with a solid bottom. Commercial-purity oxygen is blown through a water-cooled copper nozzle, mounted vertically above the surface of the slag, with sufficient force to penetrate the slag and reach the metal surface, so creating the necessary turbulence to propagate the slag-metal reactions [39–46]. The agitation of the metal is less violent, however, than in the bottom-blown practices and this is reflected in the fact that the refining time is approximately doubled. The blow can be halted at any stage and metal samples taken for rapid analysis checks without turning the converter down to prevent the metal running back into the tuyeres. In this manner some measure of control can be exercised over the metal composition; a feature which is notably absent with the bottom-blowing.

The reaction sequence is changed with top-blowing as shown in Figure 8.7. Iron oxides are formed at the metal surface and oxygen atoms are transferred across the interface from the slag into the metal, where they are brought into contact with the solutes by the turbulence of the bath. Silicon and manganese are oxidised in the same way as in the orthodox process, but the start of the carbon reaction is retarded. In the bottom-blown practices carbon and oxygen combine to form molecules of carbon monoxide at the surface of the bubbles of gas which are rising through the metal from the tuyeres but, when the blast is introduced from above, it penetrates only a short distance into the metal. The formation of carbon monoxide is then impeded by the difficulty of nucleating a gas bubble within the melt (see Chapter 13.3) and the product $[a_C].[a_O]$ rises in excess of the value in equilibrium with carbon monoxide at one atmosphere pressure. Eventually, when a sufficient degree of supersaturation is attained, bubbles of carbon monoxide start to form at crevices in the refractory surface or in the vortices of the metal turbulence. Once bubble formation has started the oxidation of carbon is vigorous and proceeds at a rate which is not far short of that which is attained with

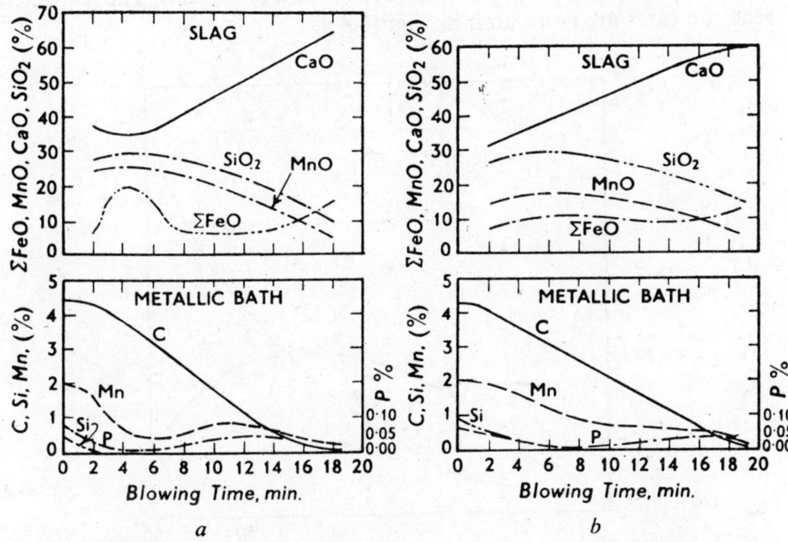

FIG. 8.7. Changes with time in the composition of the metal and the slag during refining in the LD converter process when the melt is cooled with (*a*) ore, and (*b*) metal scrap. (After Trenkler [44].)

bottom-blowing, but if the start of the reaction is delayed too long, as may happen when the crevices are covered over with a glazed layer of slag, the ensuing reaction is extremely violent, the metal is turned into a froth by the rapid formation of gas bubbles and some is ejected from the converter mouth.

The violent boiling action, when the carbon starts to oxidise, ejects metal droplets into and through the slag layer, where they are partly oxidised before returning to the metal. This increases the slag–metal contact area and oxygen is transferred to the metal at a greater rate than during the oxidation of manganese and silicon, whilst the iron oxide content of the slag is decreased and may fall below 10 per cent. The turbulence carries the oxygen-enriched metal away from the interface and downwards into the bath of molten metal, where it reacts with carbon and increases the vigour of the boil, but as the FeO content of the slag is decreased the rate of oxygen transfer to the metal falls back towards the initial level and the intensity of the boil is diminished. Thus the rate of carbon removal is initially low, passes through a maximum and decreases rapidly towards the end of the reaction. In contrast, with bottom-blowing the rate of carbon oxidation increases steadily from the start of the reaction up to a point which is close to the end of the boil. The respective reaction rates are compared in Figure 8.8.

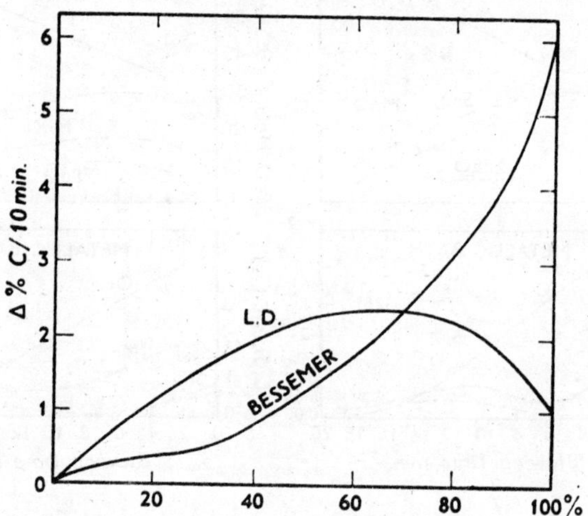

FIG. 8.8. Comparison of the rates of decarburisation in the bottom-blown Bessemer and LD converter processes. (After Matuschka [45].)

The reduction of the FeO content of the slag lowers the oxygen potential of the slag-metal system and there is a temporary reversion of manganese to the metal as the carbon content falls. The manganese peak is reached at about 0·5 to 1·0 per cent of carbon [44–48], and then, as the vigour of the carbon boil decreases, the FeO content of the slag is increased again and the manganese is re-oxidised to about the original level. This effect is shown clearly in Figure 8.7*a*. At the end of the carbon boil the slag contains from 14 to 18 per cent of FeO, which is somewhat higher than the FeO content at the same stage in the bottom-blown practices and reflects the difference in the rates of carbon elimination in the two processes. The increase is in part due also to the higher slag temperatures which are obtained in the LD process from the partial combustion of the carbon monoxide with the oxygen from the blast just above the slag surface and within the reaction vessel.

Compared with the conventional air-blown practice the LD process requires the release of only a small amount of thermal energy by the oxidation reactions, because the thermal loss in heating a large volume of nitrogen to the metal temperature is eliminated and also because additional heat from the combustion of the carbon monoxide is made available within the converter. As a result the converter process can be extended to refine iron of lower impurity content or, alternatively, fairly large quantities of iron ore or metal scrap can be added as coolants. In this respect the LD process resembles the bottom-blown practices with an oxygen-enriched blast, but rather more heat is available in the LD process and greater quantities of coolants can be added.

THE REMOVAL OF PHOSPHORUS IN TOP-BLOWING

The generation of intense heat where the blast impinges upon the metal and the rapid formation of iron oxides at the start of the blow give ideal conditions for the solution of lime in the slag and a highly-basic slag is formed almost from the start of refining. The activity coefficient of P_2O_5 in the slag is then low and significant amounts of phosphorus are found in the slag whilst manganese and silicon are still oxidising and before the start of the carbon boil. According to equation 8.10 and the equilibrium oxygen activities in the metal which are given on page 208 for a metal temperature of 1,300°C, $(\gamma_{P_2O_5})$ in the slag must be less than 10^{-10} for the phosphorus to start oxidising before the carbon, and this value becomes even smaller as the metal temperature increases. However, it is evident that the actual value of $(\gamma_{P_2O_5})$ is very much lower than this for by the time the carbon has decreased to 1·0 per cent the ratio (%P)/[%P] may be 1,000 times as great as at the same stage in bottom-blowing [38].

During the vigorous carbon boil the oxygen potential of the slag is lower and the slag temperature is higher than in the orthodox process, so the slag must be highly basic at this stage to avoid phosphorus reversion to the metal. A phosphorus content of 0·5 to 0·6 per cent in the charge metal can be lowered to about 0·04 per cent by the time the carbon has decreased to 1·0 per cent, but thereafter the phosphorus content changes very slowly with time and it appears that the slag is saturated with calcium phosphate. If this phosphorus content is suitable for the specification, the blow can be stopped before the carbon boil has ended to produce medium or high carbon steels without the uncertainty which is inherent when an after-blow and extensive recarburisation are required.

Further, the FeO content of the slag and the oxygen content of the metal increase rapidly at very low carbon contents and, therefore, when the blow is terminated with a moderate carbon content remaining in the metal, the metallic yield is improved and a smaller quantity of deoxidants is required.

Two-slag Process

Whilst the removal of phosphorus is facilitated by top-blowing, the amount which can be removed from the metal before the required carbon content is reached is limited by the time required to dissolve the necessary quantity of lime in the slag. In the normal practice it is possible to refine satisfactorily an iron containing up to about 0·5 per cent of phosphorus [49], although it is claimed [50] that larger amounts can be handled when a suitable flux such as bauxite is added to speed up the formation of the slag. The range of phosphorus contents can be extended up to about 2·0 per cent if the first slag is removed after blowing for about 8 or 10 minutes and a new slag is formed before continuing the blow. The first slag is removed when the carbon content has fallen to between 1·0 and 2·0 per cent and, starting with 1·8 per cent of phosphorus in the metal, the residual phosphorus concentration at this stage is 0·2 to 0·4 per cent [44].

The log of a typical heat has been given as follows: [41]

	Metal Composition weight-per cent			*Slag Composition weight-per cent*				
	C	Mn	Si	CaO	MgO	SiO_2	Fe	Mn
Charge	4·12	2·2	0·16					
After 8 minutes top-blowing	2·25	0·64		38·52	2·73	10·47	7·7	19·43
Slag removed; new slag formed with lime plus small additions of quartz and fluorspar								
After further blow for 10 minutes	0·54	0·39		48·20	3·08	11·32	9·8	10·37

The P_2O_5 content in the first slag can be as high as 16 to 20 per cent and is usually between 3 and 6 per cent in the second, but in the second slag the lime–silica ratio is greater and the slag is probably saturated with lime. The manganese concentration of the second slag is more than half that of the first, yet the amount of manganese oxidised into the first slag is approximately six times the amount oxidised in the second stage. The reason is that only a small amount of phosphorus remains to be removed in the second stage and hence the volume of slag is small when compared with the volume of the first slag.

At constant basicity, the amount of phosphorus which is removed under the first slag can be related to the residual carbon content as shown

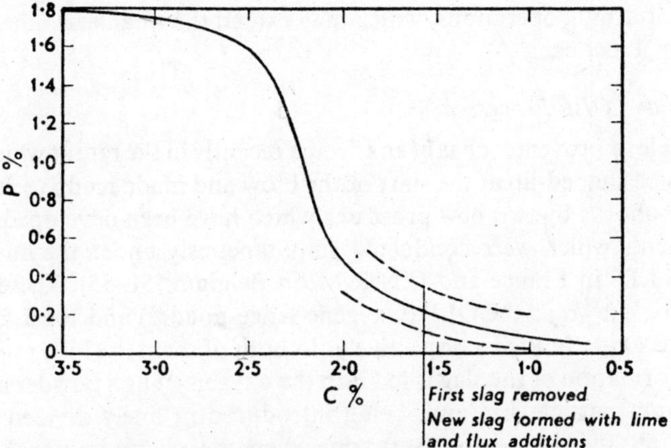

FIG. 8.9. Relation between the carbon and phosphorus contents of the metal for a two-slag process in the LD converter. (After Messin, Lambert and Helfstein [51].)

in Figure 8.9 [51]. An exact relation in terms of concentration units cannot be expected for at any time the concentrations are dependent on both the metal and slag activities and the reaction kinetics, but the diagram indicates that most of the phosphorus has oxidised by the time the carbon content has fallen to between 1·5 and 2·0 per cent. The final phosphorus content is then in the range 0·05 to 0·01 per cent. When the formation of the second slag is postponed until the carbon is below 1·0 per cent, there is little time available for the further removal of phosphorus before the carbon boil ceases, and there is only a small increase in the amount of phosphorus removed into the first slag. If the blow is continued after the drop in the carbon flame the rate of removal of phosphorus is very slow, but the ferrous oxide content of the slag and the oxygen in the metal increase rapidly with the result that the metal yield is lowered and larger quantities of deoxidisers are required.

The two-slag process involves a greater loss of iron and a lower metallic yield than when only one slag is used, but this can be improved by returning the second slag to the converter to form part of the first slag for the next heat. The phosphorus which is already present in this slag decreases the maximum amount which can be removed from the metal in the next heat, but this decrease is countered to some extent by the fact that the lime is already in solution and the first slag for the subsequent heat is formed more rapidly, so that a higher basicity can be attained before the slag-off. Economically the two-slag process is not very satisfactory, for the production rate is lowered by the additional slag removal and slag forming operations, which may extend the total heat time by as much as 50 per cent.

LD–AC and OLP Processes

Notable improvement has been effected recently in the rate at which the slag can be shaped-up at the start of the blow and made ready to receive the phosphorus by two new processes which have been developed from experiments which were conducted simultaneously under the auspices of I.R.S.I.D. in France and C.N.R.M. in Belgium [51–55]. These processes are known as the OLP (Oxygéne-lance-poudrè) and the LD–AC (*A*.R.B.E.D.–*C*.N.R.M.) respectively. In both of them the lime required for the formation of the slag is fed into the oxygen jet as a powder of less than 2 mm particle size and, being introduced in finely divided form directly into the high temperature zone where the oxygen impinges upon the metal, the lime passes rapidly into solution in the slag. The height of the lance is adjusted above the bath to ensure the initial formation of a frothy slag and to minimise ejections from the bath, but as the blow proceeds the lance is gradually lowered to the normal position for top-blowing. The phosphorus is oxidised almost from the start of refining and is removed more rapidly than in the L.D. process. After blowing for about 14 minutes the carbon is down to about 0·7 per cent and the phosphorus can be lowered from 1·8 per cent to less than 0·2 per cent, whilst the slag contains 20 to 27 per cent of P_2O_5 and 6 to 9 per cent of Fe [55]. The slag is removed at this stage and the blow is continued with further lime injection to the desired end-point, but the fresh slag is formed very rapidly and the second blow lasts for only 3 or 4 minutes. The rapid formation of the second slag allows the slag-off to be delayed to a lower carbon content in the metal than in the normal L.D. process, and both the additional blowing time and any thermal correction required can thus be estimated with greater accuracy.

This method of lime injection has proved successful also for refining iron containing up to 1·0 per cent of silicon under a basic slag [53]. In any converter process where the lime is added in lump form the first-formed iron silicate slag tends to form a froth and in extremes may pour out of the converter mouth, this tendency increasing with increasing silicon content of the metal. The frothing is due to a high surface tension in the slag which causes air bubbles to become enveloped in a thin film of liquid, but in the OLP and LD–AC processes the iron oxide content of the slag is very low during the early part of the blow and this fact, together with the rapid solution of the lime, effectively prevents the occurrence of excessive frothing. The remarks made earlier regarding the effect of silica on the slag volume and basicity are still applicable, however, and with only 0·2 per cent of phosphorus in the charge it is necessary to slag-off after the silicon oxidation period and re-form the slag.

The early formation of a highly basic slag which is replaced with a new slag at part of the way through the blow, when the FeO content is low, gives conditions which are more favourable to the removal of sulphur from the metal than in the bottom-blown practices. With the OLP and LD–AC practices a sulphur partitioning (%S)/[%S] of 14 to 18 has been realised and up to 70 per cent of the sulphur can be removed from the metal by the end of the refining [51, 54].

TOP-BLOWING WITH LOW-PRESSURE OXYGEN

Parallel with the development of the OLP and LD–AC practices attempts were being made to promote early slag formation by blowing oxygen on to the surface of the melt with insufficient force to penetrate the slag cover. The principle of this, the *buffer-slag* process, is that by flooding the surface of the slag with oxygen the ferrous oxide is converted into ferric oxide, which is a more powerful flux for the solution of the lime, and a basic, highly-oxidising slag is formed in about 4 minutes [56]. The phosphorus is oxidised by reaction at the slag-metal interface but the absence of gas bubbles in the melt retards the start of the carbon oxidation and the removal of the phosphorus precedes the carbon boil. However, the metal bath is not agitated until the carbon starts to oxidise; hence the transport of phosphorus from the bulk metal to the interface is slow and the refining time is considerably extended. It has been recorded [56] that after blowing for 25 minutes the phosphorus was lowered from 1·8 to 0·01 per cent, whilst the carbon was lowered from 3·5 to 0·9 per cent.

The reaction rate can be improved by a combination of the buffer slag and the normal, bottom-blown practice [56–58] in which a smaller volume of oxygen–carbon dioxide or oxygen–steam blast is blown through a restricted number of tuyeres in the hearth to supply part of the oxygen, but primarily to create the necessary turbulence, whilst the remainder of the oxygen is supplied in a low-pressure blast above the hearth.

One serious objection to the top-blown processes is the copious evolution of iron oxide fumes from the converter mouth. The localised temperature in the zone where the oxygen jet impinges upon the liquid metal is close to the boiling point of iron and the volume of iron vapour which is formed is considerably greater than that encountered with any of the bottom-blown processes. It is claimed that the fume emission is considerably reduced with the buffer-slag process.

COMPARISON OF THE MODIFIED CONVERTER PROCESSES

The performance of the various modifications of the basic Bessemer process is most readily compared in terms of the residual phosphorus,

Table 8.2

	Phosphorus %	Sulphur %	Nitrogen %
Normal air-blow	0·04	0·035	0·015
Air-blow with double slag	0·025	0·025	0·020
Air-blow with ore and limestone additions	0·030	0·030	0·006
Oxygen–steam blow (OV Process)	0·030	0·030	0·002
Oxygen top-blow with double slag (LD Process)	0·025	0·020	0·001
Oxygen top-blow with lime powder injection and double slag (OLP and LD–AC processes)	0·010	0·010	0·001
Buffer slag process	0·020	0·010	0·001

sulphur and nitrogen contents which can be obtained and these are in the order given in Table 8.2 for a metal containing initially 1·5 per cent of phosphorus and 0·05 per cent of sulphur.

Before concluding this subject, reference must be made to the side-blown or *Tropenas* converter practice. These converters are principally used for metal preparation in foundries and generally have relatively small capacity. The blast is introduced through tuyeres in the converter wall and impinges on the slag-metal surface at a low angle; hence the principles underlying the practice are more closely related to top-blowing than to the orthodox processes. In fact the Tropenas process pre-dates all the other top-blowing processes described in this chapter and the information gained from its operation influenced markedly the recent developments in converter practice. Phosphorus can be removed along with or before the carbon and low nitrogen contents can be attained, particularly if the blast is enriched with oxygen [59], but the agitation of the bath is not very pronounced and the refining time per ton of metal is greater than that required in the LD Process.

8.6 Bessemer Instrumentation

Although during recent years marked improvements have been made in the efficiency of the pneumatic process, the metal yield and the quality of the metal produced are still primarily dependent on the assessment of the progress of the blow from the appearance of the carbon flame. This is particularly true with bottom-blowing, where the after-blow is timed from the drop in the carbon flame. The uncertainty arising from visual assessment has been largely obviated by the development, mainly by IRSID, of instruments which are capable of recording continuously the progress of the blow. The instruments and their purposes are as follow [10, 60–63]:

Volume Debitgraph

This records the volume of air blown as a function of the time, which can be related to the changes in metal composition when the composition and temperature of the charged metal are constant. Hence the meter gives an indication of the time at which the various additions should be made.

Two-colour Pyrometer

The instrument is sighted through one of the bottom tuyeres and gives a continuous recording of the bath temperature. The temperature indicated is affected by the rate of blowing and the instrument must be calibrated accordingly.

Flame Pyrometer

This is sighted just above the nose of the converter and indicates the flame temperature, which is dependent on the temperature and composition of the bath. The reading must be corrected for the flame emissivity, and the corrected temperature is found to be about 80°C below the bath temperature at the end of the blow. The readings are less reliable with an oxygen enriched blast, owing to the partial combustion of carbon monoxide within the converter.

Opacimeter

This comprises a light source focused through the flame on to a photocell. The opacity of the flame rises to a peak midway through the afterblow, then falls to a minimum at low phosphorus contents, followed by a further rise as the rate of oxidation of the iron increases.

Although the instruments have simplified somewhat the estimation of the progress of the blow, interpretation of the readings requires considerable skill. It is improbable that the degree of control which can be exercised over the final metal composition will ever be comparable with that which obtains in the open-hearth and electric steel-making processes, where the duration of refining is such that samples can be analysed and the data made available to the melter before any marked change has occurred in the metal composition.

REFERENCES

1. Iron and Steel Inst. Special Report No. 42, 1949, Section 2, p. 13 and appendix 3.
2. Ibid., Section 3, p. 39.
3. A. RIST and J. CHIPMAN, *Rev. Met.*, 1956, Vol. 53, p. 796; *The Physical Chemistry of Steel-making*, 1958, New York, John Wiley, p. 3.
4. J. CHIPMAN, J. C. FULTON, N. GOKCEN and R. GASKEY, *Acta Met.*, 1954, Vol. 2, p. 439.
5. M. OHTANI, *Tetsu to Hagane*, 1957, Vol. 43, p. 1211.
6. R. SCHUHMANN and P. J. ENSIO, *Trans. Amer Inst. min. (metall.) Engrs.*, 1951, Vol. 191, p. 401.
7. C. BODSWORTH, *J. Iron St. Inst.*, 1959, Vol. 193, p. 13.
8. E. T. TURKDOGAN and J. PEARSON, ibid., 1953, Vol. 175, p. 393.
9. P. J. LEROY, J. G. GALEY and F. B. CAWLEY, ibid., 1956, Vol. 183, p. 208.
10. F. B. CAWLEY and D. R. WATTLEWORTH, ibid., 1956, Vol. 183, p. 198.
11. P. KOZAKEVITCH, *Rev. Met. Memoires*, 1949, Vol. 46, pp. 505, 572.
12. J. F. ELLIOTT, *Trans. Amer. Inst. min. (metall.) Engrs.*, 1955, Vol. 203, p. 485.
13. W. WEBER, T. KOOTZ and K. OBST, *Stahl u. Eisen, Düsseldorf*, 1958, Vol. 78, p. 1735.
14. J. B. BOOKEY, F. D. RICHARDSON and A. J. E. WELCH, *J. Iron St. Inst.*, 1952, Vol. 171, p. 402.
15. W. A. FISCHER and H. VOM ENDE, *Stahl u. Eisen, Düsseldorf*, 1952, Vol. 72, p. 1398.
16. G. TROMEL and W. FRITZE, *Arch. Eisenhüttenw.*, 1959, Vol. 30, p. 461.
17. P. METZ, A. DECKER and J. NEPPER, *Stahl. u. Eisen, Düsseldorf*, 1960, Vol. 80, p. 20.
18. F. C. LANGENBERG, *Trans. Amer. Inst. min. (metall.) Engrs.*, 1956, Vol. 206, p. 1099.
19. Y. KASAMOTO and S. MATOBA, *Tetsu to Hagane*, 1959, Vol. 45, p. 100.
20. L. S. DARKEN and R. W. GURRY, *The Physical Chemistry of Melts*, New York, McGraw Hill, 1953, p. 392.

21. J. D. FAST and M. B. VERRIJP, *J. Iron St. Inst.*, 1955, Vol. 180, p. 337.
22. F. W. HARBORD and T. TWYNAM, ibid., 1896, Vol. 11, p. 161.
23. H. A. DICKIE, ibid., 1948, Vol. 159, p. 360.
24. K. G. SPEITH and H. BUCKEN, *Stahl. u. Eisen, Düsseldorf*, 1954, Vol. 74, p. 1650.
25. W. OELSEN and H. J. DARMANN, ibid., 1953, Vol. 73, p. 338.
26. P. COHEUR, J. DAUBERSY, L. MARBAIS, N. NEPPER and L. DOR, *Rev. univ. Mines*, 1950, Vol. 6, p. 401.
27. R. GRAEF, *Stahl u. Eisen, Düsseldorf*, 1951, Vol. 71, p. 1189.
28. W. BERTRAM, ibid., 1952, Vol. 72, p. 243.
29. P. LEROY, M. GOMBERT and B. TRENTINI, *Rev. Met.*, 1954, Vol. 51, p. 45.
30. P. LEROY and E. DEVERNAY, ibid., 1951, Vol. 48, pp. 53, 140.
31. M. MICHAUD and M. ZEHRINGER, *Iron Coal Tr. Rev.*, 1957, Vol. 174, p. 979.
32. J. DAUBERSY, *Rev. Met.*, 1951, Vol. 48, p. 23.
33. B. KALLING, F. JOHANSSON and L. LINDSKOG, *J. Iron St. Inst.*, 1951, Vol. 168, p. 337.
34. P. COHEUR and H. KOSMIDER, *Trans. Amer. Inst. min. (metall.) Engrs.*, 1954, Vol. 200, p. 829.
35. J. E. STUKEL, *J. Metals, N.Y.*, 1956, Vol. 8, p. 763.
36. V. G. MALINOVSKII and A. S. BABII, *Stal'*, 1957, (7), p. 670.
37. H. NEUHAUS, *Stahl u. Eisen, Düsseldorf*, 1957, Vol. 77, pp. 816, 1863.
38. C. GOODEVE, *J. Iron St. Inst.*, 1956, Vol. 184, p. 113.
39. C. V. SCHWARZ, *Stahl. u. Eisen, Düsseldorf*, 1952, Vol. 72, p. 1642.
40. O. CUSCOLECA, ibid., 1952, Vol. 72, p. 989; *Trans. Amer. Inst. min. (metall.) Engrs.*, 1954, Vol. 200, p. 817.
41. O. CUSCOLECA and K. RÖSNER, *J. Metals, N.Y.*, 1958, Vol. 10, p. 673; *J. Iron St. Inst.*, 1959, Vol. 192, p. 147.
42. H. RELLERMEYER and T. KOOTZ, *Stahl u. Eisen, Düsseldorf*, 1954, Vol. 74, p. 381.
43. K. RÖSNER, ibid., 1952, Vol. 72, p. 997.
44. H. TRENKLER, *Rev. univ. Mines*, 1958, Vol. 9, p. 644; *J. Metals, N.Y.*, 1960, Vol. 12, p. 538.
45. B. MATUSCHKA, *J. Metals, N.Y.*, 1956, Vol. 8, p. 760.
46. E. C. WRIGHT, *Metal Progr.*, 1958, Vol. 73, p. 65.
47. H. TRENKLER and H. F. HAUTTMANN, ibid., 1956, Vol. 69, p. 45.
48. S. I. LIFSHITS and S. G. AFANASEY, *Metallurg.*, 1958, (5), p. 12.
49. H. TRENKLER and R. RINESCH, *Stahl u. Eisen, Düsseldorf*, 1958, Vol. 78, p. 883.
50. H. TRENKLER and R. RINESCH, Eisenhuttentag Düsseldorf meeting 1957, Oct.
51. G. MESSIN, L. LAMBERT and G. HELFSTEIN, *J. Metals, N.Y.*, 1960, Vol. 12, p. 542.
52. P. METZ, *Blast Furn.*, 1958, Vol. 46, pp. 1065, 1187; *Iron Coal Tr. Rev.*, 1958, Vol. 177, p. 931; *Rev. univ. Mines*, 1958, Vol. 9, p. 501.
53. B. TRENTINI, P. VAYSSIERE, D. JORRE and M. GOMBERT, *Rev. Met.*, 1957, Vol. 54, p. 147.
54. B. TRENTINI and M. ALLARD, *J. Metals, N.Y.*, 1958, Vol. 10, p. 466; *Rev. Met.*, 1958, Vol. 55, p. 67; *Centre Doc. Sider. Circ.*, 1957, Vol. 4, p. 2259.
55. B. TRENTINI, P. VAYSSIERE and M. ALLARD, *J. Iron St. Inst.*, 1959, Vol. 192, p. 143.

56. H. KOSMIDER, H. NEUHAUS and H. SCHENCK, *Stahl. u. Eisen, Düsseldorf*, 1957, Vol. 77, p. 1277.
57. P. COHEUR and H. KOSMIDER, *J. Metals, N.Y.*, 1956, Vol. 8, p. 754.
58. H. RELLERMEYER, H. KNÜPPEL and J. SITTARD, *Stahl u. Eisen, Düsseldorf*, 1957, Vol. 77, p. 1296.
59. C. E. SIMS and F. L. TOY, *Trans. Amer. Inst. min. (metall.) Engrs.*, 1950, Vol. 188, p. 694.
60. P. J. LEROY, *J. Iron St. Inst.*, 1956, Vol. 183, p. 56.
61. J. GALLEY, L. BEAUJARD, P. VALLET and G. URBAIN, *Journees d'Automne*, 1958, p. 10.
62. L. VON BOGDANDY and H. D. PANTHE, *Stahl u. Eisen, Düsseldorf*, 1958, Vol. 78, p. 792.
63. P. HOLLER and H. SMIT, ibid., 1958, Vol. 78, p. 799.

CHAPTER 9

The Open-Hearth Process

9.1 Introduction

As the name suggests, the open-hearth furnace is a shallow bath with a large area of slag–metal interface per unit volume of metal. The melting chamber is similar to a reverberatory furnace, to which Siemens applied his regenerator principle and thus raised the maximum temperature obtainable with coal-gas firing into the range required for the refining of steel. Today the majority of furnaces are heated by alternative fuels and, although the regenerative principle is retained, it is now employed only for heating the combustion air.

When compared with the Bessemer process the oxidation reactions in the open-hearth are slow and the heat evolved is not sufficient to compensate for the thermal losses arising from the large bath area; hence the thermal balance is restored by radiant and convective heat obtained from the combustion of a gaseous fuel above the metal. The oxidation products are transferred to a slag layer which separates the metal from the combustion zone and this feature has an important bearing on the temperature distribution. The thermal conductivity of any slag is low in relation to that of a metal at the same temperature and the rate of heat transfer to the metal is considerably retarded when the slag layer is formed, but heat is lost continuously from the metal by conduction through the lining of the hearth. This results in the average temperature of the metal being lower than that of the slag and the difference may be as great as 100°C [1], but it varies with the depth of the slag layer and the degree of agitation of the metal. On the other hand, the thermal conductivity of the steel is high and the difference in temperature between the upper and lower levels of the metal bath is only a few degrees [1, 2].

SELECTION OF THE CHARGE

Since the process is not dependent for the supply of the thermal energy on the oxidation of elements dissolved in the metal, larger quantities of scrap metal can be melted than in the Bessemer and the charge may vary from 100 per cent of hot metal to entirely cold pig iron and scrap. The use of hot metal generally requires the provision of a blast furnace plant within a reasonably close proximity to the melting shop, but many open-hearth plants are located in districts remote from those producing the iron. In whatever state the iron is supplied, however, economics generally dictate that the charge contains a proportion of scrap metal. The advantages of this are two-fold.

(1) In the major steel-producing areas scrap costs less than pig iron, so the raw material costs decrease in proportion to the amount of scrap charged. Part of this scrap arises in the processing of the metal into its final form (e.g., top and bottom discards from ingots, scrap ends from billets, used ingot moulds etc.).
(2) Scrap steel is closer to the composition ultimately required than the metal from the blast furnace and the refining time decreases in proportion to the quantity of scrap in the charge.

There is obviously a limit to this addition of scrap metal for:

(*a*) some oxidation occurs during melting, so that an all-scrap charge may melt-out with too low a carbon content, and
(*b*) the melting time increases in proportion to the amount of cold metal in the charge such that the tap-to-tap time for a cold charge may be greater than when most of the charge is molten.

The rate of removal of other elements dissolved in the iron is critically dependent on the agitation of the bath by the carbon boil, because, apart from natural convection effects, the bath is almost stagnant when carbon is not being oxidised. When a steady boil is obtained the apparent volume of the liquid metal bath is increased by 50 to 100 per cent by the bubbles of gas. This large volume of gas, with a density much lower than that of the metal, causes a very rapid circulation in the bath as the gas rises to the surface and escapes. To supply the necessary agitation it is customary to aim at a melt-carbon of about 0·5 per cent greater than that required in the product, and this usually means that there must be a minimum of 30 per cent of pig iron in the charge.

The supply of scrap metal has tended to lag behind the recent rapid increase in world steel production and, in consequence, it has become necessary to accept rather higher iron/scrap ratios in the charge. This trend favours the integrated iron- and steel-making plant where the availability of hot metal reduces the melting costs to the open-hearth, so that it is rare today to find open-hearth shops under construction remote from a supply of hot metal, and a number of existing shops have incorporated cupolas to supply molten iron and reduce the melting time in the open-hearth [3]. The cost factor is also responsible for the trend towards larger melting units. Whereas thirty years ago few furnaces exceeded 100 tons capacity, many are now operating in the 250–500 ton range without any increase in the labour force per furnace.

SUPPLY OF OXYGEN TO THE METAL

The partial pressure of oxygen in the furnace atmosphere is about 10^{-3} atmosphere, whereas the pressure in equilibrium with iron and ferrous oxide at 1,600°C is about 10^{-8} atmosphere (Figure 1.2). Thus, during the time the metal is exposed to the furnace atmosphere oxidation of the iron is rapid, but the metal is isolated from the atmosphere when a slag cover is formed. The oxygen is then supplied by the addition of ore or mill scale to the slag and by the formation of higher oxides of iron at the slag–gas interface, so that the oxygen potential of the slag is greater than that of the metal and oxygen is transferred across the slag–metal interface to take part in the refining reactions. Some oxygen is supplied by the ejection of metal droplets through the slag layer into the furnace atmosphere by the vigour of the carbon boil, and these droplets are oxidised by the atmosphere to carry oxygen back into the metal on their return. Further quantities may also be supplied by injecting gaseous oxygen through a lance which is usually located just above the slag to increase the rate of refining. The slag thus takes an active part in the process by supplying part of the oxygen to the metal and absorbing the solid and liquid reaction products.

As in the Bessemer process, the open-hearth process can be divided into acid and basic practices, depending on the nature of the lining and the slag formed. The same restrictions regarding the elimination of sulphur and phosphorus also apply, these elements being removed only under a basic slag. Hence the acid practice has declined in importance until today only a comparatively small proportion of the world's steel production is made in these furnaces, and the greater percentage of world output is made in the basic open-hearth.

9.2 The Acid Open-Hearth Process

CHARGING AND MELTING OPERATIONS

Since sulphur and phosphorus are not removed under an acid slag, the charge materials must be so selected that the average concentration of these elements is lower than the maximum limits allowed in the specification to which the steel is being made. The pig iron used is generally an acid haematite iron, containing 1·0 to 1·5 per cent of silicon.

If the entire charge is solid the iron is charged first and followed by the scrap. While the metal is exposed to the atmosphere the oxidation of the iron is quite rapid and, on an acid hearth, the iron oxides can react with silica to form a slag which is liquid at 1,200°C. When the iron is charged first it is partially molten by the time the scrap is added and the oxidation of the silicon in the iron provides the silica necessary for the initial slag formation. If the reverse procedure is adopted and the scrap is charged first, only a small amount of silica is produced until the pig iron starts to melt; during this time the iron oxides cause rapid erosion by reaction with the silica in the furnace hearth.

When the slag layer has formed the rate at which heat is transferred from the furnace atmosphere to the metal is inversely proportional to the depth of the slag, which throughout the refining period is almost saturated with silica, but the slag volume is related directly to the amount of silicon which is oxidised from the charge, plus the amount of silica which is dissolved into the slag from the furnace lining. Thus ideally the charge should contain just sufficient silicon to combine with the iron oxides which form before the metal is under a slag cover and produce a slag which is approximately saturated with silica.

When melting is just complete a typical analysis of the metal is as follows:

Element	C	Si	Mn	S	P
Weight-per cent	1·0	0·5	0·2	0·025	0·025

and the metal temperature is about 1,400 to 1,450°C. For the production of very low-carbon grades the melt-carbon may be lowered to about 0·6 per cent if sufficient scrap is available, but the concentrations of the other elements remain more or less constant. The first slag which forms consists principally of ferrous silicates, as is indicated by the dark-coloured fracture of a solidified sample, and its composition is generally in the range:

Oxide	SiO_2	FeO	MnO
Weight-per cent	40/50	30/35	10/15

REFINING PROCEDURE

Since the liquidus temperature of pure iron is depressed by most of the common solute elements, the metal temperature must be increased before the start of refining, otherwise the bath will become solid once again. As the temperature rises some oxidation of the solutes is caused by reaction with the iron oxides in the slag, but the rate is comparatively slow. The composition of the slag changes towards saturation with silica, while the FeO content is lowered slightly by dilution and by reaction with the metal. When the metal temperature is high enough (i.e.,

1,520 to 1,560°C) the refining reactions are stepped up by the addition of iron oxides to the slag, usually in the form of lump haematite ore.* There is of course a continuous transfer of oxygen from the gas phase through the slag to the metal, which can be expressed as two consecutive reactions [4]:

$$2FeO_{(slag)} + \tfrac{1}{2}O_{2(g)} \rightleftharpoons Fe_2O_{3(slag)} \text{ at the slag-gas interface} \qquad 9.1$$

$$Fe_2O_{3(slag)} \rightleftharpoons 2FeO_{(slag)} + O_{(metal)} \text{ at the slag-metal interface} \qquad 9.2$$

This change is shown schematically in Figure 9.1, and the addition of iron oxides to the slag is equivalent to the first step in the sequence. The rate of these reactions is controlled by the diffusion of oxygen across the slag

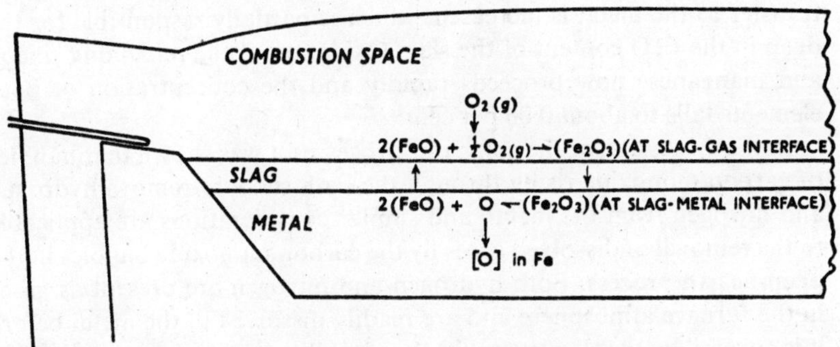

FIG. 9.1. Schematic representation of the transfer of oxygen from gas to metal via the iron oxides in the slag.

layer when the bath is quiescent, but the speed is greatly increased if the slag is stirred by turbulence or by agitation within the bath.

As the metal melts it forms an almost stagnant pool, covered by a stationary layer of slag, and the bath remains undisturbed when ore feeding is commenced, but analysis shows that the silicon and manganese are steadily oxidised while the silica content of the slag increases to between 55 and 60 per cent. At this stage the slag is saturated with silica and becomes very viscous. Whilst this condition is desirable in the acid Bessemer process, it must be avoided here, since the rate of oxygen transfer to the metal, and therefore the rate of refining, is adversely affected. Fluidity can be restored by the addition of a basic oxide such as lime but, in the presence of the solid silica lining, the unsaturated slag tends to dissolve silica from the lining to become saturated once more. The

* Consideration of the changes which are brought about by the use of gaseous oxygen to assist in refining is deferred until later in the chapter.

rate of solution of the lining is increased with increasing undersaturation, so that it is preferable to make frequent additions of small quantities of lime, rather than the occasional feeding of larger quantities. A lime feed serves a further useful purpose in the acid slag for the activity coefficient of the ferrous oxide increases with the lime content [5, 6], and so the iron oxide content of the slag (expressed as FeO) decreases to between 10 and 14 per cent when lime is added and remains in this range to the end of the heat.

The carbon starts to oxidise with continued feeding of oxides and, at about 1 hour after the completion of melting, bubbles of carbon monoxide can be seen erupting through the slag surface. From this stage onwards the carbon boil provides the turbulence which is needed to cause mixing of the slag and metal at the interface and the rate of oxygen transfer to the metal is increased, which is partially responsible for the drop in the FeO content of the slag. Oxidation of the remaining silicon and manganese now proceeds rapidly and the concentration of both elements falls to about 0·06 per cent.

In the discussion of the converter processes it was shown that bubbles of carbon monoxide rising through the bath serve to remove hydrogen and nitrogen from the metal, and similar considerations are applicable to the removal of dissolved gases by the carbon monoxide bubbles in the open-hearth process. Both hydrogen and nitrogen are present as gases in the furnace atmosphere and are readily dissolved in the metal before it is covered by the slag layer, but the solubility of these gases in the slag is very low [7]. Thus the slag serves as a barrier to their transfer into the metal, and the removal of the dissolved gases in the carbon monoxide bubbles results in a steady decrease in the concentrations remaining in the metal, so that only small amounts are present at the end of refining. The residual nitrogen constant is usually about 0·004 to 0·005 per cent[8].

Compared with the converter processes, the rate of carbon oxidation is very slow, even in relation to the LD top-blown practices, and rarely exceeds 0·3 per cent per hour. The rate decreases with decreasing carbon content and falls to between 0·10 and 0·15 per cent per hour as the carbon is oxidised to below 0·3 per cent. This is due largely to a progressive increase in the difficulty of forming bubbles of the gaseous reaction product in the melt as the concentration of carbon atoms decreases (see Chapter 13), and also to a fairly rapid increase in the metal oxygen content in equilibrium with the carbon at low carbon concentrations, so that less carbon is oxidised per unit amount of oxygen transferred into the metal. Thus at low carbon contents the carbon boil becomes less vigorous, the rate of oxygen transfer to the metal decreases, and the duration of the refining period is increased very considerably with respect to the converter processes. With a cold metal charge the tap-to-tap time is usually 10 to 14 hours, the longer time applying to the production by oxide feeding of the very low carbon steels.

As the carbon content of the metal approaches the level required in the finished steel the oxygen additions are stopped and the rate of carbon oxidation diminishes to a very low level. Metal samples taken from the bath can then be analysed and the results reported to the melter before any significant change has occurred in the bath composition.

APPLICATION OF THERMODYNAMIC DATA TO THE REACTION SEQUENCE

A more rigorous treatment can be applied to the equilibrium distributions of the elements than was possible in the converter process since, during a greater part of the refining time, the concentrations of the solute elements are low and are more amenable to the dilute-solution treatment. The slag temperature remains fairly constant throughout the refining period at about 1,600 to 1,650°C, but the metal temperature increases progressively from the time of clear melting up to the tapping time. The temperature range is small, however, and for purposes of computation it is assumed that the metal temperature varies from 1,550°C at the start of refining to 1,600°C when the metal is tapped. The effect of temperature on the activity coefficients of many of the solutes is unknown and it is assumed that the values at 1,600°C apply equally at 1,550°C. Using the activity and interaction coefficients tabulated in Appendix 1 for the solute concentrations shown at the start of refining, the activities relative to the dilute weight-per cent standard state are approximately as follow:

Element	C	Si	Mn
Weight-per cent	1·0	0·5	0·2
Activity	1·73	0·83	0·194

The estimated activities of the oxides, with respect to the pure oxide standard state, in the early refining slag at 1,600°C are as follow:

Oxide	MnO	FeO	SiO_2
Mole per cent	12	30	50
Activity	0·05 [9]	0·20 [5,6]	0·82 [5]

These activities are expected to vary with the temperature, but the temperature range is small and the variation is neglected. The partial pressure of carbon monoxide in equilibrium with the carbon is again taken as unity.

These values can be substituted in the equations for K_C, K_{Mn} and K_{Si}, which are given in Chapter 8, to obtain the activity of the oxygen in the metal in equilibrium with each element, and the results obtained for mean temperatures of 1,550°C and 1,600°C are given in the first two lines of Table 9.1.

Table 9.1

	Temperature °C	Carbon per cent	Silicon per cent	Manganese per cent	Activity of Oxygen in equilibrium with		
					Carbon	Manganese	Silicon
1.	1,550	1·0	0·5	0·2	0·0013	0·0084	0·0030
2.	1,600	1·0	0·5	0·2	0·0013	0·0126	0·0049
3.	1,550	1·0	0·1	0·1	0·0014	0·0189	0·0068
4.	1,600	1·0	0·1	0·1	0·0014	0·0286	0·0112

The activity of oxygen in pure liquid iron in equilibrium with FeO in the slag at 0·20 activity can also be estimated using the equation for K_{Fe} given in Chapter 8. Assuming that the activity of the liquid iron is unity, the values are 0·0372 at 1,550°C and 0·0454 at 1,600°C.

Comparison of the Equilibrium and the Actual Reaction Sequence

From this it is evident that if equilibrium is maintained the carbon, manganese and silicon oxidise in preference to iron, and the expected initial order of oxidation is: carbon followed by silicon and lastly manganese. In practice it is found that the silicon starts to oxidise first, followed by the manganese and by the start of the carbon boil the manganese and silicon concentrations are very low. Complete analysis reveals that the slag and metal are not in equilibrium prior to the start of the carbon boil for the slag contains some 2·5 per cent of ferric oxide, whereas in equilibrium with pure iron the ferric oxide content of the slag is less than 0·5 per cent [6]. This departure from equilibrium is due largely to the absence of agitation in the bath and the resulting low rate of diffusion of oxygen across the slag-metal interface, but this merely lowers the rate of build-up of oxygen in the metal and affects all the reactions equally; in no way can it account for the different sequence of the reactions.

As the silicon concentration decreases the departure from the sequence predicted thermodynamically becomes even more striking, for the activity of the silicon falls rapidly while the activity coefficient of carbon is lowered only slightly and the manganese coefficient is virtually unaffected.

For example, if the concentrations of silicon and manganese are lowered to 0·1 per cent while the carbon content and the slag activities remain unchanged, the activities of the solutes are as follow:

Element	C	Si	Mn
Weight-per cent	1·0	0·1	0·1
Activity	1·60	0·161	0·097

On substituting these values in the mass-action equations, the oxygen activities obtained are given in lines 3 and 4 of Table 9.1.

The further oxidation of the silicon causes the composition of the slag to move closer to saturation with silica and (a_{SiO_2}) is then approximately unity, but this raises slightly the oxygen activity in equilibrium with the residual silicon and increases the discrepancy between the equilibrium and the actual oxygen activity in the presence of the carbon.

The cause of this discrepancy is again to be found in the reaction kinetics. The manganese and silicon are removed via a reaction at the slag–metal interface and there is fairly strong evidence that these processes are at, or very close to, equilibrium during the carbon boil [10], which implies that the transfer across the interface is comparatively rapid. In contrast, carbon is removed as a gas and in the discussion of the top-blown converter processes mention was made of the fact that the oxidation of carbon is retarded when gas bubbles are not introduced mechanically into the metal. Similar considerations apply to the open-hearth process when the oxygen required for refining is fed to the bath in solid form as iron ore or iron oxides, and the metal is supersaturated with oxygen before the generation of carbon monoxide bubbles is initiated. The degree of supersaturation can be estimated from the data in Table 9.1. Generally the manganese is lowered to about 0·10 per cent and silicon to less than 0·10 per cent before a vigorous carbon boil is obtained, and the oxygen in the metal is then in equilibrium with the carbon only if the pressure of the carbon monoxide is between 3 and 4 atmospheres. A similar degree of supersaturation is needed to maintain the carbon boil once it has been initiated, and if for any reason the supply of oxygen atoms to the metal is diminished or halted the boil quickly subsides and does not start again until the necessary level of oxygen concentration is restored.

The difficulty encountered in initiating the boil is readily observed in practice. If a steel rod or rabble bar is moved fairly rapidly through the molten bath before the boil has started, bubbles of carbon monoxide are seen almost immediately to emerge from the slag surface. This is probably due to the creation of a vortex of low pressure liquid behind the moving bar, in which the carbon and oxygen atoms can unite more readily to form a bubble of gas than within the mass of liquid metal. If the rod is held stationary a few bubbles are generated as it is inserted into the bath, but the gas evolution does not continue. The boil is also initiated more readily on a heavily-fettled hearth with a rough surface than when the surface is glazed-over by partial melting of the lining or by the absorption of slag. The difference arises from the greater proportion of capillary orifices in a rough surface into which the metal cannot penetrate, owing to surface tension forces, but which are of suitable cross-section for the nucleation of a gas bubble. If a small area of the hearth surface becomes detached the newly-exposed surface contains a high concentration of nucleating sites and a very vigorous, highly localised boil is created. If this is not checked further portions of the lining may be removed from the same area by the rapid release of gas bubbles which, in extreme cases, leads to penetration of the hearth by the metal.

THE RATE OF CARBON REMOVAL

The equation for the oxidation of carbon has previously been written in the form:

$$O_{(\text{in Fe})} + C_{(\text{in Fe}} \rightleftharpoons CO_{(g)} \qquad 9.3$$

The reaction rates for the forward and back reactions can be expressed as:

$$\left(\frac{dC}{dt}\right)_{(\text{forward})} = \kappa_1[a_O][a_C] \qquad 9.4$$

and

$$\left(\frac{dC}{dt}\right)_{(\text{reverse})} = \kappa_2 p_{CO} \qquad 9.5$$

where t is the time, p_{CO} is the partial pressure of carbon monoxide in atmospheres, and κ_1 and κ_2 are the specific rate constants for the forward and back reactions. The net reaction rate (i.e., the rate of carbon removal) is then given by

$$\frac{dC}{dt} = \kappa_1[a_O][a_C] - \kappa_2 p_{CO} \qquad 9.6$$

If this equation is solved for a range of carbon contents it should be possible to predict the total time required from the taking of a metal sample up to the completion of refining, and numerous attempts have been made to adopt equations of this type to the conditions encountered in the open-hearth. The activities of the oxygen and carbon in the metal can only be determined by complete analysis and an empirical approach, replacing the activities by the weight-per cent, has often been adopted. However, it is clear from the preceding discussion that agreement between the calculated values and those found in practice is fortuitous. In the first place the values of $[a_O]$ and $[a_C]$ are not linear functions of the concentrations of these elements and they are affected by the contents of other elements present. It may be argued that after the first hour of refining the concentrations of silicon, manganese, sulphur and phosphorus are low and are relatively constant from cast to cast but, even with this simplification, the relation between per cent of carbon and per cent of oxygen is not linear and the product of the two is not constant. Schenck [11] recognised this point and tabulated values of κ_1 and κ_2 which decreased with diminishing carbon content.

Secondly, the value of p_{CO} is not constant but varies with the surface roughness of the hearth and the mobility of carbon and oxygen atoms in the metal. This mobility is, in turn, dependent on the metal composition, the temperature and the degree of turbulence in the bath. There is some confusion in the literature as to what values should be assumed for p_{CO}. Schenck assumed that the appropriate value is that of a gas bubble at approximately the mid-depth of the bath, in which case the total pressure is atmospheric, plus the hydrostatic pressure of the metal and slag layers above the bubble. A small correction was made for the partial pressure in the gas of hydrogen, nitrogen and carbon dioxide to arrive at a value of $p_{CO} = 1{\cdot}1$ atmospheres, but it is shown in the previous section that the value of p_{CO} in equilibrium with 1 weight-per cent of carbon and the oxygen in the metal is between 3 and 4 atmospheres. If the actual value is greater than 1·1 atmospheres by a constant factor it would mean simply that the values of κ_2 are too large, but it is more probable that the partial pressure of carbon monoxide is concentration dependent.

One of the most serious criticisms of this type of approach is the inherent assumption that the rate is unaffected by extraneous conditions. In practice the carbon rate is affected markedly by the rate of supply of oxygen from the slag layer to the metal (which varies with the slag composition and viscosity), by the rate of supply of oxygen to the slag from ore additions, and by a number of other factors.

Whilst these empirical formulae may give reasonable agreement with the actual conditions when the composition of the bath and the method of working the heat are more or less identical from cast to cast, it is unreasonable to expect that they will apply equally to another practice, or when any of the conditions is changed.

THE RETURN OF SILICON TO THE METAL

It is evident from the first two lines of Table 9·1 that as the value of the equilibrium constant K_{Si} decreases with increasing temperature, if (a_{SiO_2}) and $[a_O]$ remain unchanged the silicon content of the metal increases with the temperature of the bath. The activity of the silica in the slag increases during the first hour or so of refining as the silica concentration approaches nearer to the saturation value by pick-up from the silica lining of the furnace, and at constant $[a_O]$ this will also tend to increase the silicon content of the metal. The value of $[a_O]$ does, in fact, remain relatively constant after the silicon and manganese are oxidised, and only rises significantly when the carbon content is reduced to a low level. Hence it is expected that the silicon in the metal will fall initially to a low value and then increase slightly as both the metal temperature and (a_{SiO_2}) increase. This trend is found in practice, the silicon falling to about 0·06 per cent early in the refining period and subsequently increasing by 2 or 3 points (i.e. 0·02–0·03 per cent) later in the heat.

A similar, although smaller, reversion of manganese may also be observed over the same period, for the value of K_{Mn} decreases slowly as the temperature is raised and favours the return of manganese to the metal. However, the activity of the manganese oxide in the slag is very low and is scarcely affected by the change in the silica content; hence the extent of manganese reversion should in fact be small.

The silicon content of the metal may fluctuate during the heat if the rate of oxygen supply by oxide additions to the slag is irregular. As the oxygen from the last oxide addition is exhausted the carbon boil decreases in intensity, the slag and metal layers become more stagnant and the transfer of oxygen through the slag to the slag–metal interface is retarded. Meanwhile oxygen continues to transfer into the metal from the static layer of slag adjacent to the interface, so that this layer is depleted in iron oxides and eventually becomes saturated with silica. Thereafter the transfer of oxygen to the metal is extremely slow. The generation of carbon monoxide bubbles is then impeded by the low oxygen content of the metal, but this oxygen content is still appreciably greater than the level in thermodynamic equilibrium with the carbon present. Thus, although the turbulent carbon boil ceases, carbon continues to oxidise at a very low rate, the net result being a fall in the oxygen content of the metal compared with the level under a steady boil, and the silicon content of the metal must increase correspondingly to maintain K_{Si} constant. When a further oxide feed is made the steady boil conditions are resumed, $[a_O]$ is restored to the original level, and the silicon is re-oxidised to the equilibrium value.

The magnitude of the fluctuation in the silicon content is small and is generally only a few points, whilst the ultimate silicon content is unaffected, but the consequences of this fluctuation are important. During the initial oxidation of the silicon, manganese is oxidised simultaneously and the oxidation product may be a mixed oxide which is liquid at the working temperature. In this form the oxidation products are able to coalesce and form larger masses which, according to Stoke's Law (Chapter 13.6), rise rapidly into the slag. On the other hand, the oxidation of the reverted silicon produces solid silica particles which consist of only a few molecules and, as such particles do not readily coalesce, the rate of rise from the metal into the slag is correspondingly slow. A longer time is required for their removal than when the product of the oxidation is liquid, and the non-metallic content of the finished steel is greater than when a steady boil is maintained throughout the heat.

FINISHING THE HEAT

The ore additions to the slag are stopped and the bath is allowed to go 'off the boil' when the carbon in the metal is within a few points of the content ultimately required. The rate of carbon removal is then very slow and the metal can be tapped into the ladle with only negligible carbon loss. If the carbon content is inadvertently lowered too far it can be adjusted by anthracite additions to the ladle or by the addition of pig iron to the bath.

The oxygen content of the metal is appreciably greater than that in the acid Bessemer process and is commonly in the range from 0·015 to 0·020 weight-per cent; hence deoxidisers must be added in amounts determined by the required oxygen content for the class of ingot to be produced, i.e. killed, semi-killed or rimming. These additions may be made entirely to the bath, because the rate of oxygen transfer from the slag is very low when the saturated silica layer has formed at the interface and the bath does not re-oxidise appreciably in the time required for the deoxidisers to react with the metal and the subsequent tapping of the metal into the ladle. However, there is some loss of the deoxidisers by the slag reaction and the amount which must be added is greater than that theoretically required to remove the necessary amount of oxygen. In fact, the remarks made earlier regarding the deoxidation of a high-manganese, acid-Bessemer heat are to a large extent applicable to the acid open-hearth process.

9.3 The Basic Open-Hearth Process

SELECTION OF THE CHARGE

No real physical limitation to the charge composition which can be refined on the basic hearth exists, apart from the usual requirement that carbon is present in excess of that ultimately required to provide the necessary agitation. The composition is controlled only by (i) economic considerations of materials available, and (ii) the duration of the refining period. If appreciable concentrations of solid or liquid reaction products must be removed from the metal then either the volume of the slag is increased, which lowers the rate of heat transfer to the metal, or at suitable stages in the process the slag is removed and a fresh one formed to continue the refining operations. This is true of all elements which are removed as oxides, but it is particularly important when the element forms an acid radical such as P_2O_5 or SiO_2. The activities of both of these species are related inversely to the basicity of the slag and this relation in turn affects the amounts of phosphorus and silicon which remain in the metal. Therefore, for optimum dephosphorisation the silicon in the charge should be low and, as most of the silicon originates in the pig iron in the charge, there is obviously an advantage in lowering the silicon content by an active mixer practice or by an oxygen ladle treatment before charging the iron to the open-hearth. Alternatively the silicon in the iron, together with the carbon, phosphorus and other solutes, can be diluted by melting a large proportion of ferrous scrap, which is the only method available to a shop melting cold pig iron. However, even when desiliconised hot metal is charged, steel scrap is useful for diluting the concentration of other elements and reducing the duration of the refining period.

The excess carbon which is supplied for the promotion of turbulence in the bath is generally less than that in the acid process, and when sufficient scrap is available the charge is usually adjusted to give a melt carbon which is 0·3 to 0·4 per cent greater than that required in the final specification. The duration of the carbon boil is thus shortened and lasts for only about 2 or 3 hours, but the total refining time is extended if the phosphorus in the charge is high enough to necessitate a 'slag-off' and the formation of a new slag.

MELTING AND REFINING OPERATIONS

With an all-cold-metal charge the reverse charging procedure is adopted to that used in the acid open-hearth; steel scrap is charged first to protect the hearth and banks from erosion by the silica from the pig iron, and most of the lime required for the formation of the initial refining slag is added at the same time. When part of the charge is hot metal there is a further advantage in charging and partly melting the steel scrap before the hot metal is added, because the solid charge is melted more rapidly when it is exposed to the combustion gases than when it is immersed in a bath of molten iron. The charge may frequently contain large quantities of light scrap with a large surface area per unit volume, which oxidises more rapidly than heavy scrap, but melting also is more rapid and the solids are exposed to the oxidising atmosphere for a shorter time so that, with good melting conditions, the total amount of iron oxide formed in this stage remains fairly constant.

The slag which is formed first is rich in silica from the oxidation of the silicon in the charge. If the furnace roof is silica brick some silica may enter the slag also from roof spalls (caused by excessive temperature fluctuations during fettling and charging) and from partial melting of the roof if the combustion flame is deflected on to the brickwork by negligent charging of the scrap. The iron oxide content of the slag is low, for the silicon is oxidised to a low level and some manganese and phosphorus are also removed while melting is proceeding. The usual composition range of the first-formed slag is as follows:

Oxide	SiO_2	P_2O_5	CaO	MnO	FeO	Fe_2O_3
Weight-per cent	20–35	2–5	30–40	6–15	12–16	2–5

The Lime Boil

When the slag is formed by charging limestone to the furnace, the bath shows signs of turbulence towards the end of the melting period, while the slag becomes noticeably more fluid and increases in thickness. The agitation presents a very different appearance from that of the carbon boil and large bubbles of gas emerge from the slag surface. This effect, which is called the *lime boil*, is due to the decomposition of the limestone into lime and carbon dioxide and it is not encountered when the limestone is calcined before charging into the furnace. The decomposition of limestone normally starts at about 900°C and it might be expected that the reaction would start as soon as the stone is charged into the furnace, but the thermal conductivity of calcium carbonate is low and the reaction is retarded until the inner layers of each lump of stone attain the reaction temperature. When gas is liberated within the lumps the pressure is sufficient to cause their disintegration and a considerably larger surface is exposed. Consequently once dissociation starts it proceeds rapidly, and the finely divided lime which is produced is able to float out of the skeleton of partially fused scrap and rise into the slag layer. The basicity of the slag increases rapidly and by the time the lime boil has ended the slag composition lies within the following range:

Oxide	SiO_2	P_2O_5	CaO	MnO	FeO	Fe_2O_3
Weight-per cent	7–18	6–14	40–54	3–9	3–12	2–5

and the basicity ratio $(CaO)/(SiO_2+P_2O_5)$ is usually between 1·8 and 2·2.

At the start of the lime boil the metal temperature is low, but the turbulence increases the rate of heat transfer and the temperature is usually in the range from 1,520 to 1,570°C when the boil is completed. The turbulence also increases the rate of oxygen transfer from slag to metal, with the result that silicon is almost completely eliminated from the metal, together with most of the manganese, before the completion of this stage. The slag may be almost saturated with phosphorus if the initial concentration of this element is high and as much as 80 per cent of the phosphorus may be transferred into the slag before the start of the carbon boil. The thermal burden in the bath is lower when the slag is formed by additions of lime (i.e., calcined limestone) and the metal reaches the necessary temperature more rapidly, but the rate of oxygen transfer to the metal is less rapid than during a lime boil and the total time from clear melt to the start of the carbon boil is not significantly less than when limestone is used.

The Removal of Phosphorus

From this stage onwards the primary concern is the lowering of the phosphorus and sulphur contents to within the limits specified, and the lime content of the slag is adjusted by further additions to ensure that these limits are reached before the carbon is oxidised to the required level. Otherwise the refining period is prolonged and pig iron or hot metal additions are needed to maintain the carbon within the specification. However, excessive lime additions increase the slag bulk beyond the necessary minimum and more time is then required to bring the metal up to the tapping temperature. Thus at this stage the efficiency of the process is critically balanced, for the melter must 'shape up' the slag before the carbon boil starts, but use the minimum quantity of lime. The methods of slag control adopted for this purpose are discussed in Chapter 12.5.

It is common practice to add small quantities of fluorspar (calcium fluoride, CaF_2) to flux the lime and speed up the rate of slag formation, but the slag is then unsuitable for sale as a basic fertiliser. However, in basic slags alumina behaves as a very weak acid with a fairly strong fluxing action and it can be added to the slag as the mineral (bauxite) without detriment to either the refining reactions or the subsequent use of the slag as a fertiliser.

When the slag is in 'condition' to receive the necessary quantities of phosphorus and sulphur, oxygen additions are made to propagate the carbon boil. The oxygen is variously supplied, as in the acid process, by transfer via the slag phase from the atmosphere, as iron ore or mill scale to the slag or as oxygen gas to the metal. The last method is more commonly used when the carbon is below 0·5% and is particularly useful for increasing the speed of refining in the production of the very low carbon grades. If the slag is not already saturated with P_2O_5, the content of this oxide increases during the carbon boil and the total content of acidic radicals ($P_2O_5 + SiO_2$) is about 22 per cent at the end of the heat. The iron oxide content of the slag rises slowly at first, but more rapidly when the carbon content is below about 0·3 per cent, and reaches a level which is closely related to the basicity and volume of the slag and to the final carbon content of the metal, ranging from 15 to 20 per cent of ($FeO + Fe_2O_3$) at 0·3 to 0·8 per cent of carbon in the metal to 20 to 26 per cent of iron oxide at 0·1 to 0·3 per cent of carbon.

At the end of refining the P_2O_5 in the slag is in equilibrium with the phosphorus in the metal and the reduction in the oxygen content of the metal resulting from the addition of the deoxidants may displace the equilibrium and cause a reversion of phosphorus to the metal. The extent of the reverse reaction is reduced by adding only sufficient deoxidants to the bath to check the elimination of carbon, the remainder being added to the ladle where only sufficient slag is left on the metal to prevent oxidation from the air and to act as a thermal insulator. However, the ladle is generally lined with an alumino-silicate refractory, which is dissolved slowly by the basic slag and the slag basicity falls steadily with increasing time of contact of the slag with the lining. This also displaces the phosphorus equilibrium and causes a reversion to the metal but, because the volume of metal in contact with the slag is decreased when the metal is poured into the moulds, the phosphorus content of the metal remaining in the ladle increases exponentially. The extent of the reversion is not solely dependent on the increase in the silica content of the slag during this period, but mainly on the activity of the lime in the slag. If solid lime is present in suspension silica can be absorbed into the slag without any marked change in the activity of the P_2O_5 and the extent of the reversion is insignificant until all the solid lime is fluxed.

The reversion is liable to occur in any process in which a basic slag containing P_2O_5 is held in an acid-lined ladle, but it is particularly important in the basic open-hearth and arc-furnace processes when steels are being produced to a low phosphorus specification. If the whole cast is to come within the required limits, the phosphorus content of the metal at tap must be low enough to allow for the reversion, and the slag must be sufficiently basic to minimise its extent.

THERMODYNAMIC TREATMENT OF THE REACTIONS

The factors controlling the equilibrium distribution of the common impurity elements between the slag and the metal have been demonstrated in the first part of this chapter and in the previous chapter (see p. 207); the reaction sequence in the basic open-hearth can be interpreted from these considerations. It is now proposed to extend the thermodynamic treatment to determine the activity of the oxygen in the metal in equilibrium with the various elements at three stages during refining in the basic process. This treatment serves to show how closely the equilibrium conditions are approached in the various reactions. The analyses of simultaneous slag and metal samples have been made available to the author by the Lancashire Steel Co. Ltd., the Steel Company of Wales, Ltd. and The United Steel Cos. Ltd., and details of the heats are given in Appendix 2. The following cast is selected as typical of British hot-metal practice:

Table 9.2 *Log of Basic Open-Hearth Heat*

Charge

	Tons
Scrap	91·95
Hot Metal	123
Ore	12·85
Scale	2·5
Limestone	18·95

Time	
9.55	Finish charge
11.40	Metal sample: 0·51%C 0·035%S 0·035%P 0·22%Mn
12.00	Metal sample: 0·385%C 0·032%S 0·025%P 0·19%Mn
12.07	Temperature: 1,565/1,570°C
12.08	Metal sample A: 0·35%C 0·032%S 0·025%P 0·17%Mn
12.09	Slag sample A
12.24	Temperature: 1,605°C
12.25	Metal sample B: 0·155%C 0·029%S 0·025%P 0·18%Mn
12.26	Slag sample B
12.37	Metal sample: 0·06%C 0·028%S 0·025%P 0·13%Mn
12.48	Temperature: 1,605/1,610°C
12.51	Metal sample C: 0·045%C 0·028%S 0·015%P 0·09%Mn
12.52	Slag sample C
12.57	Temperature: 1,595°C
1.00	Tapped

Pit Sample

0·055/0·060%C, 0·027%S, 0·014/0·015%P, 0·36%Mn, 0·06%Ni, 0·08%Cu, 0·01%Sn, 0·01%Cr.

The time at which melting was complete is not indicated, but it can safely be assumed that the lime boil had finished before the slag and metal samples which are designated A were taken. At this stage the silicon content of the metal would be very low and the carbon boil well under way. The sample marked B corresponds roughly to the mid-point in the refining time and sample C represents the condition of the bath at the end of refining.

Activities in the Metal and in the Slag

Proceeding in similar manner to that used in the previous examples, the first step in the calculation is to find the activities of the solutes in the metal and in the slag. The pit sample shows the presence of small amounts of nickel, copper, tin and chromium in the metal. Of these the concentrations of tin and chromium are very low and can be neglected, but nickel and copper are not oxidised in the open-hearth process and it is assumed

Table 9.3 *Activities of the Solutes in the Metal*

	C	S	P	Mn
Sample A	0·41	0·034	0·25	$0{\cdot}16_8$
Sample B	0·167	0·030	0·25	$0{\cdot}17_9$
Sample C	0·046	$0{\cdot}027_5$	0·15	0·090

that the concentrations of these elements found in the pit sample were present also in the bath when samples A, B and C were taken. Neglecting the effect of the oxygen in the metal, the activities (relative to the dilute weight-per cent solution) of the other solutes can be obtained from the data in Appendix 1 and are listed in Table 9.3.

The composition of the slag must be expressed in mole fractions of the component oxides before the activities in the slag phase can be determined. When a solid slag is chemically analysed the iron oxides are usually reported in terms of the ferrous and ferric oxide contents but, for the iron oxides in the slags under consideration, the ratios of oxygen to iron are only slightly greater than unity and correspond closely to the composition of the non-stoichiometric wüstite phase in equilibrium with iron at the relevant temperature. This composition is the reference state which is usually selected for the assessment of the activity of ferrous oxide in molten slags and, therefore, for comparison with the experimental graphs of activity versus composition, the total iron content of the slag is first expressed as ferrous oxide and designated Fe_1O. The mole fractions of the oxides in the slag are then calculated and are given in Table 9.4.

Table 9.4 *Slag Compositions (Mole Fractions)*

	SiO_2	Fe_TO	Al_2O_3	MnO	CaO	MgO	S	P_2O_5
Sample A	7·03	19·10	1·08	5·49	50·4	14·2	0·42	2·55
Sample B	9·11	18·61	1·50	4·74	48·9	14·0	0·51	2·36
Sample C	8·71	26·40	1·05	3·68	42·7	15·64	0·49	1·87

The activity of the manganese oxide in the slag can be estimated by a small extrapolation of the experimental data [12] and the values obtained are approximately: 0·45 for A, 0·30 for B and 0·15 for C. The activity of P_2O_5 in steel-making slags has not been determined directly, but the activity coefficients can be calculated by the method described in Chapter 15.3 and suggested by Turkdogan and Pearson [13]. The values thus obtained are: $9{\cdot}4 \times 10^{-18}$ for sample A, $9{\cdot}6 \times 10^{-17}$ for B and $5{\cdot}6 \times 10^{-17}$ for C. The activity of the ferrous oxide in the slag can be determined from the appropriate activity-composition curves [5, 6, 14], but there are no reliable data from which the activity of the slag sulphur can be estimated.

Equilibrium Conditions

The oxygen activities in equilibrium with the residual manganese and phosphorus, and with the residual carbon and carbon monoxide at a pressure of one atmosphere, are listed in Table 9.5 for the three stages

Table 9.5 *Oxygen Activities in the metal in Equilibrium with:*

	C	Mn	P	Fe_TO	
Sample A	0·006	0·011	0·050	0·117	0·072
Sample B	0·014	0·088	0·095	0·121	0·082
Sample C	0·050	0·088	0·085	0·151	0·093

during the heat. The last two columns of the table show the oxygen activities in the metal in equilibrium with the Fe_TO in the slag. The values in column 4 are estimated with the aid of the earlier experimental measurements of (a_{Fe_TO}) at 1,600°C [5] whilst those in column 5 are based on an extrapolation of more recent lower-temperature data [6, 14].

The oxygen activities in equilibrium with the manganese and phosphorus contents in samples B and C are remarkably consistent, but show significant deviation in sample A. This implies that the manganese and phosphorus oxidation reactions had more or less reached equilibrium with respect to each other by the time sample B was taken and the equilibrium was maintained to the end of the heat. The oxygen activities in the last column of the Table suggest that these reactions are in equilibrium with the ferrous oxide in the slag, but the agreement is less satisfactory in relation to the oxygen activities in column 4. This should not be interpreted as evidence that the (a_{Fe_TO}) values used in determining the activities in column 5 are more reliable than those used with column 4, as attention was drawn earlier to the fact that the temperature of the slag may be between 50 and 100°C higher than that of the metal. This temperature difference alters both the equilibrium constants and the slag activities. However, the evidence points strongly to the fact that the partitioning of manganese and phosphorus between the slag and metal is at, or very close to, equilibrium with the iron oxides in the slag during the later stages of refining, and similar findings have been reported in a number of investigations of basic open-hearth practice [1, 15–18]. Experiments in which radioactive phosphorus is added to the bath have demonstrated that the phosphorus equilibrium is approached fairly rapidly, for small additions become uniformly distributed throughout the slag and the metal within 60 to 70 minutes [19, 20]. This conclusion is supported also by the results of calculations with the data for the heats which are listed in Appendix 2. These results are not detailed here, but the reader may try some of the calculations for practice in handling the data.

In marked contrast to the oxygen activities in equilibrium with the manganese and phosphorus, the oxygen activities in equilibrium with the carbon are very low at all three stages in the heat and it is clearly evident that the carbon–oxygen reaction deviates markedly from the equilibrium condition. In fact when sample B was taken the carbon is only in equilibrium with the residual manganese and phosphorus contents if the partial pressure of the carbon monoxide is about 6·5 atmospheres. It is unfortunate that the actual metal oxygen contents for the heat are not available. If they were, it would be found that at each stage the mean activity of the oxygen in the metal lies between the lower limit set by the carbon–oxygen equilibrium at one atmosphere pressure of carbon monoxide and the upper limit set by the slag–metal reactions for phosphorus and manganese removal. The actual values are usually closer to the lower limits, but the two limits converge at very low carbon contents, as is illustrated in Figure 9.2 where the carbon–oxygen relation obtained in a number of investigations is compared with the equilibrium values and with the metal oxygen contents in equilibrium with open-hearth slags [17, 21]. The carbon–oxygen relations for the basic Bessemer and oxygen-blown converter practices [22] are also included in the diagram to emphasise the important rôle of bath turbulence and the presence of gas bubbles within the metal in the question of closeness of approach of the carbon–oxygen reaction to the equilibrium conditions.

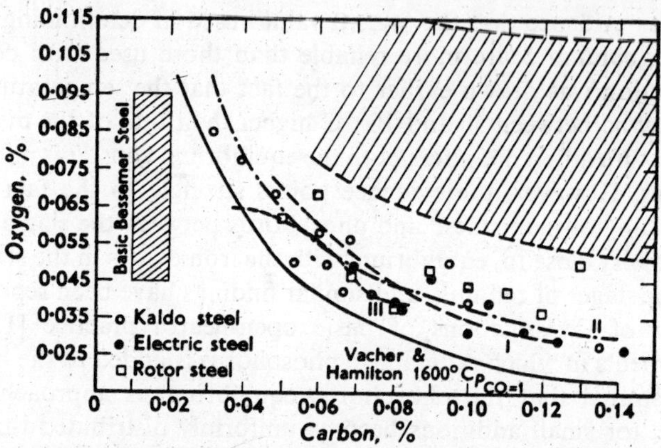

FIG. 9.2. Relation between carbon and oxygen concentrations in the metal for various steel-making processes. The solid line is the carbon–oxygen relation at equilibrium under one atmosphere pressure of carbon monoxide. The shaded band on the right shows the equilibrium oxygen concentrations calculated from the activities of ferrous oxide in open-hearth slags. (After Metz, *J. Iron St. Inst.*, 1960, Vol. 194, p. 65, and Turkdogan and Pearson, *J. Iron St. Inst.*, 1954, Vol. 176, p. 59.)

Curve I German open-hearth
Curve II American open-hearth
Curve III LD–AC

FLUSH-SLAG PRACTICE FOR PHOSPHORUS REMOVAL

At constant temperature and oxygen activity in the metal the partitioning of phosphorus can be expressed as

$$\frac{(\gamma_{P_2O_5} \cdot N_{P_2O_5})}{[f_P \cdot \%P]}$$

Hence the residual concentration of phosphorus can be lowered by two methods:

1. Increasing the basicity of the slag so as to lower $(\gamma_{P_2O_5})$, but this effect operates only when the basicity of the *liquid* slag is raised, for lime in excess of the solubility limit remains as a solid suspension which

makes little or no positive contribution to the removal of phosphorus from the metal.

2. Increasing the volume of the slag so as to lower ($N_{P_2O_5}$) by dilution; this improves the phosphorus partitioning at the expense of a thicker layer of slag and retards heat transfer to the metal.

With moderate amounts of phosphorus in the charge the residual concentration can be lowered to acceptable limits with a reasonable volume of slag, but as the initial phosphorus concentration is increased the production rate must be lowered to allow time for the large amounts of lime to dissolve in the slag before the start of the carbon boil and for the reduced rate of thermal transfer. In the latter case it is often preferable to operate a flush-slag (i.e., two-slag) process

The numerical value of the equilibrium constant for the phosphorus reaction decreases rapidly as the temperature is raised and so, for constant ($\gamma_{P_2O_5}$), the residual phosphorus in the metal increases with the temperature. Thus if a basic slag is formed while the metal temperature is still relatively low, a large percentage of the phosphorus in the charge can be transferred to the slag before the start of the carbon boil. At this stage the slag is flushed from the furnace and a new slag is formed by additions of lime and a flux, but the second slag is required to remove only small amounts of phosphorus and a smaller volume will suffice, which facilitates thermal transfer during the ensuing carbon boil. Against this advantage must be set a lower metal yield, since the total amount of iron oxides present in the two slags is often greater than in a one-slag practice. The manganese recovery in the metal is also lowered, since a large part of the manganese is oxidised into the flush slag and the remaining manganese is partly oxidised into the second slag until the equilibrium partitioning is restored.

9.4 The Removal of Sulphur in the Open-Hearth Process

The transfer of sulphur from the metal to the slag has been expressed previously by the ionic equation:

$$O^{--}_{(slag)} + S_{(in\,Fe)} \rightleftharpoons S^{--}_{(slag)} + O_{(in\,Fe)} \qquad 9.7$$

The equilibrium relation for this reaction can be used to examine the factors which affect the sulphur partitioning in the open-hearth process.

The sulphur content of an open-hearth slag rarely exceeds 0·5 per cent and it is probable that at these low concentrations the activity of the slag sulphur is directly proportional to its concentration [23]. With the exception of manganese and copper all the common solute elements raise the activity coefficient of sulphur in the iron and so as refining proceeds $[f_S]$ is lowered. However, the data in Tables 9.2 and 9.3 show that, when the sulphur content of the metal is low, $[a_S]$ is approximately equal to [%S] and the change in $[f_S]$ during refining is negligible.

There is no direct method for measuring the activity of the oxygen ions in the slag and it is arbitrarily assumed that the activity is proportional to the ionic fraction of oxygen ions in excess of the number required to satisfy the positive charges in the acid-ion polymers.

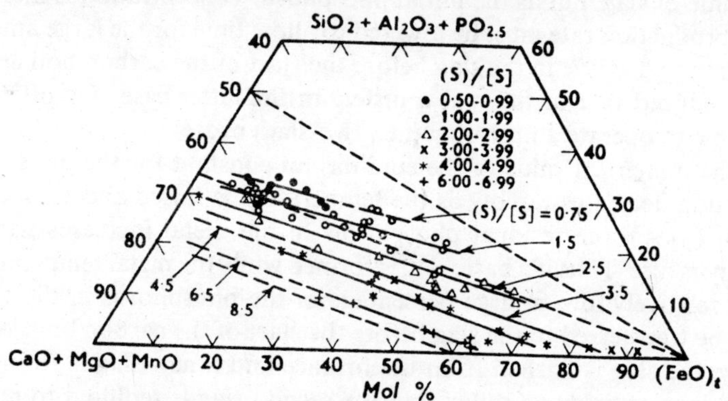

FIG. 9.3. Iso-distribution lines for sulphur between the metal and basic steel-making slags in the temperature range 1,570° to 1,630°C. (After Bishop, Lander, Grant and Chipman [26].)

Applying these approximations the apparent equilibrium constant for equation 9.7 is given by:

$$K'_S = \frac{(N_{S^{--}}).[a_O]}{(N_{O^{--}}).[\%S]} \qquad 9.8$$

which suggests that the sulphur distribution $(N_{S^{--}})/[\%S]$ is raised either as the oxygen ion content of the slag is increased or as the oxygen activity in the metal is lowered.

The value of K'_S increases as the temperature is raised (see Chapter 14.2) and favours the transfer of sulphur into the slag, but this is opposed by the corresponding increase with temperature in the oxygen content of the metal for equilibrium with the other residual elements. These two effects are frequently found to compensate each other and the sulphur distribution appears to be independent of the temperature. Thus in lime-saturated, iron oxide slags the ratio (%S)/[%S] decreases from about 11 at 1,530°C to 10 at 1,700°C [12].

In view of these considerations it appears that sulphur distribution in the open-hearth is determined primarily by the 'available' oxygen ion fraction in the slag and all other factors are of secondary importance. In acid practice this fraction is very low, the distribution ratio is rarely greater than 1·0 [24, 25, 26] and, as the slag volume is small, only a negligible amount of sulphur is removed from the metal. Ratios as high as 4·0 have been reported for acid heats [27], but these are exceptional. The 'available' oxygen ion fraction increases rapidly as the basicity is raised in the range of slag compositions typical of basic steel-making

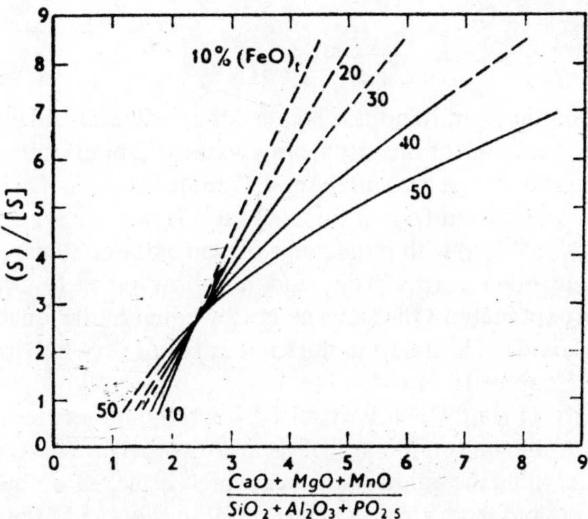

FIG. 9.4. Relation between sulphur distribution and slag basicity for various molar concentrations of iron oxide in the temperature range 1,570° to 1,630° C. (After Bishop, Lander, Grant and Chipman [26].)

practice and with basicities of 2·5 to 3·0 it is possible to obtain a sulphur distribution ratio of 10·0 to 12·0, although in open-hearth practice the ratio rarely exceeds 8·0. Figure 9.3 shows the type of variation in the ratio (%S)/[%S] with slag composition which is obtained in practice [26]. The lines for constant sulphur ratios are almost parallel to the join FeO–2CaO . SiO_2, but when the data are re-plotted in the form of sulphur ratio versus basicity (Figure 9.4), it is found that for basicities greater than about 2·4 the sulphur ratio decreases as the iron oxide content of the slag is raised. Thus when the oxygen lance is used to facilitate refining the slag may contain more iron oxides and with very basic slags the sulphur ratio is lower than when the oxygen is supplied as iron ore or mill scale [18].

The transfer of sulphur from slag to metal takes place via a reaction at the slag–metal interface; the equilibrium sulphur distribution is therefore dependent on the oxygen activity in the metal at the interface and not on the oxygen activity in the bulk metal. This means that the sulphur distribution is related to the oxygen activity in the metal which is in equilibrium with the iron oxide content of the slag, and it follows that an increase in the activity of the iron oxides in the slag should lower the sulphur distribution ratio. This effect is seen more clearly when the equilibrium constant for equation 9.7 is expressed in the following form, where f_S and f_O are assumed to be unity:

$$K''_S \frac{(a_{O^{2-}})}{(f_{S^{2-}})} = \frac{(\%S)}{[\%S]}[\%O] \qquad 9.9$$

The quantity on the right-hand side is called the *S–O ratio* and at 1,600°C the ratio decreases almost linearly from a value of about unity in lime–iron oxide slags to almost zero at about 0·35 mole fraction of acids in the slag [12, 26]. At constant (N_{SiO_2}) the S–O ratio is nearly independent of the ratio $(N_{CaO})/(N_{Fe_TO})$ within the range of iron oxide contents normally found in basic open-hearth slags, and it follows that $(a_{O^{2-}})/(f_{S^{2-}})$ is changed approximately to the same extent by equal molar quantities of lime and iron oxide. This relation does not apply at very low iron oxide contents (see Chapter 14.3).

The quantity of slag which is required for refining is determined by the amounts of silicon, sulphur and phosphorus which are to be removed from the metal, plus the amount of silica which is picked-up by the slag from roof drips and from erosion of the refractory lining of the furnace. All three elements are present as anions or anion-complexes in the slag and so, in general, their activity coefficients are lowered as the basicity of the slag is increased. The extent to which each of these elements is removed from the metal is dependent, therefore, on both the volume and the basicity of the slag and at constant basicity the minimum slag volume required is determined by the total amount of these elements which must be removed from the metal. Thus the slag volume which is necessary to remove the phosphorus may be insufficient to remove the sulphur, and vice versa, and the lime charge should be determined separately for the removal of each of these elements. The actual quantity required is determined by whichever of these two amounts is the greater.

GAS REACTIONS WITH SULPHUR

The amount of sulphur found in the tap metal and slag may differ considerably from the estimated quantity in the charge materials. This difference arises from an exchange of sulphur between the charge and the furnace gases, for the latter may contain an appreciable concentration of sulphur, particularly when oil fuel is used. Early attempts to construct a sulphur balance on gas-fired furnaces were frequently negatived, because part of the sulphur in the gas was absorbed by the accumulated iron deposits in the chequer chambers, but sufficient information has been obtained in recent years to permit at least a semi-quantitative assessment of the conditions influencing the extent of sulphur pick-up.

Sulphide and Sulphate Equilibria

It has been shown that over a wide range of oxygen pressures the exchange of sulphur between the gas and the charge is controlled by two separate equilibrium relations [23, 28, 29]. When the partial pressure of oxygen is less than 10^{-5} atmospheres the sulphur in the slag exists as a sulphide:

$$\tfrac{1}{2}S_{2(g)} + O_{(slag)} \rightleftharpoons S_{(slag)} + \tfrac{1}{2}O_{2(g)} \qquad 9.10$$

whereas when p_{O_2} is greater than 10^{-3} atmospheres the sulphur is present mainly as sulphate:

$$\tfrac{1}{2}S_{2(g)} + 1\tfrac{1}{2}O_{2(g)} + O_{(slag)} \rightleftharpoons SO_{4(slag)} \qquad 9.11$$

and the equilibrium constants for the two reactions can be written in the form:

$$K'_{10}\frac{(a_{O^{--}})}{[f_S]} = \frac{(\%S)p_{O_2}^{1/2}}{p_{S_2}^{1/2}} \qquad 9.12$$

and

$$K'_{11}\frac{(a_{O^{--}})}{[f_S]} = \frac{(\%SO_4)}{p_{S_2}^{1/2}p_{O_2}^{3/2}} \qquad 9.13$$

The right-hand sides of these two equations are termed the *sulphide* (C_S) and *sulphate* (C_{SO_4}) capacities of the melt respectively and at constant temperature and composition both C_S and C_{SO_4} are constant over the range of sulphur contents normally found in the melt. Thus, at constant partial pressure of sulphur in the gas, the sulphur content of the melt decreases with increasing oxygen partial pressure for reaction 9.10, whilst the reverse condition holds for reaction 9.11. The conditions are illustrated in Figure 9.5. The oxygen potential of an open-hearth slag corresponds to an oxygen partial pressure of about 10^{-8} atmospheres at 1,600°C, and therefore it is evident that the sulphur in the slag exists mainly as the sulphide.

The sulphide capacity varies markedly with the composition of the slag, but in general the value of C_S falls steadily with increasing silica content in the composition range of basic steel-making slags and is approximately constant for acid slags. Figure 9.6 shows the variation of C_S with silica concentration in iron oxide–lime–silica melts [30]. In the basic slag range C_S is raised as silica is replaced by alumina and lowered as silica is replaced by P_2O_5; equi-molecular proportions of alumina and P_2O_5 are about as effective as four molecular parts of silica. At constant basicity C_S is raised when lime is replaced by MnO and lowered when lime is replaced by magnesia [23, 28, 31]. Thus for a gas of constant p_{O_2} and p_{S_2} the probability of sulphur transfer from gas to slag increases as the basicity of the slag is raised whilst, from equation 9.12, for a slag of

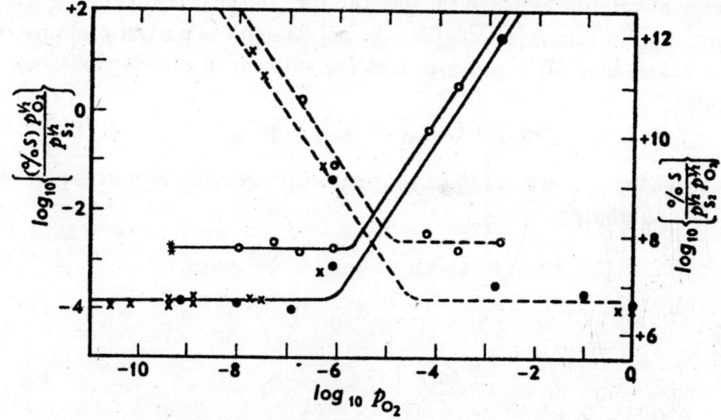

FIG. 9.5. Variation of the sulphate (broken lines) and sulphide (solid lines) capacities with the partial pressure of oxygen for slags of constant composition at 1,500°C. (After Fincham and Richardson [28].)

constant composition the probability decreases as the sulphur content of the slag is raised. Expressing this in a different way, the critical sulphur content of the gas above which sulphur is transferred from gas to slag is lowered as the basicity of the slag is increased or its sulphur content is lowered, or as the degree of combustion (and hence the partial pressure of oxygen) of the gas is lowered.

The atmosphere above an acid slag will not transfer sulphur to the slag unless the gas striking the surface of the metal is almost completely unburnt or the sulphur content of the gas is of the order of several per cent. Thus in normal acid practice sulphur should transfer continuously from the metal, via the slag, to the gas, but the quantity actually removed in this way is negligible on account of the low sulphur content and the comparatively high viscosity of the acid slag.

The situation is more complex in the basic practice for the temperature, the sulphur content and basicity of the slag increase during the refining period and these factors are inter-related in their effects on sulphur transfer between gas and liquid. Consequently, if a highly basic slag is formed rapidly it is possible for sulphur to transfer from gas to slag during the early part of the heat and then for the direction of the transfer to reverse as the sulphur content of the slag is raised by transfer from metal to slag. With moderate amounts of sulphur in the gas a fairly rapid transfer of sulphur from slag to gas, has been observed in basic practice [32, 33] and, other factors remaining constant, the critical sulphur

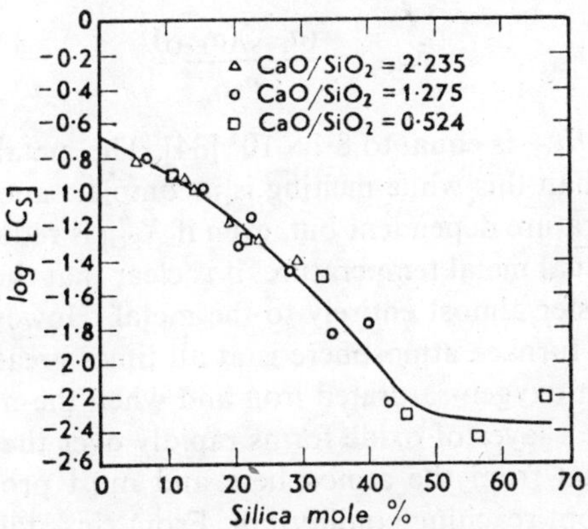

FIG. 9.6. Relation between sulphide capacity and silica content of steel-making slags. (After St. Pierre and Chipman [30].)

content above which the direction of the transfer is reversed should increase as the degree of combustion of the gas in contact with the slag is raised [23].

Sulphur Transfer During Melting

Although a significant amount of sulphur may be removed into the gas during the refining period, it is invariably found that the total sulphur content of the slag and metal has increased during the heat. This may be due in part to sulphur pick-up by metal droplets which are ejected through the slag and into the furnace atmosphere by a vigorous carbon boil and which carry sulphur back into the metal on their return [33], but it is probable that most of the sulphur is picked up by the metal during the melting period and before the slag cover is formed. It is often found that, with a fuel of constant sulphur content, the sulphur gain in the bath is greater when the melting period is extended than when the charge is quickly covered with slag, and the sulphur increment is smaller when the charge is mainly molten metal. This being so, it is interesting to apply the foregoing concepts to the first part of the heat before the start of the refining.

Sulphur absorption during the charging and melting period is probably controlled by two separate reactions. When the metal is exposed to the gas the reaction can be written as follows:

$$3Fe + SO_{2(g)} \rightleftharpoons FeS_{(l)} + FeO_{(l)} \qquad 9.14$$

$$K_{SO_2} = \frac{(a_{FeS})(a_{FeO})}{p_{SO_2}} \qquad 9.15$$

and at 1,600°C K_{SO_2} is equal to $8{\cdot}1 \times 10^4$ [34]. The metal temperature is obviously less than this while melting is incomplete and K_{SO_2} is almost certainly temperature dependent but, even if K_{SO_2} is reduced by a factor of 100 at the actual metal temperature, it is clear that the sulphur in the gas should transfer almost entirely to the metal. However, the oxygen potential of the furnace atmosphere is at all times greater than that in equilibrium with oxygen-saturated iron and when the metal is charged into the furnace a layer of oxide forms rapidly over the surface, which isolates the metal from the atmosphere and most probably prevents reaction 9.14 from reaching equilibrium. From this stage onwards the reaction mechanism is probably similar to that previously described for the reaction between the gas and the slag.

The oxide layer can be considered as almost pure wüstite, in which the activity of the oxygen ions is approximately unity and, from equation 9.12, its sulphide capacity is greater than that of a basic slag at the same temperature. The sulphide capacity decreases with falling temperature, but at 1,500°C the value for pure wüstite is still considerably greater than the usual range of values for basic open-hearth slags at 1,600°C [23, 31]. Hence, for a given partial pressure of oxygen in the furnace atmosphere, the critical sulphur content of the gas above which sulphur is absorbed is lower for a mixture of iron and iron oxides than for iron which is covered by a blanket of slag.

The introduction of the melting time as a variable probably arises from the fact that the sulphur from the gas is transferred via a pasty oxide layer into the metal. The oxide can be considered as a static layer, and the transfer of sulphur from the gas–oxide to the oxide–metal interface is then diffusion controlled, but the oxide layer is continuously melting and reforming and the oxide–metal interface is moving inwards into the metal mass. Such conditions are not readily amenable to treatment by formal diffusion theory, but it is evident that the amount of sulphur absorbed by the charge is time-dependent. Rapid melting causes the metal to be isolated from the gas by a layer of slag before the equilibrium sulphur distribution is attained, but as the melting time is increased the distribution approaches closer to the equilibrium condition and more sulphur is transferred to the charge.

Some silicon is oxidised together with the iron and combines with the iron oxides to lower the sulphide capacity of the oxide layer during melting. In view of the low silicon content of a basic charge this effect is negligible, but it may have an important bearing on the amount of sulphur which is absorbed during melting in the acid practice.

The sulphide capacity of an acid slag is low when melting is complete and is little affected by the change in composition during the heat. If the equilibrium sulphur distribution is approached during melt-down there should be a continuous loss of sulphur to the gas during refining but, for reasons discussed previously, the actual sulphur loss from the bath is negligible.

In the basic process the initial slag is silicious and the sulphide capacity is markedly lower than that of an iron oxide layer. As the melt progresses both the temperature and the basicity of the slag are increased and the sulphide capacity rises correspondingly, but this is countered to some extent by the transfer of sulphur from the metal into the slag, and the critical sulphur content of the gas is always less during refining than during the melting period.

Sulphur Transfer During Refining

Using a radioactive-sulphur tracer technique, it has been shown [32] that there is fairly rapid exchange of sulphur between the slag and the gas and theoretically one would expect that the sulphur partitioning between the slag and the gas should approach closely to the equilibrium condition by the end of the heat. If this were so then the rate of transfer from the slag to an atmosphere of fairly low sulphur content should be high at the start of refining and diminish towards zero at the end of the heat, but the available evidence indicates that sulphur is transferred to the gas at a fairly constant rate throughout the refining period. The total sulphur content of the slag and metal at tap should be in no way related to the duration of the melting period. However, it is possible that the net rate of sulphur transfer to the gas during the carbon boil is less rapid than the results of the tracer studies suggest, because of the return of sulphur to the melt via the ejected metal droplets [23], in which case the gas–slag equilibrium may not be established in the time available.

Attempts have been made to relate the sulphur pick-up by the bath to the sulphur content of the fuel and to the partial pressures of the sulphur-bearing gases in the furnace atmosphere [35–37], but it is evident that a number of variables must be considered and it is improbable that the sulphur gain or loss can be assessed with sufficient accuracy by means of a simple empirical relation.

9.5 The Use of Oxygen in the Open-Hearth Process

Gaseous oxygen has been applied extensively to the open-hearth process, both to enrich the combustion gases and to facilitate the refining reactions. Opinions differ widely regarding the merits of oxygen addition to the fuel mixture. It appears that the heat time can be decreased and the output raised by oxygen enrichment, but this applies only when bad burner and/or furnace design preclude the attainment of the maximum safe working temperature within the furnace chamber [38]. The flame direction and intensity can be improved with close control of the oxygen injection, but the consequences of an accidental deflection of the flame on to the furnace roof while charging, melting and feeding the bath are correspondingly more serious.

The oxygen lance has been applied extensively to speed up the melting process, but this is accompanied by a heavy fume emission which necessitates provision for cleaning the exhaust gases before discharge to the air and the useful chequer life is decreased by the accumulation of a heavy deposit. The fume problem is less serious when the charge is partly molten and covered by a layer of slag, but it is still serious enough to necessitate the installation of a gas-cleaning plant.

The most promising application of oxygen to the open-hearth process is to assist carbon elimination during the refining period. The maximum oxygen efficiency is obtained if the gas is blown into the bath when the lime boil is almost ended and the slag is 'in condition' to absorb the sulphur and phosphorus. The rate of carbon elimination is increased at least three-fold over that obtained with oxide feeds and both refining time and fuel consumption are markedly decreased. The fume problem is again serious with high carbon heats but the amount of fume evolved decreases rapidly with reduction in the carbon content and becomes negligible below 0·2 per cent of carbon. The oxygen efficiency decreases rapidly at low carbon contents on account of the rapid increase both in the amount of oxygen dissolved in the metal and in the iron oxide content of the slag, but the increase in production rate is more spectacular, and oxygen injection is extremely useful for the production of steels containing less than 0·1 per cent of carbon [39–45]. For example, an average reduction of 40 minutes in the total heat time has been reported [43] when oxygen is used to assist carbon elimination from 0·1 per cent to about 0·05 per cent.

The oxygen injected into the liquid metal forms gas bubbles into which the carbon atoms can diffuse and carbon monoxide is then generated more readily than at the hearth surface. As a result the carbon boil is stimulated and the bath turbulence is increased, whilst the heat generated in the bath by exothermic reaction with the solutes causes an increase in the temperature of the metal. Lime additions are fluxed more rapidly, enabling a higher basicity to be attained, and the increased turbulence increases the rate of transfer of sulphur and phosphorus across the metal–slag interface. The rate of elimination of these elements is increased proportionately to the increase in the rate of carbon removal, and slightly lower residual phosphorus contents can be achieved. However, the liquid bath becomes extremely heterogeneous when oxygen gas is injected [45]; in particular the oxygen content of the metal is high at and just beneath the slag-metal interface, and decreases with increasing depth in the metal. The iron oxide content of the slag is also higher than is the case where refining is effected with oxide feeds, but these differences can be largely eliminated if the oxygen blow is discontinued when the carbon content is about 0·02 per cent in excess of the level required at tap to allow time for the iron oxides in the slag to react with the metal. The metal yield is then increased slightly and more sulphur is transferred to the slag, but the heat time is increased and part of the saving obtained through the use of oxygen is lost.

Recently, considerable interest has been focused on the modified open-hearth or AJAX process [46], in which the principals of the LD converter practice have been applied successfully to the open-hearth and which has been developed by the Appleby-Frodingham Branch of the United Steel Companies Ltd. A jet of oxygen at high velocity is directed on to the metal surface as soon as melting is complete and very high refining rates are obtained. A hot-metal charge containing 3·2 per cent of carbon and 2·0 per cent of phosphorus can be refined to give 0·2 per cent of carbon, 0·030 per cent of phosphorus, in less than 5 hours. The process also resembles the LD converter process in the amount of heat which is liberated during oxidation, for the normal fuel supply to the furnace is discontinued while the oxygen jet is in operation and cooling additions are required to regulate the temperature of the bath.

The Mechanism of Fume Formation

The formation of a dense fume during oxygen injection is a serious problem, but it is one which has only recently begun to receive theoretical attention. Fume formation in the oxygen-blown converter processes is usually ascribed to the volatilisation of iron at the hot spots where the gas impinges on the metal [47], and similar temperatures are attained in the open-hearth where the oxygen jet is immersed in the metal. At 1,600°C the vapour pressure of iron is about 10^{-4} atmosphere, but the vapour pressure of iron equivalent to the fume in the open-hearth process is about 10^{-2} atmosphere, which is considerably greater than the true vapour pressure of iron even at temperatures of the order of 2,000°C. In fact, a dense fume is obtained from oxygen-blowing of a high carbon iron at temperatures as low as 1,100°C [48]. It thus appears that some process other than the simple volatilisation of iron is partly, if not entirely, responsible for fume formation in the open-hearth. Further, if volatilisation alone is responsible, then it is difficult to account for the effect of the carbon content of the bath on the density of the fume.

It has been suggested that a gaseous Fe–C–O complex is formed within the bubbles of carbon monoxide [49, 50] which breaks down above the slag surface to form elemental iron and carbon monoxide, and the iron is then oxidised by the furnace atmosphere. Alternatively, the formation of a gaseous iron oxide or hydroxide, with a similar sequence of dissociation and oxidation, has been proposed [50–52]. It is often observed that there is a clear space in the furnace between the slag surface and the fume layer. Analysis of gas samples taken from this clear space show that the iron is unoxidised [53, 54], which apparently supports the complex dissociation mechanisms, but others have suggested that any iron oxides present in this region are reduced to iron as the gas sample is cooled down to room temperature [48].

At any particular carbon content in the bath, the amount of fume formed is related directly to the rate of oxygen blowing. The amount is not changed significantly by changing to oxygen-air mixtures, but very little fume is formed when blowing with pure air [54], and the amount is decreased markedly when blowing with a mixture of oxygen and steam [55, 56]. These facts have been proposed as evidence in support of the high-temperature volatilisation hypothesis, but the composition of the gas in the clear space above the slag layer is dependent on the composition of the blowing mixture and it is more probable that the change in the density of the fume is brought about by a catalytic effect or some other factor such as the rapid diffusion rate of hydrogen [51]. It is evident that the mechanism of the fume formation is not yet resolved and a considerable amount of work remains to be done before a satisfactory theory can be formulated.

9.6 Foaming Slags

A condition sometimes encountered early in refining is the tendency of the slag to form a froth. Small gas bubbles become enveloped in a film of slag, which prevents their escape from the surface, the slag volume is considerably increased and in extreme cases, slag may pour out of the charging doors. Once this condition develops heat transfer through the slag to the metal virtually ceases. The condition is by no means restricted to the open-hearth process for frothing can occur in any type of vessel when a gas is bubbled through molten slag, but it is more frequently encountered here, and the resulting loss of productivity is generally greater than in any other process.

The entrapped gas bubbles are almost certainly carbon monoxide, but of smaller size than those which are normally created when the bubbles are generated at the surface of the hearth, and it has been suggested that the condition may develop from the nucleation of bubbles at the slag-metal interface by solid particles such as silica, lime, iron ore or chromic oxide [15, 57]. Numerous small droplets of liquid metal are suspended in the foam and it has also been suggested that these may react with the oxides in the slag to form small bubbles of carbon monoxide which adhere to the metal surface [58], but a more probable explanation is that, when the active boil is suppressed, bubbles rising from the hearth are not circulated in the metal and have little time for growth before they reach the slag-metal interface. It has been calculated [59] that under these conditions the bubble diameter would not exceed 1 mm by the time it reached the slag layer.

Foaming is usually experienced when the viscosity of the slag is high. At constant composition the viscosity decreases as the temperature rises and a cold melt will have a greater tendency to froth than when the bath temperature is high, but in many cases a foam is formed when the bulk slag is relatively fluid. In fact the tendency to foam is most marked when the slag is close to the neutral composition and decreases rapidly as saturation with either basic or acidic oxides is approached, yet the viscosity does not vary significantly over a wide range of compositions and only increases rapidly when the melt is almost saturated. Thus it would appear that the viscosity is of secondary importance, but under certain conditions the surface viscosity of the melt may be considerably greater than the mean viscosity; the condition most commonly encountered is the lowering of the interfacial tension by a component which becomes concentrated in the surface layer. For example, the surface tension of lime is lowered by silica [60, 61] and an adsorbed layer is formed on the surface of a calcium silicate melt in which the ratio of CaO/SiO_2 is much lower than the bulk ratio. In consequence the number of Si–O–Si bonds is increased at the surface and a more extensive silica network is formed which increases the surface viscosity. Bubbles of gas reaching the slag surface then become enveloped in the viscous liquid film and remain entrapped in the slag to form a foam. The addition of P_2O_5 to molten calcium silicates causes a more marked lowering of the surface tension [61] and increases further the foam stability.

It was suggested [62] and subsequently confirmed [58] that addition of a component such as TiO_2, which caused a breakdown of the anion network, would disperse an existing foam or prevent its formation. This observation gives strong support to the adsorbed-film theory (there is no suitable alternative explanation of the effect), but it does not explain why foams are not experienced with very acid slags where network formation is more extensive, and it is possible that the presence of some molecules or ions of species other than those forming the network is essential for stabilisation of the adsorbed surface layer [58]. Thus with increasing acidity a stage may be reached where the film becomes unstable and the network formers are mixed randomly in the bulk slag. This change is envisaged as a gradual, rather than a sudden, transition.

It is probably significant that the first-formed slags in both acid and basic practice are essentially ferrous silicates which lie close to the neutral composition, so that the addition of lime raises the surface tension and opposes the formation of an adsorbed layer. Manganese oxide is even more effective in raising the surface tension [60] and, since manganese is oxidised rapidly in the early stages of refining, the risk of foam formation should decrease inversely in proportion to the manganese content of the charge. Conversely, P_2O_5 is more highly surface active than silica and the transfer of phosphorus into an early slag of low basicity will encourage foaming, but this is often regarded as an advantage in flush-slag practice, for it facilitates the removal of the slag over the slag notches.

Foaming may still be encountered when the surface tension is fairly high if the slag contains a suspension of finely divided solids as may occur, for example, when finely divided lime and iron ore are added [63], or when a solid chromic oxide is formed by oxidation of chromium in the charge [58]. The effect of the iron oxide particles is only transient and disappears as the oxide dissolves, but the effect is more permanent with oxides such as CrO, which are almost insoluble in the slag, and the foam is very difficult to disperse. It is possible that the suspended solids act as nucleants for the carbon monoxide bubbles, as was suggested earlier; alternatively the solids may adhere to the surface of the gas bubbles and prevent their coalescence and growth. On the other hand, solids increase the viscosity of a slag and this would give slower drainage and a greater stability to the froth.

Other conditions being equal, the stability of a foam is lowered as the luminosity of the combustion flame is increased and foaming is rarely encountered with producer-gas firing which gives a very luminous flame, but is liable to occur when the furnace is fired with coke-oven gas where the luminosity is low [64]. The luminosity is due primarily to the presence of finely-divided, incandescent particles of carbon which are entrained in the gas. These particles may collapse the foam by either impacting against the bubbles and bursting the slag film [65], or by attaching themselves to the film and reacting with the iron oxides to form additional carbon monoxide molecules which extend the film and cause it to collapse [58]. Pre-heating the combustion gases to a high temperature or an increase in the gas-moisture content increases the foaming tendency [66], and it is perhaps significant that both of these treatments cause oxidation of the carbon particles and lower the flame luminosity. However, whilst there is strong evidence that the foam stability is influenced by the flame luminosity, the way in which this is effected is still uncertain.

REFERENCES

1. J. F. ELLIOTT and F. W. LUERSSEN, *Trans. Amer. Inst. min. (metall.) Engrs.*, 1955, Vol. 203, p. 1129.
2. V. S. KOCHO, *Stal'*, 1947, Vol. 7, p. 698.
3. A. RICHTER, G. COHNEN and P. JACOBI, *J. Metals, N.Y.*, 1958, Vol. 10, p. 599.
4. R. HAY, J. M. FERGUSON and J. WHITE, Iron & Steel Inst. Special Report No. 22, 1938, p. 52.
5. J. F. ELLIOTT, *Trans. Amer. Inst. min. (metall.) Engrs.*, 1955, Vol. 203, p. 485.
6. C. BODSWORTH, *J. Iron St. Inst.*, 1959, Vol. 193, p. 13.
7. J. H. WALSH, J. CHIPMAN, T. B. KING and N. J. GRANT, *Trans. Amer. Inst. min. (metall.) Engrs.*, 1956, Vol. 206, p. 1568.
8. B. L. BIGGS, *J. Iron St. Inst.*, 1959, Vol. 192, p. 361.
9. E. T. TURKDOGAN, ibid., 1953, Vol. 175, p. 393.
10. P. HERASYMENKO, *Proc. Amer. Inst. Min. (metall.) Engrs. nat. Open Hrth Comm.*, 1958, Vol. 40, p. 49.
11. H. SCHENCK, *The Physical Chemistry of Steel-making*, English translation B.I.S.R.A., London, 1945.
12. H. L. BISHOP, N. J. GRANT and J. CHIPMAN, *Trans. Amer. Inst. min. (metall.) Engrs.*, 1958, Vol. 212, p. 890.
13. E. T. TURKDOGAN and J. PEARSON, *J. Iron St. Inst.*, 1953, Vol. 175, p. 398.

14. I. M. DAVIDSON and C. BODSWORTH, ibid., 1960, Vol. 195, p. 163.
15. F. KÖRBER and W. OELSEN, *Mitt. K.-Wilh. Inst. Eisenforsch.*, 1935, Vol. 17, p. 39.
16. E. T. TURKDOGAN and J. PEARSON, *J. Iron St. Inst.*, 1954, Vol. 176, p. 59.
17. E. T. TURKDOGAN, ibid., 1955, Vol. 179, p. 147.
18. F. W. LUERSSEN, *Proc. Amer. Inst. Min. (metall.) Engrs. nat. Open Hrth. Comm.*, 1958, Vol. 40, p. 398.
19. A. I. OSIPOV, L. A. SHVARTSMAN, M. T. BULSKII and A. G. ALIMOV, *Stal'*, 1955, (8), p. 709.
20. A. M. SKREBTSOV, *Zavods Lab.*, 1959, (9), p. 1078.
21. J. CHIPMAN, Chapter in *Basic Open-Hearth Steel-making*, second edition, A.I.M.E., 1951, p. 652.
22. P. METZ, A DECKER and J. NEPPER, *Stahl u. Eisen, Düsseldorf*, 1960, Vol. 80, p. 20
23. F. D. RICHARDSON and J. C. B. FINCHAM, *J. Iron St. Inst.*, 1954, Vol. 178, p. 4.
24. H. SCHENCK, *The Physical Chemistry of Steel-making*, English translation B.I.S.R.A., London, 1945, p. 487.
25. E. MAURER and H. BISCHOF, *J. Iron St. Inst.*, 1936, Vol. 133, p. 183.
26. H. L. BISHOP, H. N. LANDER, N. J. GRANT and J. CHIPMAN, *Trans. Amer. Inst. min. (metall.) Engrs.*, 1956, Vol. 206, p. 862.
27. G. E. A. BRAMLEY, W. R. MADDOCKS and G. TATESON, Iron & Steel Inst. Special Report No. 25, 1939, p. 11.
28. C. J. B. FINCHAM and F. D. RICHARDSON, *Proc. royal Soc.*, Series A, 1954, Vol. 233, p. 40.
29. F. D. RICHARDSON and G. WITHERS, *J. Iron St. Inst.*, 1950, Vol. 165, p. 66.
30. G. R. ST. PIERRE and J. CHIPMAN, *Trans. Amer. Inst. min. (metall.) Engrs.*, 1956, Vol. 206, p. 1474.
31. K. P. ABRAHAM and F. D. RICHARDSON, *J. Iron St. Inst.*, 1960, Vol. 196, p. 313.
32. W. KOCH and W. FINK, *Arch. Eisenhüttenw.*, 1951, Vol. 22, p. 371.
33. S. N. STUPOR, *Stal'*, 1957, (8), p. 707.
34. J. CHIPMAN and TA LI, *Trans. Amer. Soc. Metals*, 1937, Vol. 25, p. 435.
35. C. A. EDWARDS, *J. Iron St. Inst.*, 1955, Vol. 181, p. 138.
36. B. TRENTINI, A. PETERS and G. HUSSON, *Soc. Francaise de Met.*, 1954, B.I.S.R.A.Translation, SM/C/189/54.
37. F. HARDERS, H. GROVE and W. OELSEN, *Stahl u. Eisen, Düsseldorf*, 1951, Vol. 71, p. 973.
38. F. J. CLARK and J. L. HARRISON, *J. Iron St. Inst.*, 1957, Vol. 186, p. 305.
39. E. B. HUGHES and F. G. NORRIS, *Trans. Amer. Inst. min. (metall.) Engrs.*, 1948, Vol. 176, p. 52.
40. D. J. O. BRANDT, *Iron Coal Tr. Rev.*, 1956, Vol. 178, p. 1377.
41. O. J. JACUNSZKAI and M. N. SZTAROVICS, *Kohâszati Lapok*, 1953, (8), pp. 194, 212.
42. J. E. HOOD, *J. Metals, N.Y.*, 1956, Vol. 8, p. 418.
43. A. J. KESTERTON, *Trans. Amer. Inst. min. (metall.) Engrs.*, 1957, Vol. 209, p. 274.
44. J. L. HARRISON, *Iron Coal Tr. Rev.*, 1956, Vol. 178, p. 15.
45. E. V. CHELISCHEV, *Kislorod*, 1958, (3), p. 11.
46. A. JACKSON, *J. Iron St. Inst.*, 1958, Vol. 190, p. 1; 1959, Vol. 191, p. 337.
47. H. KOSMIDER, H. NEUHAUS and H. KRATZENSTEIN, *Stahl u. Eisen, Düsseldorf*, 1954, Vol. 74, p. 1045.

48. C. HOLDEN, *J. Iron St. Inst.*, 1959, Vol. 193, p. 93; 1960, Vol. 195, p. 463.
49. E.J. KOHLMEYER and H. SPANDAU, *Arch. Eisenhüttenw.*, 1944–45, Vol. 18, p. 1.
50. E. T. TURKDOGAN and L. E. LEAKE, *J. Iron St. Inst.*, 1959, Vol. 192, p. 162.
51. F. D. RICHARDSON, ibid., 1960, Vol. 195, p. 461.
52. G. R. BELTON, ibid., 1960, Vol. 195, p. 463.
53. W. STRAUSS and M. W. THRING, ibid., 1959, Vol. 193, p. 216.
54. K. KNAGGS and J. M. SLATER, ibid., 1959, Vol. 193, p. 211.
55. H. KOSMIDER, H. NEUHAUS and P. E. HARDT, ibid., 1959, Vol. 193, p. 6.
56. G. S. SEL'KIN, *Kislorod*, 1958, (6) p. 3.
57. B. M. LARSEN, *Proc. Amer. Inst. Min. (metall.) Engrs. nat. Open Hrth. Comm.*, 1949, Vol. 32, p. 231.
58. P. KOZAKEVITCH, *The Physical Chemistry of Steel-making*, New York, John Wiley, 1958, p. 89.
59. L. S. DARKEN, ibid., p. 101.
60. T. B. KING, *Trans. Soc. Glass Tech.*, 1951, Vol. 35, p. 241.
61. C. F. COOPER and J. A. KITCHENER, *J. Iron St. Inst.*, 1959, Vol. 193, p. 48.
62. P. KOZAKEVITCH, *Rev. Met.*, 1949, Vol. 46, p. 505.
63. J. C. CURPHY, *J. Iron St. Inst.*, 1953, Vol. 174, p. 305.
64. R. W. EVANS, *Blast Furn.*, 1944, Vol. 32, p. 932.
65. P. T. CARTER, *Iron & Steel, Lond.*, 1953, Sept., p. 43.
66. H. M. DAVIS, *Proc. Amer. Inst. Min. (metall.) Engrs. nat. Open Hrth. Comm.*, 1949, Vol. 32, p. 238.

CHAPTER 10

The Electric-Arc Furnace Process

10.1 Introduction

Compared with the Bessemer and open-hearth processes, electric steel-making is a comparatively recent innovation. High-frequency induction melting was first introduced in 1922 and rapidly replaced the crucible process for the production of high-alloy and tool steels, but this is essentially a re-melting process which is not readily adapted for refining, and is not considered further here. The electric-arc furnace process, which is ideally suited for certain refining operations, was developed at the turn of the century and was used fairly extensively for the production of alloy steels during the First World War, but only during the past twenty years has the demand for quality steels thrust electric-arc steel-making into prominence. Today arc furnaces of more than 150 tons capacity are in use and it is widely claimed that these are competitive in cost and superior on a quality basis to the open-hearth for the production of plain carbon steels from an all-cold metal charge [1–5]. Techniques have also been developed which enable a charge consisting of up to 50 per cent of hot metal from a blast furnace to be refined in the arc furnace [5, 6], and by adding iron ore to the charge a metallic yield of 130 per cent (based on the quantity of iron and scrap in the charge) can be obtained. The production costs are again competitive with those of an open-hearth plant working with a similar hot-metal charge. Thus the economic position has changed considerably in favour of electric steel-making and it is confidently expected that the world tonnage of steel which is produced in this way will increase very rapidly in the near future.

The major advantages of the arc-furnace process over the other principal methods of steel production are: (1) the ability to control closely and to alter fairly rapidly the temperature of the metal, and (2) the ability to control the furnace atmosphere over a wider range of oxygen potentials. The heat generated by an electric arc is more easily regulated and concentrated in a chosen direction than that obtained from the combustion of a fuel or from the oxidation of elements in the metal. The heat is generated immediately above the bath and there is no risk of flame impingement on the refractory lining, but very high temperatures are reached in the arc and very close control is needed to avoid excessive heating of the entire furnace lining.

Carbon monoxide is formed by the oxidation of the electrodes adjacent to the arc and, as no other gaseous products of combustion are present, the oxygen potential of the atmosphere can be varied by regulating the amount of air or oxygen which enters the furnace. It is possible to form a slag of very low oxygen potential, such that oxygen is transferred from the metal into the slag, whereas in the open-hearth process the oxygen transfer is at all times in the reverse direction. If the later stages of refining are conducted under a reducing slag a smaller amount of deoxidants is required in finishing the heat and less non-metallic inclusions remain entrapped in the metal. Also the more readily oxidised alloy elements can be added to the bath to allow more time for mixing and homogenisation of the metal composition and a better recovery of the alloy elements in the metal than is possible when the alloys are added in the ladle to metal covered with a layer of oxidising slag. Further, the heat required to raise large alloy additions to the bath temperature is readily supplied by the electric arcs. In the other steel-making processes these alloys are added to the bath after deoxidation or to the metal as it enters the ladle and the percentage of alloy which can be added is limited by the surplus heat available in the metal.

The reduction in inclusion density in arc-furnace steels is not always an advantage, for the inclusions which are present usually have a high melting point. For example, such inclusions are not readily deformed during hot working and in high-strength steels they may act as stress-raisers under fatigue conditions. A larger density of inclusions which are deformable during hot working may give a better fatigue life.

The arc-furnace process is operated with both acid and basic hearths, but the former type is now almost entirely limited to foundry work and the principles are very similar to those of the acid open-hearth process; the differences which do exist can be inferred from an examination of the basic electric-arc process.

10.2 The Basic Arc-Furnace Process

The commonest application of the process is for the melting and refining of a cold-metal charge and, when the slag which is formed is oxidising to the metal, the sequence and the extent of the slag–metal reactions are similar to those of the basic open-hearth process. Thus the silicon and part of the manganese are oxidised rapidly during melting and at the start of refining, whereas sulphur and phosphorus are removed only after a basic slag is formed and during the carbon boil. However, the conditions are more favourable for the removal of sulphur from the metal than they are in the open-hearth, for the atmosphere in contact with the slag is free from sulphur-bearing gases.

When adequate supplies of steel scrap are available the arc-furnace charge usually contains only sufficient pig iron to ensure that the required carbon content is present when the metal is tapped; hence the volume of acid oxides ($SiO_2+P_2O_5$) which is formed during refining is small, relative to the amount formed from an open-hearth furnace charge, and the slag volume is reduced correspondingly. When the oxidation reactions with a scrap-metal charge are completed a typical composition of the slag is as follows:—

	FeO	MnO	CaO	MgO	SiO_2	P_2O_5	Al_2O_3
Weight-per cent	10·8	4·7	53·0	11·2	15·8	1·7	1·3

The basicity ratio $(CaO+MgO)/(SiO_2+P_2O_5)$ of this slag is 3·5, which is rather greater than the basicity which is obtained in the open-hearth process, owing to the higher slag temperature which can be attained and the more rapid solution of the lime in the slag under the intense heat of the electric arc. The increase in basicity favours better phosphorus and sulphur partition ratios between the slag and the metal than are obtained in the open-hearth.

The refining can be conducted with a slag which is only mildly oxidising, with the intention of conserving the alloys in the metal, when the scrap-metal charge contains a large percentage of alloy elements. Only small quantities of oxide are fed to the bath, the rate of removal of carbon, phosphorus and sulphur from the metal is low and the carbon content cannot be decreased below about 0·2 per cent by virtue of the carbon–oxygen relationship. The alloy recovery can be stepped up if oxide additions are completely eliminated and deoxidisers are added to the slag to maintain a reducing condition, but then the process degenerates into a re-melting practice, in which the only refining action is the removal of some sulphur. The problem of alloy recovery has, however, been simplified with the application of gaseous oxygen to the refining processes and arc-furnace practices based on the use of a controlled atmosphere throughout the melting and refining periods are now almost obsolete.

PROCESSING OF HOT METAL CHARGES

The processing of charges consisting of hot metal and scrap steel in the arc furnace has been made possible by the development of a modified technique. Normally during melting the electrodes burn downwards through the charge to form shallow craters with liquid metal at the bottom but, if sufficient iron ore is included in the charge to give a rapid rate of carbon removal, the slag tends to form a froth and metal and slag are ejected from the furnace. This can be avoided by reducing the amount of ore in the charge, but the rate of carbon removal is lowered correspondingly and the total heat time is increased. In the modified process the scrap metal and most of the ore required are charged on top of the molten iron, and melting is controlled to produce deep craters in the charge [6]. When these extend down to the molten iron the metal boils violently and the slag foams around the electrodes, but the walls of the craters protect the furnace refractories and reduce the risk of ejections from the furnace. When the craters collapse to form a large pool of metal most of the ore has reacted and a steady carbon boil is obtained.

REFINING UNDER A REDUCING SLAG

For the production of quality steels refining is usually completed under a reducing slag. In this case it is necessary to work the heat under the oxidising slag to a lower carbon content than the limits of the specification because carbon can be picked up by the metal under the reducing slag from the electrodes and from the slag when carbon is used as the reducing agent. Most ferro-alloys contain some carbon and the pick-up of carbon by the metal from this source can be large, particularly when large ferro-alloy additions are made in the production of high-alloy steels. The total gain in carbon by the metal is frequently in the range from 4 to 10 points (i.e., 0·040 to 0·10 per cent) and this gain determines the carbon content to which the metal must be refined at the end of the oxidising period.

A carbon-reducing slag is formed by adding lime, coke and a fluorspar flux to the bare metal after the complete removal of the oxidising slag. The ratio of the additions is usually varied according to the carbon content required in the finished steel and for moderately low-carbon steels a typical ratio is:

	CaO	Coke	CaF_2
Parts by weight	10 :	1 :	2

This slag is only mildly deoxidising and the carbon pick-up is low. Some calcium carbide is formed, but the solidified slag is white. For steels containing more than about 0·5 per cent of carbon the proportions can be varied to 6:2:1 to form a more strongly reducing slag, but the carbon pick-up is greater and 3 or 4 per cent of calcium carbide is formed, which gives the solid slag a grey colour and liberates acetylene gas when the slag is immersed in water.

Whilst the removal of carbon from the metal is comparatively straightforward when the concentration of this element is greater than about 0·10 per cent, it becomes increasingly difficult as the content is lowered beyond this level. The iron oxide content of the slag is also increased and the metallic yield is lowered. However, many of the present-day specifications for alloy steels call for carbon contents of 0·05 per cent or even lower (so as to minimise the formation of alloy carbides during fabrication and in service), and in these circumstances a large carbon gain during the reducing period cannot be tolerated. The carbon pick-up can be held to as low as one point, i.e. 0·010 per cent, by using very low-carbon ferro-alloys and by eliminating carbon from the reducing slag, but an alternative reducing agent is then required. Ferro-silicon is widely used for this purpose, but this raises the concentration and hence the activity of the silica in the slag and, if the specification calls for a low residual silicon content in the steel, it may become necessary to increase the amount of lime in the reducing slag in order to lower (a_{SiO_2}) and so decrease the silicon transfer to the metal. Aluminium is also used as the deoxidising agent when traces of aluminium in the finished steel are not detrimental to the service life.

Heat is reflected from the reducing slag on to the furnace roof, causing accelerated wear of the silica refractory and the silica content of the slag may rise to as high as 20 per cent from the roof drippings. When these slags solidify they contain large amounts of calcium ortho-silicate ($2CaO.SiO_2$), which undergoes a transition from a γ to a β structure during cooling. The transition is accompanied by a volume change of about 10 per cent, which causes the slag to fall to a powder during cooling, and the slags are commonly referred to as white-falling-slags.

During this period the transfer of oxygen is from metal to slag, but the bath is quiescent and the transfer of reactants and products between the slag and the metal is dependent entirely on transport by diffusion and convection within the two liquid layers. The reaction rates are lowered very markedly, relative to the rates which are obtained during the carbon boil, and the time taken to effect a given change in the concentration of a dissolved element is correspondingly very much greater. Thus any means of inducing turbulence and reducing the refining time is beneficial and a number of techniques have been employed to promote agitation of the bath, including hand and mechanical rabbling, rotation of the furnace body and bubbling an inert gas through the melt [7, 8]. Inductive stirrers are also used, in which a field coil located under the hearth induces a secondary current in the bath [9, 10]. The current in the primary coil is alternated at a low frequency and a strong stirring action is created in the liquid metal but, although this form of agitation appears to be ideally suited to the arc-furnace construction, opinions differ widely as to the advantages which are obtained [11–14]. It is well established, however, that marked concentration gradients exist in the metal when the bath is not agitated, and it is probable that similar gradients exist in the slag, but the reducing slag layer is usually too thin to permit adequate sampling for the detection of composition differences.

THE REMOVAL OF SULPHUR UNDER A REDUCING SLAG

The conditions which exist during the reducing period are better suited to the transfer of sulphur from metal to slag than in any of the other principal steel-making processes. In fact, one of the unique features of the arc-furnace process is the ability to make high-alloy steels of low sulphur content without the necessity for selecting materials to give a low sulphur content in the charge.

From examination of the equilibrium constant for the sulphur reaction,

$$K_S = \frac{(a_{S^{--}}).[a_O]}{(a_{O^{--}}).[a_S]}, \qquad 10.1$$

it has been shown earlier that the sulphur distribution $(a_S)/[a_S]$ is raised as the oxygen ion content (or the basicity) of the slag is increased or as the oxygen activity in the metal is lowered. The high bath temperature which is reached during the reducing stage allows the formation of a liquid slag with a basicity of about 4·0, which is higher than the optimum basicity of an entirely liquid slag that can be formed in the open-hearth and Bessemer processes. This increase in the basicity contributes to the improvement in the efficiency of sulphur removal, but the major contribution comes from the very low oxygen content of the metal.

In Chapter 14 is it is shown that the sulphur distribution is primarily a function of the slag basicity and is almost independent of the oxygen potential of the slag–metal system over a wide range of potentials, which include those of the metal and slag in the open-hearth and converter processes. Conversely the sulphur distribution is affected markedly by a small change in the oxygen potential in the range encountered during the reducing period in the electric-arc process and normal fluctuations in the basicity of the slag are almost completely masked.

The absence of natural bath agitation during this period is again important with regard to the rate of approach to equilibrium, since the sulphur distribution is directly related to the oxygen in the metal, and it follows that any acceleration in the removal of one of these elements should promote the rate of removal of the other. The lime-injection techniques which have been described in Chapter 7.6 can be used with advantage at this stage to increase the rate of sulphur removal and a fine powder of lime, calcium carbide, calcium silicide or calcium cyanamide, together with a reducing agent, can be blown into the bath in a jet of nitrogen, argon or other oxygen-free carrier gas. The best results are obtained with additions of magnesium plus silicon to give strongly reducing conditions [15], when the sulphur content can be lowered to 0·005 per cent in a few minutes [16]. It is difficult to lower the sulphur to below 0·015 per cent in the traditional practice, and it may take 1 to 2 hours under the reducing slag to reach this level. When the additions selected contain carbon or silicon there is some pick-up by the metal, which can be allowed for but which may prove troublesome in the production of steels required to contain low concentrations of these elements. For example, calcium carbide and cyanamide are undesirable for removing sulphur from very low-carbon steels.

10.3 The Behaviour of Chromium in Steel during Refining

Low-chromium-content steels can be produced by any of the principal steel-making processes, but medium- and high-chromium steels can only be made satisfactorily in the high-frequency and electric-arc furnaces. As already indicated, the high-frequency is essentially a re-melting practice and a consideration of chromium distribution is thus most pertinent to the arc-furnace process.

From examination of the oxygen potential diagram (Figure 1.2) it is evident that chromic oxide is more stable than ferrous oxide and that it is formed at oxygen potentials similar to those required to form the oxides of manganese and silicon. This suggests that the behaviour of chromium under steel-making conditions should be similar to that of manganese and silicon in that the chromium should oxidise in preference to iron, but it does not indicate the extent to which the chromium is oxidised for this is dependent on the activities of the element in the metal and in the slag. In practice it is found that the variation of the chromium recovery with the temperature and slag basicity is similar to that of manganese.

Under oxidising conditions part of the chromium is oxidised and enters the slag, but the solubility of chromium oxide in the slag is only 2 or 3 per cent and, if this amount is exceeded, the oxide remains as a solid suspension which causes a rapid increase in the viscosity. When the charge contains large amounts of chromium the viscosity may become so high that the slag approaches the 'dry' condition which is characteristic of acid Bessemer and high-frequency slags. The loss of chromium into the slag is lowered as the bath temperature is raised, but before the use of gaseous oxygen made possible the attainment of the necessary high temperatures without damage to the refractories, it was practically impossible to refine a charge containing large amounts of high-chromium, or stainless steel scrap. This situation has changed rapidly during the past few years and, by making use of the oxygen injection techniques, it is now routine practice to refine charges containing relatively high chromium contents to obtain a high chromium recovery in the metal and retain adequate fluidity in the slag.

In the early literature there is some confusion regarding the valency of the chromium oxide which is formed in the slag. At first it was thought to be present as the divalent oxide CrO [17–19], but later reports suggested that it is present as chromite ($FeO.Cr_2O_3$) or as the trivalent oxide (Cr_2O_3), depending on the chromium content of the metal [20–22]. In a more recent and thorough investigation it is concluded that a series of chromium oxides is formed in the slag [23], the range of existence of each oxide being determined by the chromium content of the metal. With up to 0·2 per cent of chromium the oxide phase is a wüstite-rich solid solution, which is replaced by the chromite spinel $FeO.Cr_2O_3$ in the range 0·2 to 3·0 per cent of chromium, and by a distorted spinel structure when the chromium exceeds 3·0 per cent. Above 9 per cent of chromium in the metal, the oxide in equilibrium is Cr_3O_4 containing some wüstite in solid solution.

EQUILIBRIUM CONSTANTS FOR CHROMIUM OXIDATION

At the completion of the oxidation stage the amount of chromium which is in equilibrium with the oxygen in the metal can be determined from the equilibrium constant for the formation of the particular chromium oxide which is present in the slag. The equilibrium constants for the formation of the wüstite-rich and the distorted spinel phases have not been evaluated, but the constants for the formation of chromite and Cr_3O_4 are given as [24]:

$$Fe_{(l)} + 2Cr_{(\%\,in\,Fe)} + 4O_{(\%\,in\,Fe)} \rightleftharpoons FeO.Cr_2O_{3(s)} \qquad 10.2$$

$$\Delta G_2^0 = -232{,}200 + 99{\cdot}1T \text{ cal mol}^{-1}{}_{(0\text{–}3\,\%Cr)}$$

and

$$3Cr_{(\%\,in\,Fe)} + 4O_{(\%\,in\,Fe)} \rightleftharpoons Cr_3O_{4(s)} \qquad 10.3$$

$$\Delta G_3^0 = -244{,}800 + 109{\cdot}6T \text{ cal mol}^{-1}{}_{(>9\,\%Cr)}$$

where the standard states for chromium and oxygen in liquid iron are selected as the dilute weight-per cent solutions. Using the activity and interaction coefficients listed in Appendix 1, when the slag is saturated with a chromium oxide the equations can be solved for the amount of chromium in equilibrium with a given oxygen content in the metal.

CHROMIUM–CARBON RELATION

The chromium oxidation reaction takes place primarily at the slag–metal interface and resembles the oxidation of manganese, silicon and phosphorus. This means that during an active carbon boil the residual chromium content of the metal is controlled by the activity of the oxygen in the metal in equilibrium with the iron oxides in the slag and *not* by the oxygen activity in the bulk metal. However, at very low carbon contents the metal oxygen activities in equilibrium with the carbon in the metal approach fairly closely to the oxygen activities in the slag (see Chapter 13.3 and Figure 9.2) and, as the majority of high-chromium steels contain not more than about 0·1 per cent of carbon, it is instructive to examine the chromium recovery relative to the oxygen in equilibrium with a given carbon content. This relation is applicable also to the equilibrium recovery of chromium in the metal when ferro-chromium is added to the bath under a carbide reducing slag. Consider equations 10.2, 10.3 and the equation given in Chapter 13.2 for the oxidation of carbon, viz.,

$$C_{(\%\,\text{in Fe})} + O_{(\%\,\text{in Fe})} \rightleftharpoons CO_{(g)} \qquad 10.4$$

$$\Delta G_4^0 = -3{,}710 - 10{\cdot}09T \text{ cal mol}^{-1}$$

If the slag is saturated with the chromium oxides their activities are unity. At equilibrium p_{CO} is taken as unity and $[a_{Fe}]$ is also assumed to be unity in the chromite range, so that the equilibrium constants for equations 10.2, 10.3 and 10.4 can be written in the following form:

For equation 10.2:

$$[a_O] = \frac{1}{K_2^{1/4}[a_{Cr}]^{1/2}} \qquad 10.5$$

For equation 10.3:

$$[a_O] = \frac{1}{K_3^{1/4}[a_{Cr}]^{3/4}} \qquad 10.6$$

For equation 10.4:

$$[a_O] = \frac{1}{K_4[a_C]} \qquad 10.7$$

If the chromium and carbon reactions are brought to equilibrium at the same oxygen contents, then combining equations 10.5 and 10.7 we obtain:

$$[a_C] = \frac{K_2^{1/4}[a_{Cr}]^{1/2}}{K_4} \quad (0\text{–}3\%\text{Cr}) \qquad 10.8$$

Similarly equations 10.6 and 10.7 are combined to give:

$$[a_C] = \frac{K_3^{1/4}[a_{Cr}]^{3/4}}{K_4} \quad (>9\%Cr) \qquad 10.9$$

Using the activity coefficients for carbon and chromium from Appendix 1, these two equations can now be solved at any temperature for the carbon in equilibrium with the residual chromium in the metal in the chromium ranges specified and Figure 10.1 shows the results which are obtained at temperatures of 1,600 and 1,800°C. The relations are represented by broken lines across the region of the distorted spinel phase to show the approximate positions.

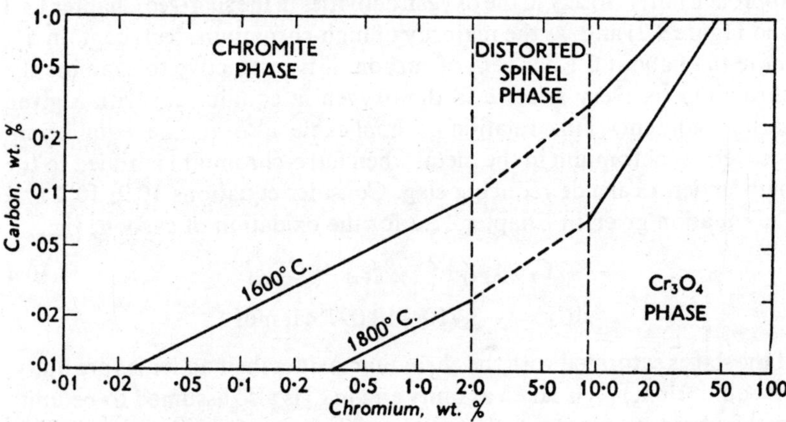

FIG. 10.1 Relation between carbon and chromium concentrations in liquid iron for equilibrium conditions at 1,600° and 1,800°C.

At 1,600°C with 0·1 per cent of carbon in the metal the residual chromium content in equilibrium with the carbon is only slightly greater than 2·0 per cent, and is independent of the chromium in the charge, provided that the slag is saturated with chromium oxides, but the residual chromium content decreases more rapidly than the carbon. For example, if the carbon content is halved to 0·05 per cent, the residual chromium is reduced by more than two-thirds to about 0·6 per cent. Conversely, if the charge contains say 10 per cent of chromium, the diagram indicates that a high chromium recovery is obtained only if the carbon content of the metal is greater than about 0·4 per cent at 1,600°C. However, the actual pressure of carbon monoxide in contact with the carbon and oxygen in the metal is greater than unity. The effect of this on the chromium recovery is not very pronounced at very low carbon concentrations, but for carbon contents greater than about 0·1 per cent the activity of the oxygen in the metal in equilibrium with the slag is greater than the value in equilibrium with carbon and carbon monoxide at one atmosphere pressure. This change in the oxygen activity lowers the chromium recovery relative to the values shown in Figure 10.1 to an extent which is determined by the ratio $p_{CO(actual)}/p_{CO(equilibrium)}$.

THE EFFECT OF OXYGEN INJECTION ON CHROMIUM RECOVERY

Comparison of the curves in Figure 10.1 for the two temperatures shows the pronounced effect of an increase in the temperature on the chromium recovery, this effect being particularly marked at very low carbon contents. Thus at 0·05 per cent of carbon an increase in the temperature of 200°C is accompanied by a ten-fold increase in the chromium recovery. The development of these high temperatures in the metal has become a practical proposition with the application of gaseous oxygen to steel-refining; it is possible to raise the metal temperature to 1,800°C with oxygen blowing solely by the heat liberated from the oxidation of iron, chromium, carbon and other oxidisable elements in the iron. The electric arcs may be cut out during this period and, as the heat is generated within the metal, the slag now acts as a thermal barrier which prevents the overheating of the refractory shell by the hot metal. At the higher temperature less chromium is transferred to the slag, the slag remains fluid with quite high chromium contents in the charge and it is possible to refine a charge containing moderate amounts of chromium to very low carbon contents (i.e. $\leqslant$ 0·05%C). However, as the chromium concentration in the charge is raised the temperature required to obtain both fluidity in the slag and a high chromium recovery in the metal is increased correspondingly. Although the slag provides some protection for the furnace lining, the rate of wear is accelerated as the metal temperature is raised and the improvement in productivity and chromium recovery resulting from the higher bath temperature must then be balanced against greater refractory costs and more frequent interruptions to the furnace relining schedule.

A considerable amount of research has been undertaken, mainly in the U.S.A., to determine the effects of oxygen blowing rate and other variables on the recovery of the chromium [25–30]. The experimental data have been treated mainly in an empirical manner to obtain a relation between the pounds of solute elements which are oxidised, the oxygen blowing rate and the bath temperature (Figure 10.2). For any particular practice there is a critical blowing rate below which the thermal losses from the bath are greater than the thermal gain from oxidation, and this critical rate must be exceeded if the bath temperature is to increase. The total amount of metal which is oxidised into the slag is affected markedly by the bath temperature at the start of the oxygen blow, and it is preferable to start the oxygen when the bath temperature is about 1,600°C, rather than at the clear-melt stage [28, 29].

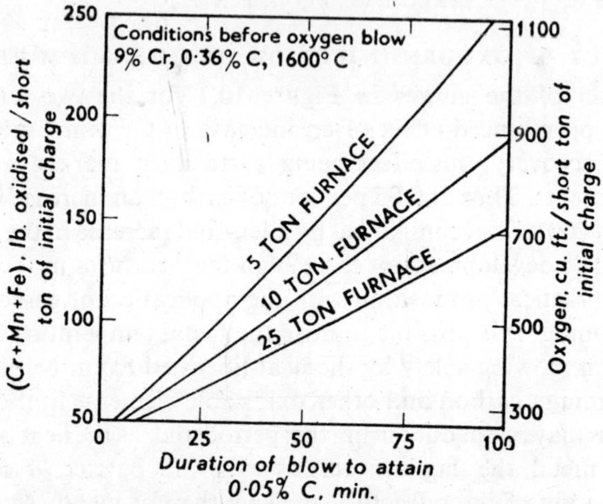

FIG. 10.2 Relation between the weight of metal oxidised, the rate of oxygen blowing and the duration of the blow for oxygen-refining of chromium steels in electric-arc furnaces of various capacities. (After Hilty, Rassbach and Crafts [28].)

CHROMIUM RECOVERY BY DEOXIDATION

As much as one-third of the charge may be oxidised into the slag during oxygen-blowing to very low carbon contents [29] but, at the end of the blow, the metallic yield can be increased by adding a strong deoxidiser to the bath to displace some of the metals from the slag. Thus silicon can displace chromium from the slag by a reaction such as:

$$Cr_3O_{4(slag)} + 2Si_{(in\ Fe)} \rightleftharpoons 3Cr_{(in\ Fe)} + 2SiO_{2(slag)} \qquad 10.10$$

Alternatively the reaction may be considered as consisting of two simultaneous steps, in one of which the oxygen potential of the bath is lowered by the silicon addition and in the other chromium is transferred from slag to metal to restore the equilibrium at the new oxygen potential.

When silicon is added as the reducing agent the amount required is determined by the amount of metal to be reduced from the slag, but a limit is set by the maximum residual silicon content allowed in the steel specification. Other deoxidisers such as aluminium or titanium may be used to supplement the silicon addition if the amount required would cause the silicon specification limit to be exceeded. The opportunity is usually taken to lower the metal temperature to about 1,650°C when the deoxidisers are added to the bath, thus increasing the efficiency of the deoxidation (the deoxidising reactions are exothermic) and protecting the refractory lining against excessive overheating. The temperature can be lowered by adding ferro-silicon with a low silicon content, to the extent that the heat consumed in melting the iron content of the alloy is considerably greater than the exothermic heat of oxidation of the silicon.

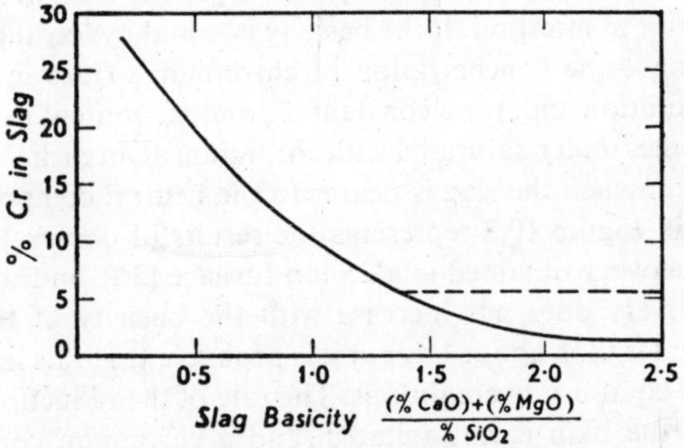

FIG. 10.3. The effect of basicity on the concentration of chromium in the slag. (After Rassbach and Saunders [26].)

The amount of iron, chromium and manganese reduced from the slag is estimated as 2·5 to 3·0 lb per lb of silicon added [26, 28, 31], but the recovery of chromium varies with the basicity and the total volume of the slag. The reason for this is readily seen in terms of the equilibrium constant for reaction 10.10

$$K_{10} = \frac{[a_{Cr}]^3 (a_{SiO_2})^2}{[a_{Si}]^2 (a_{Cr_3O_4})} \qquad 10.11$$

When the slag is saturated with the chromium oxide the ratio $[a_{Cr}]/[a_{Si}]$ varies inversely in proportion to the activity of the silica in the slag. Thus for a given value of $[a_{Si}]$ the chromium recovery is raised as the basicity of the slag is increased, as illustrated by the full line in Figure 10.3. In silica-saturated slags the recovery is very low, whereas between 90 and 95 per cent of the chromium in the charge can be recovered when the slag basicity is greater than 2·0. If the reduction is continued into the range where the slag is not saturated with chromium oxides, the recovery of chromium is dependent also on both the chromium content of the slag and the variation in the activity of the chromium oxides with the basicity. The variation with composition in the activity of chromium oxides in unsaturated slags has not been determined experimentally, but in the solid state chromic oxide forms spinel compounds with the basic oxides FeO, MgO and MnO, whereas there is extensive liquid and solid immiscibility in the system Cr_2O_3–SiO_2. On this basis it is expected that the activities are high in acid slags and decrease slowly as the basicity is raised.

The effect of the slag volume on the extent of the chromium recovery is more difficult to interpret. If the basicity is raised by feeding more lime into the furnace the concentration of chromium oxides in the slag is lowered by dilution and, for a constant chromium content in the charge, the slag becomes under-saturated with chromium at an earlier stage in the reduction than when the slag is nearer to the neutral composition. The broken line in Figure 10.3 represents the results of observations of the chromium recovery obtained in a 25-ton furnace [26], and this suggests that the recovery does not increase with the basicity of the slag for basicities greater than about 1·4, but it is probable that this line does not represent the equilibrium conditions. The rate of the reduction reactions is low when the bath is not agitated, and a chromium concentration gradient is formed in the slag in a similar manner to the sulphur concentration gradient which forms during the reducing-slag period. This premise is supported by experimental evidence [32], for a marked improvement in the extent of the chromium recovery was obtained when, after the apparent attainment of the silicon–chromium equilibrium, the metal and slag were mixed by tapping from the furnace into a ladle and were then returned to the furnace.

On comparing these opposing effects it appears that the optimum chromium recovery should be obtained with a slag basicity of about 2·0, for at high basicity the activity of the silica in the slag changes only slowly as the slag composition is altered, but it increases more rapidly with a decrease in basicity below 2·0, whereas the volume of the slag increases in proportion to the increase in the basicity. In view of the recent development of chromium–manganese stainless steels it is interesting to note that a high manganese recovery is obtained also under similar conditions.

Whilst the addition of deoxidisers to the oxidising slag improves the recovery of these elements, it must be realised that the lowering of the slag basicity and the reduction in the oxygen potential of the system decreases the phosphorus-holding capacity of the slag; some phosphorus reverts to the metal during the deoxidation and only small amounts remain in the slag after reduction. The lowering of the oxygen potential favours sulphur transfer to the slag, but this is opposed by the lower basicity of the slag, with the result that less sulphur is removed from the metal than when a highly basic and oxidising slag is maintained throughout the first part of the refining cycle.

10.4 The Recovery of other Alloy Elements

Many alloy steels contain expensive alloy elements other than chromium and the extent of the recovery of these elements in the refined metal is important, but the factors affecting the recovery of any particular element should now be clear from the preceding discussion. Briefly these are: (i) the variation with temperature of the equilibrium constant for the appropriate oxidation reaction, (ii) the oxygen activity and the activity coefficients of the elements in the slag and metal phases, (iii) how closely the equilibrium condition for the oxidation reaction is approached during the time the slag and metal are in contact.

The equilibrium constants for the oxidation of nickel, molybdenum, titanium, tungsten and vanadium, relative to the dilute weight-per cent solution reference state, can be calculated from the known thermodynamic data (see Chapter 13.5) provided that the valency of the oxide in the slag is known. Some elements may form a series of oxides, similar to the series of chromium oxides which are known to exist, but the oxidation products of most of the other alloy elements as a function of the concentration of each element in the metal have not been investigated as extensively as in the case of chromium. However, over limited concentration ranges the more common elements are known to exist in the slag as the oxides NiO, MoO_2, TiO_2, WO_3 and V_2O_3 and the values of the equilibrium constants for the formation of these oxides at 1,600 and 1,800°C are calculated as follows:

$$Ni_{(\%\,in\,Fe)} + O_{(\%\,in\,Fe)} \rightleftharpoons NiO_{(l)} \qquad 10.12$$

$$\log K_{12} = \log \frac{(a_{NiO})}{[a_{Ni}][a_O]} = \frac{7{,}526}{T} - 7{\cdot}82 \qquad 10.13$$

$$K_{12(1,600°C)} = 1{\cdot}6 \times 10^{-4} \quad K_{12(1,800°C)} = 6{\cdot}5 \times 10^{-5}$$

$$Mo_{(\%\,in\,Fe)} + O_{(\%\,in\,Fe)} \rightleftharpoons MoO_{2(s)} \qquad 10.14$$

$$\log K_{14} = \log \frac{(a_{MoO_2})}{[a_{Mo}][a_O]^2} = \frac{17{,}750}{T} - 10{\cdot}47 \qquad 10.15$$

$$K_{14(1,600°C)} = 0{\cdot}10 \quad K_{14(1,800°C)} = 1{\cdot}2 \times 10^{-2}$$

$$Ti_{(\%\,in\,Fe)}+2O_{(\%\,in\,Fe)} \rightleftharpoons TiO_{2(s)} \qquad 10.16$$

$$\log K_{16} = \log\frac{(a_{TiO_2})}{[a_{Ti}][a_O]^2} = \frac{32{,}870}{T} - 11{\cdot}92 \qquad 10.17$$

$$K_{16(1{,}600°C)} = 4{\cdot}3\times10^5 \quad K_{16(1{,}800°C)} = 8{\cdot}7\times10^3$$

$$W_{(\%\,in\,Fe)}+3O_{(\%\,in\,Fe} \rightleftharpoons WO_{3(l)} \qquad 10.18$$

$$\log K_{18} = \log\frac{(a_{WO_3})}{[a_W][a_O]^3} = \frac{25{,}720}{T} - 15{\cdot}30 \qquad 10.19$$

$$K_{18(1{,}600°C)} = 2{\cdot}8\times10^{-2} \quad K_{18(1{,}800°C)} = 1{\cdot}3\times10^{-3}$$

$$2V_{(\%\,in\,Fe)}+3O_{(\%\,in\,Fe)} \rightleftharpoons V_2O_{3(s)} \qquad 10.20$$

$$\log K_{20} = \log\frac{(a_{V_2O_3})}{[a_V]^2\,[a_O]^3} = \frac{42{,}490}{T} - 17{\cdot}19 \qquad 10.21$$

$$K_{20(1{,}600°C)} = 3{\cdot}2\times10^5 \quad K_{20(1{,}800°C)} = 3{\cdot}5\times10^3$$

The activities of the oxides can be set at unity if it is assumed that the slag is saturated with respect to each of the oxides and the residual concentration of each element can be calculated for equilibrium conditions as a function of the oxygen activity in the metal. However, it is clear from the values of the equilibrium constants that the recovery of nickel, molybdenum and tungsten in the metal should be high, whereas vanadium and titanium should be transferred almost entirely into the slag. The residual amounts of vanadium and titanium have been calculated for an oxygen potential in equilibrium with 0·10 per cent of carbon in order to compare the recoveries of these elements with that of chromium and the values obtained are as follow:

	Bath Temperature		
	1,600°C	1,800°C	
Chromium	2·06	10·6	weight-per cent
Vanadium	0·025	1·45	,, ,, ,,
Titanium	0·005	0·70	,, ,, ,,

It is again evident that the residual concentrations are increased very significantly when the bath temperature is raised, as with oxygen-blowing.

The relative amounts of the recoveries of these elements apply only to the case where the slag is saturated with the oxide product considered in the calculations. It is known that vanadium and titanium in low concentrations associate with ferrous oxide to form $FeO.V_2O_3$ and $FeO.TiO_2$

in preference to the simple oxides (the oxides which are formed are examined in more detail in Chapter 13.5). The recoveries are different from those listed when these compounds are formed.

The residual concentration of an element is also lowered if the slag is unsaturated so that the activity of the oxidation product is less than unity, which is important with regard to the recovery of tungsten and molybdenum. The solubility of the tungsten oxide (WO_3) in the slag is fairly appreciable and the oxide associates with lime to form a stable calcium tungstate compound in the solid state. Thus the oxide has an acidic character, its activity in the slag decreases with increasing basicity [33, 34], and it follows that when the slag is not saturated with the oxide the recovery of tungsten in the metal decreases as the basicity and the oxidising power of the slag are increased. Up to 20 per cent of the tungsten may be transferred into a basic slag. Molybdenum also forms a calcium molybdate compound and similar considerations apply to the recovery of this element when the slag is under-saturated, but in practice it is found that molybdenum is not oxidised in preference to iron. This is sometimes a disadvantage for, if the scrap metal contains molybdenum, it is then impossible to make a molybdenum-free steel.

Many alloy steels contain small concentrations of niobium (columbium) which apparently is present in the slag as the pentoxide [35] but the solubility and activity of the oxide as a function of the slag composition have not been determined. However, it is generally found that at 1,600°C the recovery of niobium in the metal is about 60 to 70 per cent of the concentration in the charge and so the recovery is intermediate between that of vanadium and titanium, but closer to the former element.

Since the partitioning of alloy elements between metal and slag is apparently controlled largely by the oxygen potential and composition of the slag, the factors affecting alloy recovery in the arc-furnace process apply similarly to the open-hearth and Bessemer processes. The essential difference between the processes lies in the ability to recover the alloys from the slag by deoxidation. The atmosphere is more readily adjusted and the amount of alloy recovered per pound of deoxidiser added is considerably greater in the arc furnace than in the open-hearth, and the reduction of the open-hearth and Bessemer slags is more costly on account of the very much greater slag volume. In all processes however, the recovery of alloys by reduction is not feasible when the slag is required to contain large amounts of P_2O_5.

In the second stage of the arc-furnace process, when the metal is covered by a reducing slag, the activity of the oxygen in the metal is very low, and the recovery of any alloy additions which are made during this period is correspondingly greater than the recovery which is obtained during the oxidising period. From the earlier considerations it is evident that the yield is high for the less-readily oxidised elements such as chromium, molybdenum and tungsten, but the recovery of elements such as vanadium and titanium is still comparatively poor and these are usually added either immediately before tapping and after a final deoxidation with aluminium or to the stream of metal as it enters the ladle.

10.5 Hydrogen in the Electric-Arc Furnace Process

Hydrogen in solution in solid steel has a deleterious effect on the mechanical properties. In particular, it lowers the ductility without producing a corresponding increase in strength, and it is one of the principal factors which determine the incidence of hair-line crack formation. Under one atmosphere pressure of hydrogen the solubility of hydrogen in pure iron is 0·0027 weight per cent at 1,600°C [36], in air the solubility is correspondingly lower and it is more conveniently expressed in units of millilitres per hundred grams (1 ml/100 g = 0·00009 % w/w) or in parts per million, rather than as a percentage by weight. Under steel-making conditions the hydrogen content of liquid steel varies between 2·0 and 10.0 millilitres per 100 grams, but the variation within this range has a pronounced effect on the mechanical properties. Contamination of the metal by hydrogen can occur in any steel-making process, but the loss of quality which results from the presence of the gas in solution is usually more important with alloy than with plain-carbon steels, and the control of the hydrogen concentration is thus generally of more importance for steels produced in the arc-furnace process. Long and costly heat treatments are required to lower the concentration to within safe limits in steels which are particularly susceptible to hair-line cracking or hydrogen embrittlement, but these heat treatments can be shortened, or eliminated entirely, by close control of the pick-up of the gas while the metal is molten.

Hydrogen can be absorbed by molten steel up to the saturation limit when it is exposed to an atmosphere in which hydrogen or hydrogen-containing compounds, hydrocarbons or moisture are present. One or more of these may be present in the furnace atmosphere, in the charge and the bath additions, or in the atmospheres to which the metal is exposed between tapping from the furnace and up to the time when it solidifies. For example, steam is formed when the molten metal comes into contact with damp refractory linings in the furnace launder, in the transfer ladle, or in the casting pit settings, whilst the moulds are often dressed with tar-based compounds to facilitate stripping the ingots and the hydrocarbons contained in these dressings may cause a 100 per cent increase in the hydrogen content of the metal.

Considerable quantities of moisture may be introduced with additions of solids to the bath. Of these, lime and limestone are probably the principal sources [37], but moisture may also be present in the iron ore or as an adsorbed layer on the ferro-alloys and oxidised scrap metal. Under adverse conditions more than 5 per cent by weight of the charge may consist of water [38] and pre-heating to dry the materials before feeding to the molten bath is a decided advantage.

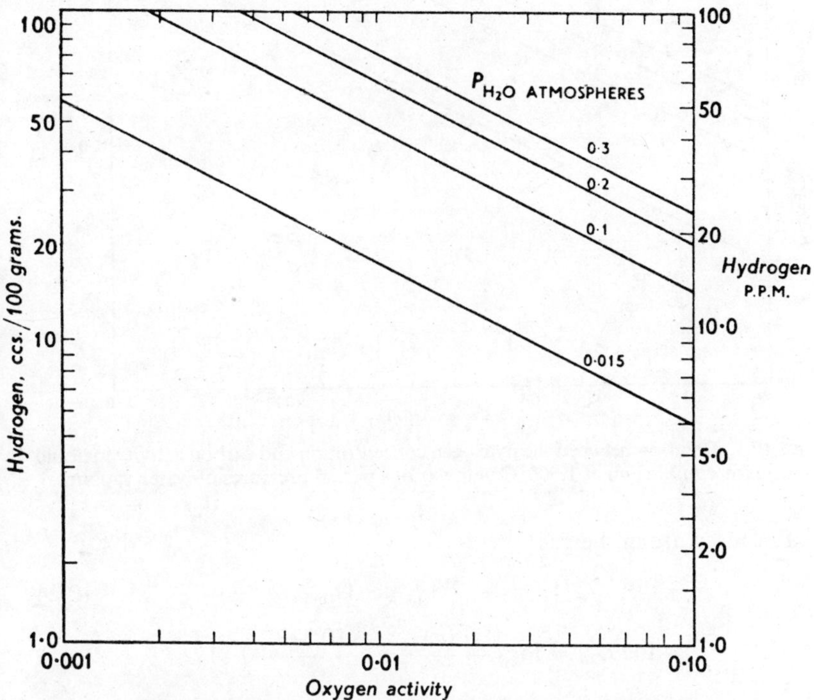

FIG. 10.4. Relation between the hydrogen concentration and oxygen activity in liquid iron for equilibrium at 1,600° C with various partial pressures of water vapour.

HYDROGEN–WATER VAPOUR EQUILIBRIA

Water vapour is present in the furnace atmosphere from the humidity in the air and when the molten metal is exposed to the atmosphere the vapour can react with the metal according to equations such as the following [38]:

For 'free' hydrogen in the gas:

$$\tfrac{1}{2}H_{2(g)} \rightleftharpoons H_{(in\,Fe)} \qquad 10.22$$

$$\log K'_{22} = \log \frac{[H]}{p_{H_2}^{1/2}} = -\frac{1670}{T} + 2{\cdot}32 \qquad 10.23$$

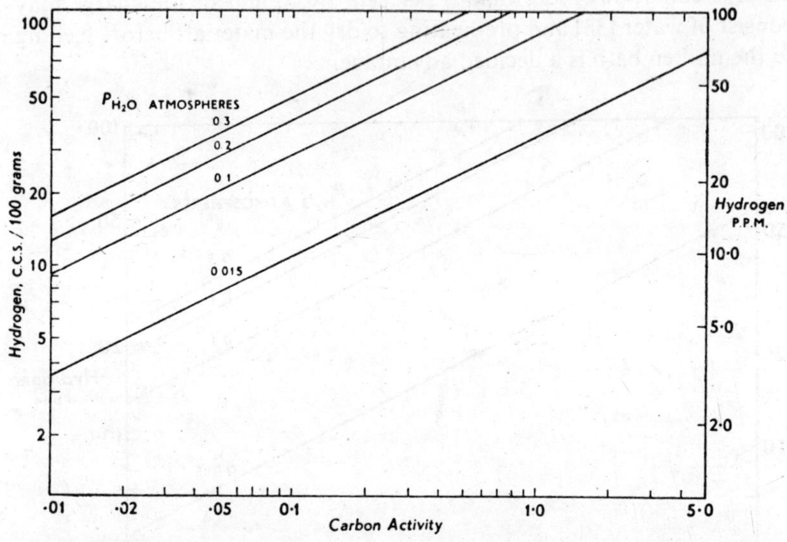

FIG. 10.5. Relation between the hydrogen concentration and carbon activity in liquid iron for equilibrium at 1,600° C with various partial pressures of water vapour.

For moisture in the gas:

$$H_2O_{(g)} \rightleftharpoons 2H_{(in\,Fe)} + O_{(in\,Fe)} \qquad 10.24$$

$$\log K'_{24} = \log \frac{[H]^2[a_O]}{p_{H_2O}} = -\frac{10{,}390}{T} + 7{\cdot}81 \qquad 10.25$$

where the hydrogen concentration in the metal is expressed as parts per million. Equation 25 is shown in graphical form in Figure 10.4 for partial pressures of steam of 0·015, 0·1, 0·2 and 0·3 atmospheres at 1,600°C. The lowest partial pressure is typical of the atmosphere in the arc furnace, whilst the highest partial pressure represents the maximum moisture contents which are normally found in the atmosphere of the open-hearth furnace. The axes are plotted on a logarithmic basis to show clearly the equilibrium at both high and low oxygen and hydrogen contents; the data are then represented by a family of straight lines, which means that as the oxygen activity is raised the equilibrium hydrogen concentration falls rapidly at first and then more slowly. Similar behaviour is shown in Figure 9.2 for the carbon-oxygen relation and it follows, therefore, that the equilibrium hydrogen content of the metal is low when the carbon content is low, and vice versa. This relationship is illustrated in Figure 10.5, where the hydrogen content is plotted against the activity of the carbon in the metal for the same partial pressures of moisture in the atmospheres as in Figure 10.4.

At very low oxygen activities, of the order of 0·001, or at the higher steam pressures, the solubility of hydrogen in the metal in equilibrium with the water vapour is greater than the value in equilibrium with pure hydrogen gas at one atmosphere pressure and the same temperature, but during the refining of medium and low-carbon steels the oxygen activity is in the range 0·01 to 0·10 and the lower oxygen activities are obtained only in very high carbon heats or after deoxidation.

SOLUBILITY OF HYDROGEN AND WATER VAPOUR IN THE SLAG

When a slag cover forms over the metal the transfer of hydrogen from the atmosphere to the metal is mainly via the slag. It appears that steel-making slags are relatively impermeable to hydrogen gas [39, 40], but they are slightly permeable to water vapour [38, 40], and there is good correlation between the humidity of the atmosphere and the hydrogen content of the metal [41]. The solubility of hydrogen in the slag is found to be proportional to the square root of the partial pressure of the water vapour in the furnace atmosphere so, from Sievert's law relation, (equation 2.22) it is suggested that the water vapour does not dissolve in the slag in molecular form, but as hydrogen or hydroxyl ions [36] according to the equations:

$$H_2O_{(g)} + O^{--}{}_{(slag)} \rightleftharpoons 2OH^{--}{}_{(slag)} \qquad 10.26$$

$$H_2O_{(g)} \rightleftharpoons 2H^{+}{}_{(slag)} + O^{--(slag)} \qquad 10.27$$

Other evidence [42] suggests that the formation of hydroxyl ions is more probable than of hydrogen ions. It is probable that the moisture in the atmosphere is dissociated into hydrogen, oxygen and hydroxyl ions in the very high-temperature zone of the electric arc, so that the partial pressures of these ion groups are greater than at steel-making temperatures and the transfer of the ions to the slag should be correspondingly greater.

The solubility of moisture is negligible in slags free from silica [40]. In lime–alumina–silica melts the activity of the hydroxyl or hydrogen ions is proportional to the square root of the activity of the lime [42] but, within the range of slag compositions which are normally encountered in basic steel-making practice, the solubility varies to only a minor extent with changes in the composition and temperature of the slag.

The partitioning of hydrogen between the slag and the metal is dependent on the activity of the hydrogen in the two phases. In the metal the activity coefficient of hydrogen is raised by aluminium, silicon, phosphorus and sulphur and it is lowered by chromium, manganese and vanadium [43], which means that for alloy steels containing large amounts of the last three elements the solubility of hydrogen is greater than it is in a plain-carbon steel. Cobalt, nickel [43] and oxygen [44] appear to have no effect on $[f_H]$. The effect of oxygen is contrary to that expected for, in view of the ease of combination of hydrogen and oxygen atoms to form water vapour or hydroxyl groups, it would normally be expected that $[f_H]$ would be lowered by oxygen.

TRANSFER OF HYDROGEN BETWEEN GAS, SLAG AND METAL

While the metal is held in the furnace hydrogen can transfer to the metal from the atmosphere either directly or via the slag until such time as the equilibrium concentration is attained, but this equilibrium is not reached in practice and equations 10.23 and 10.25 are limiting conditions. Thus it is commonly found that steels made under an acid slag have lower hydrogen contents than those made under a basic slag [37, 40, 45]. The solubility of hydrogen is lower in the acid slags but this accounts for only part of the difference and it is possible that the higher viscosity retards the rate of transfer of hydrogen from slag to metal, relative to the rate with a basic slag of lower viscosity. It has been shown that the addition of bauxite to a basic slag to increase the fluidity causes an increase in the hydrogen content of the metal [46] and heavy lime feeds might usefully be employed to thicken a basic slag and lower the rate of hydrogen transfer.

During an active carbon boil the transfer of hydrogen between slag and metal can be regarded as a two-way process. The transfer from slag to metal continues as before, but it is probable that the transfer rate is increased by the greater turbulence of the slag and metal, and particles of metal which are ejected through the slag may absorb hydrogen from the furnace atmosphere before returning to the bath. This is countered by hydrogen diffusing into the bubbles of carbon monoxide as they rise through the bath, and the change in the hydrogen content of the metal during this period is determined by the rates of the two opposing processes. The rate of transfer from gas to metal is dependent on the actual displacement from the equilibrium condition, whilst the rate of the reverse reaction is proportional to the rate at which the carbon is oxidised. Thus the minimum rate of carbon oxidation to ensure a reduction in the hydrogen content of the metal is about 0·5 per cent of carbon per hour in the basic and 0·3 per cent of carbon per hour in the acid open-hearth [36, 38, 41, 45, 47, 48]. The critical boil-rate in the basic arc-furnace process is probably close to that for the acid open-hearth, for the partial pressure of water vapour in the atmosphere of the arc furnace is considerably lower than in the open-hearth. The critical rate is lowered when gaseous oxygen is used to promote the rate of carbon removal, since hydrogen is also absorbed by the bubbles of unreacted oxygen gas which escape from the bath, and it is claimed that low carbon steels containing less than 4 parts per million of hydrogen can be produced by oxygen blowing [49].

When the oxidising slag is removed the bath is exposed to the furnace atmosphere while the deoxidisers are added and until the reducing slag is formed. The transfer of hydrogen from gas to metal is then fairly rapid and the rate should increase as the deoxidisers lower the oxygen content of the metal. It is important, therefore, that the reducing slag is formed as quickly as possible and significant decrease in the hydrogen content is found when the oxidising slag from the first stage is reduced to recover the alloy elements and the same slag is then used for the reducing period. The bath is quiescent under the reducing slag and hydrogen is transferred only from gas to metal, but the oxygen activity in the metal is lower and so the rate of transfer should be greater than when the bath is quiet under an oxidising slag. Hence any means of reducing the period during which the metal is held under the reducing slag serves also to lower the extent of the hydrogen pick-up. However, the probable main source of hydrogen contamination at this stage is the moisture contained in both the ferro-alloys and the slag-forming additions. If the additions are heated to drive-off the moisture and the time under the reducing slag is kept as short as possible, the hydrogen pick-up in the reducing period can be held to not more than 2 parts per million. Alternatively, the hydrogen content can be lowered by flushing the metal with an inert gas, such as argon or helium [50], or more effectively by flushing with one of these gases together with lithium vapour which is generated by adding lithium pellets to the bath [51].

Further reduction in the hydrogen content can be obtained after the metal leaves the furnace by either vacuum pouring or vacuum casting [52, 53]. Providing the pressure in the vacuum chamber is less than 5 millimetres of mercury, the hydrogen content of the metal can be lowered to less than 3 parts per million [54].

NITROGEN IN THE ARC-FURNACE PROCESS

The contamination of the melt with nitrogen is not usually a problem in the arc-furnace process, since it is continuously removed from the bath during the carbon boil. The final nitrogen contents are approximately equal to those obtained in the open-hearth process [55], but are usually greater when highly-alloyed steels are produced, for vanadium and chromium in particular cause a marked lowering in the activity coefficient of nitrogen and raise the solubility in iron. For example, the solubility of nitrogen in a steel of high chromium content is 2 to 3 times as great as the solubility in a plain-carbon steel of similar carbon content.

A small increase in the nitrogen content of the metal is often observed between the times when the oxidising slag is removed and when the metal is tapped from the furnace. It is probable that this is due in part to absorption of nitrogen from the atmosphere before the reducing slag covers the metal and in part to the presence of nitrogen in solid solution in the ferro-alloy additions.

Interest has recently centred on the production of alloy steels containing up to 0·2 per cent of nitrogen. The austenitic phase is stabilised by nitrogen and, as such, it has proved beneficial in the chromium–manganese stainless steels. It is also useful as a grain refiner and as a means of improving certain properties of chromium–alloy steels. The nitrogen is usually added in the form of a high-nitrogen content ferro-chromium or as nitrided electrolytic manganese, but the available thermodynamic data are not suitable for forecasting the recovery of nitrogen in the metal from these additions.

REFERENCES

1. H. W. MCQUAID, *Iron Steel Engr.*, 1952, Vol. 29, p. 139.
2. J. C. HOWARD, *J. Iron St. Inst.*, 1953, Vol. 173, p. 285.
3. J. E. WILBANKS, *J. Metals, N.Y.*, 1953, Vol. 5, p. 627.
4. E. KREBS and E. PAKULLA, *Stahl u. Eisen, Düsseldorf*, 1955, Vol. 75, p. 9.
5. A. C. OGAN and D. I. CARNEY, *J. Iron St. Inst.*, 1958, Vol. 178, p. 307.
6. R. DURRER and G. HEINTZE, ibid., 1959, Vol. 192, p. 13; *Roll'schen Mitt.*, 1958–60, Vol. 17, p. 44.
7. C. E. SIMS, *Proc. Amer. Inst. Min. (metall.) Engrs. elect. Furn. Steel Conf.*, 1949, Vol. 7, p. 302.
8. A. FAERDEN, *Metallurgia*, 1958, Vol. 58, p. 57.
9. L. DREYFUS, *Jernkontor. Ann.*, 1949, Vol. 133, p. 371.
10. F. NILSON, ibid., 1949, Vol. 133, p. 384.
11. K. G. SPEITH, H. VOM. ENDE and W. SCHNEIDER-MILO, *Stahl u. Eisen, Düsseldorf*, 1958, Vol. 78, p. 215.
12. P. E. HAMMARLUND, *Hutn. List.*, 1958, Vol. 13, p. 868.
13. H. F. WALTHER, *J. Metals, N.Y.*, 1954, Vol. 6, p. 21.

14. P. E. HAMMARLUND, and B. HANAS, *Iron Coal Tr. Rev.*, 1960, Vol. 180, p. 299.
15. W. B. BROOKS, A. C. DOUMAS and B. W. ROMEFELT, *Iron Age*, 1960, Vol. 185, p. 148.
16. L. MARPLES and J. PEARS, *J. Iron St. Inst.*, 1960, Vol. 195, p. 195.
17. F. KÖRBER and W. OELSEN, *Mitt. K.-Wilh. Inst. Eisenhforsch.*, 1935, Vol. 17, p. 231.
18. P. HERASYMENKO, *Trans. Faraday Soc.*, 1938, Vol. 34, p. 1251.
19. D. C. HILTY, *Trans. Amer. Inst. min. (metall.) Engrs.*, 1949, Vol. 185, p. 91.
20. H. M. CHEN and J. CHIPMAN, *Trans. Amer Soc. Metals*, 1947, Vol. 38, p. 70.
21. N. J. GRANT, E. C. ROBERTS and J. CHIPMAN, *Trans. Amer. Inst. min. (metall.) Engrs.*, 1954, Vol. 200, p. 145.
22. E. T. TURKDOGAN, *J. Iron St. Inst.*, 1954, Vol. 178, p. 278.
23. D. C. HILTY, W. G. FORGENG, and R. L. FOLKMAN, *Trans. Amer. Inst. min. (metall.) Engrs.*, 1955, Vol. 203, p. 253.
24. J. CHIPMAN, *J. Iron St. Inst.*, 1955, Vol. 180, p. 97.
25. D. C. HILTY, G. W. HEALY and W. CRAFTS, *Trans. Amer. Inst. min. (metall.) Engrs.*, 1953, Vol. 197, p. 649.
26. H. P. RASSBACH and E. R. SAUNDERS, ibid., 1953, Vol. 197, p. 1009.
27. E. PACHOLY, *Stahl. u. Eisen, Düsseldorf*, 1953, Vol. 73, p. 461.
28. D. C. HILTY, H. P. RASSBACH and W. CRAFTS, *J. Iron St. Inst.*, 1955, Vol. 180, p. 116.
29. G. W. HEALY and D. C. HILTY, *J. Metals, N.Y.*, 1956, Vol. 8, p. 325; *Blast Furn.*, 1956, Vol. 44, p. 1048; *Trans. Amer. Inst. min. (metall.) Engrs.*, 1957, Vol. 209, p. 95.
30. T. MAZANEK, J. NATHANIEC, Z. KOSIEWICZ and A. WOLKENBERG, *Hutnik*, 1958, Vol. 3, p. 3.
31. W. CRAFTS and H. P. RASSBACH, *Proc. Amer. Inst. Min. (metall.) Engrs. elect. Furn. Steel Conf.*, 1951, Vol. 9, p. 95.
32. A. J. TEXTER, ibid., 1953, Vol. 11, p. 73.
33. V. O. ZOTKIN, *Chern. Metallurg.*, 1959, (2), p. 27.
34. B. M. MOGUTNOV, N. N. PEREVALOV and L. A. SHVARTSMAN, *Izv. Ak. Nauk Metallurg i Topl.*, 1959, (1), p. 22.
35. M. L. SAZONOV and L. A. SHVARTSMAN, ibid., 1959, (3), p. 34.
36. D. C. CORNEY, J. CHIPMAN and N. J. GRANT, *Proc. Amer. Inst. Min. (metall.) Engrs. elect. Furn. Steel Conf.*, 1948, Vol. 6, p. 34; *Trans. Amer. Inst. min. (metall.) Engrs.*, 1950, Vol. 188, p. 404.
37. C. SYKES, H. BURTON and C. GEGG, *J. Iron St. Inst.*, 1947, Vol. 156, p. 155.
38. H. EPSTEIN, J. CHIPMAN and N. J. GRANT, *Trans. Amer. Inst. min. (metall.) Engrs.*, 1957, Vol. 207, p. 597.
39. L. S. DARKEN, *Proc. Amer. Inst. Min. (metall.) Engrs. elect. Furn. Steel Conf.*, 1948, Vol. 6, p. 55.
40. J. H. WALSH, J. CHIPMAN, T. B. KING and N. J. GRANT, *Trans. Amer. Inst. min. (metall.) Engrs.*, 1956, Vol. 206, p. 1568.
41. C. F. SIMS, G. A. MOORE and D. WILLIAMS, *Proc. Amer. Inst. Min. (metall.) Engrs. elect. Furn. Steel Conf.*, 1947, Vol. 5, p. 106.
42. L. N. BARMIN, O. A. ESIN and S. K. CHUCHMOREV, *Izv. Ak. Nauk Otdelenie Tekn.*, 1957, (9), p. 114.
43. M. OHTANI and N. A. GOKCEN, *Trans. Amer. Inst. min. (metall.) Engrs..* 1960, Vol. 218, p. 533.

44. N. N. DASTUR and J. CHIPMAN, *Trans. Amer. Inst. min. (metall.) Engrs.*, 1949, Vol. 185, p. 441.
45. H. N. KARNAUKHOV and A. K. URAZGUL'DEEV, *Izv. Ak. Nauk Otdelenie Tekn.*, 1955, (8), p. 93.
46. N. H. DOBROKHOTOV, D. Y. POVALOSKII and K. T. KAHN, *Inst. Chemoi. Met. Akad. Navi, SSSR*, 1953, p. 23.
47. N. CHUIKO, *Stal'*, 1940, (10), p. 37.
48. H. WENTRUP, H. FUCKE and O. RIEF, *Stahl u. Eisen, Düsseldorf*, 1949, Vol. 69, p. 117.
49. N. CHORINOV, *Hutn. List.*, 1958, Vol. 13, p. 402.
50. C. E. SIMS, *Proc. Amer. Inst. Min. (metall.) Engrs. elect. Furn. Steel Conf.*, 1949, Vol. 7, p. 302.
51. A. J. NIMETH, *Iron Age*, 1959, Vol. 184, p. 168.
52. F. HARDERS, H. KNÜPPEL and K. BROTZMANN, *Stahl u. Eisen, Düsseldorf*, 1956, Vol. 76, p. 1721.
53. A. TIX, *Trans. Amer. Inst. min. (metall.) Engrs.*, 1956, Vol. 208, p. 420.
54. V. SCALISE, A. DE NEGRI and V. PICCARDI, *J. Iron St. Inst.*, 1960, Vol. 195, p. 260.
55. B. L. BIGGS, ibid., 1959, Vol. 192, p. 361.

CHAPTER 11

Alternative Methods of Iron and Steel Production

In the previous five chapters the thermo-chemistry has been described of what may be called the traditional processes of iron and steel manufacture. These processes together account for the present-day production of about 99·5 per cent of the annual output of steel throughout the world. However, a large number of other processes have been suggested, some of which have been developed to the pilot-plant stage, whilst a few have progressed to full-scale production. This review of the production of iron and steel would not be complete without some reference to these alternative processes. Although the new processes often appear to be radically different from the traditional methods, the physico-chemical principles on which they are based are very similar, and the factors which affect the equilibria and the kinetics of the various reactions are also similar. It is intended, therefore, to describe briefly some of these new processes to illustrate how the reaction kinetics can be changed by altering the conditions under which a reaction occurs.

11.1 The Reduction of Iron Ores

MODIFICATIONS TO THE BLAST FURNACE PROCESS

The blast furnace has remained the principal means of reducing iron oxides for several centuries and it is probable that it will continue in this capacity within the foreseeable future. This process is, however, dependent on the availability of coals suitable for the production of blast-furnace coke and of iron ores which are suitable for reduction in the blast furnace.

The theoretical thermal requirement for the reduction of iron oxides by carbon monoxide is about 8·5 million B.Th.U. per ton of iron whereas, even under the most favourable conditions, in the blast furnace the corresponding figure is about 20 million B.Th.U. and a further 3 to 5 million B.Th.U. are consumed in processing the impure iron product into steel [1]. Part of this increase is accounted for by the heat required in melting the charge to separate the gangue materials from the iron but, even so, the process is far from being 100 per cent thermally efficient.

Any process which can perform similar functions for a smaller expenditure of thermal energy and which uses fuels other than metallurgical coke is thus advantageous from this point of view. On the other hand, there is no immediate shortage of suitable coking coals in Britain or the United States and it is understandable that the ferrous industries in these countries with a heavy capital investment in existing plant have shown less interest in alternative reduction processes than countries such as the Scandinavian bloc and Mexico, which have no indigenous coal supplies.

Many of the alternative processes are designed to operate with a finely-divided iron ore, and are therefore better suited for the reduction of concentrates. The provision of the ore in a suitable condition for reduction involves higher material costs than the preparation of the ore for the blast furnace but, with close control of the operating conditions, it is possible to produce a metal whose composition is closer to that of steel and which needs less refining than the metal from the blast furnace. The ultimate objective is to produce in one operation a steel of any required composition directly from the iron ore and several of the new processes approach closely to this ideal. Against these advantages must be set the facts that (i) none of the processes which have been proposed to date appears suitable for development to give the output per unit which can be obtained from the larger blast furnaces now in use, and (ii) at present the cost per ton of metal produced is significantly greater than in the traditional process.

Since the blast furnace process is so firmly established, it is natural that some thought should have been given to modifying the geometry of the furnace to allow other fuels to be used and/or to alter the strength and reducibility requirements of the iron ore. Electric smelting has been used on a comparatively small scale for many years, but the low-shaft blast furnace has been developed only since the end of World War II.

Electric Smelting

Carbon is necessary only for the reduction of the ore and, if the energy required to heat the charge is supplied by means other than the combustion of coke, the coke consumption per ton of iron produced is reduced to less than half that of the orthodox blast furnace. In countries with an abundance of cheap hydro-electric power the charge can be smelted in a furnace similar to the electric-arc refining unit. The earlier furnaces of the ELECTROMETAL type superimposed a blast furnace stack on the hearth with the object of partially reducing the ore by the ascending gases, but the volume of gas passing up the stack is only one-quarter to one-sixth of that produced in the normal blast furnace. The sensible heat in the gas is transferred rapidly to the charge, the ore is pre-heated for only a short distance above the hearth, and only a very limited amount of gaseous reduction is obtained before the charge reaches the hearth. Later furnaces of the TYSLAND HOLE type have dispensed with the stack and charge both the ore and reducing agent directly into the hearth. Lower grade fuels such as coke breeze can then be used for reduction, since the permeability and crushing strength requirements of the stack are obviated.

The Italian LUBATTI process [2] is designed to reduce ore fines or concentrates with fine coal or coke. The charge is fed continuously to form a layer about 6 inches thick, resting upon a liquid slag in which graphite electrodes are immersed. The ore is reduced by the solid fuel and by carbon monoxide from the slag, and the metal falls as droplets through the slag to collect on the furnace hearth. It is claimed that the waste gases are low in carbon monoxide and, since the charge acts as an insulating blanket, the heat losses are low.

The potential development of electric smelting is severely limited by the relatively high power consumption and, where cheap power is available, it is probable that the production of a sponge iron by processes using electricity will prove more economical.

The Low-Shaft Blast Furnace

The air blast of the orthodox furnace contains 79 per cent by volume of unreactive nitrogen which absorbs heat and increases the thermal burden in the combustion zone, and the high stack is necessary to recover this heat by transfer to the charge. If the oxygen content of the blast is increased a smaller blast volume is required, more heat is available in the hearth and the rate of smelting is increased, but the process then resembles electric smelting in that the total volume of gas passing through the furnace is lowered and less heat is available in the stack. For example, if the hot blast is enriched to 28 per cent of oxygen the zone of red heat extends to only about 4 feet above the tuyeres [3]. Thereafter the temperature of the gas falls rapidly to 100–200°C at about 20 feet above the tuyeres and no real thermal advantage is gained by increasing the height of the furnace beyond this level. If the height of the furnace is limited to about 20 feet the charge is not required to possess such a high crushing strength as in the normal blast furnace and lower grade ores and fuels can be handled satisfactorily. As the volume of gas is lowered by oxygen enrichment the velocity of the gas through the furnace is decreased, the dust losses from the furnace are lowered correspondingly, and a more finely divided ore can then be smelted. This is the basis on which the low-shaft blast furnace process has developed.

Pilot plants have been operating for some time at Calbe, Liege, Oberhausen, Troisdorf and elsewhere [4–9], where extensive trials have been conducted to determine performance with a variety of charge materials which are unsuitable for the orthodox furnace. The process can operate successfully without any coke in the charge, the carbon required being supplied in the form of anthracite, steam coals, or lignites, but the shortage of stack heat necessitates the decomposition of carbonates and the removal of moisture from the ore before it is charged into the furnace. Serious trouble is also encountered from hanging and irregular operation unless the sizing and the distribution of the burden within the furnace are closely controlled. The operation is more regular if the ore, fuel and flux are thoroughly mixed and charged to the furnace as briquettes.

The chemical reactions are similar to those which occur in the orthodox blast furnace process, the principal differences compared with the normal blast furnace practice being as follow:

1. The restricted hot zone above the tuyeres allows little time for gaseous reduction of the iron oxides and the amount of direct reduction is considerably greater.
2. The oxygen potential of the slag and metal in the hearth is slightly greater, as shown by the higher iron oxide content in the slag. For example, the composition of the tap slag in the Demag-Humboldt process is given as [9]:

Total Iron	CaO	MgO	SiO_2	Al_2O_3
1·1	44·7	6·6	34·1	10·4

The analysis of the slag iron content is uncertain because of the presence of entrapped metal particles, but a concentration of 1·1 per cent is greater than the value which is normally reported for a blast furnace slag.
3. The temperature of the slag and metal in the hearth is greater with an oxygen enriched blast. Thus the ratio CaO/SiO_2 in the above slag is 1·3 and with this basicity in the normal blast furnace process there is no difficulty in producing an iron which contains less than 0·8 per cent of silicon, but the range of metal compositions obtained from the Demag-Humboldt process is [9]:

C	Si	Mn	P	S
4·3	0·9–1·4	0·9–1·1	0·12–0·14	0·025–0·050

The higher silicon content is probably a result of the higher hearth temperature. With close control of the slag composition the higher temperature can be utilised to melt a slag with a basicity of about 1·8 and it is then possible to produce an iron containing about 0·2 per cent of silicon. At the same time the manganese recovery is improved and the transfer of sulphur into the slag is increased. Alternatively an endothermic gas such as steam could be added to the blast to control the hearth temperature, as described in Chapter 5.

A higher hearth temperature is useful when the furnace is operated with an acid burden. In the normal blast-furnace process the slag is extremely viscous if the basicity is lowered below about 0·85. The low-shaft process will operate satisfactorily with a slag basicity as low as 0·7, but the extension of the working range may be due only partly to the higher hearth temperatures and partly to the higher concentrations of ferrous and manganese oxides in the slag [10].

No large-scale furnaces of the low-shaft design have yet been constructed; those at present in operation have throughputs of about 60 to 150 tons per day and a true comparison with the orthodox furnace is thus not possible, but it is clear from the data available that the thermal consumption is considerably higher in the low-shaft. The increased amounts of direct reduction are reflected in the higher CO/CO_2 ratio in the exit gas, the composition of which is usually within the following range:

CO %	CO_2 %	CH_4 %	H_2 %	N_2 %	CO/CO_2	C.V. (B.Th.U./cu.ft)
29–32	2–4	1–4	7–10	52–58	8–10	130–160

The calorific value is higher than that of normal blast furnace gas, but the gas is a by-product and this represents a decrease in the thermal efficiency of the process.

There seems little probability of the low-shaft replacing the traditional process in countries with an established ferrous extraction industry, but the process is attractive in those countries where a new extraction industry is being established and where the raw materials are unsuitable for existing practices.

11.2 The Direct Production of Iron from the Ore

A large proportion of the impurity elements in pig iron is absorbed by the iron in the high-temperature bosh and hearth zones. This applies particularly to the carbon, silicon and manganese contents, which are low at the stage where gaseous reduction ceases. If the iron could be separated from the gangue within the temperature range pertinent to the indirect reduction reactions the fuel consumption would be reduced markedly and the metal produced would require less extensive refining. However, at these low temperatures neither the iron nor the gangue materials are completely liquid and the usual separation within the furnace by differences in density does not operate. A sponge iron containing about 90 per cent of metallic iron can be obtained if the ore is available as a rich concentrate and the end product of the reduction is crushed and separated by magnetic or gravity concentration. This sponge can form part of the charge in the electric-arc or open-hearth furnace [11–14], but the comparatively low density of the sponge make charging difficult unless it is previously compacted into nodules or briquettes.

Because the reduction of solid ore is diffusion controlled the process must be operated near to the upper temperature limit, where sticking and build-up commence, in order to obtain reasonably high rates of production. This maximum temperature is generally about 1,000–1,100°C, but varies with the grade of ore being reduced and for a given ore it decreases with increasing reduction, owing to the formation of low melting-point ferrous silicates and sintering of metallic iron particles. At lower temperatures difficulties are encountered from the pyrophoric nature of the reduced metal, for very finely divided metallic iron burns spontaneously in air at atmospheric temperatures and re-forms iron oxides. If the reduction is completed at temperatures in excess of about 700°C the particles of iron sinter together to form a sponge with a smaller surface area per unit volume, and a sponge which is formed at 900°C is pyrophoric at high temperatures but comparatively stable at room temperature. This means that the reduction process is either confined to the relatively narrow temperature range of 900 to 1,100°C or the reduced material must be stored in sealed containers in an atmosphere of hydrogen or in oil to avoid spontaneous combustion.

Finely divided concentrates provide ideal feed material as intimate contact between ore and reducing agent or reducing gas is essential for maximum production. High operating rates necessitate an excess of reducing agent, and provision must be made for cleaning and recirculating the gas and for separating unreacted solid reducing agents. Magnetic or similar methods are used in the latter case but the sponge inevitably becomes contaminated with some mechanically-entrapped carbon.

The gangue materials can be separated by density difference from the metallic iron if the end-product of the process is liquid, but when a carbonaceous reducing agent is used the metal is carburised to an extent which depends on the intimacy and duration of contact of the metal with carbon, the oxygen potential and the temperature. Some transfer of other elements from the ore into the iron will also occur as in the blast furnace. By close control of the operating conditions it may be possible to produce a metal which is closer in composition to that of steel and which needs less refining, but a rather impure iron must result. Strictly speaking, processes which yield an impure liquid end product should not be included under the general heading of direct reduction processes, but it is now customary to group under this heading all reduction processes other than those which are conducted either in the traditional or the modified blast furnace.

GASEOUS REDUCTION

If the process is to be competitive the reducing gas must be produced readily from abundant and fairly cheap materials; this condition limits the gaseous reducing agents to carbon monoxide and hydrogen or to mixtures of these gases. Carbon monoxide is readily produced by the controlled combustion of coal or oil, and carbon monoxide-hydrogen mixtures can be obtained by the water-gas reaction, but natural gas is only used after a preliminary reforming treatment.

If the free-energy equations for the reactions between FeO or Fe_3O_4 and carbon monoxide or hydrogen are solved at various temperatures for the gas compositions in equilibrium with the iron oxides, it is found that above 815°C hydrogen is a more efficient reducer than carbon monoxide (i.e., the equilibrium H_2/H_2O ratios are lower than the corresponding CO/CO_2 ratios), whilst the converse applies at lower temperatures (Figure 11.1). These equilibria are rarely, if ever, attained in industrial practice, because the rate of reduction becomes very slow as equilibrium is approached, and when conditions deviate markedly from equilibrium the respective reaction rates for reduction with hydrogen and carbon monoxide are in the reverse order to those which would normally be expected from the equilibrium considerations (see Chapter 6). Thus hydrogen is actually a more efficient reducer for a non-equilibrium process which is designed to operate at temperatures below 815°C and carbon monoxide is more efficient at higher temperatures.

The reduction of iron oxides to metallic iron with hydrogen is endothermic and an external source of heat is necessary to maintain the charge at the required temperature, but the corresponding reactions with carbon monoxide are exothermic and under suitably controlled conditions the reactions are thermally self-sustaining. In fact, it may be necessary to dilute the carbon monoxide with hydrogen or other heat-absorbing gas to avoid overheating the charge. Some processes have been designed to take advantage of the CO–H_2 thermal balance and use mixtures of these gases to increase the amount of reduction obtained during the heating of the ore from the ambient temperature up to the maximum reaction temperature of about 1,100°C.

Figure 11.1 shows that for all temperatures within the range where gaseous reduction is economically feasible, the equilibrium gas mixtures contain at least 60 per cent of carbon monoxide and/or hydrogen. When equilibrium is not attained the unreacted concentrations of these gases are even higher and the greater part passes unchanged through the

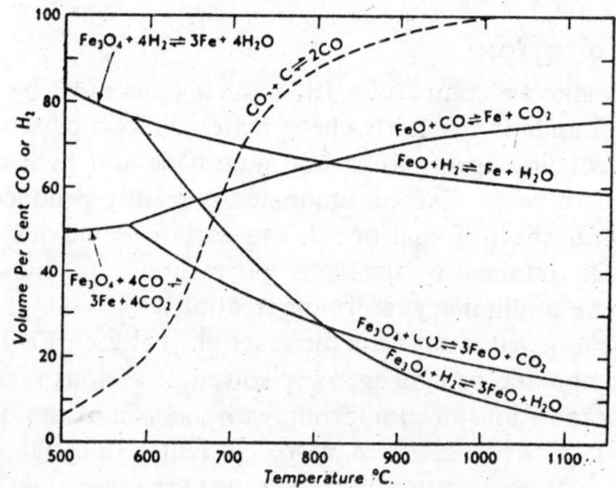

FIG. 11.1 Equilibrium relations in the systems iron–carbon–oxygen, iron–hydrogen–oxygen and carbon–oxygen.

reducing chamber. If the process is to be economical provision must be made for utilising the gas which remains after the reduction of wüstite to metallic iron for reduction of the higher oxides of iron to wüstite and/or for the regeneration of the gas mixture and the removal of the gaseous reaction products.

Figure 11.1 shows also that at temperatures below about 700°C carbon monoxide should react to form carbon dioxide and carbon rather than reduce wüstite to metallic iron. However, the rate of the latter reaction is considerably greater than the thermal breakdown of carbon monoxide and in the temperature range 600 to 700°C it is possible to convert finely divided ferrous oxide almost entirely into cementite (Fe_3C) [15]. This would not be possible if equilibrium conditions were attained.

A large number of process have been suggested for the gaseous reduction of iron ores, but they can be grouped into three main types and the basic principles are illustrated by the following brief description.

Reduction of a Solid Charge in a Kiln-Type Process

This group is most readily described by reference to the Swedish WIBERG-SÖDERFORS process, which has been in commercial production since 1941 and furnaces producing 25,000 tons of sponge iron per annum are now in operation. Although this output is very low in comparison with the present-day blast furnace it is high in relation to the productivity of most other processes. The construction and operation of the reducer have been described fully in a number of publications [16–20]. The reduction is carried out in a vertical shaft kiln into which the ore is charged from the top as fine concentrates or as a pelletised feed and a gas mixture of carbon monoxide and hydrogen is fed in near the base of the kiln at a temperature of 950-1,000°C. If the kiln is heated externally to bring the charge up to this temperature there is a risk that the finely divided and partly reduced iron oxides will adhere to the hot refractory lining of the kiln and build up accretions which interfere with the smooth descent of the charge. However, by close control of the gas ratio p_{CO}/p_{H_2} the reduction reaction can be made just sufficiently exothermic to maintain the temperature of the ore at about 1,000°C without external heating. The heat required is then generated entirely within the charge, the refractory lining is cooler than the mean temperature of the ore and the possibility of a build-up on the lining is lowered.

The reducing gas ascends the kiln counter-current to the charge and the higher oxides of iron are reduced to wüstite in the upper half of the kiln, while the final reduction to metallic sponge iron is effected in the lower half by the rich gas which is fed in continuously. At the level at which the reduction to wüstite is practically complete, most of the gas is bled-off, regenerated and re-cycled. The remaining gas continues up the shaft where the higher oxides of iron are reduced and is burned with admitted air near the top of the kiln to preheat the incoming charge.

A typical sponge analysis with a charge comprising Swedish magnetite concentrates containing about 70 per cent of iron is as follows [20]:

	Metallic Iron	*Total Iron*	C	P	S	SiO_2
Weight-per cent	82	90	0·9	0·010	0·010	3·0

If 90 per cent of the oxygen is removed from the iron oxides the energy consumption is about 7 million B.Th.U. per ton of sponge iron produced and the output per cubic foot of kiln is only slightly less than that of a modern blast furnace [19]. This degree of reduction is suitable when the sponge iron is used as part of the charge in the open-hearth or electric arc-furnace processes, but if further reduction is required the energy consumption increases rapidly and the rate of production falls.

The burden requirements are on a par with those of the blast furnace, i.e., the ore must have a relatively high crushing strength and a good reducibility. Iron ore concentrates and sinter are not regarded as suitable feed materials as the charge tends to choke in the shaft; smoother performance is obtained with a pelletised feed [19], but any further increase in the height of the kilns is dependent on the development of pellets with a higher crushing strength and better reducibility than those now available. The process appears to be a promising alternative to the blast furnace for, apart from some carbon pick-up, the iron contains only very small amounts of impurity elements, but the gangue is not separated from the iron and either very rich ore concentrates are required or the sponge must be crushed and the iron separated by magnetic, gravity or other means. The electricity consumption in the generation and purification of the gas is also very high and the process is suitable only in countries with abundant, cheap electric power.

A similar principle is applied in the O.R.F. process [21] where the ore is carried through the reducer on a travelling grate, similar to that in a sinter machine. With high-grade ore concentrates the reduced sponge iron can be fed from the reducer between pairs of steel rollers to produce a usable form of sheet metal. The Mexican H-Y-L process uses a vessel similar in shape to a steel ladle but with a closed top, in which the ore is reduced by hot, re-formed natural gas [21, 22].

Fluidised-Bed Processes

The rate of a reaction between a gas and a solid is enhanced considerably if the solids are suspended as finely divided particles in a gas atmosphere; this condition can be achieved if the velocity of the gas flowing upwards through the kiln is just sufficient to counteract the gravity settling of the ore. The particles then remain suspended and separated in the gas and the bed moves in similar manner to a liquid, a small variation in the gas velocity causing the particles to rise slowly or descend in the gas stream.

The settling rate of the ore is dependent on the particle size and the ore must be sieved to a very small range of sizes if the process is to operate efficiently. The usual size range is about 0·5–3·0 mm diameter. It is more difficult to maintain a fluidised bed with smaller particles, but it is often necessary to crush the ore to a finer size so as to release the iron oxides from the gangue content. If such concentrates are used in a fluidising process it may be necessary first to sinter and then crush to obtain the optimum size range. The settling rate is determined also by the density of the ore, and this is one of the major problems in the development of the technique, for during reduction the density of the particles changes by 20–30 per cent. Some improvement can be obtained by reducing the ore in two stages, first to wüstite in a mildly reducing atmosphere and then to metallic iron in a more strongly reducing gas.

The upper temperature limit is set by the tendency of the particles to sinter or stick together, which again interferes with the maintenance of a fluidised bed. In the H-IRON process the reduction is effected by hydrogen at about 500°C [11, 14, 23, 24], but the rate of the reaction is low at this temperature, when the pressure of the hydrogen gas is 1 atmosphere, and a more rapid rate is obtained by pressurising the gas to 300–400 lb per sq. in. The high-pressure operation also facilitates the removal of the moisture from the reacted gas before recirculation. The sponge iron product is strongly pyrophoric and must be handled accordingly. Similar processes currently under development are the NU-IRON and ESSO-A.D. LITTLE.

It is claimed that iron oxides can be converted into cementite with an atmosphere of carbon monoxide in a fluidised bed without interference from carbon deposition; the presence of cementite in the end product lowers the tendency for the reduced particles to stick together so that the process can be operated at a temperature as high as 650°C [15, 25]. The reduction is halted before all the ferrous oxide has been reduced and the mixture of oxide and carbide is heated in a neutral atmosphere at a temperature in excess of 750°C, when a rapid reaction occurs between the oxygen and the carbon. It should be possible to produce a steel to any desired carbon specification if the ratio Fe_3C/FeO can be controlled within close limits at the end of the reduction stage [15].

Gaseous Reduction with a Liquid End-product

The rate of reduction of iron oxides by a reducing gas increases rapidly as the temperature is raised, but if the metal and gangue remain solid throughout the process the maximum working temperature is limited to less than that at which the particles stick together. A significant increase in reaction rate should be obtained by raising the temperature above the fusion range of the ore and the metal, but in Chapter 6 it is shown that the rate of gaseous reduction is seriously retarded when the ore particles are coated with a layer of partly molten slag. Thus any method based on this procedure must be radically different from the orthodox process, and it is not surprising that application of this principle has not yet progressed beyond the experimental stage. Nevertheless two techniques have been proposed which are theoretically sound and appear to be promising if the operational problems can be overcome.

The CYCLO-STEEL process is described as 'an inversion' of the Bessemer process [26]. Finely-divided ore is preheated in a fluidised bed to a temperature just short of sticking and is then injected with powdered lime and solid carbon into a jet of oxygen at the top of a tower. The strongly exothermic reaction between the solids and the gas causes the ore particles to melt and form drops of liquid dispersed in the gas phase. By close control of the partial pressure of the oxygen in the gas the ore can be reduced to the metallic state without significant carbon pick-up, while the lime reacts with any sulphur in the gas to prevent the contamination of the metal. The metal and slag collect in two layers at the bottom of the tower.

In the JET SMELTING process [21], which is basically very similar, a magnetite ore concentrate is heated rapidly to about 2,000°C by oxidation to haematite in a jet of oxygen. The oxidised particles are immediately entrained in a large excess of natural gas and reduced to metallic iron which accumulates on the furnace hearth below a slag cover.

THE USE OF SOLID REDUCING AGENTS

The only suitable solid reducing agents are coal, coke and charcoal or their mixtures; the alumino-thermic and similar processes are eliminated if the process is to compare in cost with established processes. Unless the metal is molten at some stage during reduction it is contaminated with excess of reducing agent and fuel ash, in addition to the gangue materials from the ore. The reducing agent usually contains sulphur, most of which is transferred to the metal unless a sufficient quantity of lime is included with the charge. Generally the sulphur and ash contents of the charge are increased when low-grade fuels which are unsuitable for the blast furnace are used.

Reduction with a Solid End-product

Reduction by solid carbon is very much slower than the rate of gaseous reduction which can be obtained at temperatures below the limit at which the ore particles start to stick together. This is mainly due to the inadequate contact between the ore and the reducing agent and a very long time is required for complete reduction by this means. For example in the HÖGANÄS process, where the finely divided ore is heated in clay saggers surrounded by coke, complete reduction may take several days [19, 27]. This process has been in commercial operation since 1911, but it is used mainly to produce a sponge which is suitable for powder metallurgy from very high grade concentrates.

Some improvement in the reaction rate may be expected if the ore and the carbonaceous fuel are agitated during the reduction and the contact zones between the particles are continuously changing. One method of achieving this condition is to effect the reduction in a rotary kiln and this has been adopted in a number of processes such as the DOMNARVET ROTOR [28] and the R-N [21, 29]. The kiln is partly filled with ore and reducing agent and is rotated about the horizontal axis, while the charge is heated by combustion within the kiln. The reduced material is discharged continuously from one end of the kiln and fresh charge is added from a hopper at the opposite end. The working range of temperatures is similar to that for gaseous reduction with a solid end-product and the temperature must be controlled within a narrow range if a high rate of production is to be obtained without build-up on the refractory lining. The ore is heated by the combustion of gas or oil above the surface and the flame temperature must be higher than that of the charge to obtain a rapid rate of heat transfer. Consequently exposed parts of the refractory lining tend to become hotter than the charge, but this is corrected as the speed of rotation is increased. In practice it is found necessary to rotate the kiln at speeds of 20–40 r.p.m.

The end-product from these processes is separated partly from the excess fuel by gravity or magnetic concentration. After this treatment the carbon content is still greater than 1·0 per cent, most of which is mechanically entrapped in the porous sponge and is not in solution in the iron [28]. Further purification can only be effected after crushing or grinding to release the carbon.

Reduction with a Liquid End-product

Most of the processes in this group are modifications of the blast-furnace process, such as the low-shaft furnace and electric smelting which were mentioned earlier in this chapter. The product is an impure iron which contains large amounts of carbon and other impurities and which differs only in detail from the product of the blast furnace.

The KRUPP-RENN process occupies a special position and operates most satisfactorily with silicious ores which are unsuitable for the blast furnace. It achieves complete melting of the gangue materials but only partial melting of the metal [30–33]. The ore is mixed with a carbonaceous fuel and heated in a rotary kiln to a maximum temperature of 1,300–1,400°C.

The gangue constituents must form relatively low melting-point compounds and freeze over a wide temperature range if the process is to operate satisfactorily. The iron and slag are separated outside the furnace and it is desirable that the viscosity of the slag should be considerably greater than in the traditional processes, because separation is effected more readily when the slag solidifies as a vitreous mass than when it is crystalline [34]. The vitreous condition is achieved more readily when the slag is rich in silica and alumina.

The fuel consumption is comparatively low and almost any low-grade fuel can be used, but the low basicity of the slag results in almost complete transfer of the sulphur from the fuel and the ore into the metal, and either fuels of low sulphur content must be selected or the metal must be desulphurised externally before charging into the refining unit. The carbon content of the iron is again about 1 per cent, but the atmosphere is not as strongly reducing as in the blast furnace and only comparatively small amounts of manganese, silicon and chromium are reduced into the metal. Japanese ores of high titanium content, which are unsuitable feed for the blast furnace, can be reduced fairly readily in the Krupp-Renn process [35].

Up to 90 per cent reduction of the iron oxides is readily achieved in this process but some re-oxidation occurs when the product is discharged hot from the kiln. This fact, together with the high sulphur content and the gangue remaining in the sponge after gravity or magnetic separation, means that the end-product is not really suitable for charging directly to the refining processes. It is often suggested that the Krupp-Renn is best suited for up-grading iron ores which are unsuitable in the raw state for smelting in the blast furnace, and the iron obtained could be desulphurised and separated from the remainder of the slag in a low-shaft blast furnace. It is claimed that the total fuel consumption for this two-stage process is less than that observed when the same ores are smelted directly with large amounts of limestone in the blast furnace [30]. Similar considerations may apply also to the treatment of the end-product from some of the other alternative processes, but a two-stage practice for the production of pig iron does not appear to be attractive while ores are available which can be used directly in the blast furnace.

11.3 Alternative Refining Processes

Although the metallurgical efficiency and the productivity of the established refining processes have been improved considerably in recent years, they all suffer from inherent weaknesses which limit the improvement which can be obtained without drastic alteration to either the furnace or the technique. For example, the refining times in the open-hearth and electric-arc processes have been shortened appreciably by the application of gaseous oxygen, but the heat times are still reckoned in hours. At the other extreme the refining is so rapid in the bottom-blown Bessemer converter that the composition of the tap metal can be controlled within suitable limits only when low-carbon steels are produced. Even if the change in the composition with time during the blow could be estimated with greater precision, the basic Bessemer is still limited to the production of low-carbon heats by the necessity for an after-blow to remove the phosphorus. These problems have been largely surmounted by the development of the LD top-blown converter processes, but the charge composition is limited to the refining of moderately low phosphorus heats and it remained for the development of the LD–AC and OLP modifications to extend the process to the refining of a charge of any chosen composition.

None of these processes can be regarded as thermally efficient. The heat losses through the hearth are very large in the open-hearth and electric-arc processes and, except during oxygen-blowing, the temperature of the furnace atmosphere is significantly greater than that of the metal. In the converter processes there is a loss of about 780,000 B.Th.U. per ton of metal refined because the carbon is only partly oxidised to carbon dioxide within the converter [36].

Thus although in comparison with the traditional refining methods the LD–AC and OLP practices represent a very significant improvement in both metallurgical and thermal efficiency, there is still considerable scope for improvement and numerous attempts have been and are being made to develop higher-efficiency techniques.

THE ROTOR PROCESSES

Some indication of the type of process which may satisfy the various requirements can be obtained from consideration of the ROTOR processes. These were initiated at about the same time as the LD converter and were subsequently developed mainly in an attempt to extend the range of iron compositions which could be refined with oxygen-blowing. In fact they can be regarded, together with the LD–AC and OLP processes, as further modifications of the converter principles, but they introduce the novel feature to steel refining of mixing the slag and metal by rotation of the refining vessel. A vital feature of these processes is that practically the whole of the heat available from the oxidation of the carbon is recovered by combustion of the carbon monoxide with oxygen gas inside the vessel (Figure 11.2). The heat of oxidation of the silicon and phosphorus in the metal is then of secondary importance to the thermal requirements and an iron of almost any composition can be refined, provided that its carbon content is high. Excess of heat is available from the refining of a high-phosphorus, basic-Bessemer grade of iron and this can be utilised to melt fairly large quantities of scrap metal, iron ore or sponge iron. Even with an iron low in phosphorus and silicon there is a small surplus heat available for the melting of some solid metal.

The KALDO ROTOR is a cylindrical vessel, of similar shape to the Bessemer converter, which rotates at up to 30 r.p.m. on an axis inclined

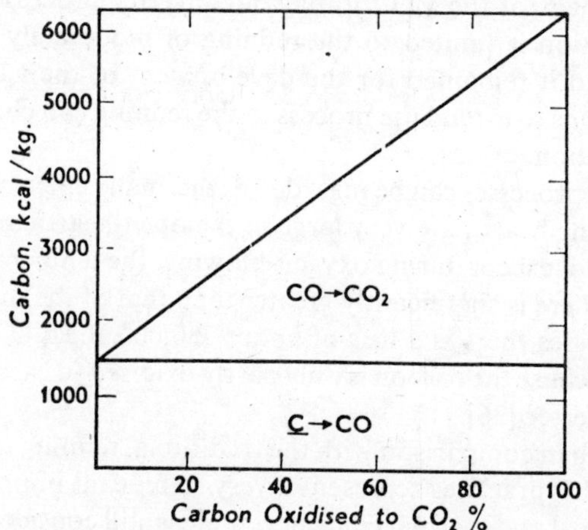

FIG. 11.2. Relation between the heat liberated per kg of carbon and the percentage of the carbon which is oxidised to carbon dioxide within the Rotor. (After Kalling and Johansson [40].)

at 17 to 20° to the horizontal [37–39]. A water-cooled lance is inserted through the nose of the vessel through which oxygen gas is blown to strike the surface of the bath at a shallow angle. The agitation of the slag and metal by rotation of the bath gives a more uniform oxygen distribution in the metal with less localised heat generation and a decrease in fume formation.

The viscosity and iron content of the slag and the rate of carbon oxidation are controlled to some extent by varying the lance position and the speed of rotation [40]. If the oxygen is blown at a small angle to the bath the rate of formation of iron oxides is low and the slag tends to be viscous, but as the angle of impingement is increased the iron oxides form more rapidly and the fluidity of the slag is raised. High speeds of rotation give intimate mixing of slag and metal, a high rate of oxygen transfer to the metal and rapid elimination of the carbon, with the result that the iron oxide content of the slag is lowered and the viscosity is raised. Lower speeds decrease the rate of carbon removal, but allow more opportunity for the complete combustion of the carbon monoxide within the vessel. Thus high rotation speeds are desirable to shorten the refining time and,

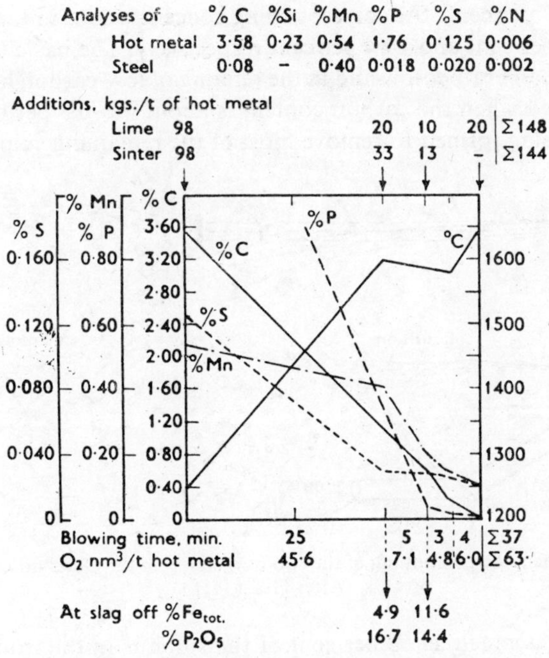

FIG. 11.3. Variation with time of the temperature and composition of the metal for a high-phosphorus heat in the Kaldo rotor process. (After Kalling and Johansson [37].)

particularly towards the end of the heat, to lower the iron oxide content of the slag and increase the metal yield, but even at the maximum speed the actual refining time is about double the time required in the LD converter process.

With a basic lining in the vessel, sulphur and phosphorus can be made to transfer into the slag almost from the start of refining, for the lime is fluxed very rapidly by the oxygen jet and phosphorus can be removed to a low residual level while the carbon content is still high. When high-phosphorus irons are refined it is usually necessary to remove the slag once or twice during the course of refining and a small volume of new slag is formed with lime additions. Figure 11.3 shows the change in metal composition with time during a heat in which the slag was removed after 25 minutes and again after a further 5 minutes of refining. The effect of the rotational action on the iron oxides in the slag, relative to the corresponding concentrations in the converter process, is clearly evident, for the total iron content of the first slag is only 4·9 per cent and the P_2O_5 content is 16·7 per cent; the corresponding concentrations in the second slag are 11·6 per cent and 14·4 per cent respectively. The basicity of the first slag is usually rather low and in the refining of low-carbon heats this slag is removed when the carbon content is about 0·2–0·3 per cent and a more basic slag formed to remove most of the remaining sulphur and

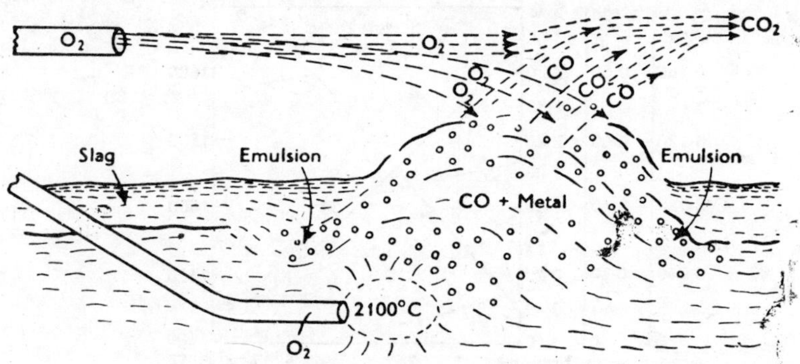

FIG. 11.4. Schematic positioning of the two oxygen jets in the Oberhausen rotor process. (After Graef [41].)

phosphorus. As much as 50 per cent of the sulphur in the iron can be removed into this second slag and it is possible to refine an iron containing 1·5–1·8 per cent of phosphorus to meet a specification of 0·025 per cent phosphorus maximum in the steel [37]. For the production of higher-carbon steels the first slag is removed at an earlier stage in refining and the residual phosphorus content is higher but, even in high-carbon steels, it can be held to not greater than 0·035 per cent [40].

The oxygen content of the refined metal corresponds closely to that in open-hearth furnace steels at similar carbon contents [36] and the nitrogen content of the steel is about 0·002–0·003 per cent when the oxygen gas is not less than 95 per cent pure [40].

A second Rotor process which differs only in detail has been developed at OBERHAUSEN [41, 42]. This process employs a long, cylindrical vessel rotating on a horizontal axis. Oxygen for refining is supplied through a jet which is immersed in the metal below the slag and the carbon monoxide produced is ignited above the bath by oxygen supplied through a second nozzle situated just above the slag surface (Figure 11.4). The tap-to-tap time and the iron oxide content of the slag are similar to those in the Kaldo process, but the speed of rotation of the vessel is only about 0·5 r.p.m. and the additional turbulence is created by the immersed jet, which causes a vigorous mixing of the slag and metal layers. The sequence of elimination of the solute elements and the final phosphorus, sulphur, nitrogen and oxygen contents in the metal are similar in the two processes.

The Rotor processes extend the principles of the pneumatic processes to the refining of an iron of almost any composition and the reactions can be halted at any stage to produce a steel of any chosen carbon content. There is no theoretical limiting size to which the Rotor vessel can be developed and 100-ton capacity units are already in operation. Furthermore, the capital costs are considerably lower than for an open-hearth or electric arc-furnace plant of similar capacity. The major drawbacks at present are the rather high refractory consumption per ton of steel and the low production rate, relative to the converter processes, but Rotors appear to be the most promising competitors to the established methods of refining iron into steel.

Recently preliminary trials for the direct production of steel from iron ore in the Kaldo rotor have been reported [40]. It is suggested that iron-ore concentrates and coke breeze should be charged into a steel bath containing a suitable carbon content and the heat generated by the combustion of the coke breeze and the carbon in the iron utilised to melt the ore. The oxygen in the ore can then react with the carbon to form carbon monoxide and metallic iron, which is refined to give a steel of any selected composition while still in the Rotor. If this technique can be developed successfully future application of Rotor processes is assured.

REFERENCES

1. S. MITCHELL, *J. Iron St. Inst.*, 1953, Vol. 174, p. 1.
2. C. RESCHE, *Stahl u. Eisen, Düsseldorf*, 1954, Vol. 74, p. 291.
3. Anon., *J. Iron St. Inst.*, 1955, Vol. 179, p. 36.
4. P. COHEUR, *J. Metals, N.Y.*, 1955, Vol. 7, p. 872.
5. F. W. STARRATT, ibid., 1957, Vol. 9, p. 1432.
6. H. SCHUHMACKER, *Stahl u. Eisen, Düsseldorf*, 1953, Vol. 73, p. 257.
7. W. SCHILLING and D. REIMANN, *Neue Hutte*, 1958, Vol. 3, p. 683.
8. A. B. CHATTERJEA, *Iron Coal Tr. Rev.*, 1956, Vol. 173, pp. 1255, 1329.
9. H. REINFELD, ibid., 1956, Vol. 173, p. 1139.
10. G. VON STRUVE, *J. Iron St. Inst.*, 1960, Vol. 196, p. 50.
11. A. M. SQUIRES and C. A. JOHNSON, *J. Metals, N.Y.*, 1957, Vol. 9, p. 586.

12. P. E. CAVANAGH, ibid., 1956, Vol. 8, p. 1642.
13. J. L. STÅLHED, *Stahl. u. Eisen, Düsseldorf*, 1952, Vol. 72, p. 459.
14. P. M. UNTERWEISER, *Iron Age*, 1956, Vol. 178, p. 71.
15. O. STELLING, *J. Metals, N.Y.*, 1958, Vol. 10, p. 290.
16. A. AMÉEN, *Iron Age*, 1944, Vol. 153, p. 55.
17. B. KALLING and J. L. STALHED, *Steel*, 1949, Vol. 125, p. 72.
18. S. EKETORP, *Tekn. Tidskr., Stockh.*, 1950, Vol. 80, p. 1131.
19. S. FORNANDER, *J. Iron St. Inst.*, 1954, Vol. 177, p. 1.
20. J. L. STÅLHED, *J. Metals, N.Y.*, 1957, Vol. 9, p. 246.
21. P. E. CAVANAGH, ibid., 1958, Vol. 10, p. 804.
22. Anon., *Steel*, 1958, Vol. 143, p. 78.
23. E. C. WRIGHT, *Metal Progr.*, 1957, Vol. 72, p. 99.
24. R. A. LUBKER and K. W. BRULAND, *J. Metals, N.Y.*, 1960, Vol. 12, p. 321.
25. O. STELLING and L. P. MORATH, *Jernkontor. Ann.*, 1951, Vol. 141, p. 237.
26. C. GOODEVE, *J. Iron St. Inst.*, 1956, Vol. 184, p. 113.
27. D. J. GIRARDI, *Jernkontor. Ann.*, 1945, Vol. 129, p. 703.
28. B. KALLING and F. JOHANSSON, *J. Iron St. Inst.*, 1954, Vol. 177, p. 76.
29. A. STEWART and H. K. WORK, *J. Metals, N.Y.*, 1958, Vol. 10, p. 460.
30. Anon., ibid., 1956, Vol. 8, p. 849.
31. E. SCHÜRMANN, *Stahl u. Eisen, Düsseldorf*, 1958, Vol. 78, p. 1297.
32. R. GERLACH, *Tech. Mitt. Krupp*, 1958, Vol. 16, p. 27.
33. S. HENKEL, *Met. Giesserei-Techn.*, 1954, Vol. 4, p. 101.
34. F. JOHANSSON and H. BRUNION, *Z. Erzbergb Metallhüttenw.*, 1959, Vol. 12, pp. 211, 272.
35. K. KUWADA, Proc. First World Met. Congress 1951, A.S.M., p. 188.
36. W. O. PHILBROOK, *J. Metals, N.Y.*, 1958, Vol. 10, p. 477.
37. B. KALLING and F. JOHANSSON, *Stahl u. Eisen, Düsseldorf*, 1957, Vol. 77, pp. 1308, 1885. *Iron Coal Tr. Rev.*, 1956, Vol. 173, p. 1497; *J. Internat. De Siderurgie*, 1958, June, p. 1.
38. F. JOHANSSON, *J. Metals, N.Y.*, 1957, Vol. 9, p. 972.
39. D. J. O. BRANDT, *Iron Coal Tr. Rev.*, 1957, Vol. 174, pp. 325, 437.
40. B. KALLING and F. JOHANSSON, *J. Iron St. Inst.*, 1959, Vol. 192, p. 330.
41. R. GRAEF, *Stahl u. Eisen, Düsseldorf*, 1957, Vol. 77, p. 1; *Iron Coal Tr. Rev.*, 1957, Vol. 174, p. 213.
42. L. VON BOGDANDY, *Z. Elektrochemie*, 1957, Vol. 1, p. 222.

PART IV

SLAG-METAL REACTIONS

CHAPTER 12

Refining Slags

Experimental studies of liquid slags can be divided into four major groups. These are:

1. The determination of the structure of the slag.
2. The location of the liquidus surfaces.
3. Measurements of the thermodynamic properties of the various components of the slag system.
4. Examination of the factors which determine the rate of chemical reactions within the slag and between the slag and the metal.

The first two groups are discussed in Chapter 3 and have received occasional consideration in the chapters which deal with the production of iron and steel; the last two groups remain to be considered here and in succeeding chapters. During the past decade or so a fairly considerable amount of data has accumulated on these aspects, but a very extensive programme of work remains to be undertaken and present knowledge covers only a limited field. Variation in the thermodynamic activity with composition has been examined in a number of binary systems and in some of the more important of the ternary systems. A few measurements have been made with quaternary and more complex mixtures, but it is not yet possible to predict accurately the activity of any one component in the multi-component slags which occur in iron and steel refining. The kinetic aspects have received less attention than the thermodynamic properties and the data are very limited. With the aid of the available data it is, however, possible to obtain some idea of the mixing behaviour in liquid slags and the probable effects of adding other components to the mixture.

12.1 Thermodynamic Properties

ACTIVITIES IN TWO-COMPONENT MELTS

The experimental problems in this field of study are numerous and it is not unusual to find that in any particular investigation the results are reproducible but are not consistent with those obtained in other investigations. For example, in Figure 12.1 some of the curves which have been published for the activities of silica versus concentration are compared for liquid mixtures in the system $CaO–SiO_2$. Each of these curves has been determined by closely-controlled, laboratory experiments, yet the only feature on which there is agreement is that the activities of silica tend towards a negative deviation from Raoult's law as the concentration of

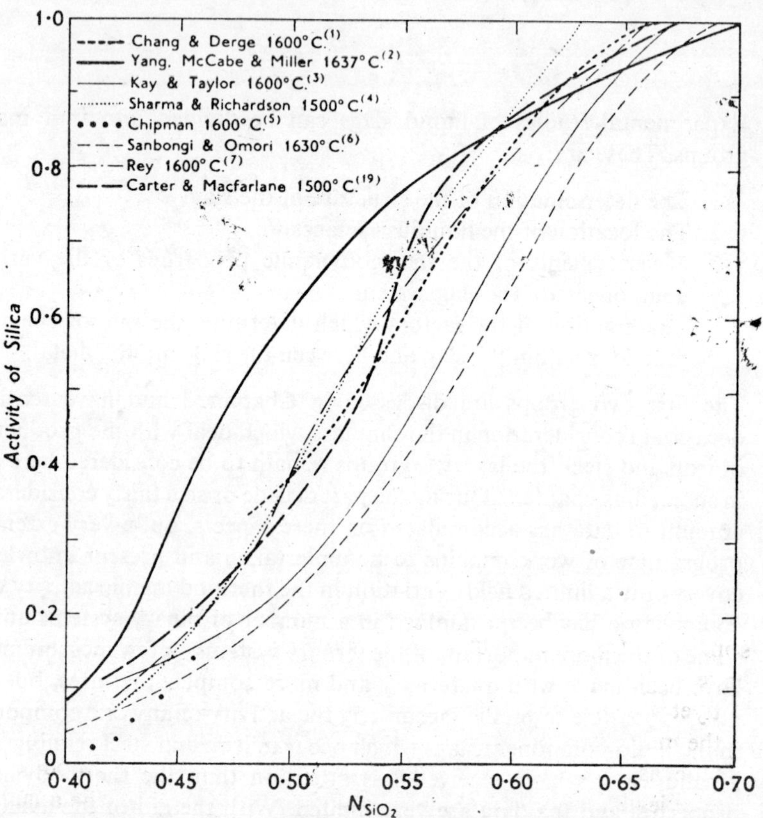

FIG. 12.1. Comparison of some measurements of the activities of silica in the binary system $CaO–SiO_2$, relative to pure solid silica.

silica is lowered. At constant composition the activity values vary over a fairly wide range and they include the curve which has been calculated by application of the regular-solution relations [7] to the liquidus lines for the binary system and extrapolated to 1,600°C. There is good agreement at 1,500°C between the data of Sharma and Richardson [4] and those of Kay and Taylor [3] for mole fractions of silica greater than 0·5. Both sets of data show that the activities of silica are lowered as the temperature is raised and extrapolation to 1,630°C gives a fair measure of agreement with the curve due to Sanbongi and Omori [6] at that temperature. The agreement is less satisfactory for compositions close to the lime saturation surface, but Chipman's data [5]* are reasonably consistent with those of Sharma and Richardson in this range.

In Figure 12.2 the results are shown of two independent determinations of the activities of manganese oxide in liquid $MnO–SiO_2$ mixtures [8, 9]. Agreement between the two sets of results is poor, but the broken-line curves have been obtained by extrapolation to the binary side of data for

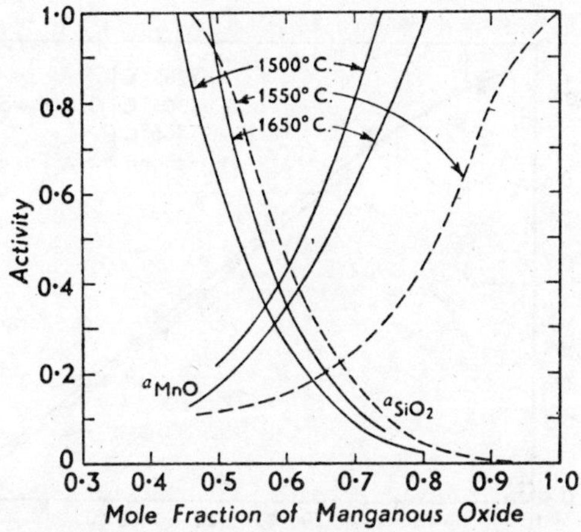

FIG. 12.2. The activities of manganous oxide and silica at various temperatures in binary $MnO–SiO_2$ melts, relative to pure solid manganous oxide (solid lines) [9] and pure supercooled liquid manganous oxide (broken lines) [8].

the ternary system $MnO–FeO–SiO_2$. These last data were obtained by making the very doubtful assumption that the activity coefficient of ferrous oxide is unity in the ternary melts and the upper curves in the diagram are regarded as more probably correct. At constant composition the activity coefficient of manganese oxide is lowered as the temperature is raised from 1,550 to 1,650°C.

* The curve attributed to Chipman in Figure 12.1 differs slightly from that given in reference 5. Chipman's values were calculated using free-energy data for equation 12.7 which have since been shown to be incorrect [59] and it is necessary to multiply the activities of silica given in reference 5 by a factor of 2·7 to bring them into line with the more recent free-energy equation. This correction has been applied in Figure 12.1 and all the activity data for silica and lime taken from reference 5 and reproduced in this chapter have been adjusted similarly.

It has been suggested that at 1,600°C the activities of ferrous oxide are almost ideal in the binary system FeO–SiO_2, but this conclusion is reached from the extrapolation of data for ternary melts to intercept the binary side [10, 11], and it is also possible to perform the extrapolation to show a small negative deviation from ideality in the binary system. The results from two investigations of the activities of ferrous oxide in this system in the temperature range 1,250–1,380°C are shown in Figure 12.3 [12, 13]. The two curves agree well and show marked deviation from ideality, although the deviation is less pronounced than that for manganese oxide in the silicate melts. Within the experimental temperature

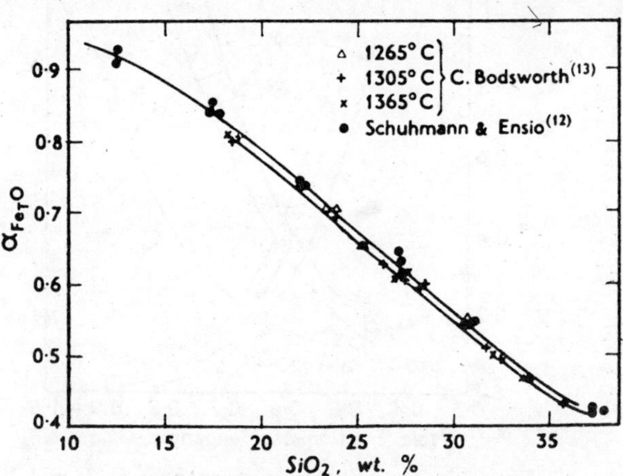

FIG. 12.3. The influence of composition on the activities of wüstite (Fe_TO) in binary silicate melts at equilibrium with pure iron.

range the activities are found to be independent of the temperature. The recommended activity/composition curves for lime, manganous oxide and ferrous oxide in these binary melts are compared in Figure 12.4.

When the activity-composition curves for lime and manganous oxide are recalculated, relative to the pure, supercooled liquid oxides as the standard states, it is found that as the concentration of silica is raised the activities of lime and manganese oxide decrease almost linearly at first, whereas the activities of ferrous oxide show positive deviation from ideality at low silica contents. The shape of the ferrous oxide activity-composition curve is not unusual in binary systems, but it should be noted that the activities in this system were measured with reference to the cation-deficient wüstite phase and not relative to the stoichiometric ferrous oxide composition. It is probable that the positive deviation from ideality is due to the presence of the triple-charged ferric ions which are required to preserve electro-neutrality in the wüstite phase and, by

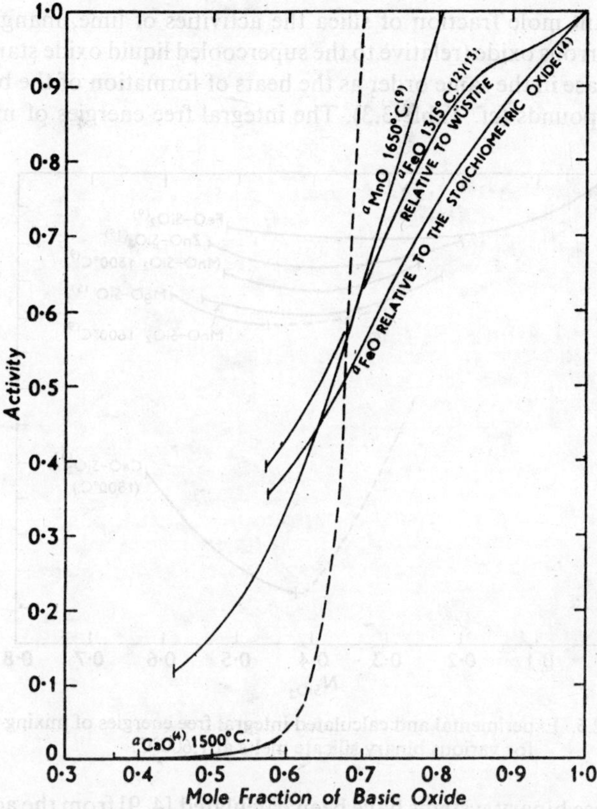

FIG. 12.4. Comparison of the activities of ferrous oxide, lime and manganous oxide in binary silicate melts. The lines for lime and manganous oxide are for the pure solid oxide standard state. The lines for ferrous oxide are drawn relative to pure supercooled liquid wüstite and for the stoichiometric FeO standard states.

application of the ionic theories of melts, it has been shown [14] that the activities of the stoichiometric ferrous oxide in the liquid binary silicates may show a negative departure from ideality at all concentrations. The calculated curve based on the stoichiometric oxide is included in Figure 12.4. As the temperature is raised the concentration of ferric ions is decreased in wüstite which is in equilibrium with pure iron and, since the activities of wüstite are apparently independent of temperature, this change in the concentration of ferric ions could cause the activities of the stoichiometric phase to increase and move closer to ideality as the temperature is raised.

At constant mole fraction of silica the activities of lime, manganese oxide and ferrous oxide (relative to the supercooled liquid oxide standard states) increase in the same order as the heats of formation of the binary silicate compounds (cf. Table 3.3). The integral free energies of mixing

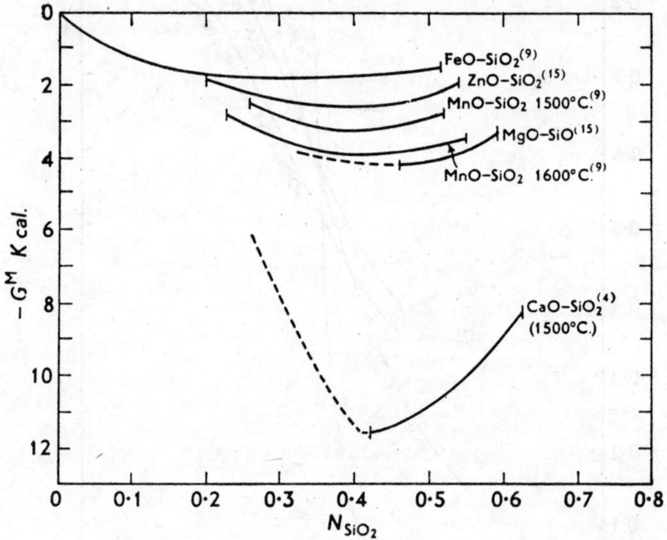

FIG. 12.5. Experimental and calculated integral free energies of mixing for various binary silicate melts at 1,600°C.

(G^M) for these binary systems have been calculated [4, 9] from the activity data and are shown in Figure 12.5 together with the G^M-composition curves for liquid binary zinc and magnesium silicates which have been calculated [15] from binary phase diagram data, heats and free energies of formation and entropies of fusion for the silicate compounds. The mixing free-energies are grouped closely together for those oxides which form extensive non-ionic bonding in the solid state and which are believed to retain some element of non-ionic bonding in the liquid silicates. They show much smaller negative excess free energies of mixing than the calcium silicates, where the bonding is essentially ionic.

The activity-composition relationships have been examined in some binary systems which are of interest and where silica is not present as one of the components. In the system FeO–CaO the activities of ferrous oxide show a marked negative deviation from Raoultian behaviour [10, 11, 16] (Figure 12.6*a*), owing probably to a tendency for the ferric ions in the non-stoichiometric wüstite to form oxy-acid complexes such as $Fe_2O_4^{2-}$ and associate with the calcium ions. A stable compound, $CaO.Fe_2O_3$, is formed in the solid state, but no corresponding compound is formed by

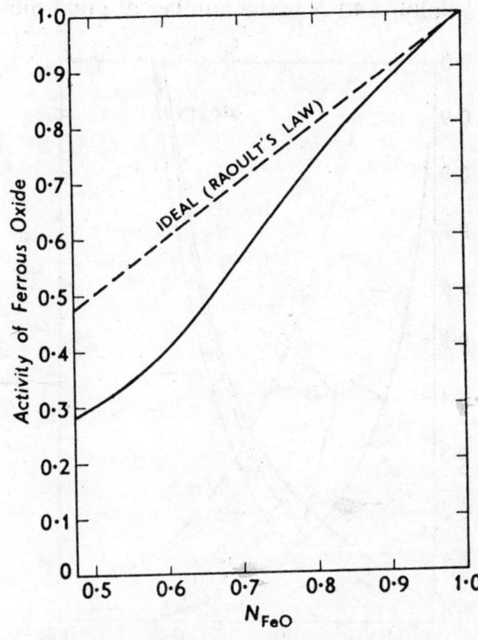

FIG. 12.6*a*. The influence of composition on the activities of ferrous oxide in binary FeO–CaO melts at 1,530° to 1,600°C. (After Bishop, Grant and Chipman [16]).

solid mixtures of manganese and iron oxides and the system FeO–MnO forms a continuous range of solutions which are approximately ideal in both the solid [17] and the liquid [18] phase fields.

The activities of lime at 1,500°C in $CaO–Al_2O_3$ melts have been estimated from studies of sulphur distribution between the melt and a gas phase containing sulphur and oxygen [4, 19]. The reaction may be represented as:

$$CaO_{(slag)} + \tfrac{1}{2}S_{2(g)} \rightleftharpoons CaS_{(slag)} + \tfrac{1}{2}O_{2(g)} \quad 12.1$$

Hence:

$$(a_{CaO}) = \frac{(a_{CaS})}{K}\left(\frac{p_{O_2}}{p_{S_2}}\right)^{1/2} \qquad 12.2$$

or

$$(a_{CaO}) = C_S . \frac{\gamma_{CaS}}{32NK} \qquad 12.3$$

where C_S is the sulphide capacity of the melt (see Chapter 9), 32 is the atomic weight of sulphur and N is the number of gram-moles of CaO,

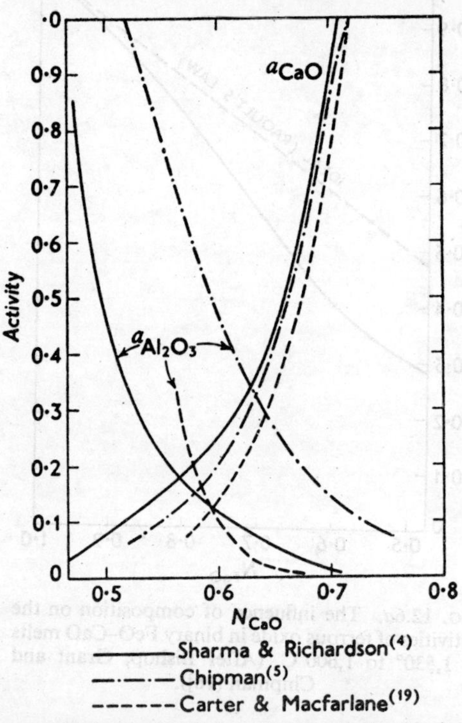

FIG. 12.6*b*. The activities of lime and alumina in binary $CaO–Al_2O_3$ melts at 1,500°C.

CaS and Al_2O_3 in the melt [4]. The activities have also been estimated [5] by combining the free-energy data for the formation of the calcium aluminate compounds with the measured [19] C_S values. The three sets of data show similar variations with composition for the activities of lime (Figure 12.6*b*), but the agreement is rather poor and this can be traced to the assumptions made regarding the variation of (γ_{CaS}) with (N_{CaS}) in the experimental [4, 19] studies and to the uncertainties in the thermal data used to derive the calculated [5] activities. At the higher lime concentrations the error band for any one of the investigations includes the other two sets of data.

The activities of both lime and alumina show pronounced negative deviations from Raoultian behaviour, but the deviations are less marked than in the case of the lime–silica melts. This is consistent with the fact that a series of aluminate compounds is formed in the solid state, similar to the CaO–SiO_2 system, but the bond strengths and the degree of non-ionic bonding of the aluminate–ion complexes are less than those for the silicate–ion complexes.

ACTIVITIES IN MULTI-COMPONENT SLAGS

The limited data which are available for more complex melts can be divided into two groups. The constituents of the first group are FeO, CaO, MnO and SiO_2, viz., the oxides which form the major part of the slags encountered in the steel-making processes. The second group is pertinent to the blast furnace slags and is based on the ternary system CaO–Al_2O_3–SiO_2.

The system FeO–CaO–SiO₂

The activities of ferrous oxide in this system were first examined at 1,600°C in the classic experiments of Chipman and co-workers [10, 11], who determined the activities from measurements of the oxygen content of liquid iron in equilibrium with the slag, relative to the oxygen in the iron in equilibrium with pure iron oxide at the same temperature. The melts contained small amounts of MgO and MnO, but from their data it is not possible to distinguish between the effects of these oxides and the effect of lime on the activities. Consequently they assumed that the three basic oxides are interchangeable on a molar basis and that the iso-activity contours for ferrous oxide could be plotted on triangular co-ordinates with the three apices Fe_TO, SiO_2 and (CaO + MnO + MgO), where Fe_TO signifies that the total concentration of iron in the slag is calculated as ferrous oxide. For constant mole fraction of ferrous oxide, the activity coefficient of this oxide in the ternary melts is greater than in the binary system FeO–SiO_2 (i.e. (γ_{Fe_TO}) is raised by the addition of lime to the binary melts) and the maximum values for (γ_{Fe_TO}) lie along the join Fe_TO–2(CaO + MnO + MgO). SiO_2. The activities are apparently independent of the temperature, but the experimental temperature range is restricted.

The data have since been re-plotted [20] and the intercepts of the iso-activity curves on the binary side FeO–SiO_2 adjusted to coincide with the values obtained by extrapolation of the lower-temperature data for the binary system [12], assuming a linear temperature dependence of the logarithm of the binary activities. Using Schuhmann's solution for the ternary Gibbs-Duhem relation (Chapter 1.9) the activity coefficient contours for lime and silica have also been calculated from the ferrous oxide data and the relationships are shown in Figure 12.7.

These ferrous oxide activity data have also been combined with data from other sources for melts containing up to 0·1 mole fraction of

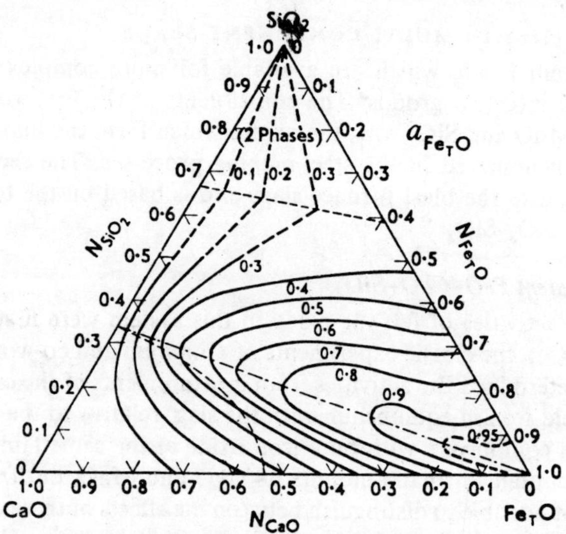

FIG. 12.7*a*. The activities of ferrous oxide in the ternary system Fe_TO–CaO–SiO_2 at 1,600°C. The broken lines are probable liquidus limits. (After Elliott [20].)

phosphorus pentoxide and it was concluded [21] that within this concentration range the effect of phosphorus pentoxide on the activities of ferrous oxide is equivalent to that of silica. However, the iso-activity contours differ markedly from those shown in Figure 12.7 and in particular the maximum values of $(\gamma_{Fe_TO})_{II}$ do not lie on a straight line, but on a smooth curve which is displaced to higher basicities between the limits Fe_TO and $2(CaO+MgO+MnO).(SiO_2+P_2O_5)$. This fact could be adduced as evidence that the activities of ferrous oxide are lowered as silica is replaced by phosphorus pentoxide, but the scatter of the data is too large to justify any quantitative conclusions.

Three other experimental studies have been made of the activities of

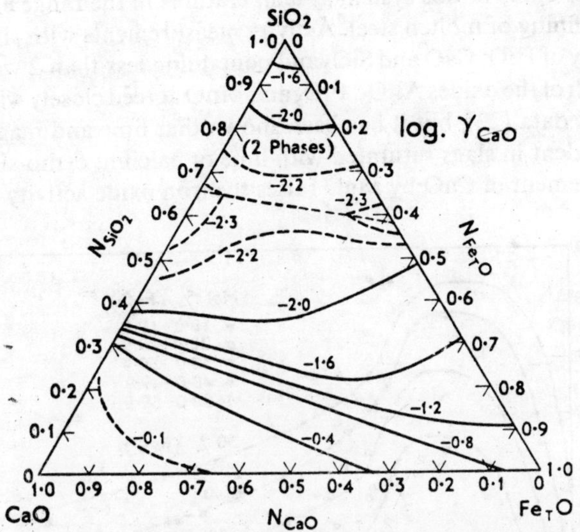

FIG. 12.7*b*. Variation of the activity coefficient of lime with composition in the ternary system Fe_TO–CaO–SiO_2 at 1,600°C. (After Elliott [20].)

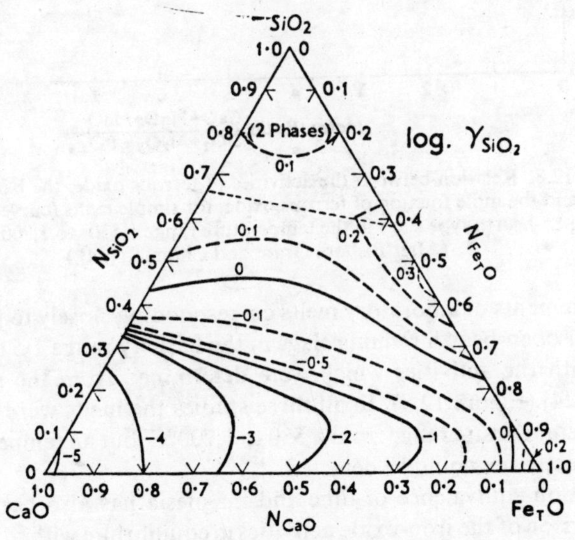

FIG. 12.7*c*. Variation of the activity coefficient of silica with composition in the ternary system Fe_TO–CaO–SiO_2 at 1,600°C. (After Elliot [20].)

ferrous oxide in this system at temperatures in the range encountered in the refining of molten steel. Activity measurements with slags consisting mainly of FeO, CaO and SiO_2 and containing less than 2 weight-per cent of each of the oxides Al_2O_3, P_2O_5 and MnO agreed closely with Chipman's earlier data [22], but it has been shown that lime and magnesia are not equivalent in slags saturated with lime or calcium ortho-silicate, for the replacement of CaO by MgO raises the iron oxide activity [16]. Activity

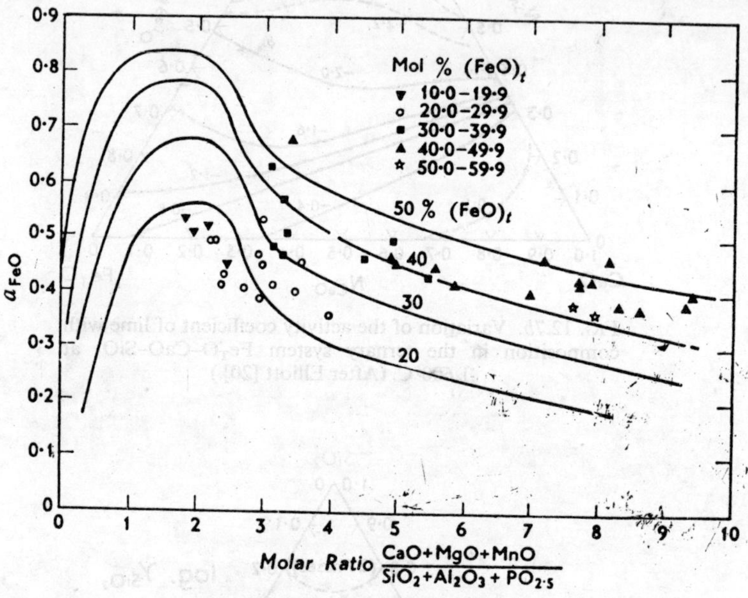

FIG. 12.8. Relation between the activities of ferrous oxide, the basicity of the slag and the mole fraction of ferrous oxide, for simple melts (curved lines) and for open-hearth type slags in the temperature range 1,530° to 1,700°C (points). (After Bishop, Grant and Chipman [24].)

measurements of laboratory melts corresponding closely to the compositions of open-hearth refining slags in the basicity range 1·5 to 4·0 agreed well with the activities which were determined from the more simple melts [24] (Figure 12.8). In all three studies the melts were equilibrated in the temperature range from 1,530 to 1,700°C, but no temperature effect on the activities could be detected.

The non-equivalence of lime and magnesia has also been shown by comparison of the iron-oxide activities in equilibrium with solid lime with those for equilibrium with solid dolomite (CaO . MgO) [23]. The replacement of lime by magnesia in the saturated melts causes a marked increase in (γ_{Fe_TO}) (Figure 12.9) and this trend is in the direction to be expected from a comparison of the heats or free energies of formation of calcium and magnesium ferrites. There is a smaller tendency for association between magnesium and ferrite ions than between calcium and ferrite ions and consequently the activities of the non-stoichiometric ferrous oxide should be increased as lime is replaced by magnesia. However, the effect of magnesia in more acid melts cannot be predicted from these results, for it is improbable that ferrite-ion complexes are formed in melts which also contain silicate ions with shared bonds and any change in the activity of the wüstite which is effected by a change in the ferrite-iron association should cease to be significant at basicities less than about 2·0.

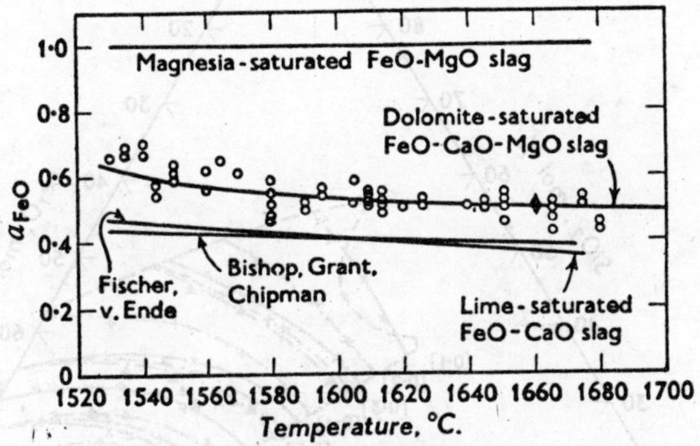

FIG. 12.9. The temperature dependence of the activities of ferrous oxide in slags saturated with various basic oxides. (After Fischer and Spitzer [23].)

One of the major problems encountered in the determination of ferrous oxide activities is the selection of a suitable material for containing the slag. No known refractory material is completely inert both to liquid iron and to liquid iron oxide slags, and the melt is progressively contaminated with the container material. Attempts have been made to circumvent this problem by using a container which does not react to any appreciable extent with liquid iron. The container is rotated rapidly and the slag is contained within the paraboloid meniscus of the iron, so reducing the contact between the slag and the container, but it is not possible to eliminate completely the contamination of the melt. Alternatively the activities can be determined for equilibrium with one of the slag components when that component is used for the container, but the field of study is confined to a restricted composition range. Contamination can be eliminated by using a solid iron crucible to contain the molten slag, but the maximum temperature which can bé studied is limited by the softening of the iron as the melting point is approached and the results obtained are not applicable directly to steel-making conditions.

The last technique has been used to determine the activities of ferrous oxide in pure FeO–CaO–SiO_2 melts in the temperature range 1,265–1,365°C [13] (Figure 12.10). The composition range which could be

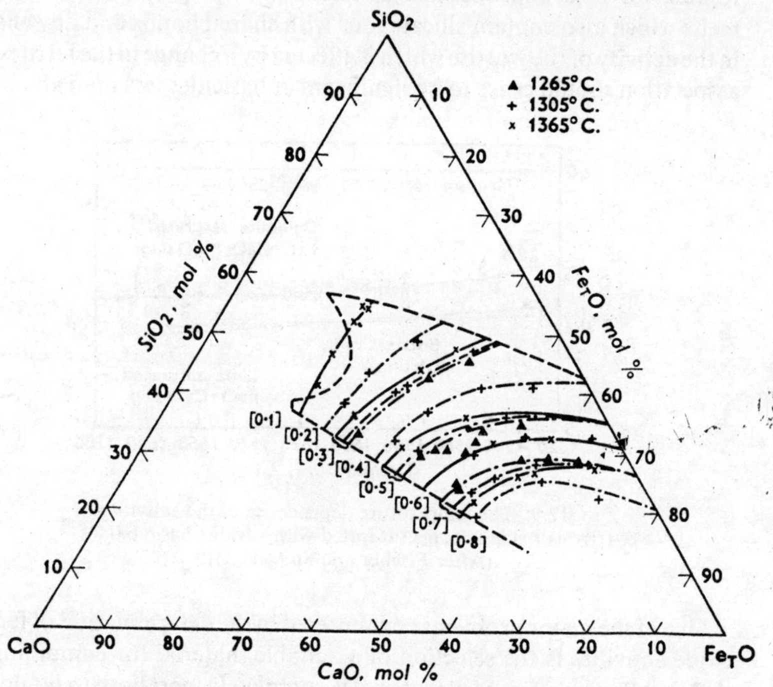

FIG. 12.10. The activities of ferrous oxide in the ternary system Fe_TO–CaO–SiO_2 at 1,265°, 1,305° and 1,365°C. (After Bodsworth [13].)

studied is restricted by the liquidus surfaces of the ternary system to slags corresponding to those which form at the start of refining and to those which are encountered in acid practices. Within this range the iso-activity curves are similar in shape to those shown in Figure 12.7, but for all compositions (γ_{Fe_TO}) is lower than at 1,600°C. In contrast to the conclusions from the earlier investigations, (γ_{Fe_TO}) is temperature dependent in the ternary melts and, for constant composition within the experimental temperature range, the logarithm of the activities is a linear function of the temperature. Over most of the composition range (γ_{Fe_TO}) is increased as the temperature is raised, but the reverse applies when (a_{Fe_TO}) is greater than 0·6. The data can be extrapolated to 1,600°C for comparison with the data of other workers and with up to 20 mole per cent of lime the extrapolated iso-activity curves are closely coincident with those in Figure 12.7, but at higher lime contents the extrapolated curves show lower values for the activity coefficient.

This investigation has been extended to determine the relative effects of lime and manganese oxide on the iron-oxide activities [25], and within

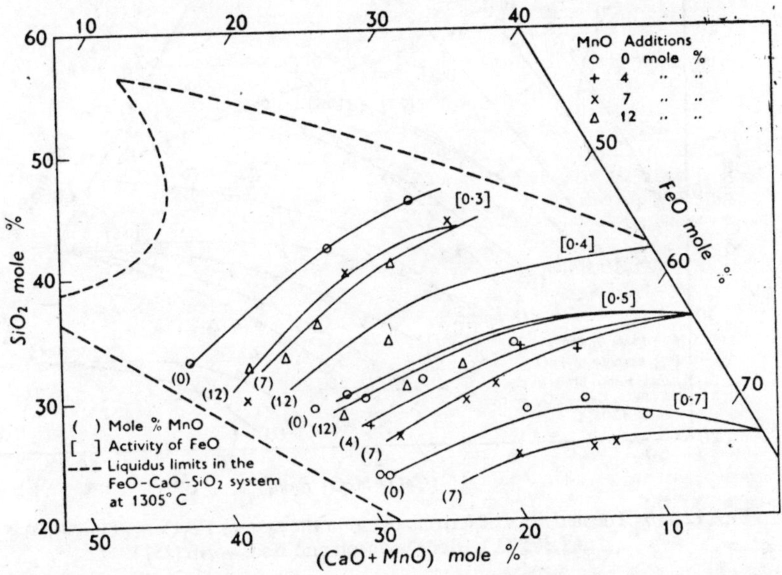

FIG. 12.11*a*. The activities of ferrous oxide in the system Fe_TO–($CaO+MnO$)–SiO_2 at 1,265°C. (After Davidson and Bodsworth [25].)

the composition range studied it is found that the replacement of lime by up to 12 mole per cent of manganese oxide lowers (γ_{Fe_TO}), the iso-activity curves move closer to ideality, and the temperature dependence of the activities is lowered (Figure 12.11).

To summarise this section, it can be stated that there is now fairly strong evidence that the basic oxides CaO, MnO and MgO are not equivalent in their effects on the activities of ferrous oxide. Theoretically an equivalence is not expected on the basis of the binary activity data or the free energies of formation of the solid compounds. Similarly there is no reason for assuming that the acid oxides SiO_2, Al_2O_3 and P_2O_5 are equivalent in this respect. If the activities of ferrous oxide can be related to the relative cation-oxygen bond energies of the various oxides in the melt then (a_{Fe_TO}) should be affected to a decreasing extent by the basic oxides in the order $CaO > MgO > MnO$ and by the acid oxides in the order $P_2O_5 > SiO_2 > Al_2O_3$. The equivalence which has been reported from the earlier studies can probably be attributed to small random variations from melt to melt in the concentrations of more than one of the

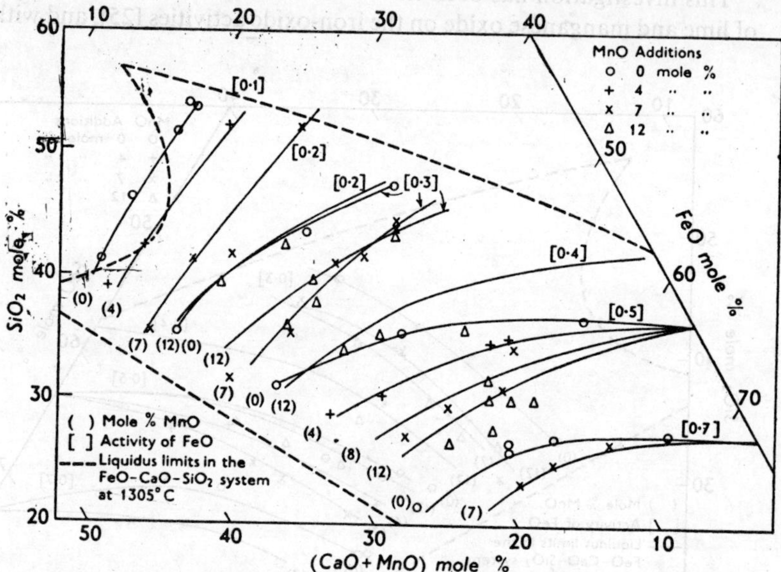

FIG. 12.11*b*. The activities of ferrous oxide in the system Fe_TO–$(CaO+MnO)$–SiO_2 at 1,365°C. (After Davidson and Bodsworth [25].)

minor components, and this may also account for the failure to detect a temperature-dependence of the activities.

The Activities of Iron Oxides at Higher Oxygen Potentials

For a slag which is in equilibrium with liquid iron at 1,600°C the partial pressure of oxygen dissolved in the slag at the slag–metal interface is about 10^{-8} atmosphere, and the corresponding pressure at the slag–gas interface varies from near atmospheric to about 10^{-5} atmosphere, depending on the process. The activities of ferrous oxide which are discussed in the preceding section relate only to a slag in equilibrium with iron, whereas in certain cases the mean oxygen potential of the slag may be considerably greater than this equilibrium value. Such conditions may obtain, for example, when the slag is very viscous and reactions between the slag and metal proceed at a low rate, or when the bath is off the boil and immediately after large ore feeds.

Several investigations have been conducted to determine the effect of the oxygen potential and slag composition on the activities of the iron oxides [26–29], using air or mixtures of carbon monoxide and carbon

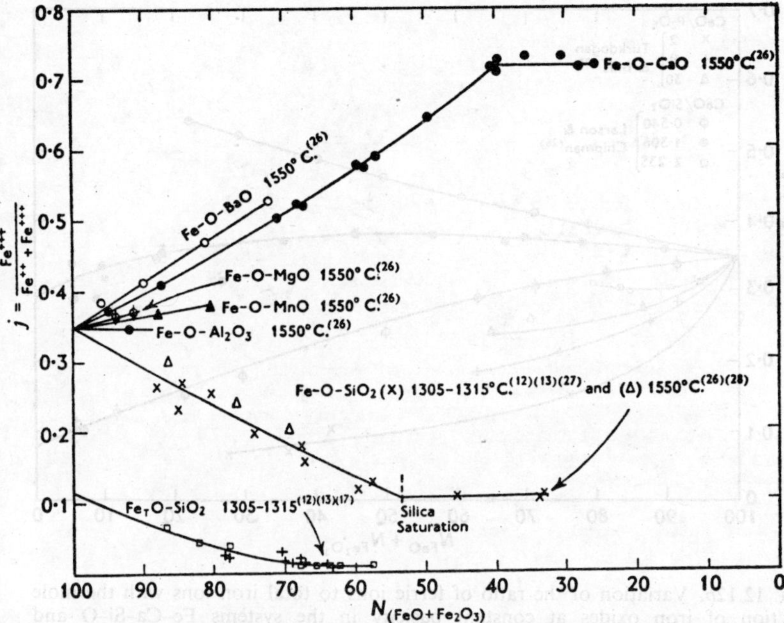

FIG. 12.12*a*. Variation of the ratio of ferric ions to total iron ions with the mole fraction of iron oxides for mixtures of iron oxides with various other oxides. The data shown are for equilibrium with carbon dioxide and carbon monoxide in the ratio 11·4:1 at 1,550°C. The data for iron–oxide–silica melts in equilibrium with solid iron at 1,305–1,315°C are included for comparison.

dioxide to vary the oxygen potential over a wide range. In one of the earlier investigations the results are presented in terms of a coefficient j which is defined as the ratio of the ionic fractions of ferric iron to the total iron in the slag, viz., $(Fe^{3+})/(Fe^{2+}+Fe^{3+})$ [26]. These data, together with those from other sources [12, 27, 28], are summarised in Figure 12.12 for a temperature of 1,550°C and $p_{CO}/p_{CO_2} = 11{\cdot}4$ which corresponds to an oxygen partial pressure of $9{\cdot}1\times10^{-6}$ atmosphere. The zero-point intercepts are located from the results of an extensive study of the pure iron-oxygen system over wide ranges of temperature and oxygen potential [29], in which it is shown that, for pure iron oxides at a fixed partial pressure of oxygen, the *j* ratio varies only slightly with the temperature.

The *j* ratio is raised by additions of the basic oxides BaO, CaO, MnO and MgO and is lowered by silica additions, but alumina up to the saturation limit apparently has no effect. These results are shown in

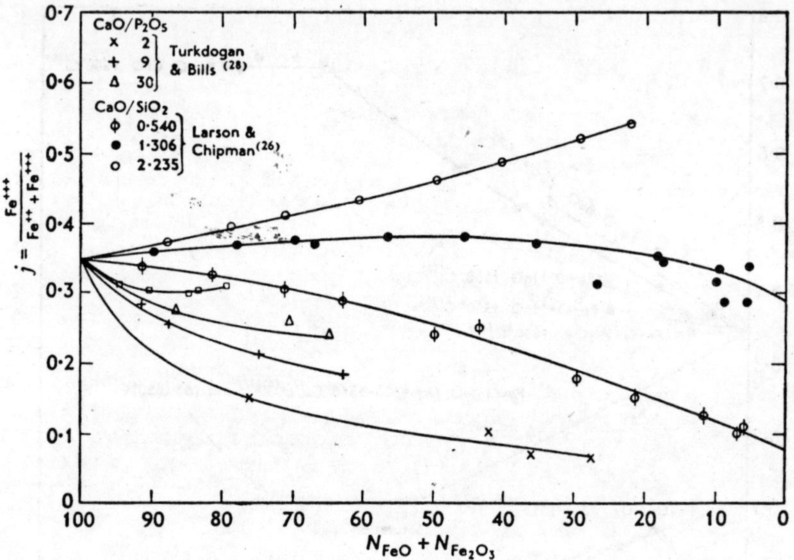

FIG. 12.12*b*. Variation of the ratio of ferric ions to total iron ions with the mole fraction of iron oxides at constant basicity in the systems Fe–Ca–Si–O and Fe–Ca–P–O. The data shown are for equilibrium with carbon dioxide and carbon monoxide in the ratio 11·4:1 at 1,550°C.

Figure 12.12*a*, and the *j* ratios for iron oxide–silica melts in equilibrium with solid iron are included for comparison. Figure 12.12*b* shows that in the presence of lime the *j* ratio is lowered to a greater extent by P_2O_5 than by silica.

These changes correspond to the sequence which is predicted from the relative bond energies of the pure oxides and the tendency to form ferrite ion complexes. The change in the *j* ratio with composition can be interpreted in terms of the equation:

$$2Fe^{2+}{}_{(slag)} + \tfrac{1}{2}O_{2(g)} \rightleftharpoons 2Fe^{3+}{}_{(slag)} + O^{2-}{}_{(slag)} \qquad 12.4$$

The equilibrium constant for the reaction is given by:

$$K = \frac{(a_{Fe^{3+}})(a_{O^{2-}})}{(a_{Fe^{2+}}) \cdot p_{O_2}^{1/2}}$$

and from this relation it is evident that, at any given temperature and partial pressure of oxygen, the ratio of the activities of the ferric and ferrous ions varies inversely with the activities of the oxygen ions in the slag. For example, the addition of the anion-formers Si^{4+} and P^{5+} lowers the activity of the oxygen ions and increases the activity ratio, but the P—O bonds are stronger than the Si—O bonds and the ratio is raised to a greater extent by phosphorus than by silicon. Conversely in basic melts the ferric ions associate with the oxygen ions to form ferrite-ion complexes; this lowers the activity of the ferric ions to an extent which should increase as the bond energies of the basic oxides are lowered. However, the bond energies between the ferrous ions and the ferrite-ion complexes are very much weaker than those between the ferrous ions and the phosphate or silicate anions. Hence the activity of the ferrous ions is raised as the network-forming anions are replaced by the basic oxides, with the overall effect that the ratio of the ferric and ferrous ion activities should be lowered to an increasing extent by additions of the basic oxides in the order MnO < MgO < CaO < BaO.

The relation between the activities and the concentrations of the iron ions is obtained by expressing the equilibrium constant for equation 12.4 in terms of the activity coefficients, i.e.,

$$\log K_2 + \tfrac{1}{2}\log p_{O_2} = 2\log\frac{(N_{Fe^{3+}})}{(N_{Fe^{2+}})} + \left[2\log\frac{(\gamma_{Fe^{3+}})}{(\gamma_{Fe^{2+}})} + \log(a_{O^{2-}})\right] \qquad 12.5$$

Experimental work has established that the mole fraction ratio of the iron ions is lowered as the oxygen ion activity is lowered at constant oxygen partial pressure, but the preceding discussion indicates that the activity ratio of the iron ions is increased. It follows, therefore, that as the oxygen ion activity is changed the Fe^{3+}/Fe^{2+} activity-coefficient ratio is changed more rapidly than the corresponding activity ratio, with the result that the mole fraction ratio varies inversely with the activity ratio. However, the variation of the activities of the oxygen and iron ions is not related simply to the composition of the melt. The ratio $N_{Fe^{3+}}/N_{Fe^{2+}}$ can be varied by altering the partial pressure of oxygen in equilibrium with a melt of otherwise constant composition and if, under these conditions, the bracketed terms on the right hand side of equation 12.5 are constant, then a plot of $\tfrac{1}{2}\log p_{O_2}$ versus $\log N_{Fe^{3+}}/N_{Fe^{2+}}$ should yield a straight line with a slope of 2 [30]. When the experimental data for the iron oxide–silica melts are plotted in this way it is found that there are wide deviations from the simple relationship and, to account for them, three causes have been suggested [28]:

1. $(a_{O^{2-}})$ changes with the ratio of ferric and ferrous ions;
2. $(\gamma_{Fe^{3+}})/(\gamma_{Fe^{2+}})$ changes also with this ratio;
3. some of the ferric ions are present as network-forming ferrite complexes.

The Activities of Manganese Oxide in Silicate Melts

The effect of the slag composition on the activity of manganese oxide has been investigated for limited ranges of composition in the ternary systems FeO–MnO–SiO_2 and CaO–MnO–SiO_2 and some measurements have also been made with more complex melts. The bond energies of

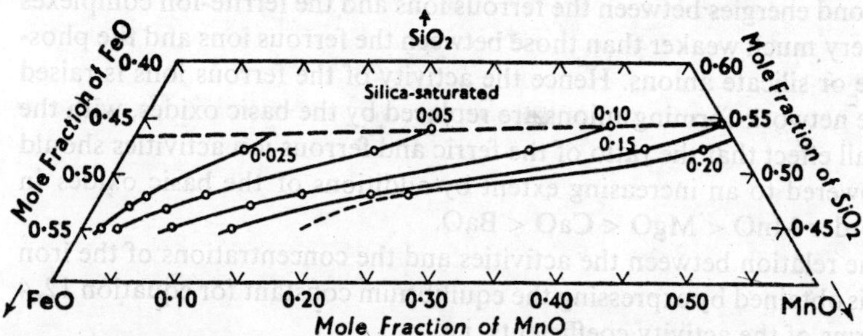

FIG. 12.13. The activities of manganous oxide in silica-rich FeO–MnO–SiO_2 melts. (After Turkdogan and Pearson [32].)

manganese oxide are intermediate between those of ferrous oxide and lime, but are closer to the former, and this is reflected in the thermodynamic behaviour of manganese oxide in these melts. Thus in the binary silicate melts manganese oxide interacts with silica more strongly than does ferrous oxide, but much less strongly than lime. In the silica-rich region of the ternary system FeO–MnO–SiO_2 the activity of manganese oxide is less than ideal (Figure 12.13) and, with increasing silica concentration, (γ_{MnO}) is lowered more rapidly than (γ_{Fe_TO}). Moreover, in the silica-saturated melts the equilibrium 'constant' for the reaction:

$$FeO_{(slag)} + Mn_{(l)} \rightleftharpoons MnO_{(slag)} + Fe_{(l)} \qquad 12.6$$

$$K'_{Mn} = \frac{(\%MnO).[\%Fe]}{(\%FeO).[\%Mn]}$$

is 6 times greater than the value of K'_{Fe} for simple FeO–MnO melts [31] and, as the MnO–FeO melts are approximately ideal, this means that at silica saturation the ratio $(\gamma_{MnO})/(\gamma_{Fe_TO})$ is approximately equal to 1/6.

Similar behaviour is found in the silica-rich region of the ternary system $CaO–MnO–SiO_2$ [9] where (γ_{MnO}) is consistently lower than the corresponding values of (γ_{Fe_TO}) in the system $CaO–FeO–SiO_2$ (Figure 12.14). In the temperature range 1,500–1,650°C (γ_{MnO}) is lowered as the temperature is raised and, at constant mole fraction of manganese oxide, the temperature dependence of the activities is lowered as the mole fraction of silica is raised. A similar variation of the temperature dependence of the activities with composition was found in the binary system $MnO–SiO_2$, but the temperature dependence in the ternary melts is

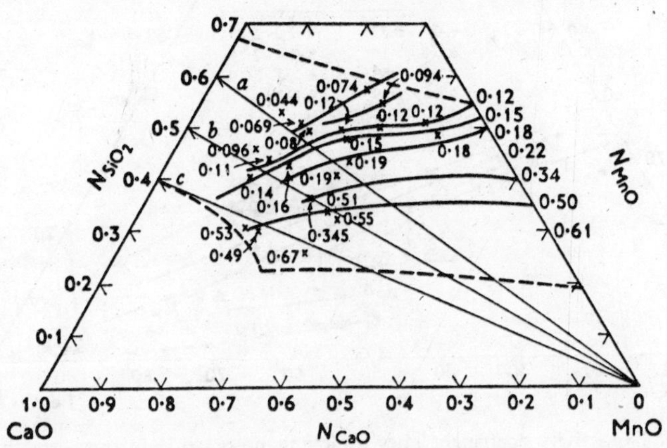

FIG. 12.14. The activities of manganous oxide in the system $MnO–CaO–SiO_2$ at 1,650°C. (After Abraham, Davies and Richardson [9].)

opposite in sign to that of the activities of ferrous oxide in the silica-rich region of the system $FeO–CaO–SiO_2$.

In complex melts containing FeO, MnO, MgO, CaO, SiO_2 and P_2O_5, (γ_{MnO}) has a maximum value along the join from the MnO apex to the 'orthosilicate' composition [24, 32] (Figure 12.15) and, as the silica concentration is raised above the values for this join, (γ_{MnO}) decreases more rapidly than (γ_{Fe_TO}), whereas the reverse applies to compositions below the join. This has been interpreted as indicative of greater association between MnO and SiO_2 and between CaO and FeO than of FeO with SiO_2 or of CaO with MnO [24]. At constant mole fractions of manganese oxide and (lime plus magnesia) (γ_{MnO}) is lowered as the mole fraction of ferrous oxide is raised [32], and this is similar to the effect of manganese oxide on (γ_{Fe_TO}) in the quaternary melts $FeO–CaO–MnO–SiO_2$. No temperature dependence of the manganese oxide activities has been detected in the multi-component melts but, in view of the variation with temperature of the activities in the binary and ternary systems, it is probable that the activities in the more complex melts are dependent also on the temperature. As with the activities of ferrous oxide in the complex melts, the experimental observation may be influenced by the masking effect of the simultaneous variation of the temperature and the concentration of one or more of the minor components of the slag.

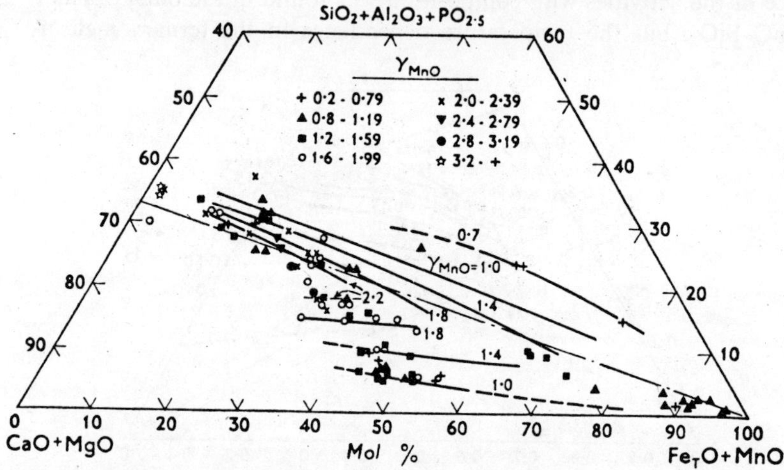

FIG. 12.15. Iso-activity coefficient lines for manganous oxide in open-hearth type slags at equilibrium with liquid, carbon-free iron in the temperature range 1,530° to 1,700°C. (After Bishop, Grant and Chipman [24].)

ACTIVITIES IN BLAST FURNACE SLAGS

Agreement is poor among the measured activities of silica in the ternary system CaO–Al_2O_3–SiO_2, but some of these data were obtained from electromotive-force measurements with electrolytic cells [6, 33] in which the back E.M.F.'s were appreciable, and it is probable that the true thermodynamic equilibria were not obtained. The results obtained from two gas equilibrium investigations in the temperature range 1,450 to 1,700°C are reasonably consistent and appear to be more reliable. In one of these investigations [5, 34] the activities have been obtained from measurements of the composition and temperature dependence of the reaction:

$$SiO_{2(\text{slag})} + 2C_{(\text{graphite})} \rightleftharpoons Si_{(1 \text{ in Fe})} + 2CO_{(g)} \qquad 12.7$$

The silicate slag was formed over carbon-saturated iron, in which the activity of the carbon is unity, and under carbon monoxide at a pressure of 1 atmosphere, so that the equilibrium constant for equation 12.7 can be simplified to:

$$K_7 = \frac{(a_{Si})}{(a_{SiO_2})}$$

and the activities of silica are obtained from a knowledge of K_7 and the activities of silicon in carbon-saturated iron.* The composition range which has been investigated extends over the liquid phase field from the

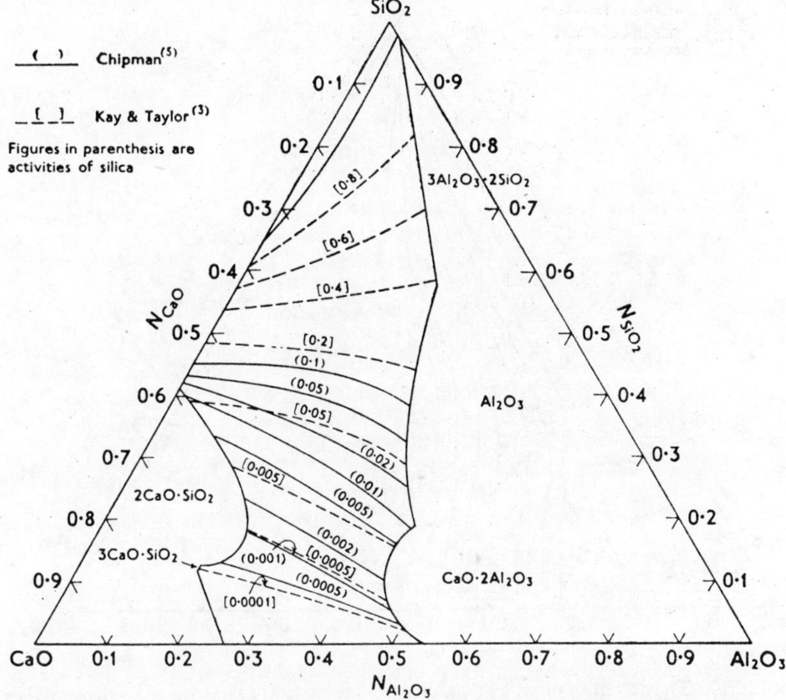

FIG. 12.16. The activities of silica in the ternary system $CaO–Al_2O_3–SiO_2$ at 1,600°C.

binary side $CaO–Al_2O_3$ towards silica saturation, but the latter surface is not intersected owing to the formation of silicon carbide at high silicon contents in the liquid iron.

In the second investigation [3] the activities have been determined from measurement of the equilibrium pressures for the reaction:

$$SiO_{2(slag)} + 3C_{(s)} \rightleftharpoons SiC_{(s)} + 2CO_{(g)} \qquad 12.8$$

and the composition range studied extends to the silica saturation surface.

* See footnote on page 323.

The iso-activity contours at 1,600°C which are shown in Figure 12.16 indicate that at low silica concentrations the activity coefficient of silica is raised progressively as the alumina concentration is increased, but this trend is reversed and (γ_{SiO_2}) is lowered by alumina additions when the silica concentration is greater than about 60 weight-per cent. For any composition the activity is raised only slightly as the temperature is increased from 1,500 to 1,700°C.

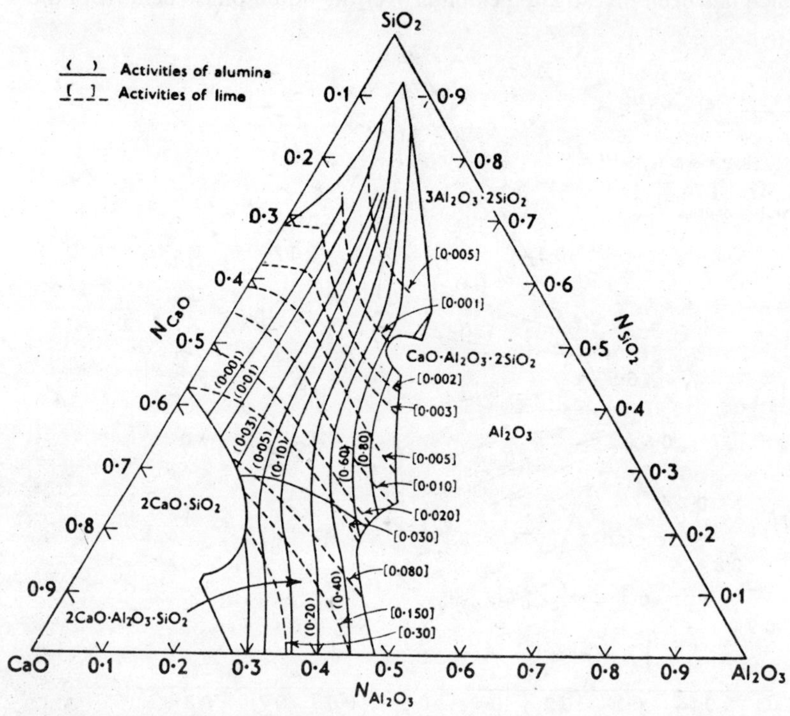

FIG. 12.17. The activities of lime and alumina in the ternary system $CaO–Al_2O_3–SiO_2$ at 1,600°C. (After Kay and Taylor [3].)

The Schuhmann solution of the ternary Gibbs-Duhem relation has been used to derive the iso-activity contours for lime and alumina from these data for silica [3] and, as shown in Figure 12.17, the activities of all three oxides deviate markedly in a negative direction from Raoultian behaviour, but the rate of change of the activity with composition is less marked for alumina than for the other two oxides.

With up to 10 weight-per cent of magnesia present in the melt the effect of magnesia on the activities of silica is approximately equivalent to the effect of an equal weight-per cent of lime and therefore on a molar basis the effect of magnesia is approximately two-thirds that of an equal concentration of lime. The effects of other oxide additions have not been examined, but in normal blast-furnace practice the concentrations of oxides other than those considered here is normally very low and the activities of the major constituents are not expected to vary appreciably from those found in the simple melts.

The activities of manganese oxide in blast-furnace slags are considered in Chapter 7 and require no further elaboration, whilst the activities of sulphur and phosphorus in liquid slags are discussed in Chapters 14 and 15 respectively.

12.2 Ideal Mixing in Silicate Melts

The phase diagrams for mixtures of the ortho or meta-silicates of the divalent basic oxides show extensive ranges of solubility in the solid state. At room temperature the heats of formation for solid solutions of the ortho or meta-silicates of both magnesium and iron and of calcium and iron are negligible [35], and this suggests that the activities in the solid state are approximately ideal. If it is assumed that the heats of mixing are also very small in the liquid state and that the structure of the melt is unchanged on mixing the two binary compounds, then the activities can be calculated in terms of the configurational (or positional) entropy of mixing. Further, the anion species and concentrations are constant between the composition limits; hence the entropy of mixing is simply that for the mixing of the cations. The calculation of this entropy is discussed in the review of the ionic treatments in Chapter 4, where it is shown that 'ideal' mixing of the cations may occur although the Raoult's law relation is not obeyed, for the 'ideal' activities are proportional to a power of the concentrations of the ions, which power is determined by the valencies of the ionic species present. From such considerations Richardson [15, 36, 37] has derived the following relations for ideal mixing:

$$a_{(\text{MaSiO}_4,\ \text{MbSiO}_4)} = N^2_{(\text{MaSiO}_4,\ \text{MbSiO}_4)} \tag{12.9}$$

and

$$a_{(\text{MaSiO}_3,\ \text{MbSiO}_3)} = N_{(\text{MaSiO}_3,\ \text{MbSiO}_3)} \tag{12.10}$$

where Ma and Mb are the divalent basic oxide cations, Ca, Mg, Mn, Fe, etc., and

$$a_{(x\,\text{MaO}\,.\,\text{SiO}_2)} = N^x_{(x\,\text{MaO}\,.\,\text{SiO}_2)} \tag{12.11}$$

$$= \frac{(a^x_{\text{MaO}})(a_{\text{SiO}_2})_{\text{ternary}}}{(a^x_{\text{MaO}})(a_{\text{SiO}_2})_{\text{binary}}} \tag{12.12}$$

The denominator of equation 12.12 is evaluated from the data for the binary system and at constant mole fraction of silica the activities of silica in the ternary melts are calculated by linear interpolation of $\log(a_{SiO_2})$ between the two binary limits. The only unknown is (a^x_{MaO}) in the ternary system and this quantity can be evaluated as a function of the composition along the lines of constant silica concentration.

The activities of the silicates, which can be calculated from the binary phase diagrams and the entropies of fusion assuming that the solid solutions are ideal, agree well with the values calculated using these equations for mixtures of $MgO.SiO_2$ with $FeO.SiO_2$; $CaO.SiO_2$ with $FeO.SiO_2$; $CaO.SiO_2$ with $MgO.SiO_2$; and $2MgO.SiO_2$ with $2FeO.SiO_2$; it is evident that these liquid mixtures conform closely to Richardson's 'ideal mixing' model. However, the activities which are calculated on this basis for liquid mixtures of $2CaO.SiO_2$ and $2FeO.SiO_2$ deviate markedly from the 'ideal' relation and this deviation is interpreted as evidence for a strong interaction in the liquid solutions [36].

The model has been used to calculate the 'ideal mixing' activities for ferrous oxide in the ternary system $FeO–CaO–SiO_2$ and for manganese oxide in the ternary system $MnO–CaO–SiO_2$ [9, 36, 37]. The pattern of behaviour is similar in both systems and the calculated activities correspond closely with the activities which have been measured at 1,600°C [9–11] for compositions at, or more acid than, the meta-silicate ratio but, as the basicity is raised beyond this ratio, the experimental activity-composition curves show negative deviation from the 'ideal mixing' curve. At the ortho-silicate join the measured activities show large positive deviation from Raoultian behaviour, but there is more marked negative departure from the 'ideal mixing' curve. These trends are shown in Figure 12.18 [9, 36].

The activities of silica in the binary systems $CaO–SiO_2$ and $MnO–SiO_2$ and the activities of manganese oxide in the system $MnO–SiO_2$ are temperature dependent. It follows from this that the 'ideal mixing' curves for the system $MnO–CaO–SiO_2$ are also dependent on the temperature, but within the range 1,500–1,650°C the deviations of the measured activities of manganese oxide from the 'ideal mixing' curves are similar to those at 1,600°C which are described above [9]. The ideal mixing curves for the system $FeO–CaO–SiO_2$ have been recalculated therefore, using the activity data for the binary system $FeO–SiO_2$ at 1,250 to 1,380°C [12, 13] to compare with the activities of ferrous oxide which have been measured at this temperature [13] in the ternary system. The more recent measurements of the activities of silica in the binary system $CaO–SiO_2$ [4, 5] have been used for these calculations in preference to the

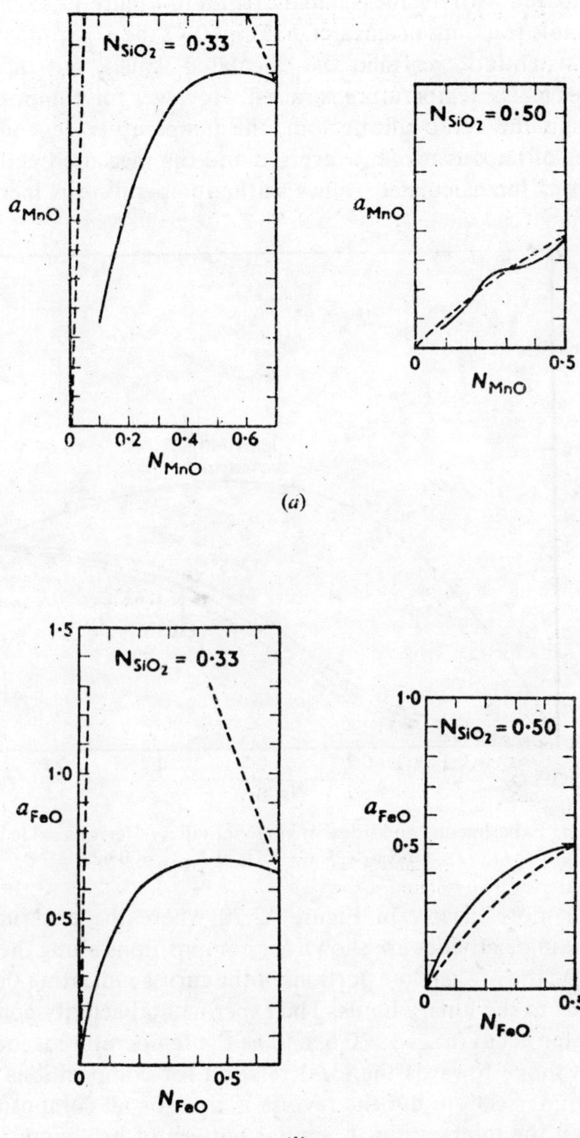

FIG. 12.18. Comparison of measured and 'ideal mixing' activities at 1,600°C of (*a*) manganous oxide, and (*b*) ferrous oxide, in ternary silicate melts. (After Richardson *et al*, [9, 36].)

data used in Richardson's original calculations. The recalculated curves are compared with the measured activities in Figure 12.19, which shows that at mole fractions of silica of 0·33 and 0·42 the experimental activities are consistently lower than the calculated values, but the divergence decreases as the temperature is raised. However for compositions more basic than the ortho-silicate join, the temperature dependence of the activities of ferrous oxide is reversed and the measured activities move away from the calculated values as the temperature is increased. This

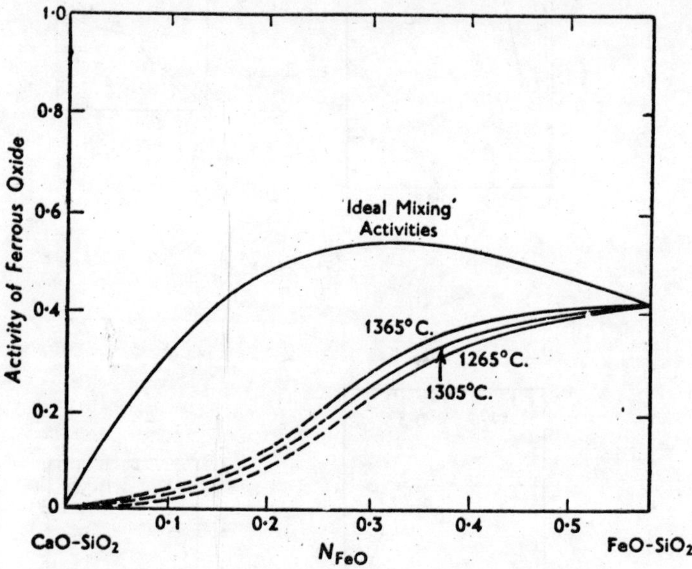

FIG. 12.19*a*. Experimental and 'ideal mixing' activities of ferrous oxide in the ternary system Fe_TO–CaO–SiO_2 at $N_{SiO_2} = 0{\cdot}42$.

is shown more clearly in Figure 12.20 where the experimental and 'ideal mixing' activities are shown for compositions along the join FeO–CaO . SiO_2, the broken line portions of the curves indicating the probable extensions to the binary limits. The experimental activity-concentration curves intersect at $(a_{Fe_TO}) = 0{\cdot}6$, and as the temperature is increased the activities move towards the ideal relation for compositions more acid than the intersection, but the reverse is true for all compositions more basic than the intersection. A similar pattern of behaviour is obtained along any join through the FeO apex to intercept the CaO–SiO_2 side of the system within the composition range investigated. Thus for all compositions within the liquid phase field of the ternary system in the temperature range 1,250–1,380°C, the measured activities of ferrous oxide are lower than those which are calculated from the 'ideal mixing' model and the extent of the deviation is increased as the silica concentration is raised.

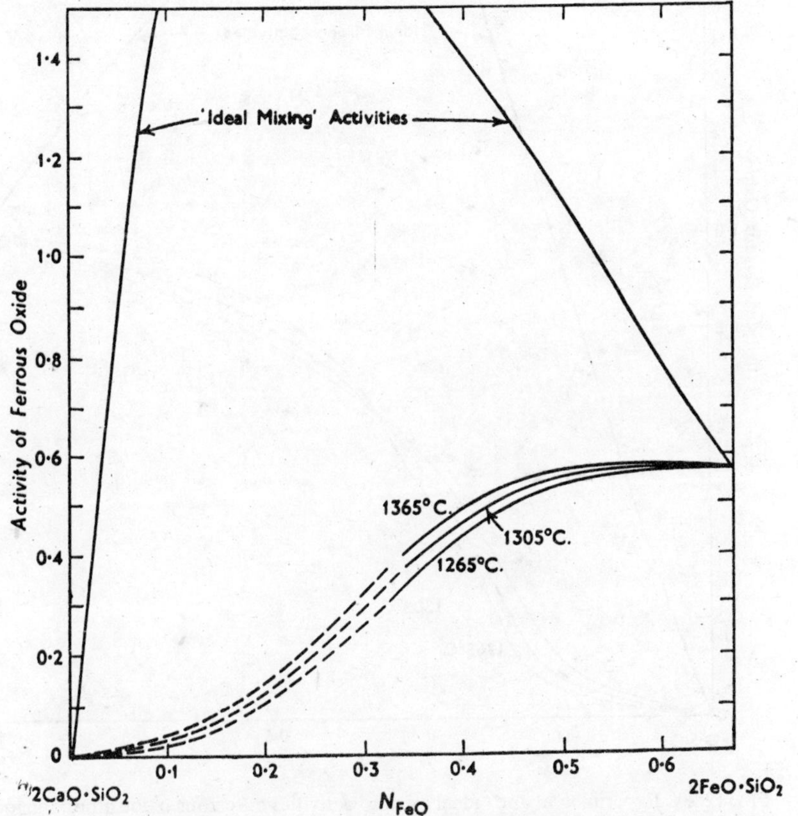

FIG. 12.19*b*. Experimental and 'ideal mixing' activities of ferrous oxide in the ternary system Fe_TO–CaO–SiO_2 at $N_{SiO_2} = 0{\cdot}33$.

When the integral free energies of mixing for the ternary systems calculated from the model are compared with the experimental data at 1,600°C they show excess negative free energies which are somewhat greater in the FeO–CaO–SiO_2 system than for MnO–CaO–SiO_2 melts, but at 1,600°C the maximum value is less than 2·5 kcal mol^{-1} [9, 35]. It has been suggested that these excess free energies in the more basic melts could arise from a tendency for the liquid solution to separate into two

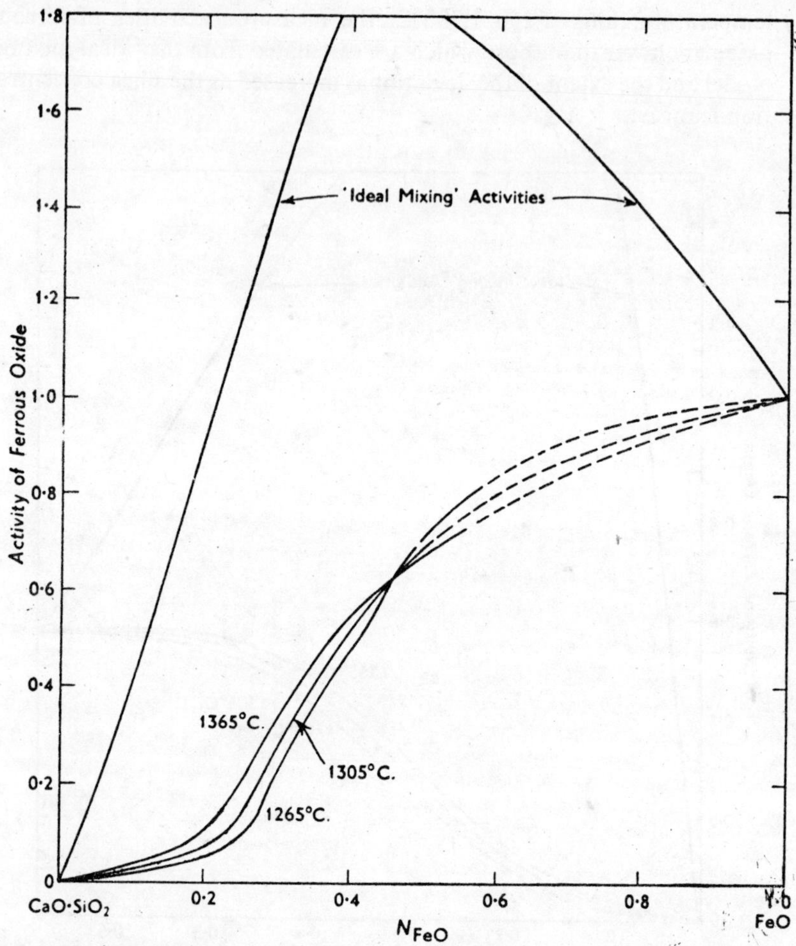

FIG. 12.20. Experimental and 'ideal mixing' activities of ferrous oxide along the join Fe_TO–CaO . SiO_2 in the ternary system Fe_TO–CaO–SiO_2

liquids (i.e., liquid immiscibility) and this premise is supported by free energy considerations [35]. For a melt consisting of 0·30 moles of CaO, 0·20 moles of SiO_2 and 0·50 moles of FeO the free energy of formation of an 'ideal mixing' solution is

$$(0{\cdot}30\,N_{CaO}+0{\cdot}075\,N_{SiO_2})+(0{\cdot}50\,N_{FeO}+0{\cdot}125\,N_{SiO_2}) \rightleftharpoons (0{\cdot}30\,N_{CaO}+0{\cdot}20N_{SiO_2}+0{\cdot}50\,N_{FeO}) \qquad 12.13$$

$$\Delta G^0 = -2{\cdot}0 \text{ kcal mol}^{-1}$$

whereas the free energy of formation of two separate liquids, one of which consists of lime and silica and the other of ferrous oxide, is:

$$(0{\cdot}30\,N_{CaO}+0{\cdot}075\,N_{SiO_2})+(0{\cdot}50\,N_{FeO}+0{\cdot}125\,N_{SiO_2}) \rightleftharpoons (0{\cdot}30\,N_{CaO}+0{\cdot}20\,N_{SiO_2})+(0{\cdot}50\,N_{FeO}) \qquad 12.14$$

$$\Delta G^0 = -3{\cdot}0 \text{ kcal mol}^{-1}$$

This result shows that the free energy of the two liquids is more negative and the mixture is more stable than the homogeneous liquid solution and implies that in the single-phase solution which is actually formed the Ca^{2+} ions are associated preferentially with $Si_2O_7^{6-}$ and other complex silicate anions, whilst the Fe^{2+} ions are associated preferentially with the excess O^{2-} ions. This tendency to form a non-random distribution of the ions lowers the configurational entropy of mixing relative to that for a random solid solution. The free energies of formation which are calculated on the basis of two separate solutions, plus an empirical factor which is equal to 40 per cent of the configurational free energy of mixing for an ideal random solution, correspond closely to the free energies of formation which have been measured experimentally at 1,600°C in the more basic composition ranges of the two ternary systems [35].

The hypothesis of non-ideal mixing and the tendency for two-liquid formation is supported by the data for the lower-temperature activities of ferrous oxide at low silica concentrations in the FeO–CaO–SiO_2 system. As noted earlier, when (a_{Fe_TO}) is greater than 0·60 the activities are lowered as the temperature is raised, and the relative partial molar enthalpy of ferrous oxide is positive in this composition range, which is usually associated with a tendency towards unmixing and separation into two phases.

The foregoing interpretation is consistent also with the pattern of behaviour which is expected from consideration of the bond energies and polarisation of the pure oxides. The polarisation of the Si—O bond is opposed more strongly by Fe^{2+} than by Ca^{2+} ions with the result that when calcium and ferrous silicates are mixed the calcium and silicate ions tend to separate from the ferrous ions, but to preserve electro-neutrality the ferrous ions must associate with an equal number of O^{2-} ions. For compositions at, or more acid than, the ortho-silicate join it is generally considered that most of the oxygen ions are associated with the silica, and therefore the separation of Fe^{2+} and O^{2-} ions in the more acid melts can be accomplished only by the sharing of more oxygen bonds between the silicate anions in order to release the requisite number of oxygen ions. This is implied in equation 12.14. However, the lower-temperature data for the FeO–CaO–SiO_2 system show that the relative partial molar enthalpy of ferrous oxide is negative for all compositions where (a_{Fe_TO}) is less than 0·6 and the change in the sign of the enthalpy occurs at compositions which are slightly more basic than the ortho-silicate join. Thus it appears that the ferrous ions are associated only with the oxygen ions in excess of the number required to satisfy the positive charges on the Si^{4+} ions and that in the more acid melts the calcium and ferrous ions remain mixed together with the silicate anions. The negative value of H^M_{FeO} in the acid melts suggests that some type of ordering is present in the liquid; this could arise from a tendency for the Fe^{2+} ions to remain in close proximity to certain preferred oxygen ions in each silicate group in an endeavour to retain a balanced electron field around each ionic grouping.

12.3 Calculation of Slag Activities from Ionic Theory

In the application of the ionic theories to the determination of the activities of the constituents of liquid slags, which is reviewed in Chapter 4, it is frequently assumed that either the cations or the anions are mixed randomly in a constant lattice of the ions of opposite charge. However, from the activity data which are presented in this chapter, it is clear that in melts containing ferrous silicates excess free energies of mixing are the rule, rather than the exception, and any treatment which presupposes ideal mixing of either cations or anions is unlikely to yield data which compare favourably with the measured activities.

An attempt has been made to extend the ionic approach to allow for departures from ideality by computing the free-energy change as the sum of the free energies which arise from both the entropy of mixing and the changes in the number and types of cation-anion bonds [38, 39]. For example, when ferrous oxide and calcium ortho-silicate are mixed to form a liquid solution, new Fe—O—Si and Ca—O bonds are formed, while some Ca—O—Si and Fe—O bonds are destroyed, and the change in free energy resulting from the changes in bonding is equated to the free-energy change for the reaction:

$$FeO + 0{\cdot}5(2CaO\,.\,SiO_2) \rightleftharpoons CaO + 0{\cdot}5(2FeO\,.\,SiO_2) \qquad 12.15$$

The activity of ferrous oxide in a slag containing the ionic species Fe^{2+}, Ca^{2+}, Mn^{2+}, Mg^{2+}, SiO_4^{4-} and O^{2-} is given by the relation [39]:

$$\begin{aligned} G^M_{FeO} = RT\ln(a_{Fe_TO}) = {} & RT\ln N_{Fe^{2+}}\,.\,N_{O^{2-}} + RT\ln(\gamma_{Fe^{2+}})(\gamma_{O^{2-}}) \\ & + N'_{Ca^{2+}}\,.\,N'_{SiO_4^{4-}}\,.\,\Delta G_{Fe\text{-}Ca} \\ & + N'_{Mn^{2+}}\,.\,N'_{SiO_4^{4-}}\,.\,\Delta G_{Fe\text{-}Mn} \\ & + N'_{Mg^{2+}}\,.\,N'_{SiO_4^{4-}}\,.\,\Delta G_{Fe\text{-}Mg} \end{aligned} \qquad 12.16$$

where N' is the electrically equivalent ion fraction (see Chapter 4.2) and the ΔG terms are the free energies of the cation-exchange reactions of the type illustrated by equation 12.15. The term $(\gamma_{Fe^{2+}})(\gamma_{O^{2-}})$ cannot be evaluated from the existing data and is usually neglected.

The activities computed with this equation for the range of slag compositions encountered in basic steel-making practices agree well with the activities which have been measured at 1,600°C [38], but the treatment cannot be extended to compositions at, or more acid than, the ortho-silicate join for then the calculated oxygen ion fraction is zero.

In this treatment the assumption is made that the cations are distributed randomly in relation to other cations of similar charge, but the large free-energy values for the various exchange reactions are probably a result of non-random ionic distributions. The restriction on mixing lowers the mixing entropy to less than that for an ideal solution, but the effect of this on the free energy of mixing may be balanced by a corresponding increase in the enthalpy of mixing from the greater number of high-energy bonds which are formed. This balance appears to hold for the calculation of the ferrous oxide activities in the basic melts which have been studied experimentally, but the degree of non-randomness may vary with the slag composition and the ionic species which are present. Hence the agreement which is obtained in the one system to which the equations have been applied does not justify their use to determine activities in other composition ranges or in the presence of other species. The error introduced by non-random mixing would be eliminated if the activity coefficient term in equation 12.15 could be evaluated.

An attempt has been made to modify the treatment and extend it to the silica-rich region of the ternary system FeO–CaO–SiO_2 [13, 25]. The oxygen ion fractions in the composition range between the ortho-silicate and the meta-silicate joins can be determined empirically by assuming that all the silica is present as SiO_3^{2-} ions. Equation 12.16 does not admit values of (γ_{Fe_TO}) below unity, but for all compositions within the range considered the measured values of (γ_{Fe_TO}) are less than unity, and to take account of this it is necessary to introduce into the equation an empirical term which includes the value of (γ_{Fe_TO}) in the binary system FeO–SiO_2 at the same *silica* concentration as that of the ternary composition for which the activity of ferrous oxide is calculated. The modified expression is:

$$RT \ln (a_{Fe_TO}) = RT \ln N_{Fe^{2+}} N_{O^{2-}} + RT \ln (\gamma_{Fe_TO})_{(binary)} + xN'_{Ca^{2+}} . N'_{SiO_3^{2-}} \qquad 12.17$$

The coefficient x cannot be identified with the free-energy change for the exchange reaction in the meta-silicate liquids:

$$FeO + CaO.SiO_2 \rightleftharpoons FeO.SiO_2 + CaO$$

because the activities of ferrous oxide are increased appreciably as the temperature is raised, whereas the free-energy change for this reaction varies only slightly with the temperature, and it is necessary to assume values for x of 19,000 at 1,365°C and 30,000 at 1,600°C in order to fit the equation to the measured activities. The agreement between the calculated and the actual activities is then reasonably good over a fairly wide range of compositions and the activities can also be calculated for acid slags containing small concentrations of manganese oxide if the latter is considered as an inert diluent. However, the additional term in the equation and the values for x cannot be justified on a theoretical basis and it is obvious that this approach can yield no useful information regarding the ionic behaviour of silicate melts or predict the effect on the activities of other ionic species present in the melts. The only useful feature is the ability to express in the form of a mathematical relation the measured activities within a specified concentration range. If the experimental studies could be extended to cover the effects of all the constituents which may be present in a steel-making slag, the summarisation of the data by a single equation would be most useful.

12.4 The Rates of Slag–Metal Reactions

Direct experimental evidence of the rates of the various reactions which occur in iron and steel-making is extremely limited. The experimental difficulties with which the investigator is faced at the relatively high operating temperatures are even more numerous than in the determination of activities, and erroneous conclusions may be reached through the neglect of unforeseen factors which affect the reaction rates. Perhaps the most serious problem lies in the fact that the kinetic data which are obtained from a laboratory investigation are pertinent only to the precise conditions under which the experiments are conducted. These conditions may not apply in practice, where the magnitude of the turbulent and convective flow of the slag and the metal layers varies from one practice to another and from time to time in any one heat.

There is some evidence of the factors which affect the reaction rates in the practical observations which are discussed in the chapters dealing with the iron and steel-making processes. Compare for example the very rapid rate of carbon removal under the strongly turbulent conditions which exist in the bath in the converter process (about 0·30 per cent of carbon per minute) with the very much slower rate (0·15 to 0·60 per cent of carbon per hour) normally obtained in the open-hearth and electric-arc processes. Even more striking is the slow rate of transfer of sulphur into a reducing slag relative to the rate at which other elements are removed into an oxidising slag in the electric-arc furnace, or the marked departure from the equilibrium distributions of the elements between the slag and the metal in the hearth of the blast furnace. The rapid refining in the Bessemer process shows clearly that the various reactions can proceed very rapidly, if the conditions are suitable for bringing the reactants into contact and for the removal of the reaction products. When the bath is not agitated the reaction rates are controlled by thermal convective flow and by the diffusivities of the reactants and products in the slag or metal layer, but they are increased very considerably when the bath is stirred either by mechanical means or by the carbon boil. Hence in practice the kinetics are dependent on artificial mixing to a greater extent than on diffusion processes.

THE SIGNIFICANCE OF DIFFUSION RATES

At steel-making temperatures the diffusivities in liquid iron of carbon, manganese, silicon, sulphur and phosphorus are of the order of 10^{-4} $cm^2\ sec^{-1}$ [40–42] and the self-diffusion of liquid iron is of the same order [43], but the diffusion rates decrease as the temperature is lowered and at 1,350–1,400°C they are about one-tenth of this value [44, 45]. Few reliable diffusion measurements have been made with liquid slags. The diffusion rate of ferrous ions in silica-saturated, ferrous silicate is $5{\cdot}8 \times 10^{-5}$ $cm^2\ sec^{-1}$ at 1,250°C [46]. The diffusivity of the calcium ions in a blast-furnace type slag containing 40 per cent CaO, 40 per cent SiO_2 and 20 per cent Al_2O_3 varies from 3×10^{-7} at 1,350°C to $1{\cdot}3 \times 10^{-6}$ at 1,450°C, whilst the corresponding diffusivities of silicon ions are lower by a factor of ten [47, 48]. These diffusivities correspond to penetration distances in the liquid metal and slag of about 1 and 0·1 mm per hour respectively, but the bath of molten metal is usually about 30 to 40 cm deep in the open-hearth furnace and refining would not be a practical proposition if the reactions were dependent entirely on diffusion.

This is demonstrated convincingly by the following data. The rate of exchange of iron between the liquid metal and an iron silicate slag 1 cm thick has been measured with radioactive tracers [49] and it is found that 1 gram-mole of iron can be transferred in each direction across an interface area of 71·5 sq. cm in 115 seconds when the metal and slag are almost completely stationary. For an open-hearth furnace containing 100 tons of metal the area of the slag-metal interface is about 500 sq. ft so, at the same transfer rate, 0·288 ton of iron can transfer each way across the interface in one hour. Assuming that the two electrons which are released by the transfer of each iron atom into the metal are consumed by the simultaneous transfer of an oxygen atom across the interface into the metal, this oxygen flow is sufficient to oxidise only 0·058 per cent of carbon per hour. The actual rate of carbon removal would be considerably lower, for some of the electron charges are consumed by the oxidation into the slag of elements such as manganese, silicon and phosphorus, and the net rate of transfer of oxygen from the slag into the bulk metal to react with the carbon is less than the rate of transfer of iron atoms in the same direction. In addition some of the oxygen remains dissolved in the iron as the carbon is oxidised and the activity of the dissolved oxygen is lowered.

These transfer data have been obtained with a slag saturated with silica, where the viscosity is relatively high; a more rapid rate of transfer may obtain with a fluid slag. A refining slag is usually about 10 cm thick and the transfer rate may also increase slightly with the thickness of the slag, but these factors are not sufficient to account for the very large difference between the actual and the predicted rates of carbon removal. In fact during a steady-state boil the actual rate of oxygen transfer to the metal must be between 5 and 10 times as great as the rate for the diffusion-controlled transfer reaction.

This does not mean that diffusion is unimportant in the reaction kinetics. The mechanical mixing which is caused by the carbon boil or thermal convective flow can be regarded as dividing the metal and slag layers into a series of thin strata and diffusion occurs over relatively short distances between these layers. Some success has been obtained in predicting the rate of oxidation of various elements from the metal into the slag by assuming the existence of a thin stagnant film of slag and metal on either side of the slag–metal interface (see Chapter 5.5) and the rate of oxidation of the elements in the iron is then dependent on the rate of supply of oxygen from the bulk slag, through the static layers and across the interface, into the bulk metal. This aspect is considered further in relation to the various elements in the following chapters.

THE EFFECTS OF TEMPERATURE AND AGITATION

The reactions proceed more rapidly as the temperature is increased since the diffusion rates increase exponentially whilst at constant composition the fluidity of both the metal and the slag is raised with increasing temperature and so, for a constant rate of carbon oxidation, the mixing of the metal and the slag by the carbon boil is increased. Thus the total refining time can be shortened by increasing the rate at which the bath is brought up to the optimum refining temperature and by adjustment of the composition to avoid saturation with one or other of the constituent oxides. For example, lime is added to acid slags to avoid saturation with silica, whilst fluorspar or bauxite is added to basic slags to flux undissolved lime.

The other necessary condition is a sufficient degree of turbulence in the bath to secure thorough mixing in the slag and metal layers. However, it is evident that the agitation created by a deep-seated carbon boil is not in itself sufficient to secure the optimum refining rates, for the rate of carbon elimination in the open-hearth furnace is accelerated considerably (relative to the speed of removal with ore feeding) when gaseous oxygen is injected into the metal. The injection of the oxygen gas into the metal promotes a greater degree of turbulence and mixing than is obtained from the normal carbon boil and the rates of carbon removal in the 'Ajax' oxygen open-hearth process are closer to those obtained with the top-blown oxygen converter processes. Although both of these rates are considerably greater than are normally obtained in the open-hearth process, they still fall short of the optimum conditions, for the refining time in the top-blown converter processes is approximately double the time required for the bottom-blown Bessemer processes.

When the refining reactions are not accompanied by the evolution of a gas the reaction rates can be increased by artificially-induced agitation, and examples of this are the use of electro-magnetic stirrers or bubbling an inert gas through the bath to speed the reactions under the reducing slag in the electric-arc furnace process. Similarly a more rapid approach to the equilibrium distributions could be obtained in the blast furnace by continuously bubbling an oxygen-free gas through the slag and metal.

To summarise, the necessary conditions for high rates of refining are a fluid slag with the required composition to absorb the elements which are to be removed as solids or liquids from the metal, a rapid rate of heat transfer to the metal at the start of refining, and the maintenance of the necessary degree of turbulence in the bath. Whilst the last factor may or may not be under the control of the operator, the other two factors are certainly amenable to adjustment. This is commonly referred to as *slag control*, which is simply the production of a slag with the most suitable composition and the minimum volume required to absorb the reaction products at the earliest possible stage in the process. Any increase over the minimum slag volume which is necessary for the refining which has to be performed increases the thermal load on the process and, in those processes where the heat is transmitted via the slag to the metal, the rate of heat transfer is lowered as the slag thickness is increased.

12.5 Slag Control Techniques

REQUIREMENTS AND LIMITATIONS

Practical slag control involves obtaining rapidly the answers to the four following questions:

1. What is the chemical composition of the metal charge? This is obtained readily when the charge is all hot metal, but when part or all of the charge is solid the composition usually can be ascertained only after melting is complete and some oxidation of the elements has already occurred.
2. What is the most suitable slag composition for effecting the required elimination of the elements from the iron? The slag must be fluid if it is to take an active part in the refining and the activities in the slag should be low for the elements which are to be transferred from the iron.
3. What is the minimum slag volume needed to absorb the liquid and solid reaction products? The correct answer can be obtained only from a knowledge of the activities of the elements in the slag and in the metal and from the equilibrium constants for the various reactions.
4. What are the total quantity and composition of the oxides which are picked up from the refractory lining of the furnace and dissolved in the slag during the course of the heat? This information cannot be predicted theoretically, but an estimate can be made on the basis of previous experience or the amounts can be determined from frequent analyses of the slag during the heat and by preparing a material balance.

The above details require time to collate and examine; hence slag control methods are not strictly applicable to processes in which the refining time is short. For example, slag control in the Bessemer process is more in the nature of a post-mortem; the composition of the slag in the first heat made is based on the hot-metal composition and in subsequent heats adjustments are made on the basis of the data from the previous casts. To some extent this applies also in the blast furnace process, where several hours may elapse from the time when the composition of the burden is adjusted to the time when the hearth slag is affected by this change. In the acid steel-making practices the slag volume is determined largely by the volume of the acid oxides which are formed by oxidation of the metal bath and by the extent of the lining erosion, for the only adjustments which are made to the slag composition are small feeds of lime to avoid over-saturation with silica and iron oxides to propagate the oxidation reactions. Hence the principal applications of the slag control methods are with the basic practices in the open-hearth, the electric-arc and certain of the new refining techniques.

In the early days of steel-making, before the slag–metal reactions were understood properly, it was common practice to add an excess of slag-forming materials to the charge to ensure that the metal would be refined to within the specified limits. Generally the metal satisfied the specification but the slag volume was often grossly in excess of theoretical requirements and the heat time was extended by the low rate of heat transfer to the metal. A reduction in the slag volume often resulted in the carbon being removed to the required level before the phosphorus and sulphur limits were reached. This necessitated adding pig-iron to raise the carbon content and maintain the boil while adjusting the slag composition to allow for the further removal of these elements. Alternatively only part of the slag materials was added with the charge and further quantities were added from time to time during the heat, the additions being estimated by frequent sampling of the metal and examination of the rate of change of the concentrations of the impurities; but the refining time was extended by the time taken to analyse each metal sample and for the slag-forming additions to work through the bath before the next sample was taken. In all cases 'rogue' casts were occasionally encountered which could not be refined to the required limits and the cast had to be rejected.

Some improvement in the control of the processes is obtained from examination of slag 'pancakes' and metal button samples which are drawn from the bath at various times during the refining period. With experience a close estimation of the basicity and the iron content of the slag can be obtained from the surface appearance and fracture colour of the slag pancake [50, 51], whilst the appearance of the button sample gives some indication of the carbon content and the degree of oxidation of the metal. For many years these were the principal methods of obtaining rapidly indication of the progress of the refining and they are still used extensively.

The first systematic attempts to devise slag control procedures were made during the early 1930's. In America a number of papers dealing with this theme were published by Herty. In Germany Schenck derived equilibrium 'constants' containing empirical terms from examination of a wide range of operating data and graphs were prepared from which the 'free' concentrations of the slag components were determined as a function of the composition [52]. The equations which he proposed for determining the equilibrium partitioning of phosphorus between the slag and the metal are reviewed briefly in Chapter 15.3. The 'free' concentrations can be regarded as an attempt to obtain thermodynamic activities from operational data and it is perhaps fortunate that a marked improvement in performance was often obtained from the application of these equations, for at last the need was recognised for more scientific examination of the refining reactions. Many years elapsed, however, between the time when the 'free' concentrations were derived and the publication of the first measurements of thermodynamic activities in molten slags, but today the determination of the most suitable composition and volume of the slag is developing gradually into an exact science as the data for activities in the slag and metal increase and the true mass-action constants are evaluated.

If the slag control methods are to be fully effective not only should they eliminate the production of reject casts which do not meet the composition requirements, but the time taken to refine the charge should be as short as possible and this means that the compositions of the metal and slag must be determined rapidly at suitable intervals during the heat.

DETERMINATION OF THE METAL COMPOSITION

An approximate value of the carbon content of the metal can be obtained within a few minutes of taking a sample from the furnace by measuring some suitable physical property; this is most commonly measurement of the magnetic properties with an instrument such as the carbometer. A wet assay of the other elements in the metal is time consuming and, even when rapid methods are used to give approximate values, a complete analysis may take at least twenty minutes; this time has, however, been reduced very considerably with the introduction of the mass quantometer and a complete analysis is now available within a few minutes from the time when the sample is drawn.

ESTIMATION OF THE VOLUME AND COMPOSITION OF THE SLAG

The next problem is the rapid estimation of the volume and composition of the slag at clear-melt and during refining. Ideally the amount of lime added with the charge is a little less than the estimated requirement and additional quantities are added as soon as the composition of the liquid slag can be determined, but wet assay methods are not suitable for determining the slag composition because of the time which is required to obtain reliable values. Measurements of the physical properties of solid slag samples have been considered, therefore, as a means of supplementing the information obtained from visual examination of the slag pancakes. Attempts have been made to relate the basicity of the slag to its pH value, which is obtained by measurement of the hydrogen-ion concentration in an aqueous solution [53–55]. Theoretically the pH increases as the basicity is raised and hydrolysis of the uncombined lime is increased, but in practice it does not vary markedly over a fairly wide range of compositions (Table 12.1) and, when the pH of the predominant compound in the slag is reached, no further change occurs with change in the composition. Measurements have also been made of the electrical conductivities of aqueous solutions of the slag [53, 56], but these suffer from similar disadvantages. In particular the conductivity of

Table 12.1 *The pH and Conductance of Various Slag Compounds* (55)

	pH	Conductivity microhm at 20°C
$CaO.Fe_2O_3$	9·0	46
MgO	10·1	98
$2CaO.Fe_2O_3$	10·5	124
$CaO.SiO_2$	10·7	123
$3CaO.2SiO_2$	11·0	192
$2CaO.SiO_2$	11·4	681
$3CaO.SiO_2$	11·9	1,810
CaO	12·6	7,800

calcium oxide is very high in relation to the conductivities of other compounds in the slag, and small quantities of undissolved lime cause a very marked change in the measurements. These methods are more time-consuming and yield little or no additional information than can be obtained from the examination of slag pancake samples, but recently a major step forward has been made with the successful application of the mass quantometer to the analysis of slag samples and a complete slag analysis can now be obtained within a few minutes of taking the sample from the bath.

Traditionally the slag volume has been estimated on the basis of the amount of lime charged to the furnace, the average lime content of the slag at clear-melt or before tapping and the volume of oxides formed by oxidation of the impurity elements in the slag, but the actual volume could be determined accurately only after the slag was tapped from the furnace. A more positive assessment can now be obtained by adding a controlled quantity of a radioactive isotope to the slag and monitoring the radioactivity of a measured volume of the slag. The isotope should have a sufficiently long half-life, disperse uniformly through the slag in a short time and remain entirely in the slag without suffering loss by volatilisation or by partitioning to the liquid metal or refractory lining. The mixed isotope ^{140}Barium plus ^{140}Lanthanum is claimed to satisfy these requirements [57] and a radioactivity of 0·1 microcuries per lb of slag is adequate for accurate measurements, but is sufficiently low to avoid any health hazard. The isotope ^{45}Ca has been tried, but this disperses rather slowly in the slag. More rapid dispersion is obtained with ^{32}P [58] but there is a high probability that some of this enters the metal either by partitioning or by an exchange interaction with the non-radioactive phosphorus in the metal.

Only a passing reference has hitherto been made to slag control in the blast furnace. With adequate burden preparation and uniformity of the charge composition, sufficient control can usually be obtained through variation of the hot blast composition and/or temperature to ensure that the slag is fluid. When the composition of the ore is not constant the burden must be adjusted to ensure that the fluctuations in the composition of the slag do not cause saturation with one or other of the constituent oxides. In this case with a basic practice some improvement in performance could be obtained by lowering the ratio CaO/SiO_2 in the burden and adjusting the composition of the hearth slag by lime injection at the tuyere level.

REFERENCES

1. L. C. CHANG and G. DERGE, *Trans. Amer. Inst. min. (metall.) Engrs.*, 1947, Vol. 172, p. 90.
2. L. YANG, C. L. MCCABE and R. MILLER, *Physical Chemistry of Steel-making*, New York, John Wiley, 1958, p. 63.
3. D. A. R. KAY and J. TAYLOR, *Trans. Faraday Soc.*, 1960, Vol. 56, p. 1372.
4. R. A. SHARMA and F. D. RICHARDSON, *J. Iron St. Inst.*, 1961, Vol. 198, p. 386; ibid., 1962, Vol. 200, p. 373.
5. J. CHIPMAN, *Physical Chemistry of Process Metallurgy*, Interscience, 1961, p. 27.
6. K. SANBONGI and Y. OMORI, *Physical Chemistry of Metallic Solutions and Intermetallic Compounds*, London. N.P.L., Vol. 2, 1959, Paper 6D.
7. M. REY, 'Physical Chemistry of Process Metallurgy', *Disc. Faraday Soc.*, 1948, No. 4, p. 257.
8. H. B. BELL, A. B. MURAD and P. T. CARTER, *Trans. Amer. Inst. min. (metall.) Engrs.*, 1952, Vol. 194, p. 718.
9. K. P. ABRAHAM, M. W. DAVIES and F. D. RICHARDSON, *J. Iron St. Inst.*, 1960, Vol. 196, p. 82.

10. K. L. FETTERS and J. CHIPMAN, *Trans. Amer. Inst. min. (metall.) Engrs.*, 1941, Vol. 145, p. 95.
11. C. R. TAYLOR and J. CHIPMAN, ibid., 1943, Vol. 154, p. 228.
12. R. SCHUHMANN and P. J. ENSIO, ibid., 1951, Vol 191, p. 401.
13. C. BODSWORTH, *J. Iron St. Inst.*, 1959, Vol. 193, p. 13.
14. C. BODSWORTH and I. M. DAVIDSON, *Physical Chemistry of Process Metallurgy*, Interscience, 1961, p. 233.
15. F. D. RICHARDSON, *Physical Chemistry of Melts*, London, Inst. Min. Met., 1953, p. 75.
16. H. L. BISHOP, N. J. GRANT and J. CHIPMAN, *Trans. Amer. Inst. min. (metall.) Engrs.*, 1958, Vol. 212, p. 185.
17. P. K. FOSTER and A. J. E. WELCH, *Trans. Faraday Soc.*, 1956, Vol. 52, p. 1636.
18. J. CHIPMAN, J. B. GERO and T. B. WINKLER, *Trans. Amer. Inst. min. (metall.) Engrs.*, 1950, Vol. 188, p. 341.
19. P. T. CARTER and T. G. MACFARLANE, *J. Iron St. Inst.*, 1957, Vol. 185, p. 54.
20. J. F. ELLIOTT, *Trans. Amer. Inst. min. (metall.) Engrs.*, 1955, Vol. 203, p. 485.
21. E. T. TURKDOGAN and J. PEARSON, *J. Iron St. Inst.*, 1953, Vol. 173, p. 217.
22. H. L. BISHOP, H. N. LANDER and J. CHIPMAN, *Trans. Amer. Inst. min. (metall.) Engrs.*, 1956, Vol. 206, p. 862.
23. W. A. FISCHER and H. SPITZER, *Arch. Eisenhüttenw.*, 1958, Vol. 29, p. 611.
24. H. L. BISHOP, N. J. GRANT and J. CHIPMAN, *Trans. Amer. Inst. min. (metall.) Engrs.*, 1958, Vol. 212, p. 890.
25. I. M. DAVIDSON and C. BODSWORTH, *J. Iron St. Inst.*, 1960, Vol. 195, p. 163.
26. H. LARSEN and J. CHIPMAN, *Trans. Amer. Inst. min. (metall.) Engrs.*, 1953, Vol. 197, p. 1089; 1954, Vol. 200, p. 759.
27. R. SCHUHMANN and P. J. ENSIO, ibid., 1952, Vol. 194, p. 723.
28. E. T. TURKDOGAN and P. M. BILLS, *J. Iron St. Inst.*, 1957, Vol. 186, p. 329; 1958, Vol. 188, p. 143.
29. L. S. DARKEN and R. W. GURRY, *J. Amer. chem. Soc.*, 1945, Vol. 67, p. 1398; 1946, Vol. 68, p. 798.
30. F. D. RICHARDSON and J. C. BILLINGTON, *Trans. Inst. Min. Met.*, 1955–56, Vol. 65, p. 273.
31. F. KÖRBER and W. OELSEN, *Mitt. K.-Wilh. Inst. Eisenforsch.*, 1933, Vol. 15, p. 271.
32. E. T. TURKDOGAN and J. PEARSON, *J Iron St. Inst.*, 1953, Vol. 175, p. 393.
33. O. A. ESIN and B. M. LIPINSKIKH, *Izv. Ak. Nauk Otdelenie Tekn.*, 1954, (2), p. 60.
34. F. C. LANGENBERG and J. CHIPMAN, *Trans. Amer. Inst. min. (metall.) Engrs.*, 1959, Vol. 215, p. 958.
35. K. P. ABRAHAM and F. D. RICHARDSON, *Physical Chemistry of Process Metallurgy*, Interscience, 1961, p. 263.
36. F. D. RICHARDSON, *Trans. Faraday Soc.*, 1956, Vol. 52, p. 1312.
37. F. D. RICHARDSON, *Physical Chemistry of Steel-making*, New York, John Wiley, 1958, p. 68.
38. H. FLOOD, T. FORLAND and K. GRJOTHEIM, *Physical Chemistry of Melts*, London, Inst. Min. Met., 1953, p. 46.
39. H. FLOOD and K. GRJOTHEIM, *J. Iron St. Inst.*, 1952, Vol. 171, p. 64.
40. W. F. HOLBROOK, C. C. FURNAS and T. L. JOSEPH, *Industr. Engng. Chem.*, 1932, Vol. 24, p. 993.

41. M. PASCHKE and A. HAUTMANN, *Arch. Eisenhüttenw.*, 1935, Vol. 9, p. 305.
42. B. V. STARK, E. V. CHELISCHEV and E. Z. KAZACHKOV, *Izv. Ak. Nauk Otdelenie Tekn.*, 1951, (11), p. 1689.
43. L. YANG, M. T. SIMNAD and G. DERGE, *Trans. Amer. Inst. min. (metall.) Engrs.*, 1956, Vol. 206, p. 1577.
44. Y. KAWAI, *Sci. Rep. Res. Inst. Tohoku Univ.*, 1957, Series A, 9, p. 520.
45. R. F. GRACE and G. DERGE, *Trans. Amer. Inst. min. (metall.) Engrs.*, 1958, Vol. 212, p. 331.
46. M. T. SIMNAD, L. YANG and G. DERGE, ibid., 1956, Vol. 206, p. 690.
47. H. TOWERS, M. PARIS and J. CHIPMAN, ibid., 1953, Vol. 197, p. 1455.
48. H. TOWERS and J. CHIPMAN, ibid., 1957, Vol. 209, p. 769.
49. G. DERGE and C. E. BIRCHENALL, ibid., 1953, Vol. 197, p. 1648.
50. W. O. PHILBROOK, *Basic Open-Hearth Steel-making*, second edition, A.I.M.E., 1951, p. 190.
51. J. D. SHARP, W. JOHNSON and K. W. ANDREWS, *J. Iron St. Inst.*, 1960, Vol. 195, p. 83.
52. H. SCHENCK, *Physical Chemistry of Steel-making*, English Translation by B.I.S.R.A., London, 1945.
53. W. A. SMITH, J. MONAGHAN and W. HAY, *J. Iron St. Inst.*, 1948, Vol. 160, p. 121.
54. C. W. SHERMAN and N. J. GRANT, *Trans. Amer. Inst. min. (metall.) Engrs.*, 1949, Vol. 185, p. 898.
55. P. D. S. ST PIERRE, ibid., 1953, Vol. 197, p. 41.
56. Y. KASAMATU, *Tetsu to Hagane*, 1957, Vol. 43, p. 925.
57. G. R. CHURCH, W. C. HESELWOOD and G. A. NICHOLSON, *Nature, Lond.*, 1957, Vol. 179, p. 1294.
58. A. M. SKREBTSOV, *Zavods. Lab.*, 1959, p. 1078.
59. H. F. RAMSTAD and F. D. RICHARDSON, *Trans. Amer. Inst. min. (metall.) Engrs.*, 1961, Vol. 221, p. 1021.

CHAPTER 13

Oxidation and Deoxidation Reactions

13.1 The Activities and Solubilities of Oxygen in Liquid Iron

The partial pressure of oxygen which is in equilibrium with the oxygen dissolved in liquid iron varies normally between 10^{-8} and 10^{-10} atmosphere, and depends on the temperature and the purity of the iron. These pressures are too low for direct measurement; hence the solubility is determined by bringing the metal to equilibrium with controlled mixtures of either carbon monoxide and carbon dioxide or hydrogen and steam. At equilibrium the partial pressure of oxygen in equilibrium with the metal is the same as that in equilibrium with the gas mixture. The method whereby the data are evaluated is desoribed with respect to the oxygen potential diagrams in Chapter 1.6.

The equilibrium constants for these reactions have been determined as follows, where the standard state for oxygen in iron is the 1 weight-per cent solution. For the reaction:

$$CO_{(g)} + O_{(\%\,\mathrm{in\,Fe})} \rightleftharpoons CO_{2(g)} \qquad 13.1$$

the equilibrium constant is

$$K_{CO} = \frac{p_{CO_2}}{p_{CO}[a_C]}$$

and the equations are:

$$\Delta G^0 = -38{,}050 + 20{\cdot}72T \text{ cal mol}^{-1}; \log K_{CO} = \frac{8{,}316}{T} - 4{\cdot}53 \text{ [1]} \qquad 13.2$$

$$\Delta G^0 = -38{,}630 + 20{\cdot}72T \text{ cal mol}^{-1}; \log K_{CO} = \frac{8{,}443}{T} - 4{\cdot}53 \text{ [2]} \qquad 13.3$$

$$\Delta G^0 = -37{,}010 + 20{\cdot}31T \text{ cal mol}^{-1}; \log K_{CO} = \frac{8{,}088}{T} - 4{\cdot}438 \text{ [3]} \qquad 13.4$$

$$\Delta G^0 = -27{,}810 + 14{\cdot}77T \text{ cal mol}^{-1}; \log K_{CO} = \frac{6{,}028}{T} - 3{\cdot}228 \text{ [4]} \qquad 13.5$$

$$\Delta G^0 = -38{,}650 + 20{\cdot}95T \text{ cal mol}^{-1}; \log K_{CO} = \frac{8{,}448}{T} - 4{\cdot}58 \text{ [5]} \qquad 13.6$$

The data are shown graphically in Figure 13.1 and it can be seen that equations 13.2, 13.4 and 13.6 are in very good agreement.

For the reaction:

$$H_{2(g)} + O_{(\%\text{ in Fe})} \rightleftharpoons H_2O_{(g)} \qquad 13.7$$

$$K_{H_2O} = p_{H_2O}/p_{H_2} \cdot [a_O]$$

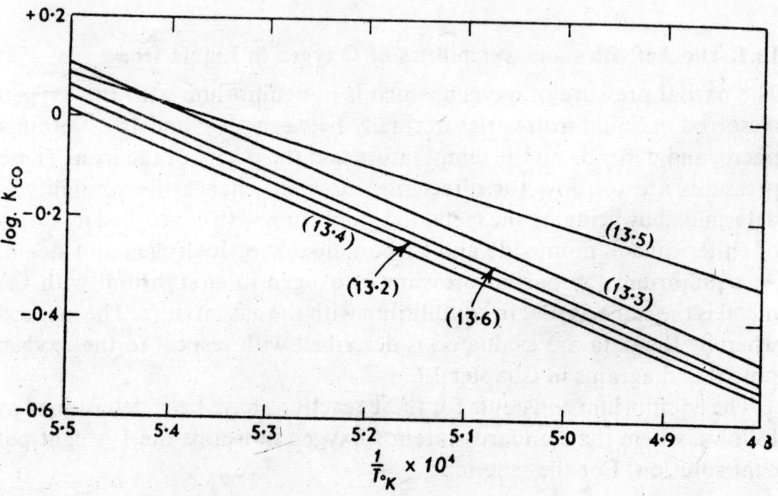

FIG. 13.1. The effect of temperature on the equilibrium between carbon monoxide, carbon dioxide and oxygen in liquid iron. The numbers against each line refer to the equation numbers given in the text.

and the equations are:

$$\Delta G^0 = -32{,}250 + 14{\cdot}50T \text{ cal mol}^{-1}; \log K_{H_2O} = \frac{7{,}050}{T} - 3{\cdot}17 \text{ [6]} \qquad 13.8$$

$$\Delta G^0 = -30{,}520 + 13{\cdot}95T \text{ cal mol}^{-1}; \log K_{H_2O} = \frac{6{,}670}{T} - 3{\cdot}05 \text{ [3]} \qquad 13.9$$

$$\Delta G^0 = -32{,}200 + 14{\cdot}63T \text{ cal mol}^{-1}; \log K_{H_2O} = \frac{7{,}050}{T} - 3{\cdot}20 \text{ [7]} \qquad 13.10$$

The data are shown graphically in Figure 13.2. All three lines have a similar slope, but the scatter is fairly appreciable.

At any temperature the solubility of oxygen in pure liquid iron is determined as the concentration which is in equilibrium with liquid iron oxide. At the melting point of pure iron the solubility is approximately

0·16 weight-per cent and the variation of the solubility with the temperature has been expressed by the following relations:

$$\log [\%O] = -\frac{6{,}320}{T} + 2{\cdot}734 \ [8, 9] \qquad 13.11$$

and

$$\log [\%O] = -\frac{5{,}762}{T} + 2{\cdot}439 \ [3] \qquad 13.12$$

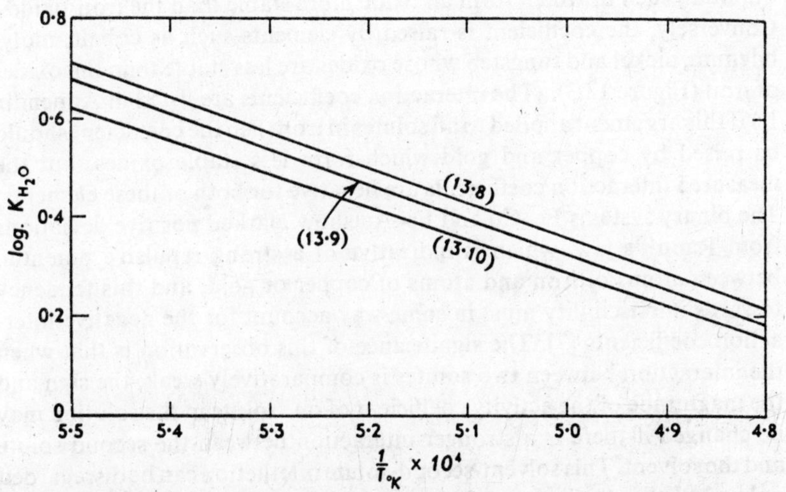

FIG. 13.2. Variation with temperature of the equilibrium between hydrogen, steam and oxygen in liquid iron. The numbers against each line refer to the equation numbers given in the text.

The two equations agree very closely at 1,600°C, but the variation of the solubility with the temperature from equation 13.11 is slightly greater than from equation 13.12:

Temperature °C	1,550	1,600	1,650	1,700
Weight-per cent of oxygen from equation 13.11	0·185	0·229	0·281	0·340
Weight-per cent of oxygen from equation 13.12	0·190	0·230	0·277	0·330

From earlier investigations it was concluded [9] that oxygen in pure iron obeys Henry's law and that the equilibrium constant for equation 13.7 could be expressed as follows:

$$K'_{H_2} = \frac{p_{H_2O}}{p_{H_2} \cdot [\%O]} \qquad 13.13$$

but a recent and more precise investigation has revealed that the activities do not vary linearly with the concentration of oxygen and the activity coefficient is given by the relation [7]:

$$\log f_{O} = -0{\cdot}20[\%O] \qquad 13.14$$

Generally the activity coefficient of oxygen in iron is lowered by solutes such as aluminium, boron, carbon, chromium, silicon, vanadium and titanium, each of which form an oxide more stable than the iron oxides. Conversely, the coefficient is raised by elements such as cobalt, molybdenum, nickel and tungsten whose oxides are less stable than the oxides of iron (Figure 13.3). (The interaction coefficients are listed in Appendix 1.) If this argument applied to all solutes in iron then the coefficient should be raised by copper and gold which form less stable oxides, but the measured interaction coefficients are negative for both of these elements. The binary systems Fe–Cu and Fe–Au show marked positive deviations from Raoult's law, which is indicative of a strong repulsive potential between atoms of iron and atoms of copper or gold, and this tendency towards immiscibility must in some way account for the negative interaction coefficients [7]. The significance of this observation is that when the interaction between two solutes is comparatively weak, the sign and the magnitude of the activity coefficient of one solute in the solution may be changed if there is a stronger interaction between the second solute and the solvent. This solvent-second-solute interaction can be disregarded only when there is evidence for strong attractive or repulsive interaction between the two solutes.

The concentration of oxygen in pure liquid iron which is in equilibrium with pure liquid ferrous oxide can be calculated for any temperature from equations 13.11 or 13.12. Hence the oxygen content of the metal for equilibrium with any slag for which the activity of ferrous oxide is known can be determined from the relation:

$$\frac{[a_{O}]_{(actual)}}{[a_{O}]_{(saturated)}} = \frac{a_{FeO(actual)}}{a_{FeO(pure)}} = \frac{a_{FeO(actual)}}{1} \qquad 13.15$$

where the standard state for ferrous oxide is chosen as the pure liquid oxide. Re-arranging and expressing $[a_{O}]_{(saturated)}$ in terms of equation 13.11, for example, we obtain:

$$\log[\%O]_{(actual)} = \log\frac{[a_{O}]}{[f_{O}]_{(actual)}}$$

$$= \frac{(\log a_{FeO(actual)})\left(-\frac{6{,}320}{T}+2{\cdot}736\right)}{\log[f_{O}]} \qquad 13.16$$

Conversely the activity of the iron oxide in the slag can be determined from measurement of the oxygen content of pure liquid iron in equilibrium with the slag.

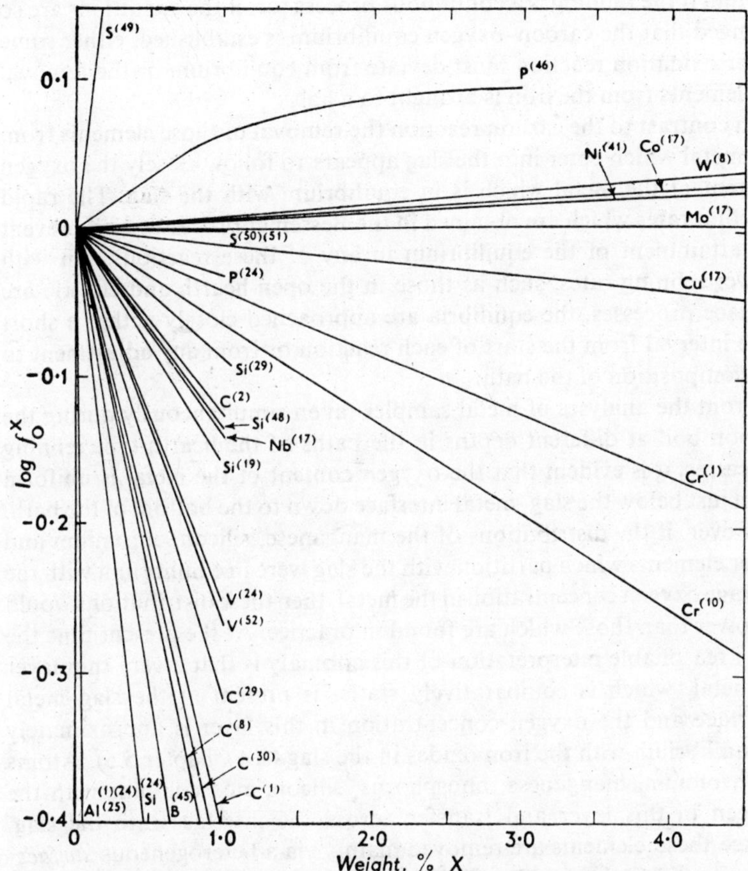

FIG. 13.3. The effect of various elements on the activity coefficient of oxygen in dilute solution in liquid iron. The numbers in parentheses against each line relate to the references for the interaction coefficients given in Table 10 of Appendix 1.

SURFACE AND VOLUME ZONE REACTIONS

It is now widely recognised that in steel refining the oxygen content of the metal is less than that for equilibrium with the slag and the available evidence suggests that this is due to the departure from equilibrium in the carbon-oxidation reaction. Thus the mean concentration of oxygen in the metal is intermediate between the values for equilibrium with the slag and for equilibrium with the dissolved carbon. On reflection it should be obvious that at least one of the reactions cannot be at equilibrium if the refining is a continuous process for, if the conditions are so changed that the carbon–oxygen equilibrium is established, either some other oxidation reaction must deviate from equilibrium or the removal of elements from the iron is brought to a halt.

In contrast to the carbon reaction the removal of those elements from the metal which enter into the slag appears to follow closely the oxygen content of the metal which is in equilibrium with the slag. The rapid reaction rates which are obtained in the Bessemer processes may prevent the attainment of the equilibrium in any of these reactions, but with slower refining rates, such as those in the open-hearth and electric-arc furnace processes, the equilibria are approached closely within a short time interval from the start of each reaction or from any adjustment to the composition of the bath.

From the analyses of metal samples taken simultaneously during the carbon boil at different depths in the baths of the hearth-type refining processes, it is evident that the oxygen content of the metal is uniform from just below the slag–metal interface down to the bottom of the bath. However, if the distributions of the manganese, silicon, chromium and other elements which partition with the slag were in equilibrium with the average oxygen concentration in the metal, then these distributions would be lower than those which are found in practice. At the present time the most reasonable interpretation of this anomaly is that a very thin layer of metal, which is comparatively static, is present at the slag–metal interface and the oxygen concentration in this layer is approximately in equilibrium with the iron oxides in the slag (see Chapter 5.5). Atoms of chromium, manganese, phosphorus, silicon, etc., combine with the oxygen in this layer and transfer across the interface into the slag. Hence these elements are removed mainly via a heterogeneous *surface-zone reaction*, whereas the oxidation of the carbon takes place mainly away from the slag–metal interface and can be described as a homogeneous *volume-zone reaction* [10]. It seems reasonable to assume that the only times at which solid or liquid oxides are formed within the metal are (*a*) before the start of the carbon boil, (*b*) when a 'dry' slag is formed, and (*c*) following the addition of large amounts of deoxidisers. When the equilibrium distributions are established for any one reaction the reaction ceases until such time as the activities of the elements are changed in the slag or metal, or until the activities of the iron oxides in the slag are changed. The distributions are not affected directly by the change in the mean oxygen content of the metal as the carbon concentration is lowered.

13.2 The Oxidation of Carbon

EQUILIBRIUM CONSTANTS

The reaction between carbon and oxygen dissolved in liquid iron is one of the most important of those which occur in iron- and steel-making and it has been studied in a large number of investigations. The reaction is usually written as:

$$C_{(\text{in Fe})}+O_{(\text{in Fe})} \rightleftharpoons CO_{(g)} \qquad 13.17$$

and the equilibrium constant is given by:

$$K_{\text{C-O}} = \frac{p_{CO}}{[a_C].[a_O]} = \frac{p_{CO}}{[\%C].[\%O].f_C.f_O} \qquad 13.18$$

In most of these investigations the value of $K_{\text{C-O}}$ has been obtained by combining data for equation 13.1 with those for the reaction:

$$CO_{2(g)}+C_{(\%\text{ in Fe})} \rightleftharpoons 2CO_{(g)} \qquad 13.19$$

$$K_C = \frac{p_{CO_2}}{p_{CO}[a_C]}$$

A number of evaluations of the free-energy change in the last reaction have been published, including the following:

$$\Delta G^0 = 26{,}400-26{\cdot}46T\,\text{cal mol}^{-1};\ \log K_C = -\frac{5{,}778}{T}+5{\cdot}784\ [11] \qquad 13.20$$

$$\Delta G^0 = 29{,}550-28{\cdot}24T\,\text{cal mol}^{-1};\ \log K_C = -\frac{6{,}460}{T}+6{\cdot}175\ [2] \qquad 13.21$$

$$\Delta G^0 = 34{,}500-31{\cdot}07T\,\text{cal mol}^{-1};\ \log K_C = -\frac{7{,}541}{T}+6{\cdot}79\ [12] \qquad 13.22$$

$$\Delta G^0 = 33{,}300-30{\cdot}40T\,\text{cal mol}^{-1};\ \log K_C = -\frac{7{,}280}{T}+6{\cdot}65\ [13] \qquad 13.23$$

$$\Delta G^0 = 27{,}400-27{\cdot}07T\,\text{cal mol}^{-1};\ \log K_C = -\frac{5{,}986}{T}+5{\cdot}917\ [4] \qquad 13.24$$

Agreement among these equations is fairly good at temperatures close to the melting point of pure iron, but there is increasing divergence as the temperature is raised (Figure 13.4). Equations 13·22 and 13·23 show very good agreement in the relevant temperature range and are considered to be the most reliable for the evaluation of the equilibrium constant.

The equilibrium constant for the carbon–oxygen reaction 13·17 has

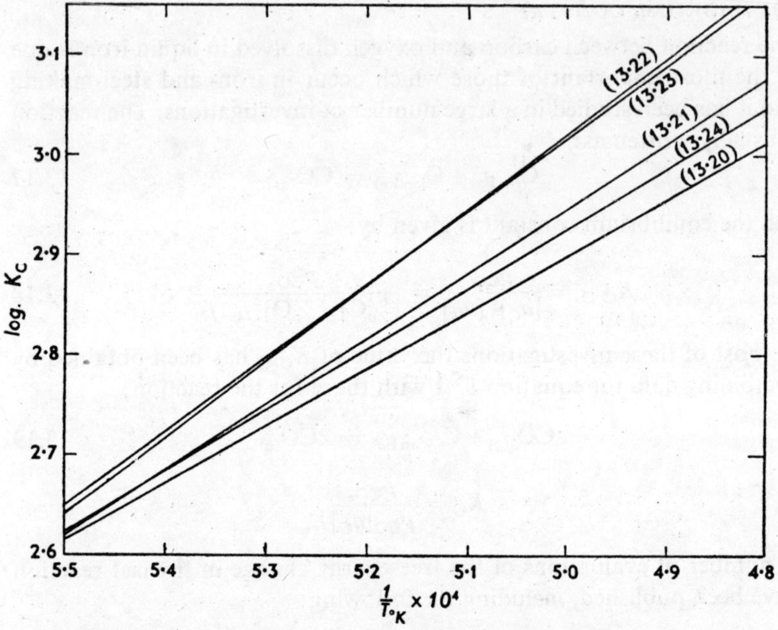

FIG. 13.4. Variation with temperature of the equilibrium between carbon monoxide, carbon dioxide and carbon in liquid iron. The numbers against each line refer to the equation numbers given in the text.

been expressed by the following relations when the standard states are chosen as the dilute weight-per cent solutions in liquid iron:

$$\Delta G^0 = -8{,}510 - 7{\cdot}52T \text{ cal mol}^{-1}; \log K_{\text{C-O}} = \frac{1{,}860}{T} + 1{\cdot}643 \text{ [11]} \quad 13.25$$

$$\Delta G^0 = -4{,}830 - 9{\cdot}75T \text{ cal mol}^{-1}; \log K_{\text{C-O}} = \frac{1{,}056}{T} + 2{\cdot}131 \text{ [14]} \quad 13.26$$

$$\Delta G^0 = -4{,}780 - 9{\cdot}75T \text{ cal mol}^{-1}; \log K_{\text{C-O}} = \frac{1{,}045}{T} + 2{\cdot}131 \text{ [15]} \quad 13.27$$

$$\Delta G^0 = -5{,}340 - 9{\cdot}47T \text{ cal mol}^{-1}; \log K_{\text{C-O}} = \frac{1{,}168}{T} + 2{\cdot}07 \text{ [5]} \quad 13.28$$

In previous chapters a constant is used which is obtained by combining equations 13·4 and 13·23:

$$\Delta G^0 = -3{,}710 - 10{\cdot}09T \text{ cal mol}^{-1}; \log K_{C\text{-}O} = \frac{811}{T} + 2{\cdot}205 \qquad 13.29$$

These relations are shown graphically in Figure 13.5. The data show fairly wide scatter and at 1,600°C the values of K range from 433 to 496.

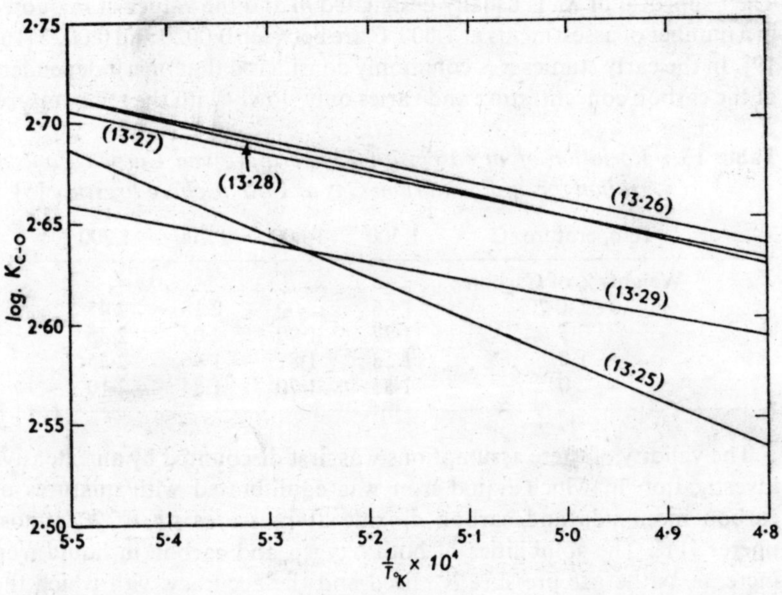

FIG. 13.5. Comparison of the values of equilibrium constants for the reaction between carbon and oxygen in liquid iron. The numbers against each line refer to the equation numbers given in the text.

Equations 13.26, 13.27 and 13.28 show very good agreement, but the first two of these have been obtained from re-evaluations of the data used to derive equation 13.25 and should therefore be consistent. Equation 13.28 is obtained by combining equations 13.6 and 13.23 whereas, as noted previously, equation 13.29 is obtained by combining 13.4 and 13.23. Equation 13.6 has been obtained indirectly whereas equation 13.4 has been determined from actual experimental measurements of the composition of atmospheres in equilibrium with oxygen in liquid iron and so equation 13.29 is probably a more reliable assessment of the equilibrium constant.

THE ACTIVITY OF CARBON IN DILUTE SOLUTION IN IRON

Prior to the experimental determination of the activities of carbon and oxygen in liquid iron it was frequently assumed that the activities are directly proportional to their mole fractions so that, when the partial pressure of carbon monoxide is 1 atmosphere, the equilibrium constant could be simplified to:

$$K' = \frac{1}{[\%C].[\%O]} \qquad 13.30$$

The reciprocal of K' is usually designated m and the values of m quoted in a number of assessments at 1,600°C are between 0·0020 and 0·0025 [16–19]. In the early studies it is commonly considered that m is independent of the carbon concentration and varies only slowly with the temperature.

Table 13.1 *Variation of $m \times 10^3$ with Temperature and Carbon Concentration for an Equilibrium Gas at 1 atmosphere Pressure* [5]

Temperature °C	1,500	1,600	1,700	1,900
Weight-% of Carbon				
0·02–0·20	1·86	2·00	2·18	2·45
0·5	1·77	1·90	2·08	2·35
1·0	1·68	1·81	1·96	2·25
2·0	1·55	1·70	1·84	2·10

The validity of these assumptions was first discounted by an extensive investigation in which liquid iron was equilibrated with mixtures of carbon monoxide and carbon dioxide at pressures up to 20 atmospheres [11]. The solubilities of both oxygen and carbon in liquid iron increase as the gas pressure is raised and the accuracy with which the concentrations of these elements in solution can be determined is raised accordingly. This study indicated that m increases as the carbon concentration is raised, but there is some doubt about the accuracy of the oxygen analyses at the higher carbon concentrations. The experiments have since been repeated [5] with the object of obtaining more accurate data at high carbon contents and the results of this re-examination, which are summarised in Table 13.1, indicate that at constant temperature m decreases as the carbon concentration is raised.

Two factors may account for this decrease. Firstly, the gas in equilibrium with carbon dissolved in liquid iron is not pure carbon monoxide because some carbon dioxide is also present. At 1,600°C the carbon dioxide concentration in the equilibrium gas is less than 1 per cent by volume when the carbon content of the metal is greater than 0·2 weight-per cent, but it increases to about 14 per cent at 0·1 per cent of carbon, which means that p_{CO} varies markedly with the carbon content at low carbon concentrations and this in turn affects the value of m. Secondly, it has already been shown that the activities of oxygen in dilute solution in liquid iron do not conform to Henry's law and the activities of carbon in iron vary likewise. From equation 13.18 it is apparent that the product $[a_C] \times [a_O]$ is constant only when the carbon monoxide content of the

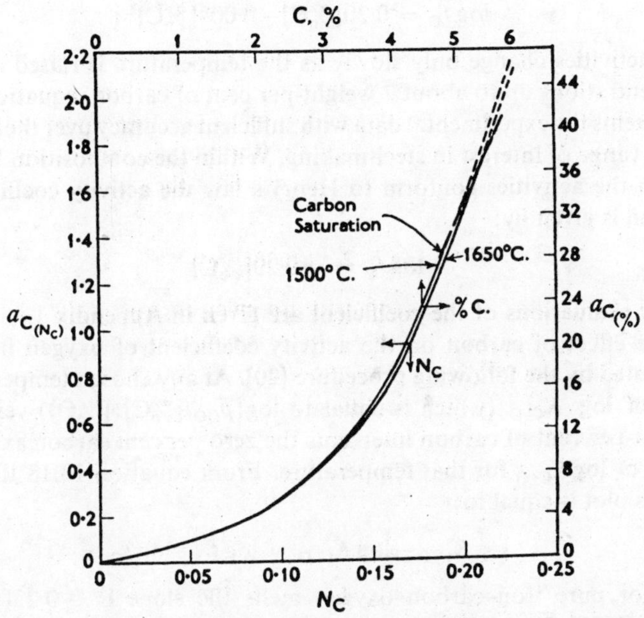

FIG. 13.6. The activities of carbon in pure liquid iron–carbon alloys with reference to pure graphite and dilute, weight-per cent carbon, standard states. (After Elliott [20].)

equilibrium gas is constant and hence it is improbable that the quantity m is independent of the concentrations of carbon and oxygen at equilibrium.

The variation with temperature of the activity of carbon in pure liquid iron–carbon alloys has been determined for all concentrations of carbon up to saturation of the melt with graphite [12, 13]; the combined data from these investigations are represented by the relation [13]:

$$\log f_C = \frac{4,350}{T}[1+4\times10^{-4}(T-1,770)](1-N_{Fe})^2 \qquad 13.31$$

where the reference state for carbon is the infinitely dilute (atom-fraction) solution of carbon in liquid iron. That is $a_C/N_C \rightarrow 1$ as $N_C \rightarrow 0$.

The variation of the activities with the concentration of carbon in terms of both the infinitely dilute atom-fraction and weight-per cent solutions is given in Figure 13.6 [20], which shows that the activities deviate markedly from Henry's law when the carbon content exceeds about 0·2 weight-per cent. At 1,540°C the data are represented by [20]:

$$\log f_C = 0{\cdot}20[\%C] - 0{\cdot}008[\%C]^2 \qquad 13.32$$

The activities change only slowly as the temperature is raised and, for concentrations up to about 2 weight-per cent of carbon, equation 13.32 represents the experimental data with sufficient accuracy over the temperature range of interest in steel-making. Within the composition range in which the activities conform to Henry's law the activity coefficient of carbon is given by:

$$\log f_C = +0{\cdot}20[\%C] \qquad 13.33$$

Other evaluations of the coefficient are given in Appendix 1.

The effect of carbon on the activity coefficient of oxygen has been evaluated by the following procedure [20]. At any chosen temperature a plot of log $K'_{C\text{-}O}$ (which is equal to $\log\{p_{CO}/[\%C].[\%O]\}$) versus the weight-per cent of carbon intercepts the zero per cent carbon axis at the value of $\log K_{C\text{-}O}$ for that temperature. From equation 13.18 the slope of this plot is equal to:

$$\log K'_{C\text{-}O} - \log K_{C\text{-}O} = \log f_O + \log f_C \qquad 13.34$$

and for pure iron–carbon–oxygen melts the slope is $-0{\cdot}22[\%C]$ at 1,600°C [21]. Expanding the activity coefficients for carbon and oxygen and substituting this value gives the relation:

$$\log f_O + \log f_O^C + \log f_C + \log f_C^O = -0{\cdot}22[\%C] \qquad 13.35$$

For all except very low carbon contents, the values of $\log f_O$ and $\log f_C^O$ are very small and can be neglected, and when the value of $\log f_C$ is substituted from equation 13·32 we obtain [20]:

$$\log f_O^C = -0{\cdot}42[\%C] + 0{\cdot}008[\%C]^2 \qquad 13.36$$

In Figure 13.7 the variation of this coefficient with the carbon concentration (line I) is compared with other published values. With the exception of line VI the equations are reasonably consistent, but line VI was determined from the most recent investigation of the carbon–oxygen reaction [5] and should be the most reliable. In the last case chromium was added to the melts to increase the oxygen solubilities and improve the accuracy of the oxygen analyses, but the value of log f_O^C is then dependent on the accuracy of the interaction coefficient for chromium with oxygen which is used to convert the data to the pure iron–carbon–oxygen system.

At one time the effect of carbon on the activity coefficient of oxygen was attributed to a small solubility of carbon monoxide in liquid iron [11,

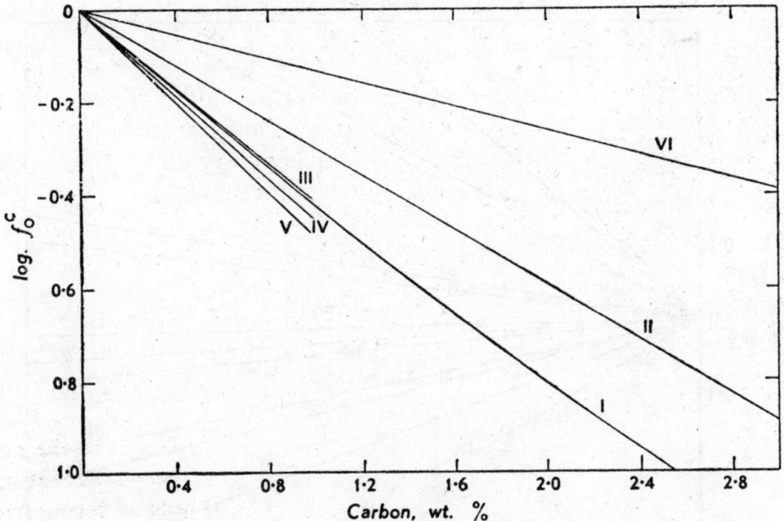

FIG. 13.7. The effect of carbon on the activity coefficient of oxygen in pure iron–carbon–oxygen alloys.

Curve I Elliott [20]
Curve II Matoba and Banya [22]
Curve III Chipman [21]
Curve IV Schenck and Gerdom [4]
Curve V Turkdogan *et al.* [14]
Curve VI Fuwa and Chipman [5]

23]. The solubility was thought to increase with the carbon content, but this hypothesis is not substantiated by the experimental evidence and it is now recognised that the effect is due probably to an interaction between the carbon and oxygen atoms [24, 25] which lowers the activity coefficients of both carbon and oxygen. Generally the concentration of carbon is very much greater than that of oxygen and the effect of oxygen on the activity of carbon is significant only at very low concentrations of carbon.

The effects of various solutes on the activity coefficient of carbon, relative to the dilute weight-per cent solution, are given in Figure 13.8, which shows that the activities of carbon are lowered by the addition of elements which form carbides more stable than cementite (Fe_3C) and vice versa, but there are some exceptions. For example, manganese carbide (Mn_3C) is slightly more stable than Fe_3C and e_C^{Mn} is small but negative [21]. The free energy of formation of silicon carbide (SiC) is more negative than that of Mn_3C and on this basis it is expected that e_C^{Si}

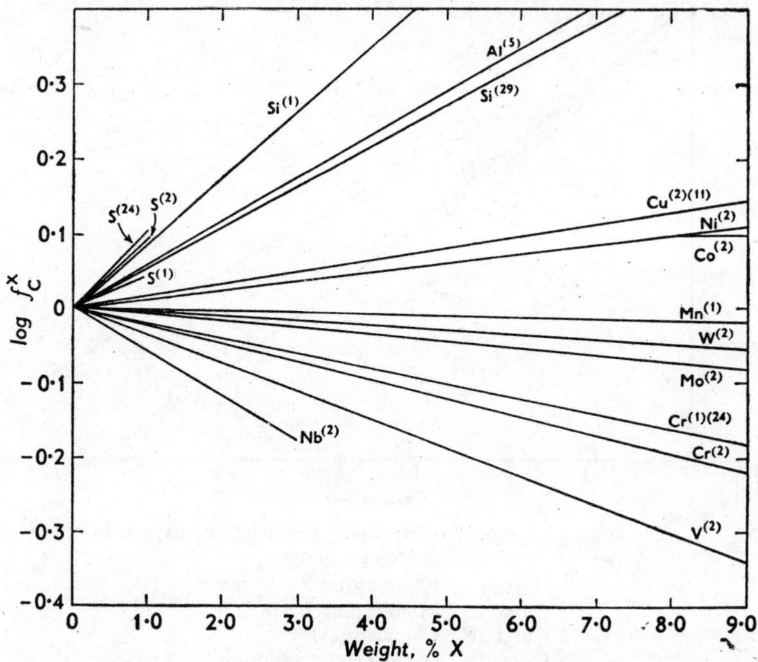

FIG. 13.8. The effect of various elements on the activity coefficient of carbon in dilute solution in liquid iron. The numbers in parentheses against each line relate to the references for the interaction coefficients given in Table 3 of Appendix 1.

should be more negative than e_C^{Mn}, but in fact e_C^{Si} has a large positive value [12, 21]. This is probably due to a stronger interaction between iron and silicon atoms than between silicon and carbon atoms, in comparison with the iron–manganese and manganese–carbon interactions [26]. In support of this contention it is notable that pure iron–manganese solutions are approximately ideal, whereas pure iron–silicon melts show large positive deviations from ideality [27] and the activity of carbon is raised by silicon to a greater extent in Fe–Si–C than in Mn–Si–C melts [28]. The activity coefficient of carbon is raised markedly by additions of sulphur [29, 30] and, in this case, the repulsion between the carbon and sulphur atoms may arise from a difference in the structural co-ordination of the two species in liquid iron; the sulphur atoms prefer an ionic bonding, whereas the carbon atoms contribute to the metallic bond [31].

CARBON ACTIVITIES IN GRAPHITE-SATURATED SOLUTIONS

The application of the thermodynamic data to the blast-furnace process frequently requires knowledge of the activity coefficients of carbon in solutions which are almost or completely saturated with carbon (i.e., in equilibrium with β graphite), but the interaction coefficients for the effects of various solutes on the activity coefficient of carbon which are determined from data for dilute solutions do not apply automatically at carbon saturation. Apart from any uncertainty, which is magnified as the concentration is increased, in the values of the interaction coefficients, the effect of the carbon content on these coefficients cannot be neglected at high concentrations of carbon. However, the interaction coefficients for the carbon-saturated solutions are readily obtained from measurements of the change in carbon solubility with the concentration of the other solutes, for at constant carbon activity

$$\partial \ln f_C = -\partial \ln N_C \qquad 13.37$$

where N_C is the atom fraction of carbon in the saturated solution. In Chapter 7 it is stated that at low concentrations of a second solute X the rate of change of the carbon solubility is a linear function of the concentration of X; hence the rate of change of the activity coefficient of carbon with the concentration of the solute X can be evaluated by the following procedure [29].

The effect of the concentration of X on f_C is expressed in the form:

$$d \ln f_C = \left(\frac{\partial \ln f_C}{\partial N_C}\right)_{N_X=0} . d \ln N_C + \left(\frac{\partial \ln f_C}{\partial N_X}\right)_{N_C} . dN_X \qquad 13.38$$

where the first term on the right-hand side is the change in f_C due to the change in the carbon solubility and the second term is the effect of the concentration of X on f_C at constant N_C. Dividing through by N_X to specify constant carbon activity:

$$\left(\frac{\partial \ln f_C}{\partial N_X}\right)_{a_C} = \left(\frac{\partial \ln f_C}{\partial N_C}\right)_{N_X=0} \cdot \left(\frac{\partial \ln N_C}{\partial N_X}\right)_{a_C} + \left(\frac{\partial \ln f_C}{\partial N_X}\right)_{N_C} \qquad 13.39$$

Re-arranging and using equation 13.37 gives:

$$\left(\frac{\partial \ln f_C}{\partial N_X}\right)_{N_C} = -\left[1 + \left(\frac{\partial \ln f_C}{\partial \ln N_C}\right)_{N_X=0}\right]\left(\frac{\partial \ln N_C}{\partial N_X}\right)_{a_C} \qquad 13.40$$

The term in square brackets has been evaluated from data for graphite solubility in pure iron–carbon melts and at 1,560°C is equal to 2·79. Hence:

$$\left(\frac{\partial \ln f_C}{\partial N_X}\right)_{N_C \text{ saturated}} = -2{\cdot}79\left(\frac{\partial \ln N_C}{\partial N_X}\right)_{a_C} \qquad 13.41$$

Values for $(\partial \ln N_C/\partial N_X)_{a_C}$ are given in Table 1, Chapter 7 for a number of solute elements and the interaction coefficients are listed in Appendix 1.

Generally the elements which lower the solubility and increase the activity coefficient of carbon in iron are found to behave similarly with regard to the solubility and activity of nitrogen in liquid iron. The exceptions are consistent with the differences in the relative stabilities of the carbides and nitrides of these elements.

13.3 The Carbon Boil

From the data which are presented in this and the preceding chapter it is possible to calculate the oxygen content of the metal which is in equilibrium with both a given carbon concentration in the metal and with the iron oxides in the slag. In hearth-refining practices the mean oxygen content of the metal is intermediate between these limiting values and attempts have been made to relate the change with time of the actual carbon–oxygen product to the change in either of the two equilibria.

One of the most informative treatments has been developed by Larsen [32] who assumes a constant value for m and from this estimates the difference $\Delta[\%O]$ between the actual oxygen content of the metal $[\%O]$ and the content in equilibrium with the carbon in solution and carbon monoxide at a pressure of 1 atmosphere $[\%O_{CE}]$. During a steady boil $\Delta[\%O]$ varies between 0·015 and 0·025, which is increased to between 0·023 and 0·035 after feed-ore additions, but returns to the range for a steady boil within 10 to 15 minutes after the ore has worked through. Conversely, during a vigorous lime boil $\Delta[\%O]$ falls to between 0·009 and 0·015 and increases to the steady-boil values when the lime boil subsides. From this it is implied that the carbon reaction tends towards a non-equilibrium, steady-state condition where the oxygen content of the metal is only slightly in excess of the carbon–oxygen equilibrium level. Values for $\Delta[\%O]$ calculated in terms of the metal oxygen content in equilibrium with the iron oxides in the slag are considerably larger and show a wide scatter. The actual carbon–oxygen relation in the open-hearth process is compared with the two oxygen-equilibrium relations in Figure 2 of Chapter 9 which shows that all three relations converge at very low carbon contents. For carbon contents below 0·05 the actual product of

[%C].[%O] is often less than the value of the constant for the carbon–oxygen equilibrium. This is probably due to the fact that the constant is calculated for a constant pressure of carbon monoxide whereas, as shown earlier, at very low concentrations of carbon the partial pressure of carbon dioxide is high in the equilibrium gas and approaches that of the carbon monoxide.

In many of the earlier investigations it is assumed that the non-attainment of the carbon–oxygen equilibrium is probably caused by an inherently slow reaction rate. However, a simple reaction of this type, which requires the combination of one atom only of each species to form a molecule of the reaction product, should be rapid unless the activation energy for the reaction is greater than about 100 kcal mol^{-1} [33]. The actual activation energy has been estimated from the temperature coefficient of the reaction rate constant and is only about 28 kcal mol^{-1} [34].

NUCLEATION OF CARBON MONOXIDE BUBBLES

Some of the more recent attempts to explain the departure from the equilibrium condition have considered the problem in terms of the removal of the reaction products from the metal bath. Since the product is a gas which is desorbed from the metal surface at a very slow rate, it is necessary for bubbles of gas to form below the slag–metal interface. However, the nucleation of a bubble within a homogeneous liquid is very difficult, for the liquid exerts a pressure on the walls of the bubble which is inversely proportional to the diameter of the bubble. This pressure in atmospheres can be calculated from the relation [23, 34]:

$$p_{CO} = p_A + p_f + 9{\cdot}87 \times 10^{-7}\left(\frac{2\pi r\sigma}{\pi r^2}\right) \qquad 13.42$$

where p_A is the atmospheric pressure and p_f is the ferrostatic pressure of the liquid metal at the point where the bubble is nucleated which, for liquid steel, is 0·25 atmosphere at about 14 inches below the surface. The bracketed term is the product of the circumference of the bubble of radius r and the coefficient of surface tension (σ) divided by the cross-sectional area of the bubble. At 1,600°C the surface tension of liquid steel is given by the relation [35]:

$$\sigma_{\text{dynes cm}^{-1}} = 1{,}600 - (100 \times \%C) \qquad 13.43$$

In liquid steel containing 0·5 per cent of carbon the pressure on a bubble 1 mm in diameter just below the metal surface is approximately 0·06 atmosphere. If the diameter is halved the pressure is doubled and for a diameter of one micron (0·001 mm) the pressure is about 58 atmospheres. When first formed the bubbles contain only a few molecules of gas and it is unlikely that the diameter is much larger than one micron. Hence, if they are nucleated within the liquid, the supersaturation of the melt with carbon would be very considerably greater than that which is found in practice.

In view of this it is now considered that the bubbles are nucleated at small crevices in the surfaces of the refractory lining and any other solids with which the metal is in contact. Surface tension prevents the liquid metal from flowing into very small crevices, but it is probable that any hole which is greater than about 0·1 mm diameter would be filled with liquid, so the nucleation of the gas is confined to the very small cavities. At the start of the reaction these cavities are probably filled with inert gases remaining from the air which is entrapped when the crevices are covered with liquid metal. Carbon and oxygen atoms can then combine to form molecules of carbon monoxide at these gas–metal interfaces more readily than elsewhere in the melt and a pocket of gas grows outwards from the cavity. The growth can continue until the pressure of carbon monoxide within the bubble is in equilibrium with the carbon and oxygen in the metal but, before this stage is reached, the radius of curvature of the bubble may become hemispherical. Further growth then causes the bubble to become unstable and it detaches from the crevice and rises through the metal, absorbing more carbon and oxygen during its transit to the slag–metal interface. Meanwhile a fresh bubble starts to grow from the gas remaining in the crevice and the nucleation of the bubbles becomes continuous.

There is considerable factual and experimental evidence in support of this picture of bubble formation. The effect of the roughness of the refractory surface has been demonstrated convincingly by melting an iron–carbon alloy in a refractory crucible [36]. When the surface of the crucible is coated with a layer of viscous slag so as to seal most of the pores, the rate of decarburisation is very low and the amount of carbon in solution in the iron is up to 15 times the concentration in equilibrium with the dissolved oxygen. The metal reacts violently and a vigorous carbon boil is obtained when the surface of the crucible is scratched or an iron rod is inserted in the melt, but immersion of a silica rod with a glazed surface has no effect. When the surface of the crucible is not coated with slag the carbon boil starts with a much smaller excess of carbon. Similarly in the open-hearth and arc-furnace practices it is well recognised that it is more difficult to start the carbon boil on a glazed hearth which has become impregnated with slag than on a clean and correctly fettled hearth and more violent and localised boiling action is caused when a metal rod is plunged below the level of the slag–metal interface or if solid metal is added to the bath. The surface of the metal provides additional nucleating sites and the rate of gas evolution subsides rapidly as the solid starts to melt.

Further supporting evidence is inferred from the uniformity of the composition and temperature of the metal at various depths in the bath during a steady carbon boil. The oxygen content of the metal is a maximum adjacent to the slag–metal interface, but the concentration is lower by only 0·002 to 0·010 per cent immediately above the surface of the hearth [37–39], and the temperature varies by only 10–15° over the depth of the bath. This uniformity is due most probably to the agitation of the bath by the deep-seated carbon boil, for marked thermal and composition gradients would arise if the carbon reaction took place only in the vicinity of the slag–metal interface. It is also significant that the rate of decarburisation is accelerated greatly when a gas is bubbled through the metal. Thus, in the Bessemer process the carbon monoxide is absorbed by the stream of nitrogen gas passing through the bath, so that the separate nucleation of gas bubbles is not required, and the rate of decarburisation is many times faster than in the open-hearth process. A smaller but still significant increase in the rate is obtained when oxygen gas is jetted into the metal from above. In this case the rate is altered mainly by the more rapid supply of oxygen to the metal than is obtained from ore feeding, but the bubbles of oxygen entering the bath can act as nuclei above the critical size for the growth of a carbon monoxide bubble. A small increase in the rate is obtained by bubbling an inert gas such as argon or nitrogen through the bath, which can be due only to the greater facility for the formation of carbon monoxide bubbles. During a lime boil the excess oxygen in the metal is lower than when a steady carbon boil is in progress, owing probably to the absorption of carbon monoxide into the bubbles of carbon dioxide as they rise through the bath.

The minimum size of the crevice which can nucleate a gas bubble in the metal is determined by the requirement that the bubble must become hemispherical before the pressure of carbon monoxide within it reaches the value for equilibrium with the carbon and oxygen in solution in the metal. The gas pressure within the bubble as it passes through the critical condition to become hemispherical can be obtained by identifying r in equation 13.42 with the radius of the cavity. Inspection of the equation shows that, because the ferrostatic pressure at the crevice increases with the bath depth, the minimum size of crevice which can function as a gas nucleating site is increased as the bath depth is increased. For example, the critical gas pressure is about 1·55 atmospheres for a bubble which forms at a cavity of 0·1 mm radius and at a depth of 14 inches below the surface of liquid steel containing 0·5 per cent of carbon at 1,600°C. For the same critical pressure the radius of the crevice is increased to about 0·6 mm if the depth of the metal bath is doubled.

The difference $\Delta[\%O]$ between the actual concentration of oxygen dissolved in the metal $[\%O]$ and that which is in equilibrium with the carbon $[\%O_{CE}]$ can be related to the minimum size of the nucleating crevice by substituting for p_{CO} in equation 13.42 the relation:

$$[\%O_{CE}] = \frac{p_{CO}}{K'_{C\text{-}O}[\%C]}$$

We thus obtain:

$$\Delta[\%O] = [\%O] - \frac{1}{K'_{C\text{-}O}}\left[1 + p_f + \frac{2\sigma}{r}\right] \qquad 13.44$$

from which it is apparent that $\Delta[\%O]$ should vary inversely with the size of the smallest crevice which can nucleate a bubble. According to the data in Table 13.1 $K'_{C\text{-}O}$ increases slowly as the carbon content of the metal is lowered, but from equation 13.43 the surface tension (σ) increases more rapidly than K' as the carbon is oxidised, with the net result that $\Delta[\%O]$ should increase slowly with decreasing carbon content of the metal. In practice $\Delta[\%O]$ remains fairly constant over a wide range of carbon contents and appears to decrease at very low concentrations of carbon; this suggests that the difficulty of nucleating the gas bubbles is not the only factor which causes the non-attainment of the carbon–oxygen equilibrium. The evidence is not conclusive and it could be argued that the discrepancy arises from the neglect of the activity coefficients for carbon and oxygen in the metal, but a more positive indication of a second controlling mechanism is obtained from the following approach.

APPLICATION OF THE BOUNDARY LAYER DIFFUSION THEORY TO CARBON OXIDATION

Since for hearth-refining processes $\Delta[\%O]$ attains a constant, non-equilibrium value under steady-boil conditions and returns rapidly to this value after any adjustment to the composition of the bath, Darken [10, 40] has considered the carbon–oxygen reaction as a normal steady-state process with two slow steps which determine the reaction rate, viz.,

$$O_{(slag)} \rightleftharpoons O_{(in\,Fe)} \qquad 13.45$$

and

$$O_{(in\,Fe)} + C_{(in\,Fe)} \rightleftharpoons CO_{(g)} \qquad 13.46$$

These are both assumed to be surface reactions so that the reaction rates are proportional to the areas of the slag–metal and the gas bubble–metal interfaces respectively. Using the two-film theory described in Chapter 5.5 and applying the restriction that the reaction rate is dependent on diffusion across the boundary layer in the phase with the lower concentration [41], the rate of oxygen transfer from slag to metal per unit area of interface can be determined by inserting the appropriate units in equation 5·35:

$$\frac{d[\%\text{O}]}{dt} = D\rho\left(\frac{[\%\text{O}_{\text{SE}}]-[\%\text{O}]}{100\Delta x}\right) \qquad 13.47$$

where t is the time in seconds, D is the diffusivity of oxygen in liquid iron of density ρ, Δx is the film thickness, $[\%\text{O}_{\text{SE}}]$ is the oxygen content of the metal at equilibrium with the iron oxides in the slag and $[\%\text{O}]$ is the actual oxygen content of the metal. If each atom of oxygen transferred across the boundary layer combines with an atom of carbon the rate of carbon removal is 12/16 of this. Each unit mass of carbon which reacts with this oxygen is removed from a volume of metal of unit cross-section extending from the slag–metal interface to the hearth, so the carbon concentration is lowered by an amount equal to $100/l\rho$ where l is the depth of the metal in the bath. Substituting this in equation 13.47 gives:

$$-\frac{d[\%\text{C}]}{dt} = D\left(\frac{12([\%\text{O}_{\text{SE}}]-[\%\text{O}])}{16l\Delta x}\right) \qquad 13.48$$

Assuming the diffusivity of oxygen in liquid steel at 1,600°C is 10^{-4} $\text{cm}^2\ \text{sec}^{-1}$, an average value of 0·04 for $([\%\text{O}_{\text{SE}}]-[\%\text{O}])$ during a steady boil and a boundary film thickness of 0·003 cm, then for a bath depth of 35 cm the rate of carbon removal during a steady boil is calculated to be 0·11 per cent of carbon per hour [40]. Following heavy ore feeds the quantity $([\%\text{O}_{\text{SE}}]-[\%\text{O}])$ is increased and a similar calculation shows that the carbon rate is increased about five-fold. Both of these calculated rates agree remarkably well with the rates which are obtained in practice.

When a bubble of carbon monoxide detaches from its nucleating site and rises through the bath, according to equation 13.46, during its transit through the metal further quantities of carbon monoxide are absorbed from reaction between carbon and oxygen atoms at the gas–metal interface. The volume of gas which is absorbed during this period must be appreciable, for the initial size of the stable bubble is probably between 2 and 10 microns in diameter [34], yet the average diameter of the bubbles which break through the slag surface is probably greater than 5 cm. During the ascent of the bubble it is assumed that the rate of growth is dependent on the diffusion of atoms across a thin boundary layer of metal which surrounds the gas. Because the concentration of carbon in the metal is generally very much larger than the concentration of oxygen, whilst the diffusivities of carbon and oxygen in liquid iron are of similar magnitude, it can be readily shown that the concentration gradient for carbon across the boundary layer is very small, relative to the oxygen gradient. Hence the rate of carbon transfer into the bubble during its ascent through the bath may be assumed to be dependent on the rate of oxygen transfer across the boundary layer and the rate of reaction is given by [40]:

$$-\frac{d[\%C]}{dt} = D\left(\frac{12([\%O]-[\%O_{CE}])}{16l\Delta x'}\right) \qquad 13.49$$

where $\Delta x'$ is the thickness of the boundary film at the gas–metal interface and the other symbols have the same significance as earlier defined. During the steady-boil conditions the rate at which oxygen atoms are transferred across the slag–metal interface is approximately equal to the rate at which oxygen atoms enter the gas bubbles, so that the rates for equations 13.48 and 13.49 are approximately the same and:

$$\frac{D.S_{S\text{-}M}([\%O_{SE}]-[\%O])}{\Delta x} \simeq \frac{D.S_{G\text{-}M}([\%O]-[\%O_{CE}])}{\Delta x'} \qquad 13.50$$

where $S_{S\text{-}M}$ and $S_{G\text{-}M}$ are the areas of the total slag–metal and gas–metal interfaces respectively. As an approximation it is reasonable to assume that Δx and $\Delta x'$ are equal and D is constant on both sides of the equation, so that the equation can be re-arranged in the form:

$$\frac{S_{S\text{-}M}}{S_{G\text{-}M}} \simeq \frac{([\%O]-[\%O_{CE}])}{([\%O_{SE}]-[\%O])} \qquad 13.51$$

For a normal boil the bath depth is approximately twice that of a stagnant bath and, assuming an average bubble diameter of 5 cm at the metal surface, the ratio $S_{S\text{-}M}/S_{G\text{-}M}$ is approximately 3 [40]. Substitution of the average values for [%O], [$\%O_{CE}$] and [$\%O_{SE}$] for corresponding conditions in equation 13.49 yields a value for the interface ratio of about 5, and in view of the approximations and assumptions made this can be regarded as a satisfactory agreement. On this evidence it is concluded [40] that the rate of carbon removal in the open-hearth, electric arc and similar hearth-refining processes is dependent on two diffusion reactions with almost equal reaction rates, and the overall rate is altered if the conditions are changed for either reaction.

The rate at which the bubble grows in the melt can be calculated from the rate of diffusion across the boundary layer and this shows that the radius in cm is roughly one-quarter of the time in seconds [40], but for an average growth from 10 microns to 5 cm the calculated time the bubble takes to escape from the bath is many times that which is expected for a free bubble to rise through the liquid metal. Darken shows that this is consistent with the fact that during a steady boil each cc of metal liberates about 0·04 cc, or 1/25th of its own volume, of carbon monoxide per second although the metal contains approximately its own volume of gas; hence the average transit time of a bubble through the bath is about 25 seconds. This result suggests that the bubbles are circulated within the metal by the turbulence of the boil and they do not traverse the slag–metal interface until they have circulated on average a large number of times between the upper and lower levels of the metal bath.

When the bubble leaves the metal and rises through the slag it is possible that the carbon monoxide reacts with the iron oxides there present to form some carbon dioxide, while most of the remaining carbon monoxide combines with the oxygen in the atmosphere immediately above the slag surface and is oxidised to carbon dioxide. This may be the reason why the oxygen potential of the iron oxides in the slag is closer to equilibrium with the liquid iron than with the bulk furnace atmosphere.

13.4 The Oxidation of Other Elements

In contrast to the oxidation of carbon, all the other elements which are removed during the refining of liquid iron are changed into either liquid or solid reaction products, so that the oxygen dissolved in the metal in equilibrium with any chosen concentration of an element is dependent on the activity of the reaction product, and this may vary over a wide range. The activity of the product is constant in the slag and in the metal when it is a solid which is immiscible and has the same composition in both the metal and in the slag. This is probably the case, for example, during the later stages of the oxidation of chromium from a charge with a high-chromium content and when silicon is oxidised under a silica-saturated slag. However, it is found more frequently that the reaction product is immiscible with the metal, but is soluble in the slag, in which case the activity is usually lower in the slag than it is in the metal and the reaction can proceed further at the slag–metal interface than within the bulk of the metal. Thus an oxidation reaction may proceed further at the interface since metal cations and oxygen anions can transfer across the interface to form an oxide whose activity is determined immediately by the composition of the slag.

During the carbon boil the oxygen activity is lower in the bulk metal than in the diffusion layer at the slag–metal interface. This fact also favours the interface reactions for the elements other than carbon, since the solubility product of any one of these elements and the oxygen in the iron is exceeded at the interface before it is reached elsewhere in the bulk metal.

Assuming that diffusion across a boundary layer controls the rates of these oxidation reactions, the rate of approach to the equilibrium distributions can be formulated similarly to equation 13.48 for the carbon reaction, but one of the major problems encountered in this approach is the determination of the thickness of the boundary layers. The thickness (Δx) is not necessarily the same for all elements in iron, for the available evidence suggests that it tends to increase in depth as the diffusivity of the atomic species is raised [41]. Also in some cases it is difficult to decide whether the rate is dependent on diffusion across the boundary layer within the slag or within the metal, for it is frequently found that the concentration of the reacting species in the metal is similar to, or greater than, the concentration of that species in the slag. It has been suggested that diffusion across the metal boundary layer is rate determining only when the concentration of the reacting element in the metal is between one-tenth and one-hundredth of its concentration in the slag [41].

One important feature does emerge from these considerations. When the reactants are brought rapidly into contact by agitation of the metal the speed of a reaction within the bulk metal should be high and, in comparison, the rate of the interface reactions should be lower, since these reactions are dependent on diffusion through the stagnant layer and transfer of the products across the interface into the slag. Once the equilibrium conditions are established for the interface reactions in a hearth-type refining process, it appears that the reactions can normally proceed at a rate which maintains the equilibrium distributions during the oxidation of the carbon. However, if the equilibrium distributions at the interface are displaced by large additions to the metal or slag it may require up to half an hour to re-establish the equilibria whereas, from inspection of the known activation energies for the various reactions, it appears that the equilibria within the bulk metal should be restored within a few minutes from the time the additions are made. Thus when additions of the more readily oxidised elements are made to the bath for the purpose of controlled deoxidation, it is probable that the saturation products for these elements and oxygen are exceeded within the bulk metal, and oxides are formed by volume-zone reactions to relieve the supersaturation. Oxides are also formed by surface-zone reactions at the slag–metal interface, but these reactions may continue after the corresponding reactions within the metal have ceased and until the activities of the deoxidising elements and the oxygen in the metal are restored to the values in equilibrium with the slag. In this case the slower rates of the interface reactions are beneficial for, if the rates were similar to those for the reactions in the bulk metal, the deoxidation would be rapid but the oxygen content of the metal would return rapidly to the original level and there would be insufficient time to tap the metal into the ladle and separate it from the slag.

13.5 Deoxidation

The solubility of oxygen in pure liquid iron is 0·16 weight-per cent at the melting temperature. On solidification this decreases to a very low level; in γ-iron it is estimated as $0{\cdot}003 \pm 0{\cdot}003$ per cent [42]. In liquid steel the amount of oxygen in solution is lowered by the presence of the other solute elements and the actual concentration is usually between these two limits, but when the metal solidifies the excess of oxygen is rejected from solution and combines with the other elements to form oxides. If the carbon content of the metal is not greater than about 0·1 weight-per cent most of the excess of oxygen combines with the carbon to form carbon monoxide; some of this carbon monoxide escapes as a gas, but some bubbles remain entrapped by the solid metal which compensates for the normal volume contraction on freezing. With careful control the solidification shrinkage is balanced exactly by the entrapped gas, so that the formation of a shrinkage cavity or pipe is eliminated. Such steels are called *rimming steels*; the name arises from the appearance of a characteristic rim of low-carbon metal at the surface of the ingot, beneath which is a layer of metal contaning numerous entrapped pockets of gas or 'blowholes'. Rimming steels are suitable for many purposes, but they are not acceptable where uniformity of properties is important, because the rimming action is accompanied by a marked increase in the extent to which the solute elements are segregated.

The gas evolution during solidification is small when the carbon content is greater than about 0·1 weight-per cent and diminishes as the carbon content is raised until, with about 2·0 weight-per cent of carbon, the equilibrium amount of oxygen dissolved in the liquid metal is similar to the solubility in the solid state. In the steel-making furnace the actual oxygen content of the metal is greater than that in equilibrium with the carbon, but generally it is found that the equilibrium is approached more closely during the transfer of the metal from the furnace into the ingot moulds. This is probably due partly (*a*) to the reduction of the ferrostatic pressure and the increased turbulence promoting bubble formation as the metal flows into the ladle, and partly (*b*) to the flushing action of bubbles of steam formed by the moisture in the casting pit refractories which rise through the metal and absorb some carbon monoxide in transit.

The excess of oxygen which does not combine with carbon forms oxides of the other elements which are present in the metal; some of these oxides escape to the surface of the solidifying casting but the majority are entrapped and remain within the metal as *inclusions*. Apart from the detrimental effects of non-metallic particles on the mechanical properties and the fatigue life of the metal, some of the oxides are semi-molten at the hot working temperatures and form grain boundary films which give rise to poor ductility and hot-shortness during fabrication. Thus, with the exception of rimming steels and steels of very high carbon content, the amount of oxygen dissolved in the metal must be lowered prior to solidification and while the deoxidation products have time to escape from the metal into the slag. The elements which are added for this purpose must form oxides which are more stable than iron and those most commonly selected are aluminium, manganese and silicon, alone or in combination with each other. Other elements such as niobium, vanadium and titanium are also effective, but are more costly. In fact deoxidation is often carried out prior to the addition of the expensive or more readily oxidised alloy elements so as to conserve the additions and minimise the losses to the slag.

CALCULATION OF THE AMOUNT OF DEOXIDANT REQUIRED

The extent to which a given concentration of any element can deoxidise the metal is readily determined from the equilibrium data for that element and for the oxide product with the oxygen dissolved in the iron. The amount of the element which is required to remove a given concentration of oxygen from the metal can be determined from the same data. Consider a simple reaction in which an element M reacts with oxygen to form the pure solid oxide MO:

$$\mathrm{M}_{(\%\,\mathrm{in\,Fe})} + \mathrm{O}_{(\%\,\mathrm{in\,Fe})} \rightleftharpoons \mathrm{MO}_{(s)} \qquad 13.52$$

In this case the activity of the oxide is unity and the equilibrium constant may be expressed as:

$$\begin{aligned} K_{\mathrm{M\text{-}O}} &= [a_{\mathrm{M}}].[a_{\mathrm{O}}] \\ &= [\%\mathrm{M}].[\%\mathrm{O}].[f_{\mathrm{M}}].[f_{\mathrm{O}}] \qquad 13.53 \end{aligned}$$

The residual concentration of the element M in equilibrium with the required concentration of oxygen in the iron is determined from data for $K_{M\text{-}O}$, $[f_M]$ and $[f_O]$ at the temperature and composition of the liquid bath, to which must be added the amount of the element M required to combine with the oxygen to form the oxide. This quantity is determined by first calculating the value of [%M] in equation 13·53 in equilibrium with the oxygen before the addition is made and, as 1 weight-per cent of M combines with $x/16$ weight-per cent of oxygen, (where x and 16 are the atomic weights of M and oxygen respectively) the amount of M which must be added is given by:

$$[\%M]_{final}+\frac{x}{16}\left([\%M]_{final}-[\%M]_{initial}\right) \qquad 13.54$$

It should be noted, however, that this quantity of deoxidiser will effect the required removal of oxygen only under the condition where no transfer of oxygen to the metal is possible during the time taken for the deoxidiser to react with the metal. In practice, when the metal is covered with a layer of slag, some oxygen is always transferred across the slag–metal interface, the amount per unit time varying with the viscosity, temperature and composition of the slag, and the area of the slag–metal interface per unit volume of metal. The calculated quantity of deoxidiser must be adjusted accordingly.

DEOXIDATION BY SILICON

The equilibrium constant for the reaction between silicon and oxygen dissolved in molten iron can be calculated by the following procedure [43].

For the reaction:

$$Si_{(l)}+O_{2(g)} \rightleftharpoons SiO_{2(cristobalite)} \qquad 13.55$$

the standard free energy of reaction is given by [44]:

$$\Delta G^0 = -222{,}800+47{\cdot}62T\,\text{cal mol}^{-1} \qquad 13.56$$

The free energy of solution of oxygen gas in liquid iron relative to the dilute weight-per cent solution, i.e.,

$$\tfrac{1}{2}O_{2(g)} \rightleftharpoons O_{(\%\text{ in Fe})} \qquad 13.57$$

has been estimated in a number of investigations and some of the more recent evaluations are as follow:

$$\Delta G^0 = -28{,}220-0{\cdot}57T\,\text{cal mol}^{-1}\ [27] \qquad 13.58$$

$$\Delta G^0 = -29{,}570-0{\cdot}12T\,\text{cal mol}^{-1}\ [3] \qquad 13.59$$

$$\Delta G^0 = -27{,}790-0{\cdot}79T\,\text{cal mol}^{-1}\ [7] \qquad 13.60$$

At constant temperature the values of ΔG^0 from these equations vary over a range of about 1 kcal, which corresponds to an uncertainty of approximately 3·8 per cent, but the values obtained from equation 13·60 lie midway within this range and this equation is selected, therefore, to represent the free-energy change. Combining equations 13·56 and 13·60, for the reaction:

$$Si_{(l)}+O_{(\%\,in\,Fe)} \rightleftharpoons SiO_{2(cristobalite)} \qquad 13.61$$

the free-energy change is:

$$\Delta G^0 = -167{,}220+49{\cdot}2T\,\text{cal mol}^{-1} \qquad 13.62$$

The free-energy change accompanying the transfer of pure silicon to the dilute weight-per cent solution in liquid iron is obtained by combining this equation with the constant for the reaction under a silica-saturated slag, (i.e., $a_{SiO_2} = 1$) when:

$$K'_{Si\text{-}O} = [\%Si][\%O]^2 \qquad 13.63$$

At 1,600°C $K'_{Si\text{-}O}$ has been evaluated as $3{\cdot}6\times10^{-5}$ [45] $3{\cdot}0\times10^{-5}$ [46] and $2{\cdot}8\times10^{-5}$ [43]. These three values show reasonably good agreement, but the last value is probably the most reliable and from this the free-energy change for equation 13.63 becomes:

$$\Delta G^0_{(1{,}878^\circ K)} = -39{,}000\text{ cal mol}^{-1} \qquad 13.64$$

The free-energy change for equation 13·62 at 1,600°C is:

$$\Delta G^0_{(1{,}873^\circ K)} = -75{,}070\text{ cal mol}^{-1} \qquad 13.65$$

Subtracting equation 13·64 from 13.65, we obtain for the reaction:

$$Si_{(l)} \rightleftharpoons Si_{(\%\,in\,Fe)} \qquad 13.66$$

$$\Delta G^0_{(1{,}873^\circ K)} = -36{,}070\text{ cal mol}^{-1} \qquad 13.67$$

The heat of solution of silicon in liquid iron at 1,600°C is estimated as $-28{,}500$ cal mol^{-1} [46]. Subtracting this quantity from equation 13.67 gives the entropy of solution and the temperature dependence of reaction 13·66 is obtained as:

$$\Delta G^0 = -28{,}500-4{\cdot}05T\text{ cal mol}^{-1} \qquad 13.68$$

Hence for the reaction:

$$Si_{(\%\,in\,Fe)}+2O_{(\%\,in\,Fe)} \rightleftharpoons SiO_{2(cristobalite)} \qquad 13.69$$

the free-energy change is obtained by combining equations 13.62 and 13.68:

$$\Delta G^0 = -138{,}720 + 53{\cdot}27T \text{ cal mol}^{-1} \qquad 13.70$$

and

$$\log K_{\text{Si-O}} = \log \frac{(a_{\text{SiO}_2})}{[a_{\text{Si}}][a_{\text{O}}]^2} = \frac{30{,}350}{T} - 11{\cdot}64 \qquad 13.71$$

By a similar combination of free-energy data the equilibrium constant has been evaluated elsewhere as:

$$\Delta G^0 = -129{,}440 + 48{\cdot}44T \text{ cal mol}^{-1}; \log K = \frac{28{,}300}{T} - 10{\cdot}59 \text{ [2]} \qquad 13.72$$

$$\Delta G^0 = -133{,}340 + 50{\cdot}37T \text{ cal mol}^{-1}; \log K = \frac{29{,}150}{T} - 11{\cdot}01 \text{ [43]} \qquad 13.73$$

The constant has been evaluated also from measurement of the H_2/H_2O ratios in equilibrium with pure silica and silicon dissolved in liquid iron to obtain:

$$\Delta G^0 = -135{,}940 + 51{\cdot}43T \text{ cal mol}^{-1}; \log K = \frac{29{,}700}{T} - 11{\cdot}24 \text{ [47]} \qquad 13.74$$

The data are compared in Figure 13.9 which shows that there is reasonable agreement between the experimental evaluation and the equations obtained from the thermodynamic relations, but equation 13.74 is obtained by a more direct approach and is considered to give the most reliable assessment.

Using the activity coefficients listed in Appendix 1 the silicon concentration can be calculated for equilibrium with any given oxygen content of the metal. However, for pure iron–silicon–oxygen alloys it is found experimentally that the change in the activity coefficient of silicon with the silicon and oxygen content of the metal is almost exactly balanced by the corresponding change in the square of the activity coefficient of oxygen for silicon concentrations up to about 2·0 weight-per cent [4, 43, 47]. That is, the product $[f_{\text{Si}}] \times [f_{\text{O}}]^2$ is approximately unity and

$$K_{\text{Si-O}} \simeq K'_{\text{Si-O}} = \frac{(a_{\text{SiO}_2})}{[\%\text{Si}].[\%\text{O}]^2} \qquad 13.75$$

where $K'_{\text{Si-O}}$ is the 'apparent' equilibrium constant. The solubility of silica in the binary liquids FeO–SiO_2 is approximately 50 per cent at 1,600°C and the solubility of oxygen in iron in equilibrium with a silica-saturated, ferrous silicate slag at this temperature is 0·088 weight-per cent [43]. From equations 13.74 and 13.75 the silicon in the metal in equilibrium with this concentration of oxygen is 0·004 weight-per cent.

Hence if the metal contains not more than 0·088 weight-per cent of oxygen, the addition of silicon in excess of 0·004 weight-per cent produces a solid reaction product within the metal. For higher oxygen contents in the metal the concentration of silica in the deoxidation product is less than the saturation limit (i.e., the ratio ferrous oxide/silica is increased) and the activity of the silica is less than unity in the reaction product. In practice a steel which is fully deoxidised or killed contains greater than 0·1 weight-per cent of residual silicon and so for pure

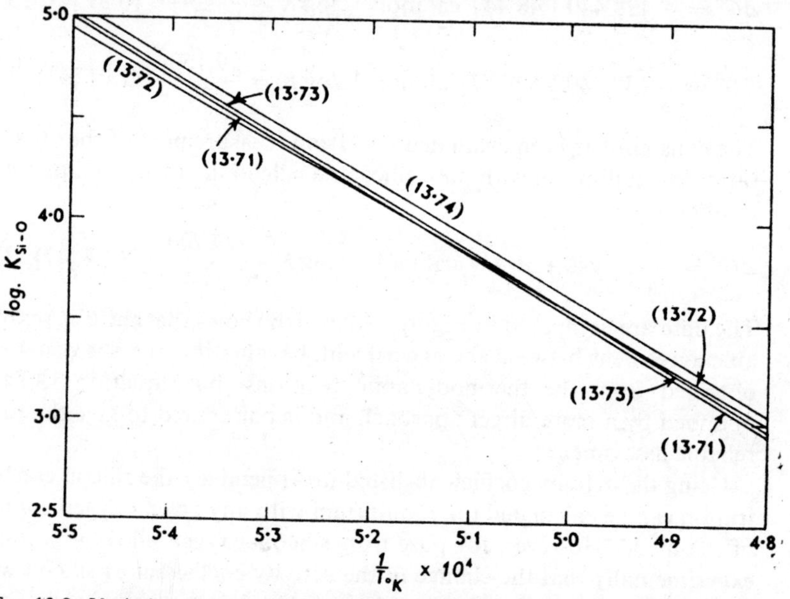

FIG. 13.9. Variation with temperature of the equilibrium between pure solid silica and silicon and oxygen in liquid iron. The numbers against each line refer to the equation numbers given in the text.

iron–oxygen–silicon melts the silicon and oxygen in the metal are in equilibrium with silica at approximately unit activity. The product $[\%Si].[\%O]^2$ is then approximately equal to the reciprocal of equation 13.74 and the equilibrium relation between these elements is shown in Figure 13.10.

These considerations apply to deoxidation by reaction within the bulk metal but, after the equilibrium is attained within the metal, the reaction may continue at the slag–metal interface. When the slag is saturated with silica the residual silicon and oxygen contents at equilibrium for the interface reaction are identical with those for the homogeneous reaction within the bulk metal. However, if the slag is not saturated the activity of the silica which is formed at the interface is controlled by the composition of the slag and may be very much lower than the activity of the product which is formed by a homogeneous reaction, in which case the deoxidation can proceed further by the interface reaction and a lower residual oxygen content can be obtained. For example, at 1,600°C the

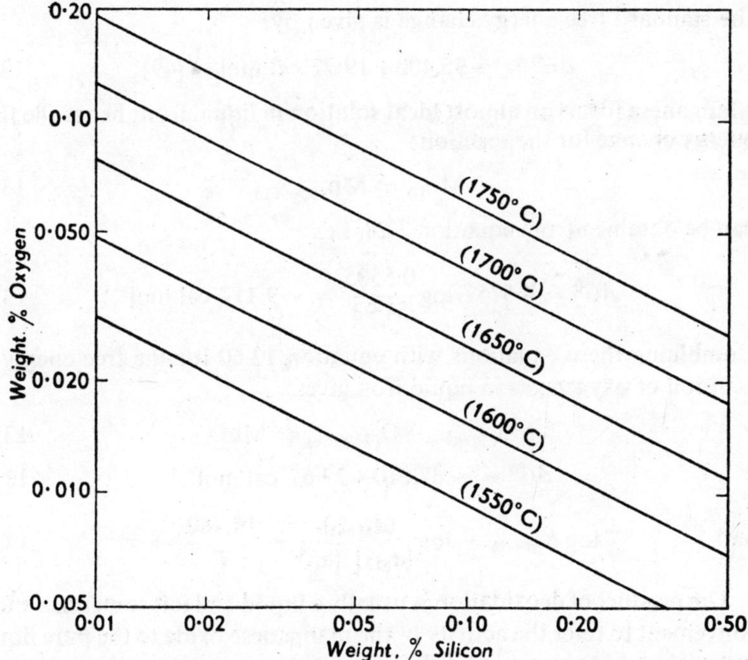

FIG. 13.10. The relation between the concentrations of oxygen and silicon in pure iron–oxygen–silicon alloys in equilibrium with pure solid silica at various temperatures.

product $[\%Si].[\%O]^2$ is equal to $2{\cdot}8 \times 10^{-5}$ when the slag is saturated with silica, but this is lowered to $5{\cdot}8 \times 10^{-6}$ for equilibrium with a slag containing 90 per cent of lime and 10 per cent of calcium fluoride [48]. This result means that silicon is a more efficient deoxidiser for basic than for acid steel-making processes.

The equilibrium product of the silicon and oxygen concentrations decreases as the temperature is lowered, so that the reaction continues after the metal is tapped from the furnace and further quantities of silica are produced right up to the time when the last metal solidifies, but the oxides formed during this period have little opportunity to escape from the metal and become entrapped to form inclusions.

DEOXIDATION BY MANGANESE

The equilibrium constant for the reaction between manganese and oxygen in liquid iron is calculated from the following data. For the reaction:

$$Mn_{(l)} + \tfrac{1}{2}O_{2(g)} \rightleftharpoons MnO_{(s)} \qquad 13.76$$

the standard free-energy change is given by:

$$\Delta G^0 = -95{,}400 + 19{\cdot}7T \text{ cal mol}^{-1} \text{ [49]} \qquad 13.77$$

Manganese forms an almost ideal solution in liquid iron; hence the free-energy change for the reaction:

$$Mn_{(l)} \rightleftharpoons Mn_{(\%\,\text{in Fe})} \qquad 13.78$$

can be obtained from equation 2.28, i.e.,

$$\Delta G^0 = 4{\cdot}575T \log \frac{0{\cdot}5585}{54{\cdot}93} = -9{\cdot}11T \text{ cal mol}^{-1} \qquad 13.79$$

Combining these equations with equation 13.60 for the free energy of solution of oxygen gas in liquid iron gives:

$$Mn_{(\%\,\text{in Fe})} + O_{(\%\,\text{in Fe})} \rightleftharpoons MnO_{(s)} \qquad 13.80$$

$$\Delta G^0 = -67{,}610 + 29{\cdot}6T \text{ cal mol}^{-1} \qquad 13.81$$

and $$\log K_{Mn\text{-}O} = \log \frac{(a_{MnO})_s}{[a_{Mn}]\cdot[a_O]} = \frac{14{,}780}{T} - 6{\cdot}5 \qquad 13.82$$

The product of deoxidation is usually a liquid and it is sometimes more convenient to refer the activity of the manganese oxide to the pure liquid rather than to the pure solid. The equilibrium constant is then obtained by combining the above equations with:

$$MnO_{(s)} \rightleftharpoons MnO_{(l)} \qquad 13.83$$

$$\Delta G^0 = 10{,}700 - 5{\cdot}2T \text{ cal mol}^{-1} \text{ [50]} \qquad 13.84$$

to obtain: $$Mn_{(\%\,\text{in Fe})} + O_{(\%\,\text{in Fe})} \rightleftharpoons MnO_{(l)} \qquad 13.85$$

$$\Delta G^0 = -56{,}910 + 24{\cdot}4T \text{ cal mol}^{-1} \qquad 13.86$$

and $$\log K_{Mn\text{-}O} = \log \frac{(a_{MnO})_l}{[a_{Mn}][a_O]} = \frac{12{,}440}{T} - 5{\cdot}33 \qquad 13.87$$

The effect of manganese on the activity coefficient of oxygen is negligible up to 0·6 weight-per cent of manganese [21, 51] and, for pure iron–oxygen–manganese melts containing low concentrations of oxygen, the equilibrium constant can be written as:

$$K_{MnO} \simeq K'_{Mn-O} = \frac{(a_{MnO})_l}{[\%Mn].[\%O]} \qquad 13.88$$

From comparison with the corresponding equations for the silicon–oxygen reaction it is evident that, when the activity of manganese oxide is unity in the reaction product, manganese is a very much weaker deoxidising element than silicon for liquid iron. For the simple ternary melts in equilibrium with pure liquid manganese oxide at 1,600°C, a residual manganese concentration of 0·5 weight-per cent is in equilibrium with approximately 0·1 weight-per cent of oxygen.

Under normal conditions the reaction product is not pure manganese oxide but is either a liquid or solid solution of manganese and ferrous oxides, the composition of which changes continuously with the temperature and the manganese content of the metal. At 1,600°C the product is liquid with up to about 0·2 weight-per cent of residual manganese. This means that the activity of the manganese oxide is less than unity, but the pure liquid and solid mixtures of ferrous and manganese oxides are approximately ideal and, when the reaction product consists only of these two oxides, the mole fraction of manganese oxide may be substituted for the activity in equation 13.82 or 13.88 without introducing any appreciable error. This procedure has been adopted in most of the experimental investigations in which the equilibrium constant has been determined for the reaction:

$$FeO_{(l)} + Mn_{(\% \text{ in Fe})} \rightleftharpoons MnO_{(l)} + Fe \qquad 13.89$$

The apparent equilibrium constant is usually expressed as:

$$K'_{MnO} = \frac{(N_{MnO})_l}{(N_{FeO})[\%Mn]} \qquad 13.90$$

and some of the published values are as follows:

$$\log K'_{MnO} = \frac{6,234}{T} - 3{\cdot}03 \ [46] \qquad 13.91$$

$$\log K'_{MnO} = \frac{6,440}{T} - 2{\cdot}95 \ [50] \qquad 13.92$$

$$\log K'_{MnO} = \frac{6,350}{T} - 3{\cdot}03 \ [52] \qquad 13.93$$

$$\log K'_{MnO} = \frac{6,350}{T} - 2{\cdot}99 \ [53] \qquad 13.94$$

The relation between the manganese and the oxygen in the metal is obtained by combining these equations with 13.11, which may be approximated to the following form for the simple solutions considered here:

$$\log \frac{[\%O]}{N_{FeO}} \simeq -\frac{6{,}320}{T} + 2{\cdot}734 \qquad 13.95$$

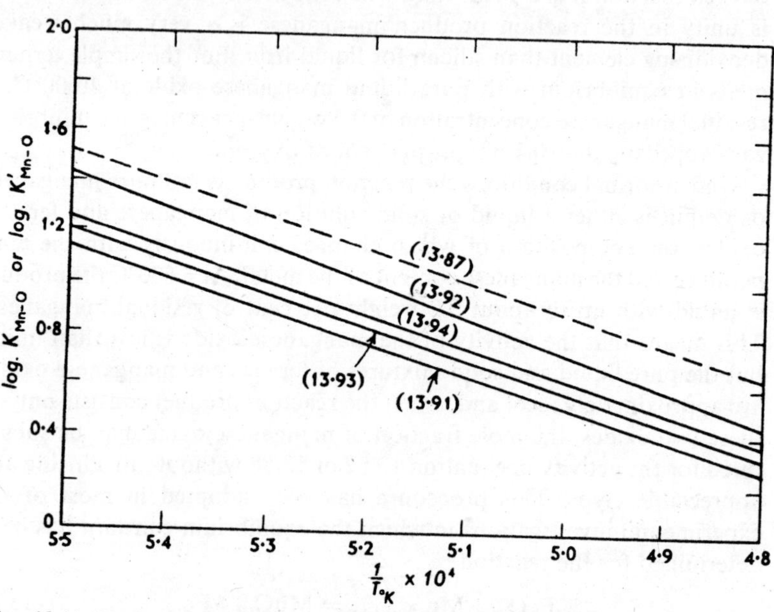

FIG. 13.11. Variation with temperature of the equilibrium constant for the reaction between pure liquid manganese oxide and manganese and oxygen in liquid iron. Solid lines determined by combining the equation numbers indicated on each line with equation 13.95. Broken line is for the calculated equilibrium constant, equation 13.87.

The apparent equilibrium constant is then written as:

$$K'_{MnO} = \frac{(N_{MnO})_l}{[\%Mn].[\%O]} \qquad 13.96$$

which is similar in form to equation 13.88. The temperature-dependence of the K'_{Mn-O} relations which are obtained by combining the equations is shown in Figure 13.11. The scatter appears to be large, but this is due mainly to the extended scale and it is actually less than that for the silicon–oxygen constant. It is noticeable that the calculated values for the constant are slightly larger than those determined experimentally; the difference may be due both to small errors in the data used to compute equation 13.87 and to neglect of the activity coefficients in the experimental relations.

Several diagrams have been prepared to show the relationship between the residual manganese and oxygen contents of the metal and the ratio of MnO/FeO in the deoxidation product [49, 54, 55]; Figure 13.12 shows

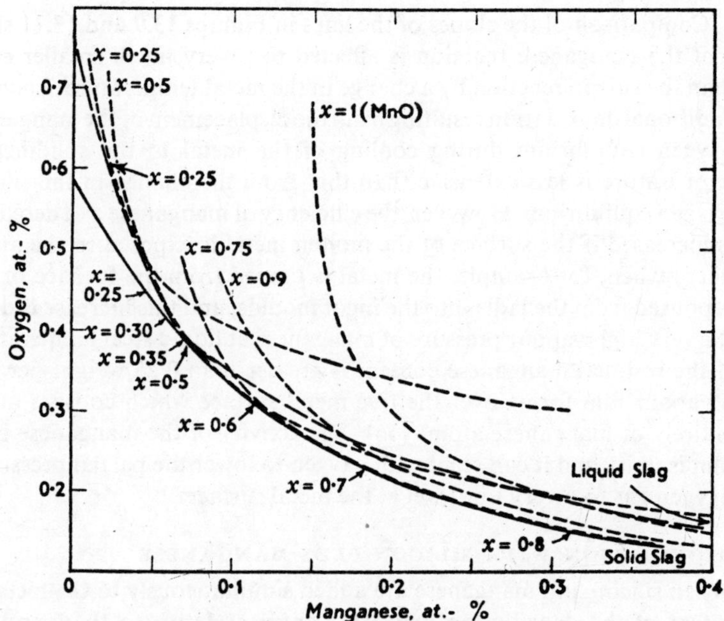

FIG. 13.12. Equilibrium between manganese and oxygen in pure iron–manganese–oxygen alloys and FeO + MnO melts at 1,600°C. Each broken line represents the conditions for equilibrium with the deoxidation product (x MnO.FeO) at the value of x given against the line. The solid line, which forms the envelope of all the broken lines, indicates the composition of the deoxidation product and the oxygen content of the metal in equilibrium with a given residual manganese concentration. (After Sloman and Evans [54].)

one of these relations. A similar diagram could be constructed from the data which are used to derive equations 13.82 and 13.87 using the methods of solution detailed in the original papers, but the curve obtained does not deviate markedly from that shown in Figure 13.12 and the diagram can be used with reasonable reliability for most practical purposes.

The earlier remarks concerning the continuance of the silicon slag–metal interface reaction after equilibrium has been attained for the homogeneous reaction apply equally to deoxidation with manganese and to any other element whose reaction product dissolves in the slag. The principal difference between the manganese and silicon reactions is that the activity of manganese oxide is lowered as the basicity of the slag is decreased, whereas the converse applies to silica. In fact, the deoxidising efficiency of manganese is raised approximately six-fold for equilibrium with a silica-saturated slag, relative to the efficiency in silica-free melts [45].

Comparison of the slopes of the lines in Figures 13.9 and 13.11 shows that the manganese reaction is affected to a very much smaller extent than the silicon reaction by a change in the metal temperature; hence the additional deoxidation resulting from the displacement of the manganese–oxygen equilibrium during cooling of the metal to the solidification temperature is less extensive than that from the change in the silicon–oxygen equilibrium. However, the efficiency of manganese as a deoxidiser is increased if the surface of the molten metal is exposed to the atmosphere when, for example, the metal is tapped from the furnace or as it is poured from the ladle into the ingot moulds, and this increase is due to the very high vapour pressure of manganese at liquid steel temperatures. If the residual manganese content is greater than 0·25 weight-per cent, a vapour film forms over the free metal surface which consists almost entirely of manganese atoms [56]. The activity of the manganese in the film is unity and it can react with oxygen to lower the partial pressure of oxygen gas to a very low level at the metal surface.

DEOXIDATION WITH SILICON PLUS MANGANESE

When silicon and manganese are added simultaneously to the metal the extent of the deoxidation may be changed, relative to the amount of oxygen which is removed when the same amount of silicon is added alone. A probable explanation of this effect can be obtained by combining equations 13.69 and 13.85 to give

$$Mn_{(\%\,in\,Fe)} + SiO_{2(s)} \rightleftharpoons Si_{(\%\,in\,Fe)} + O_{(\%\,in\,Fe)} + MnO_{(l)} \qquad 13.97$$

The equilibrium constant is:

$$K_{Mn\text{-}Si} = \frac{K_{Mn\text{-}O}}{K_{Si\text{-}O}} = \frac{[a_{Si}][a_O](a_{MnO})}{[a_{Mn}](a_{SiO_2})} \qquad 13.98$$

and the deoxidation product is a mixture of silica, iron and manganese oxides. If this is saturated with silica the activity of the silica is not affected by the ratio MnO/SiO_2 in the deoxidation product and the improvement in the deoxidation performance is negligible, but when the manganese/silicon ratio in the deoxidising addition is adjusted suitably the deoxidation product is a liquid and the activity of the silica is lowered. This effect allows the silicon reaction to proceed further in the bulk metal and the deoxidation is more effective.

Two experimental studies have shown very considerable improvement resulting from the simultaneous addition of the two elements [45, 46]; with 0·1 weight-per cent of silicon in the metal the deoxidising power of the silicon is improved by almost 30 per cent when 0·25 weight-per cent of manganese is added and is almost doubled by addition of 0·50 per cent of manganese. However, this affects only the homogeneous reaction within the bulk metal and the activities of the oxides at the slag–metal interface are not affected. This observation confirms that in practice the equilibrium is not attained for the interface reaction during the time taken to establish the equilibrium in the bulk metal; otherwise no improvement in the extent of the deoxidation would be found when the two elements are added simultaneously to metal which is covered by a basic slag.

DEOXIDATION WITH ALUMINIUM

The product of deoxidation with aluminium is either solid alumina or the spinel $FeO.Al_2O_3$. The spinel is formed only when the aluminium in the iron is not in stoichiometric excess of the amount of dissolved oxygen [57] and generally the deoxidation product is alumina at unit activity.

The equilibrium constant can be calculated from the following data. For the reaction:

$$2Al_{(\%\,in\,Fe)} + 3H_2O_{(g)} \rightleftharpoons Al_2O_{3(s)} + 3H_{2(g)} \qquad 13.99$$

the standard free-energy change is given by:

$$\Delta G^0 = -196{,}000 + 47{\cdot}44T \text{ cal mol}^{-1} \text{ [58]} \qquad 13.100$$

Combining this with equation 13.9 for the free-energy change for the reaction between hydrogen, steam and oxygen in liquid iron yields:

$$2Al_{(\%\,in\,Fe)} + 3O_{(\%\,in\,Fe)} \rightleftharpoons Al_2O_{3(s)} \qquad 13.101$$

$$\Delta G^0 = -287{,}560 + 89{\cdot}32T \text{ cal mol}^{-1} \qquad 13.102$$

and the equilibrium constant is given by:

$$\log K_{Al\text{–}O} = \log \frac{(a_{Al_2O_3})}{[a_{Al}]^2\,[a_O]^3} = \frac{62{,}870}{T} - 19{\cdot}53 \qquad 13.103$$

Other determinations of the equilibrium constant are as follow:

$$\Delta G^0 = -292{,}800 + 93{\cdot}7T \text{ cal mol}^{-1}; \log K = \frac{64{,}000}{T} - 20{\cdot}48 \text{ [2]} \quad 13.104$$

$$\Delta G^0 = -291{,}300 + 93{\cdot}6T \text{ cal mol}^{-1}; \log K = \frac{63{,}462}{T} - 20{\cdot}50 \text{ [59]} \quad 13.105$$

$$\Delta G^0 = -290{,}500 + 93{\cdot}7T \text{ cal mol}^{-1}; \log K = \frac{63{,}500}{T} - 20{\cdot}48 \text{ [58]} \quad 13.106$$

The constants are compared in Figure 13.13 which shows that there is a fairly appreciable scatter, but equations 13.105 and 13.106 are revisions of equation 13.104, using more recent data for the value of γ^0_{Al} in liquid iron, and are considered to be reliable.

The apparent equilibrium constant:

$$K'_{Al\text{-}O} = \frac{(a_{Al_2O_3})}{[\%Al]^2[\%O]^3} \quad 13.107$$

has been determined in a number of investigations for equilibrium with pure alumina and some of the equations for the apparent constant are as follow:

$$\log K'_{Al\text{-}O} = \frac{71{,}200}{T} - 27{\cdot}98 \text{ [60]} \quad 13.108$$

$$\log K'_{Al\text{-}O} = \frac{58{,}600}{T} - 18{\cdot}90 \text{ [61]} \quad 13.109$$

$$\log K'_{Al\text{-}O} = \frac{58{,}600}{T} - 22{\cdot}75 \text{ [62]} \quad 13.110$$

$$\log K'_{Al\text{-}O} = \frac{38{,}025}{T} - 7{\cdot}55 \text{ [63]} \quad 13.111$$

These equations are also compared in Figure 13.13 which shows that there is a very wide scatter among the data and they all indicate values for the apparent deoxidation constant which are very much smaller than the range of values for the true constant. This difference is partly accounted for by the strong negative deviation from ideal behaviour of aluminium in liquid iron ($\gamma^0_{Al} = 0{\cdot}031$ at 1,600°C [58]) and by the fact that the activity coefficients of aluminium and oxygen are lowered considerably by the strong interaction between the two elements. These effects do not however account entirely for the difference between the experimental and the equilibrium relations and it is probable that part of the scatter is due to difficulties in sampling the melt and in determining the very small concentrations of aluminium and oxygen. The residual aluminium concentrations are very low in some of the experimental studies and it is possible that the reaction product is the spinel compound, in which case the assumption of unit activity for the alumina is incorrect. Some recent and carefully controlled experimental measurements of the equilibrium constant agree very closely with the calculated constants [58].

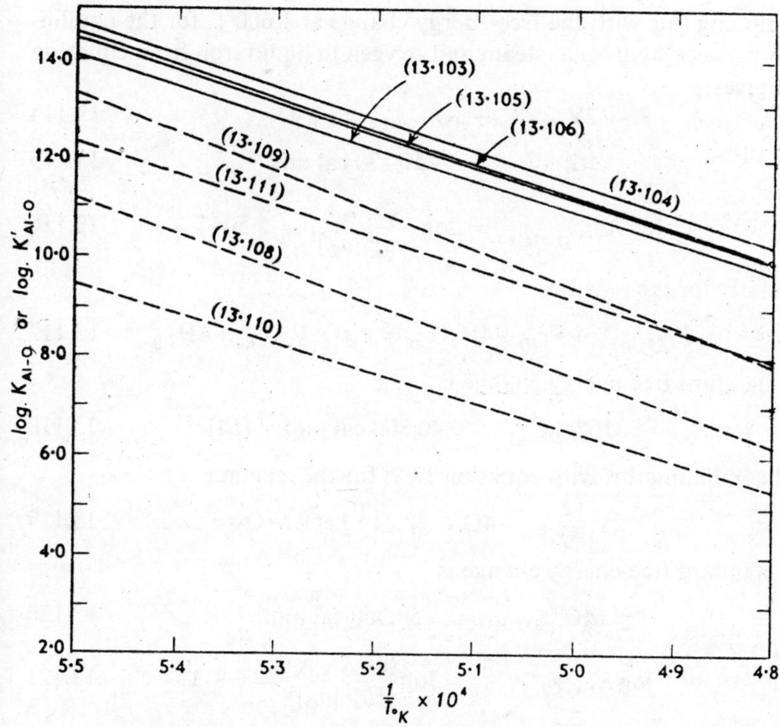

FIG. 13.13. Comparison of the true and apparent equilibrium constants for the reaction between aluminium and oxygen in liquid iron to form pure solid alumina. The numbers against each line refer to the equation numbers given in the text.

DEOXIDATION WITH VANADIUM

Vanadium is not a strong deoxidising element and if it is added to rimming steels it combines with the nitrogen in preference to oxygen and reduces the tendency to age-harden. The deoxidation product is either the spinel $FeO.V_2O_3$ or the oxide V_2O_3, but at 1,600°C the spinel is formed only when the vanadium content of the metal is less than 0·17 weight-per cent [64]. Both products are solid at steel-making temperatures.

The deoxidation constant is obtained as follows.

For the reaction:

$$2V_{(\%\,\text{in Fe})} + 3H_2O_{(g)} \rightleftharpoons V_2O_{3(s)} + 3H_{2(g)} \qquad 13.112$$

The standard free-energy change at 1,600°C is [64]:

$$\Delta G^0_{(1,873°K)} = -34{,}000 \text{ cal mol}^{-1} \qquad 13.113$$

Combining this with the free-energy change at 1,600°C for the equilibrium between hydrogen, steam and oxygen in liquid iron from equation 13.9 gives:

$$2V_{(\%\,\text{in Fe})} + 3O_{(\%\,\text{in Fe})} \rightleftharpoons V_2O_{3(s)} \qquad 13.114$$

$$\Delta G^0_{(1,873°K)} = -47{,}170 \text{ cal mol}^{-1} \qquad 13.115$$

and $$\log K_{V\text{-}O(1,873°K)} = \log \frac{(a_{V_2O_3})}{[a_V]^2\,[a_O]^3} = 5{\cdot}504 \qquad 13.116$$

Similarly for the reaction

$$2V_{(\%\,\text{in Fe})} + Fe_{(l)} + 4H_2O_{(g)} \rightleftharpoons FeO\,.\,V_2O_{3(s)} + 4H_{2(g)} \qquad 13.117$$

the standard free-energy change is:

$$\Delta G^0_{(1,873°K)} = -40{,}800 \text{ cal mol}^{-1} \text{ [64]} \qquad 13.118$$

and combining this with equation 13·9, for the reaction

$$2V_{(\%\,\text{in Fe})} + 4O_{(\%\,\text{in Fe})} \rightleftharpoons FeO\,.\,V_2O_{3(s)} \qquad 13.119$$

the standard free-energy change is

$$\Delta G^0_{(1,873°K)} = -58{,}360 \text{ cal mol}^{-1} \qquad 13.120$$

and $$\log K_{FV\text{-}O(1,873°K)} = \log \frac{(a_{FeO.V_2O_3})}{[a_V]^2\,[a_O]^4} = 6{\cdot}811 \qquad 13.121$$

Equation 13.118 applies when the residual vanadium concentration is greater than 0·17 weight-per cent, whilst equation 13·121 applies for lower residual contents. This means that the vanadium–oxygen relation in liquid iron is represented not by one line but by two lines which intersect at 0·17 weight-per cent of vanadium as shown in Figure 13.14. Similar behaviour is shown by aluminium, chromium and other elements which form a series of oxides when added as deoxidisers, each pair of lines intersecting at that concentration of the element which is in simultaneous equilibrium with the deoxidation products to which each of the two lines relates.

The vanadium–oxygen equilibria have been determined only at 1,600°C and there are no data suitable for calculating accurately the temperature dependence of the equilibrium constants, but a rough estimate can be made in the following manner [64].

For the reaction:

$$2V_{(s)} + \tfrac{3}{2}O_{2(g)} \rightleftharpoons V_2O_{3(s)} \qquad 13.122$$

the standard free-energy change is given by:

$$\Delta G^0 = -293{,}550 + 56{\cdot}15T \text{ cal mol}^{-1} \text{ [49]} \qquad 13.123$$

This equation has been determined from measurements at temperatures below the melting point of pure iron, but it can be extended to the

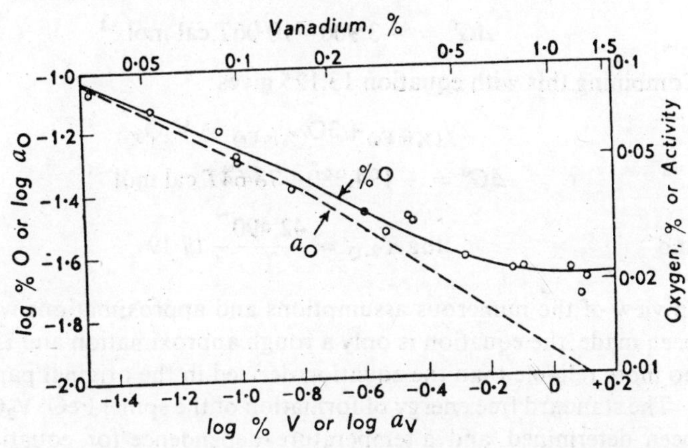

FIG. 13.14. Equilibrium between the weights-per cent (solid line) or activities (broken line) of vanadium and oxygen in liquid iron at 1,600°C. (After Dastur and Chipman[6].)

temperature range of interest probably without serious error. Combining this with equation 13.60 gives:

$$2V_{(s)} + 3O_{(\%\,in\,Fe)} \rightleftharpoons V_2O_{3(s)} \qquad 13.124$$

$$\Delta G^0 = -210{,}180 + 58{\cdot}52T \qquad 13.125$$

and $$\Delta G^0_{(1,873^\circ K)} = -100{,}580 \text{ cal mol}^{-1} \qquad 13.126$$

Subtracting equations 13.114 from 13.124 and 13.115 from 13.126 gives:

$$V_{(s)} \rightleftharpoons V_{(\%\,in\,Fe)} \qquad 13.127$$

$$\Delta G^0 = -26{,}725 \text{ cal mol}^{-1} \qquad 13.128$$

The free-energy change accompanying the formation of an ideal dilute solution of vanadium in liquid iron has been estimated [64] as −16,700 cal mol^{-1} and the difference of $-10{,}025$ calories between these quantities is a measure of the deviation from Raoult's law of vanadium in liquid iron. Assuming that the solution is regular:

$$-10{,}025 = RT \ln \gamma^{0} \qquad 13.129$$

and so at 1,600°C γ^0 is 0·068.

The temperature dependence of reaction 13.126 is obtained by inserting this value, together with the standard free-energy of fusion of vanadium, into equations 2.30 and 2.31 and we obtain the relations:

$$V_{(s)} \rightleftharpoons V_{(\%\,\mathrm{in\,Fe})}$$

$$\Delta G^{0} = -3{,}900 - 10{\cdot}06T \text{ cal mol}^{-1} \qquad 13.130$$

Combining this with equation 13.125 gives:

$$2V_{(\%\,\mathrm{in\,Fe})} + 3O_{(\%\,\mathrm{in\,Fe})} \rightleftharpoons V_2O_{3(s)}$$

$$\Delta G^{0} = -194{,}380 + 78{\cdot}64T \text{ cal mol}^{-1} \qquad 13.131$$

and

$$\log K_{V\text{-}O} = \frac{42{,}490}{T} - 17{\cdot}19 \qquad 13.132$$

In view of the numerous assumptions and approximations which have been made, the equation is only a rough approximation and is certainly no more reliable than the equation derived in the original paper [64].

The standard free energy of formation of the spinel $FeO.V_2O_3$ has not been determined and a temperature dependence for equation 13·119 cannot be estimated.

DEOXIDATION BY TITANIUM

A series of oxides is formed by reaction between titanium and oxygen in iron and the deoxidation equilibria are correspondingly more complex than for the elements which have already been considered. The true equilibrium constants have not been determined experimentally but, where suitable data exist, approximate values can be calculated.

Over the range 0·5–1·0 weight-per cent of titanium in iron the deoxidation product is titanium dioxide, TiO_2, and the free energy of formation of this oxide from the elements is given by:

$$Ti_{(s)} + O_{2(g)} \rightleftharpoons TiO_{2(s)} \qquad 13.133$$

$$\Delta G^{0} = -218{,}458 + 41{\cdot}78T \text{ cal mol}^{-1} \text{ [49]} \qquad 13.134$$

The activity coefficient of titanium at infinite dilution in liquid iron at 1,627°C is estimated as 0·011 [65] and the heat of fusion of titanium is

4,500 cal mol^{-1}. Inserting these values in equations 2.30 and 2.31, for the reaction:

$$Ti_{(s)} \rightleftharpoons Ti_{(\%\,in\,Fe)} \qquad 13.135$$

the standard free-energy change is given by:

$$\Delta G^0 = -12{,}520 - 11{\cdot}2T \text{ cal mol}^{-1} \qquad 13.136$$

Combining equations 13.57, 13.133 and 13.135 gives:

$$Ti_{(\%\,in\,Fe)} + 2O_{(\%\,in\,Fe)} \rightleftharpoons TiO_{2(s)} \qquad 13.137$$

The standard free-energy change is given by:

$$\Delta G^0 = -150{,}358 + 54{\cdot}56T \text{ cal mol}^{-1} \qquad 13.138$$

and $$\log K_{Ti\text{-}O} = \log \frac{(a_{TiO_2})}{[a_{Ti}][a_O]^2} = \frac{32{,}870}{T} - 11{\cdot}92 \qquad 13.139$$

For equilibrium with pure titanium oxide at unit activity the product $[a_{Ti}][a_O]^2$ is $2{\cdot}3 \times 10^{-6}$ at 1,600°C. This is considerably lower than the experimental value of the product $[\%Ti][\%O]^2$, which is given as approximately equal to 4×10^{-4} in the range 0·5 to 1·0 weight-per cent of titanium [66], and which suggests that there is a strong interaction between titanium and oxygen in liquid iron, although the interaction is less strong than that between aluminium and oxygen.

The solubility of oxygen in iron is lowered as the titanium concentration is increased, but the solubility passes through a minimum at about 0·5 weight-per cent of titanium and increases again as the titanium content is raised further [66]. Similar behaviour has been noted with carbon, chromium and other elements which lower the activity coefficient and raise the solubility of oxygen in iron. The solubility of oxygen is higher in pure liquid titanium than in pure liquid iron and it has been suggested that the minimum in the solubility should correspond to the composition where, on the average, there are just sufficient iron atoms present to fill all the nearest-neighbour sites of each titanium atom. Taking the mean co-ordination number as eleven for the liquid state [67], this composition corresponds to an Fe/Ti atom ratio equal to 11/1 or to a titanium content of 5 weight-per cent, whereas the actual minimum is only about 0·5 weight-per cent.

DEOXIDATION BY BORON

Boron is a comparatively weak deoxidising element and about as effective as vanadium, for it is also added to rimming steels to control the tendency to strain-ageing. The equilibrium between boron in iron and hydrogen–steam mixtures has been measured at 1,600°C in B_2O_3 crucibles [68]:

$$2B_{(\%\,in\,Fe)} + 3H_2O_{(g)} \rightleftharpoons B_2O_{3(l)} + 3H_{2(g)} \qquad 13.140$$

and the equilibrium constant is given as:

$$K_{B\text{-}H_2O(1{,}873°K)} = \frac{P^3_{H_2}(a_{B_2O_3})}{P^3_{H_2O}[a_B]^2} = 4{\cdot}284 \times 10^6 \qquad 13.141$$

Combining this with equation 13.9 gives:

$$2B_{(\%\,in\,Fe)} + 3O_{(\%\,in\,Fe)} \rightleftharpoons B_2O_{3(l)} \qquad 13.142$$

$$K'_{B\text{-}O(1{,}873°K)} = \frac{(a_{B_2O_3})}{[a_B]^2\,[a_O]^3} = 1{\cdot}2 \times 10^8 \qquad 13.143$$

When the melts are in equilibrium with the pure oxide, the activity of the oxide is unity and the product $[a_B]^2[a_O]^3$ is equal to $8{\cdot}6 \times 10^{-9}$, but this increases to $2{\cdot}4 \times 10^{-8}$ in a silica crucible and to $3{\cdot}5 \times 10^{-7}$ in a magnesia crucible [69]. The change is probably due to a lowering of the activity of the B_2O_3 in the presence of silica or magnesia, for the oxide is acidic and it appears that its activity is not very different from unity in the presence of silica but is lowered markedly in the presence of a basic oxide.

DEOXIDATION BY NIOBIUM (COLUMBIUM)

The equilibrium for niobium in liquid iron and niobium oxide NbO_2 with hydrogen–steam atmospheres has been measured at 1,600°C using silica crucibles [70].

For the reaction:

$$Nb_{(in\,Fe)} + 2H_2O \rightleftharpoons NbO_{2(s)} + 2H_2 \qquad 13.144$$

the standard free-energy change is:

$$\Delta G^0_{(1{,}873°K)} = -20{,}400 \text{ cal mol}^{-1} \qquad 13.145$$

The value of γ^0_{Nb} in iron is unknown but, assuming regular-solution behaviour, the value of the deoxidation constant has been estimated by a similar procedure to that described above (see p. 405) for the vanadium deoxidation constant and is given by [70]:

$$Nb_{(\%\,in\,Fe)} + 2O_{(\%\,in\,Fe)} \rightleftharpoons NbO_{2(s)} \qquad 13.146$$

$$\Delta G^0 = -131{,}600 + 54{\cdot}1T \text{ cal mol}^{-1} \qquad 13.147$$

and

$$\log K = \log\frac{(a_{NbO_2})}{[a_{Nb}][a_O]^2} = \frac{28{,}780}{T} - 11{\cdot}83 \qquad 13.148$$

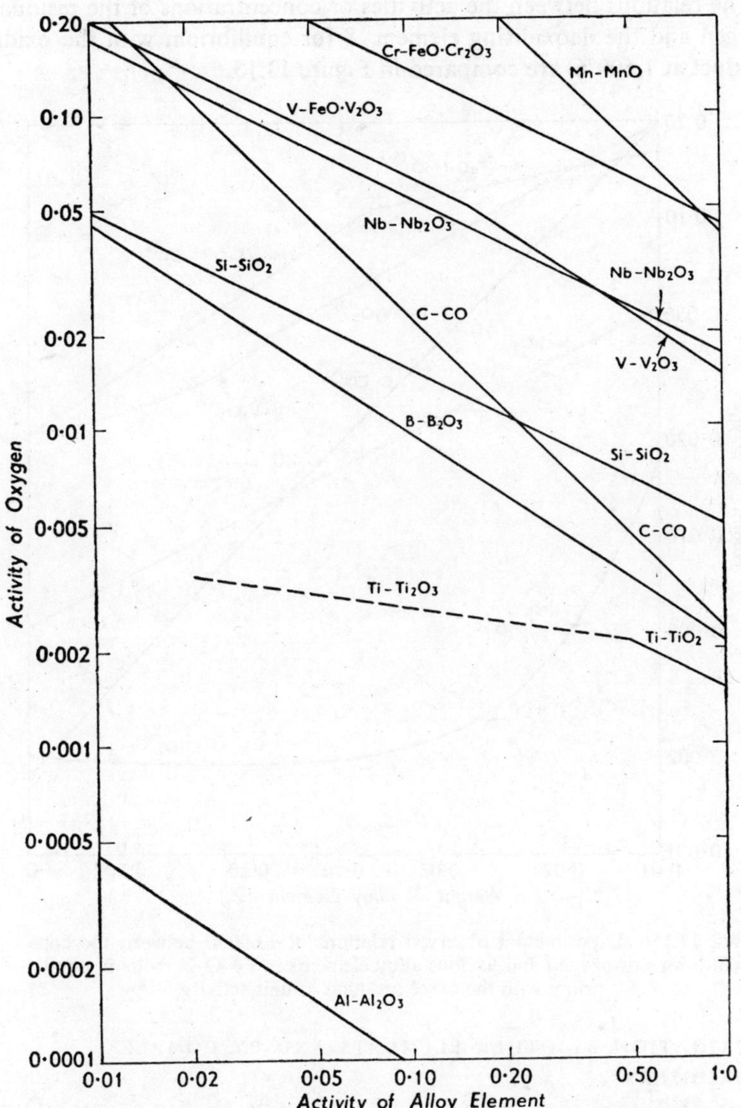

FIG. 13.15*a*. Relation at 1,600°C between the activities of oxygen and various alloy elements in liquid iron in equilibrium with the oxide products at unit activity. The line for chromium–spinel is determined from equation 10.2.

The relations between the activities or concentrations of the residual oxygen and the deoxidising element X for equilibrium with the oxide product at 1,600°C are compared in Figure 13.15.

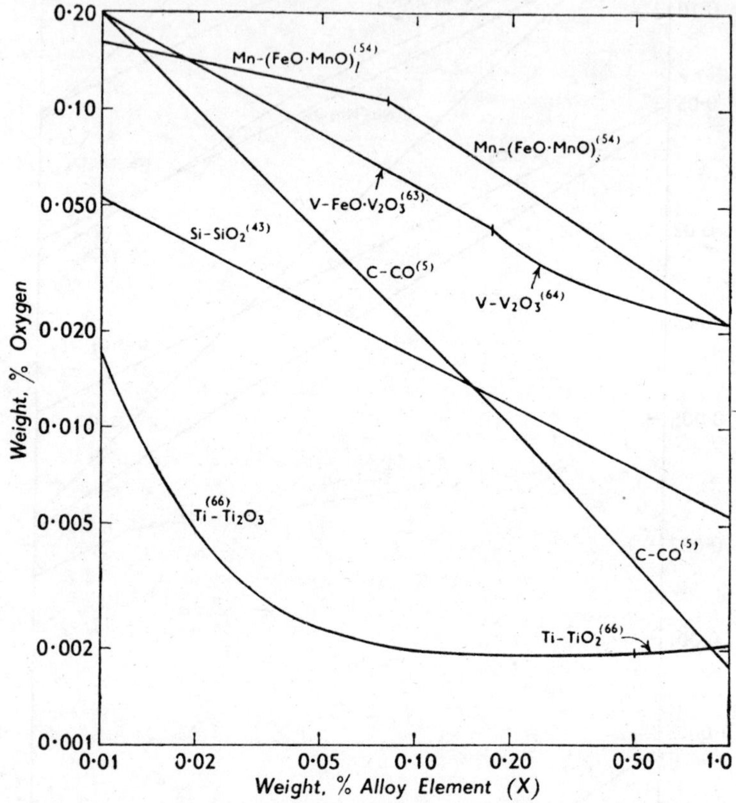

FIG. 13.15*b*. Experimental observed relations at 1,600°C between the concentrations of oxygen and various alloy elements in Fe–O–X melts in equilibrium with the oxide products at unit activity.

DEOXIDATION BY OTHER ELEMENTS AND BY COMPLEX DEOXIDISERS

Apart from chromium, which is considered in Chapter 10, the deoxidising powers of other elements have not been studied in any detail and the thermodynamic data are not sufficient to allow a reasonably accurate assessment of the equilibrium constants. Nevertheless a rough indication of the effectiveness of an element as a deoxidiser can often be obtained from approximate calculations and from indirect observations. For example, the apparent constant for deoxidation with zirconium to produce the pure oxide ZrO_2:

$$Zr_{(\%\ in\ Fe)} + 2O_{(\%\ in\ Fe)} \rightleftharpoons ZrO_{2(s)} \qquad 13.149$$

has been calculated [2] as

$$\log K'_{Zr-O} = \log \frac{(a_{ZrO})}{[\%Zr][\%O]^2} = \frac{41{,}340}{T} - 12{\cdot}07 \qquad 13.150$$

and for equilibrium with pure zirconia at 1,600°C the product $[\%Zr].[\%O]^2$ is about 10^{-10}. This suggests that zirconium is a more powerful deoxidising element than aluminium and the conclusion is invalidated only if the calculated constant is in error by several kilo-calories. From similar considerations it has been suggested [71] that on a weight-for-weight basis tantalum is about as effective as silicon.

Magnesium and calcium form oxides which are relatively more stable than the oxides of the other elements which have been considered and both of these elements are very powerful deoxidisers for liquid iron. Thus for equilibrium with pure magnesia and magnesium vapour at one atmosphere pressure, the oxygen content of pure iron at 1,600°C is $1{\cdot}5 \times 10^{-7}$ [2]. Unfortunately neither of these elements is soluble in molten iron, but complex deoxidising compounds are available which contain one or both of the elements in a form in which they can be finely dispersed in the metal and very low residual oxygen contents can be achieved. Elements such as copper, nickel, molybdenum, tin and tungsten form oxides whose stabilities are similar to, or less than, that of ferrous oxide and are not usually considered as deoxidisers.

It is common practice to use more than one element to achieve the necessary deoxidation and, especially for 'killed' steels, this is often essential to ensure that the residual concentrations of the deoxidising elements are within the specification limits. The improvement in the extent of the deoxidation which can be obtained by adding manganese and silicon together has already been considered, and a similar variation in the deoxidation efficiency by homogeneous reaction within the bulk metal is to be expected when the activities of either the oxide products or the solutes in the metal are changed by adding simultaneously two or more deoxidisers. Thus the deoxidising power of aluminium is increased considerably when either manganese or silico–manganese is added with the aluminium [62, 72].

The activity of the oxides produced within the bulk metal is lowered only if they form compounds or liquid or solid solutions which differ in composition from those formed when the deoxidising elements are added individually. For example, when aluminium is added alone the deoxidation product is often in the form of very small particles of pure solid alumina which melt at temperatures above 2,000°C, and these particles may not be fluxed if manganese or silicon is added after the aluminium. Any improvement in the deoxidation performance is obtained only when the elements are added in the correct sequence to form liquid oxides or solids which are readily fluxed when the next element is added. Similarly the deoxidising additions should follow each other with sufficient rapidity to allow the oxide products from each element to form and react with the other oxide products before those formed by the first additions have had time to escape from the metal into the slag.

REACTION BETWEEN THE DEOXIDISERS AND THE SLAG

Oxygen atoms are transferred across the interface and into the metal for as long as the metal remains in contact with an oxidising slag. The addition of the deoxidisers suppresses the carbon boil, so that the bath is no longer turbulent and the rate of transfer of oxygen across the interface is less rapid than under the normal boil conditions, but the deoxidation reactions continue at the interface after the reactions in the bulk metal have ceased. As a result the residual concentrations of the deoxidising elements in the metal are lowered progressively, the oxygen content of the metal starts to increase again as soon as the additions have worked through the bath and with time the oxygen content is restored to the level at which the carbon boil can recommence. For similar conditions the *amount* of oxygen transferred across the interface in unit time is dependent on the total area of the interface per unit mass of steel. Thus when the interface area is smaller in the ladle than it is in the refining furnace, there is seemingly an advantage in adding just sufficient of the deoxidiser in the furnace to 'block' the carbon boil and adding the remainder to the metal as it enters the ladle.

The interface area per unit volume of metal is even smaller in the ingot moulds and this suggests that it would be more profitable to add the deoxidisers to the metal as it enters the moulds. However, time is required for the deoxidiser to dissolve in the metal and react with the dissolved oxygen and for the oxide product to escape into the slag. The first two requirements need only a few minutes but separation of the oxides is less rapid and the inclusion density in the solid metal is invariably increased by this practice. Hence it is customary to add only small corrective additions of aluminium to the metal stream as it enters the moulds. It is common practice to add a large proportion of the deoxidisers to the metal as it enters the ladle, except where large additions of chromium or other readily-oxidised alloy elements are required, when the thermal requirements of the alloy additions may necessitate completion of the deoxidation within the furnace.

13.6 Removal of the Deoxidation Products

If addition of the deoxidisers results in the formation of a non-metallic particle which remains suspended within the metal, the oxygen content of the bath is not lowered but is converted into a form other than oxygen dissolved in iron. These oxides have a less serious effect on the mechanical properties of the solidified metal than the formation of grain-boundary films of ferrous oxide, but non-metallic particles entrapped in the metal do adversely affect certain mechanical properties and are undesirable when a high surface finish is required, for example when the surface is to be polished or plated. It is important, therefore, that the deoxidation products are removed from the metal as completely as possible before solidification commences. The deoxidation reactions are continuous up to the time when the last drop of metal solidifies, owing partly to the change with temperature in the numerical values of the equilibrium constants and partly to the concentration of oxygen and other solute atoms in the liquid in advance of the liquid–solid interface as the metal freezes. There is little opportunity for the reaction products to escape if they are formed while the metal is solidifying and ideally the oxygen content of the metal is lowered below the limit of solubility in the solid metal before it is transferred into the ingot mould.

The rate of rise or fall of an immiscible volume of liquid or solid which is suspended within a second liquid is given by Stoke's law:

$$V = \frac{2}{9}\frac{gr^2(\rho_1 - \rho_2)}{\zeta} \qquad 13.151$$

where V is the rate in cm sec^{-1} when g is the force of gravity, r is the radius of the particle (assumed to be spherical), ρ_1 and ρ_2 are the densities of the liquid and the particle and ζ is the viscosity of the liquid, all expressed in c.g.s. units. At 1,600°C the density of liquid steel is about 7·0 grams cm^{-3}, whilst the densities of the deoxidation products range from about 2·3 for solid silica to slightly less than 6·0 for liquid ferrous oxide; hence the density difference is positive and the inclusions rise to the slag–metal interface where they are absorbed by the slag. Within the normal range of liquid steel viscosities and particle densities the time taken to reach the interface is determined primarily by the radius of the particle and the distance from the interface at which the particle is formed; the other factors are only of secondary importance. Thus when the liquid viscosity is 0·1 dynes, the times required for a spherical particle with a density of 4·0 to rise from the bottom of a stagnant open-hearth bath 15 inches deep, or from the bottom of a ladle containing 15 feet of liquid metal are as follow:

Particle radius $cm \times 10^{-3}$	*Rate of rise* *in.* min^{-1}	*Time to rise through* 15 *in.* *min*	*Time to rise through* 15 *ft* *min*
0·1	0·013	1,154	13,848
1·0	1·30	11·54	138·48
10·0	130·2	0·12	1·39

For reasons discussed previously the metal is held in the furnace for only a short time after the deoxidisers are added and there is little probability that particles less than about 0·1 mm in diameter are removed entirely from the metal before it is tapped into the ladle. Similarly the metal cannot be held in the ladle for more than about 20 minutes because of the heat losses and only particles greater than about 0·1 mm diameter are cleared completely in this time. Assuming that the deoxidation products are formed randomly between the slag–metal interface and the lowest point in the liquid, this means that relatively few large inclusions are retained in the metal and the proportion of particles of each size which are retained varies inversely with the square of the inclusion radius.

In practice the metal is not stagnant even in the absence of the carbon boil, for the convective currents cause some circulation of the metal. Also the particles are not necessarily spherical; for example, pure solid alumina usually forms as thin needles or plates which rise less rapidly than spherical particles of similar volume, but generally it is found that these predictions of the distribution of the sizes of the non-metallic particles retained in the metal are confirmed by inclusion counts on the solid steel.

If the product of the deoxidation reaction is a solid, there is only small probability that it will grow from the initial size which may contain only a few molecules of the oxide. Liquid products can coalesce to form a larger particle with a more rapid rate of escape when the surface tension is favourable and it is clearly advantageous to control the sequence of the deoxidation additions so as to encourage the formation of liquid reaction products. For example, when silicon is added before manganese solid or very viscous silicates are formed which are not fluxed by the manganese addition, but if the manganese is added first the (MnO + FeO) reaction product combines with the silica as it forms to give an iron–manganese silicate. The deoxidising power of silicon is about 6 times as great as that of manganese and the addition of manganese, followed by silicon, in the ratio 6/1 gives a deoxidation product containing about equal parts of manganese oxide and silica; this gives a composition which is liquid at 1,600°C and the particles have a high probability of coalescence and growth. Similarly the addition in the correct proportions of manganese and/or silicon followed by aluminium, results in the formation of liquid alumino-silicates. In practice it is found that the correct selection of the sequence and the amounts of the various additions may cause a marked reduction in the inclusion density in the solid metal [72–75]. It is interesting to note that the conditions which favour the formation of a liquid reaction product correspond closely to those which have been considered previously for improving the efficiency of the deoxidising additions.

13.7 Deoxidation by Vacuum Melting and Casting

With proper control of the melting and casting it is possible to produce steels which are suitably clean for most purposes, but where high strength and fatigue life are important it is obviously desirable to produce steels which are free from non-metallic particles. This condition is never attained in the normal practices, but a very big improvement in the cleanliness of the metal can be obtained if it is vacuum treated before solidification. This is achieved by lowering the partial pressure of the carbon monoxide in equilibrium with the carbon and oxygen in the metal (equation 13.17) to pressures in the range from 1 mm to 1 micron (10^{-3} mm Hg), which causes the reaction to go further to the right. Theoretically at this pressure it is possible to lower the oxygen content of the metal to between 10^{-8} and 10^{-10} weight-per cent, depending on the carbon content of the metal, but this level is never attained in practice (Figure 13.16).

The departure from the equilibrium condition is due largely to the difficulties associated with the nucleation of a carbon monoxide bubble. From equation 13.44 it is apparent that the minimum radius of a cavity which can nucleate a gas bubble is decreased as the atmospheric pressure is lowered. Thus if the pressure is lowered to about 1 micron the conditions are more favourable for the carbon boil and the metal boils for a short time, but the excess supersaturation is rapidly removed and the boil quickly subsides. At this stage the value of the product [%C].[%O] is about 10^{-4}. However, it is found that the product [%C].[%O] continues to decrease after the boil has subsided and it has been suggested that this is due to the desorption of carbon monoxide gas from the exposed metal surface [76]. This hypothesis is supported by the relatively low rate at which oxygen is removed from the metal during the later stages of the vacuum treatment, for the overall rate at which the oxygen is removed is increased only slightly if the metal is agitated mechanically or by blowing argon gas through the melt. In view of the earlier considerations of the factors which affect the carbon boil it is improbable that this improvement is due entirely to the promotion of the boil and

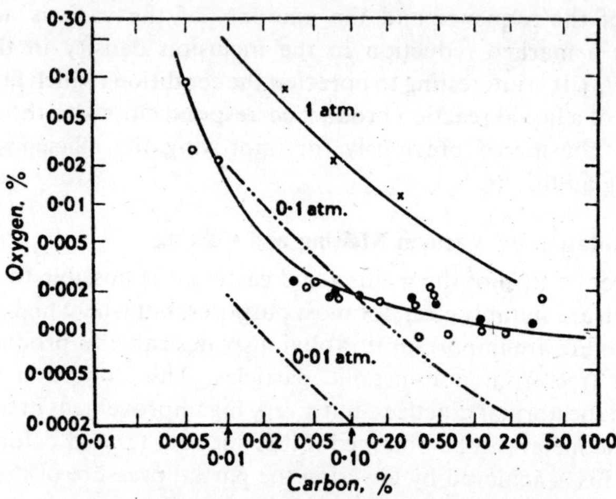

FIG. 13.16. Relation between the carbon and oxygen contents of liquid iron at 1,590°C for various atmospheric pressures. The equilibrium carbon–oxygen relations at pressures of 0·01, 0·10 and 1·0 atmosphere are shown by the three labelled curves. Points indicate the actual carbon–oxygen relation obtained under 1·0 atmosphere pressure of argon (crosses) and at pressures of $1{\cdot}5 \times 10^{-2}$ (full circles) and $5\text{–}7 \times 10^{-6}$ (open circles) mm Hg. (After Samarin, *J. Iron St. Inst.*, 1961, Vol. 198, p. 131.)

it is more likely that the circulation of the metal serves mainly to reduce the diffusion paths for the surface desorption process.

It is generally found that even after times approaching one hour under a vacuum of about 1 micron, the product [%C].[%O] in the metal is not lowered below 10^{-6} and when the metal is melted in a magnesia crucible the carbon-oxygen product approaches a limiting value between 10^{-5} and 10^{-6} [76–79], which is almost certainly due to the reduction of the magnesia and the transfer of oxygen from the crucible into the metal. Magnesium is not soluble in iron and when the pressure above the melt is atmospheric the extent to which magnesia is reduced is negligible, but at liquid steel temperatures the vapour pressure of magnesium is appreciable. Under vacuum the magnesium vapour can be removed as rapidly as it is produced and the reduction of the magnesia is then a continuous process which is limited only by the rate at which the reduction proceeds. If carbon is present in the melt the supply of oxygen atoms from the refractory is countered by the surface desorption of carbon monoxide and a steady-state condition is reached [76, 78] in which the rate of change of the oxygen in the metal is determined by the respective rates of the two reactions. The rate of reduction of magnesia is dependent on the area per unit volume of contact between the metal and the crucible, whilst the rate of desorption is dependent on the area per unit volume of the exposed metal surface. Hence, depending upon the actual conditions during the steady-state period, the oxygen content of the metal may increase, decrease or remain constant but, if this process is allowed to continue until the carbon is almost entirely removed, the oxygen content of the metal increases and ultimately rises to the level at which it is in equilibrium with liquid ferrous oxide.

Rather more rapid initial rates of oxygen removal can be obtained if the metal is caused to flow as a fine stream or spray through an evacuated chamber. This can be achieved, as in the original Bochumer-Verein practice, by allowing the metal to flow from the bottom of a ladle into a second ladle or ingot mould contained in an evacuated chamber. The time for which the metal can be held molten in the second ladle is however limited by the thermal losses and the ultimate oxygen content of the metal is higher than can be attained by vacuum melting.

With careful control of the vacuum treatment conditions the residual oxygen content of the liquid metal is less than the solubility limit after solidification, so that no deoxidising additions are required and the metal is free from those inclusions which originate from the deoxidation products. If the oxygen content of the metal is high at the start of the vacuum treatment, some adjustment is required to the carbon content of the melt to compensate for the loss by the carbon–oxygen reaction and to ensure that the treated metal complies with the specification limits.

REMOVAL OF OTHER ELEMENTS UNDER VACUUM

The vacuum treatments serve useful purposes other than the removal of oxygen. For example, the concentration of any element which has a reasonably high vapour-pressure at liquid steel temperatures is lowered by the treatment and with vacuum induction melting elements such as copper, lead, tin and manganese are almost completely eliminated. There is some loss also of chromium from the melt [80]. The following analysis of the flue dust from a vacuum casting installation [81] shows that these elements can be volatilised also during ladle de-gassing:

	FeO	MnO	Zn	Cu	Sn	Pb
Weight-per cent	17·9	47·0	1·4	2·6	0·2	1·0

Some removal of hydrogen and nitrogen from the metal can be obtained with a vacuum treatment. The partial pressure of hydrogen gas in equilibrium with the hydrogen dissolved in the metal is very low and only small amounts of this element are removed, but it is possible to lower the residual concentration to less than 2 parts per million [81, 82] and residual nitrogen concentrations of less than 0·001 weight-per cent can be obtained. The compositions of the exhaust gases from two ladle treatment plants are given as follow:

	CO	CO_2	H_2	N_2	
Per cent by Volume	80	1	15	5	[82]
	40	3·8	23	33·2	[81]

Vacuum treatments are costly processes and the increase in cost is justified only for high quality steels but, for many applications of these grades, the improvements in quality and metal yield which are obtained by a good vacuum treatment far outweight the additional cost.

REFERENCES

1. L. S. DARKEN and R. W. GURRY, *J. Amer. chem. Soc.*, 1947, Vol. 68, p. 798.
2. J. CHIPMAN, *Basic Open-Hearth Steel-making*, second edition, A.I.M.E., 1951, p. 621.
3. N. A. GOKCEN, *Trans. Amer. Inst. min. (metall.) Engrs.*, 1956, Vol. 206, p. 1558.
4. H. SCHENCK and G. H. GERDOM, *Arch. Eisenhüttenw.*, 1959, Vol. 79, p. 451.
5. T. FUWA and J. CHIPMAN, *Trans. Amer. Inst. min. (metall.) Engrs.*, 1960, Vol. 218, p. 887.
6. M. N. DASTUR and J. CHIPMAN, *Trans. Amer. Inst. min. (metall.) Engrs.*, 1949, Vol. 185, p. 313.
7. T. P. FLORIDIS and J. CHIPMAN, *Trans. Amer. Inst. min. (metall.) Engrs.*, 1958, Vol. 212, p. 549.
8. J. CHIPMAN and K. L. FETTERS, *Trans. Amer. Soc. Metals*, 1941, Vol. 29, p. 953.
9. C. R. TAYLOR and J. CHIPMAN, *Trans. Amer. Inst. min. (metall.) Engrs.*, 1943, Vol. 154, p. 228.
10. L. S. DARKEN, *Basic Open-Hearth Steel-making*, second edition, A.I.M.E., 1951, p. 867.
11. S. MARSHALL and J. CHIPMAN, *Trans. Amer. Soc. Metals*, 1942, Vol. 30, p. 695.

12. F. D. RICHARDSON and W. E. DENNIS, *Trans. Faraday Soc.*, 1953, Vol. 49, p. 171.
13. A. RIST and J. CHIPMAN, *Rev. Met.*, 1956, Vol. 53, p. 796; *Physical Chemistry of Steel-making*, New York, John Wiley, 1958, p. 3.
14. E. T. TURKDOGAN, L. S. DAVIS, L. E. LEAKE and G. C. STEVENS, *J. Iron St. Inst.*, 1955, Vol. 181, p. 123.
15. G. H. J. BENNETT, H. T. PROTHEROE and R. G. WARD, ibid., 1960, Vol. 195, p. 174.
16. H. C. VACHER and E. H. HAMILTON, *Trans. Amer. Inst. min. (metall.) Engrs.*, 1951, Vol. 95, p. 80.
17. G. PHRAGMEN and B. KALLING, *Jernkontor. Ann.*, 1939, Vol. 123, p. 199.
18. H. SCHENCK, W. RIESS and E. O. BRÜGGEMAN, *Z. Elektrochem.*, 1938, Vol. 38, p. 562.
19. I. M. MACKENZIE, *J. Iron St. Inst.*, 1946, Vol. 154, p. 55.
20. J. F. ELLIOTT, *Physical Chemistry of Steel-making*, New York, John Wiley, 1958, p. 37.
21. J. CHIPMAN, *J. Iron St. Inst.*, 1955, Vol. 180, p. 97.
22. S. MATOBA and S. BAN-YA, *Tech. Reps. Tohoku Univ.*, 1955, Vol. 20, p. 131; *Tetsu to Hagane*, 1958, Vol. 44, p. 643.
23. C. E. SIMS, *Trans. Amer. Inst. min. (metall.) Engrs.*, 1947, Vol. 172, p. 176.
24. J. CHIPMAN, *Metal Progr.*, 1949, Vol. 56, p. 211.
25. F. D. RICHARDSON, *Disc. Faraday Soc.*, 1948, No. 4, p. 114.
26. F. D. RICHARDSON, *J. Iron St. Inst.*, 1953, Vol. 175, p. 33.
27. J. CHIPMAN, J. C. FULTON, N. A. GOKCEN and G. R. CASKEY, *Acta Met.*, 1954, Vol. 2, p. 439.
28. M. S. PETRUSHESKII and P. V. GEL'D, *Zhur. Priklad Khim.*, 1959, Vol. 32, p. 86.
29. T. FUWA and J. CHIPMAN, *Trans. Amer. Inst. min. (metall.) Engrs.*, 1958, Vol. 212, p. 549.
30. J. PEARSON and E. T. TURKDOGAN, *J. Iron St. Inst.*, 1954, Vol. 176, p. 19.
31. J. A. KITCHENER, J. O.'M BOCKRIS and D. A. SPRATT, *Trans. Faraday Soc.*, 1952, Vol. 48, p. 608.
32. B. M. LARSEN, *Trans. Amer. Inst. min. (metall.) Engrs.*, 1941, Vol. 145, p. 67.
33. C. GOODEVE, *Disc. Faraday Soc.*, 1948, No. 4, p. 9.
34. P. VALLET, *Iron & Steel, Lond.*, 1955, Vol. 28, p. 463.
35. P. KOZAKEVITCH, S. CHATEL and M. SAGE, *C. R. Acad. sci., Paris*, 1953, Vol. 236, p. 2064.
36. F. KÖRBER and W. OELSEN, *Mitt. K.-Wilh. Inst. Eisenforsch.*, 1935, Vol. 17, p. 39.
37. T. E. BROWER and B. M. LARSEN, *Trans. Amer. Inst. min. (metall.) Engrs.*, 1945, Vol. 162, p. 712.
38. J. F. ELLIOTT and F. W. LUERRSEN, ibid., 1955, Vol. 203, p. 1129.
39. E. V. CHELISHCHEV, *Kislorod*, 1958, (3), p. 11.
40. L. S. DARKEN, *Trans. Amer. Inst. min. (metall.) Engrs.*, 1947, Vol. 172, p. 190; *Physical Chemistry of Steel-making*, New York, John Wiley, 1958, p. 101.
41. C. WAGNER, ibid., p. 237.
42. J. A. KITCHENER, J. O'M BOCKRIS, M. GLEISER and J. W. EVANS, *Acta Met.*, 1953, Vol. 1, p. 93.
43. J. CHIPMAN and N. A. GOKCEN, *Trans. Amer. Inst. min. (metall.) Engrs.*, 1952, Vol. 194, p. 131; 1953, Vol. 197, p. 1017.

44. H. F. RAMSTAD and F. D. RICHARDSON, *Trans. Amer. Inst. min. (metall.) Engrs.*, 1961, Vol. 221, p. 1021.
45. F. KÖRBER and W. OELSEN, *Mitt. K.-Wilh. Inst., Eisenforsch.*, 1933, Vol. 15, p. 271.
46. D. C. HILTY and W. CRAFTS, *Trans. Amer. Inst. min. (metall.) Engrs.*, 1950, Vol. 188, p. 425.
47. J. CHIPMAN and T. C. M. PILLAY, *Trans. Amer. Inst. min. (metall.) Engrs.*, 1961, Vol. 221, p. 1277.
48. W. A. FISCHER and H. ENGELBRECHT, *Stahl u. Eisen, Düsseldorf*, 1955, Vol. 75, p. 70.
49. O. KUBASCHEWSKI and E. LL. EVANS, *Metallurgical Thermochemistry*, third edition, Pergamon, 1958.
50. J. CHIPMAN, J. B. GERO and T. B. WINKLER, *Trans. Amer. Inst. min. (metall.) Engrs.*, 1950, Vol. 188, p. 341.
51. B. V. LINCHEVSKII and A. A. SAMARIN, *Izv. Ak. Nauk Otdelenie Tekn.*, 1957, (2), p. 9.
52. H. B. BELL, A. B. MURAD and T. P. CARTER, *Trans. Amer. Inst. min. (metall.) Engrs.*, 1952, Vol. 194, p. 1173.
53. H. SCHENCK, N. G. SCHMAHL and A. K. BISWAS, *Arch. Eisenhüttenw.*, 1957, Vol. 28, p. 517.
54. H. A. SLOMAN and E. LL. EVANS, *J. Iron St. Inst.*, 1951, Vol. 169, p. 145.
55. E. T. TURKDOGAN and J. PEARSON, ibid., 1953, Vol. 175, p. 393.
56. V. V. AVERIN, R. A. KARASEV, A. Y., POLYAKOV and A. M. SAMARIN, *Izv. Ak. Nauk Otdelenie Tekn.*, 1956, (11), p. 52.
57. H. A. SLOMAN and E. LL. EVANS, *J. Iron St. Inst.*, 1950, Vol. 165, p. 81.
58. N. A. GOKCEN and J. CHIPMAN, *Trans. Amer. Inst. min. (metall.) Engrs.*, 1953, Vol. 197, p. 173.
59. H. SAWAMURA, *Suiyokwai Shi*, 1956, Vol. 13, p. 142.
60. H. WENTRUP and G. HIEBER, *Tech. Mitt. Krupp*, 1939, Vol. 1, p. 47.
61. W. GELLER and K. DICKE, *Arch. Eisenhüttenw.*, 1943, Vol. 16, p. 431.
62. D. C. HILTY and W. CRAFTS, *Trans. Amer. Inst. min. (metall.) Engrs.*, 1950, Vol. 188, p. 414.
63. T. YAGIHASHI and Z. SHIBATA, *Sci. Reps. Res. Inst. Tohoku Univ.*, Series A, 1954, Vol. 6, p. 50.
64. J. CHIPMAN and M. N. DASTUR, *Trans. Amer. Inst. min. (metall.) Engrs.*, 1951, Vol. 191, p. 111.
65. J. CHIPMAN, *Trans. Amer. Inst. min. (metall.) Engrs.*, 1960, Vol. 218, p. 767.
66. R. L. HADLEY and G. DERGE, *Trans. Amer. Inst. min. (metall.) Engrs.*, 1955, Vol. 203, p. 55.
67. A. SAUERWALD, *Z. Erzbergb. Metallhüttenw.*, 1950, Vol. 41, p. 214.
68. T. YAGIHASHI and T. SATO, *Tetsu to Hagane*, 1953, Vol. 39, p. 575.
69. G. DERGE, *Trans. Amer. Inst. min. (metall.) Engrs.*, 1946, Vol. 167, p. 93.
70. M. ELLE and J. CHIPMAN, *Trans. Amer. Inst. min. (metall.) Engrs.*, 1961, Vol. 221, p. 701.
71. F. D. RICHARDSON, *J. Iron St. Inst.*, 1950, Vol. 166, p. 137.
72. W. KOCH, H. WENTRUP and O. RIEF, *Arch. Eisenhüttenw.*, 1951, Vol. 22, p. 15.
73. W. KOCH and F. WEVER, *Stahl u. Eisen, Düsseldorf*, 1954, Vol. 75, p. 264.
74. R. ROSEGER, *Radex Rundschau*, 1958, p. 292.

75. I. P. KOVYRYALOV, S. I. POPEL, G. F. KONOVALOV and A. M. POLZUNOV, *Stal'*, 1960, (4), p. 305.
76. G. H. J. BENNETT, H. T. PROTHEROE and R. G. WARD, *J. Iron St. Inst.*, 1960, Vol. 195, p. 174.
77. T. B. KING, Chapter in *Thermodynamics in Vacuum Metallurgy*, New York, Univ. Press, 1960.
78. O. WINKLER, *Metallurgical Rev. Inst. Metals, London*, 1960, Vol. 5, p. 1.
79. P. TURILLON, *Rev. Met.*, 1960, Vol. 57, p. 649.
80. G. M. GILL, E. INESON and G. W. AUSTIN, *J. Iron St. Inst.*, 1959, Vol. 191, p.172.
81. A. TIX, *J. Metals, N.Y.*, 1956, Vol. 8, p. 420.
82. F. HARDERS, H. KNÜPPEL and K. BROTZMANN, *Stahl u. Eisen, Düsseldorf*, 1956, Vol. 76, p. 1721.

CHAPTER 14

The Distribution of Sulphur between the Slag and the Metal

14.1 Introduction

At iron and steel-making temperatures elemental sulphur is stable only as a gas, but it can be dissolved to form a liquid solution in both the slag and the metal. This means that the control of the sulphur content in the metal must be considered in terms of (*a*) the partitioning between the gaseous compounds containing sulphur and the sulphur in solution at the gas–liquid interface, and (*b*) the partitioning of sulphur between the slag and the metal at the slag–metal interface.

The conditions which affect the exchange of sulphur between the gas and the liquid are examined in some detail in Chapter 9.4. These considerations can be summarised in the general statement that sulphur is transferred to the liquid (slag or metal) when the partial pressure of the sulphur in the gas exceeds a critical level which is dependent on the temperature and on the compositions of the liquid and the gas. The transfer takes place in the reverse direction when the partial pressure of the sulphur in the gas is below this level. Thus in processes such as the electric-arc and the Bessemer, where the atmosphere above the melt is free from sulphur-bearing combustion gases, there should be a continuous transfer of sulphur from the liquid into the atmosphere, but the indications are that the overall rate of loss to the atmosphere is low and the sulphur content of the metal is determined primarily by the partitioning between the slag and the metal. In the blast furnace the iron is exposed to the gas during the descent through the stack and bosh zones and the absorption of sulphur by the metal is rapid. However, the final sulphur content of the metal is determined by the slag–metal reactions in the furnace hearth and it is probable that the partial pressure of sulphur in the atmosphere immediately above the slag is lower than the level required for transfer from gas to slag. Hence it is only with processes such as the open-hearth, where measurable amounts of sulphur are contained in the furnace combustion products, that sulphur pick-up by the metal from the atmosphere via the slag is important.

In Chapter 9.4 it is shown that sulphur may be present in a liquid slag as either sulphide (S^{2-}) or sulphate (SO_4^{2-}) ions, but the sulphate form is stable only when the partial pressure of oxygen in equilibrium with the slag is greater than 10^{-4} atmosphere and in the steel-making processes the partial pressure of oxygen is very much lower than this at the slag–metal interface. The partial pressure is higher at the slag–gas interface and is usually in the range 10^{-1} to 10^{-3} atmosphere, but the data presented in Chapter 12 show that the mean value for the slag is close to that at the slag–metal interface. This means that in the refining of steel the sulphur in the slag is present principally as sulphide ions; the concentration of sulphate ions is very small and it has been suggested [1] that the small fraction of sulphate ions which is found in a solidified slag is produced by a reaction between sulphide and oxygen ions during cooling. Only sulphide ions can exist at the relatively low oxygen potentials of the slag in the blast furnace process.

14.2 Equilibrium Considerations

The partitioning of sulphur between the slag and the metal is sometimes considered in terms of three simultaneous reactions [2] which can be expressed as:

$$MS_{(in\,Fe)} \rightleftharpoons MS_{(slag)} \qquad 14.1$$

$$MS_{(slag)} + CaO_{(slag)} \rightleftharpoons CaS_{(slag)} + MO_{(slag)} \qquad 14.2$$

$$MO_{(slag)} + X_{(in\,Fe)} \rightleftharpoons M_{(in\,Fe)} + XO_{(slag\,or\,gas)} \qquad 14.3$$

where M represents iron, manganese and possibly silicon, and X is an element such as aluminium, carbon, manganese and silicon which combines with the oxygen dissolved in the iron. Treatment of the partitioning in this way suffers from the disadvantage that only lime is considered to take part in stabilising the sulphur in the slag and the true mechanism of the process is not clear; a more satisfactory interpretation is obtained when the reactions are expressed in ionic form.

The transfer of elemental sulphur from the metal into the slag can be written as:

$$S_{(in\,Fe)} + 2e^- \rightleftharpoons S^{--}{}_{(slag)} \qquad 14.4$$

where $2e^-$ represents the acquisition of two negative electron charges by each atom of sulphur which is transferred across the slag–metal interface. In order to maintain electro-neutrality a simultaneous transfer is required across the interface of some element which releases electron charges as it enters the slag. Thus the transfer of an iron atom into the slag to form a ferrous ion is accompanied by the liberation of two electron charges:

$$Fe_{(metal)} \rightleftharpoons Fe^{2+}{}_{(slag)} + 2e^- \qquad 14.5$$

Similarly the transfer into the slag of one atom of silicon releases four electron charges and one atom of aluminium donates three electron charges. Alternatively the two electrons absorbed by the sulphur atom can be regarded as transferred from an oxygen ion which crosses the interface in the reverse direction and enters the metal to take part in the normal oxidising reactions represented by equation 14.3. The sulphur transfer reaction in terms of the ionic species in the slag can then be written as:

$$S_{(\text{in Fe})} + O^{--}_{(\text{slag})} \rightleftharpoons O_{(\text{in Fe})} + S^{--}_{(\text{slag})} \qquad 14.6$$

The standard free-energy change for this reaction is given by [3]:

$$\Delta G^0 = 17{,}200 - 9{\cdot}12T \text{ cal mol}^{-1} \qquad 14.7$$

and

$$\log K_S = \log \frac{(a_{S^{--}})[a_O]}{(a_{O^{--}})[a_S]} = -\frac{3{,}750}{T} + 1{\cdot}996 \qquad 14.8$$

The activities of the sulphur in the slag have been determined experimentally only for slags of the blast furnace type [4] and there is no reliable method for calculating the activities of the oxygen ions; hence the equation cannot be solved for the actual sulphur distribution between the metal and the slag, but the general trends can be predicted from consideration of the data which are presented in previous chapters. Thus the activity of the oxygen ions is raised as the basicity of the slag is increased whereas, with the low sulphur contents which are normal in a liquid slag, it is probable that the activity of the sulphur ions is lowered as the basicity is raised. It follows from equation 14.8 that the sulphur partitioning between the slag and the metal is raised by increasing either the basicity of the slag or the activity of the sulphur in the metal, or by lowering the activity of the oxygen in the metal.

The sulphur partitioning can be expressed also in terms of the ratio of the weight-per cent concentrations of sulphur in the slag and in the metal (i.e. (%S)/[%S]) and in the steel-making processes, where the atmosphere is oxidising throughout the refining period, this ratio varies from approximately 1 under an acid slag to about 10–14 when the slag is saturated with basic oxides. The ratio is considerably greater for conditions typical of those encountered in the blast furnace where, for equilibrium at a metal temperature of 1,500°C, it varies from about 50 for a silica-saturated slag to about 350 for a lime-saturated slag [5]. As the transfer of sulphur from metal to slag is an endothermic process, these ratios should be even larger at steel-making temperatures. The ratios obtained under the reducing slags in the electric-arc furnace process are intermediate, but lie closer to the range of values for steel-making under an oxidising slag than to those for the blast furnace.

Part of the enormous difference between the ratios which are obtained in the blast furnace and those for the refining processes can be attributed to the effect of the concentrations of the solutes in the iron on the activity coefficient of the sulphur in the metal. In pure liquid iron at 1,600°C

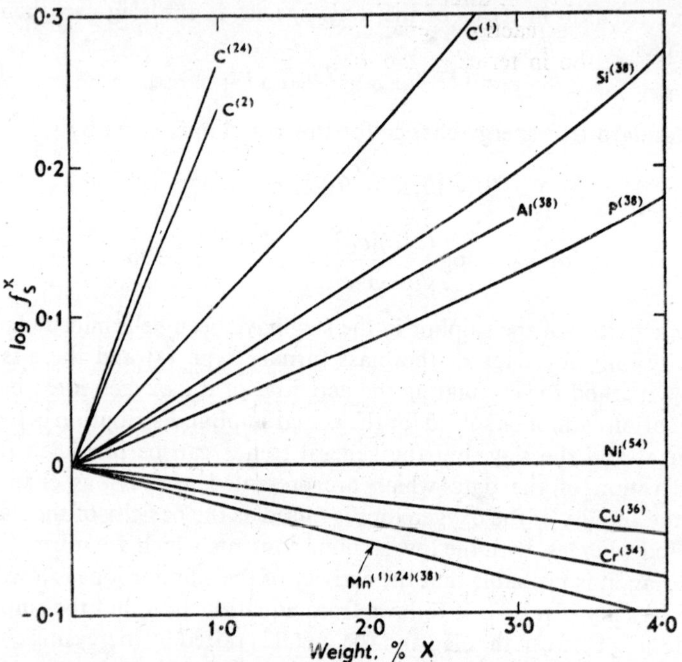

FIG. 14.1. The effect of various elements on the activity coefficient of sulphur in liquid iron. The numbers against each line relate to the references for the interaction coefficients given in Table 13 of Appendix 1.

sulphur obeys Henry's Law for concentrations up to about 0·5 weight-per cent and within this range log[f_S] is $-0{\cdot}028$ [%S] [6, 7]. The activity coefficient is raised by aluminium, carbon, nitrogen, phosphorus and silicon; nickel and oxygen have no effect and it is lowered by chromium, copper and manganese. The data are summarised in Figure 14.1 and the interaction coefficients are tabulated in Appendix 1. The concentration of solutes in the metal is low towards the end of the refining in the steel-making processes and the activity coefficient of sulphur is usually between 0·95 and 1·5, but the presence, particularly of the large concentrations of carbon and silicon or phosphorus in the blast furnace metal, raises the activity coefficient to within the range 4·0–6·0. This means that, if all other conditions remain constant, the changing of the metal composition from that obtained at the blast furnace to the normal range for plain carbon steels accounts for a decrease in the sulphur partition ratio by a factor of about 5. However, if the blast furnace data are adjusted to allow for this difference in metal compositions, the partition ratios are still between 5 and 10 times as great as those which are obtained in the open-hearth process and some other factor must be responsible for this differencc.

14.3 The Effect of Slag Composition

The distribution of sulphur ((%S)/[%S]) at 1,600°C between pure liquid iron–sulphuı alloys and almost pure liquid ferrous oxide is 3·6 [8]; this increases to about 7·8 for liquid iron in equilibrium with a slag containing 28 to 33 mol per cent of CaO and 4·8 mol per cent of MgO [9] and to about 10·5 in equilibrium with a lime-saturated, ferrous oxide slag [3, 8]. Since these compositions are approximately saturated with basic oxides it is reasonable to assume that the activity of the oxygen ions is constant. The activity coefficient of sulphur is unknown in these solutions, but it has been suggested that the value of the coefficient is unchanged as ferrous oxide is replaced by lime [10], in which case the change in the distribution ratio with the slag composition is due entirely to the change in the activity of the oxygen in the metal. As the ferrous oxide in the slag is replaced by lime the activity of the ferrous oxide is lowered and (if the equilibrium between the slag and the metal is established) the activity of the oxygen in the iron is lowered correspondingly, this change being in the direction required to account for the change in the sulphur distribution ratio. Further support for this argument is obtained from the fact that the solubility of ferrous oxide in the slag is increased and the sulphur ratio is lowered as the temperature is raised [3]. It is expected that at constant composition the activity of the ferrous oxide is lowered as the solubility is raised and the corresponding change in the oxygen activity is again in the correct direction to explain the variation in the sulphur distribution ratio.

Magnesia in the slag is generally considered to have a negligible effect on the sulphur partitioning [11, 12] and in lime-saturated slags the distribution ratio is not changed in the presence of up to 10 mole per cent of magnesia [3]. This contention is supported by a comparison of the free-energy changes for the reactions between sulphur in the metal and the oxides in the slag [13].

$$CaO_{(slag)}+S_{(in\,Fe)} \rightleftharpoons CaS_{(slag)}+O_{(in\,Fe)};\ \Delta G^0_{(1,873°K)} = +12,700 \text{ cal} \quad 14.9$$

$$FeO_{(slag)}+S_{(in\,Fe)} \rightleftharpoons FeS_{(slag)}+O_{(in\,Fe)};\ \Delta G^0_{(1,873°K)} = +14,800 \text{ cal} \quad 14.10$$

$$MnO_{(slag)}+S_{(in\,Fe)} \rightleftharpoons MnS_{(slag)}+O_{(in\,Fe)};\ \Delta G^0_{(1,873°K)} = +17,600 \text{ cal} \quad 14.11$$

$$MgO_{(slag)}+S_{(in\,Fe)} \rightleftharpoons MgS_{(slag)}+O_{(in\,Fe)};\ \Delta G^0_{(1,873°K)} = +33,400 \text{ cal} \quad 14.12$$

These free-energy values are approximate, but they are in the same order as the values which can be calculated from recent measurements at temperatures below 1,500°C [14] of the equilibria between the oxides and sulphides with oxygen and sulphur in the gas phase. These relations show that the desulphurising power of the basic oxides decreases slightly in the series CaO, FeO and MnO but magnesia is very much less powerful when compared with these three oxides.

From experimental data it is estimated that lime is about 1,000 times more effective than magnesia for holding sulphur in basic oxide-saturated melts [11], but the difference is very much less pronounced with slags in the composition range normally encountered in the refining processes. The activity of the oxygen ions in the slag is lowered when oxides which form the complex oxy-acidic radicals are added to the slag and the composition moves away from the saturation surfaces for the basic oxides. Under these conditions the oxygen ions added with magnesium and calcium ions act together to neutralise the charges on the acid groups and if magnesia is added to the melt, while holding constant the ratio CaO/SiO_2, the activity of the oxygen ions is raised and the sulphur distribution ratio is increased.

In the earlier investigations of the activities of ferrous oxide in silicate melts it was commonly assumed that lime, magnesia and manganese oxides are equivalent on a molar basis in their effects on the activities, and a similar equivalence of these oxides on the sulphur partitioning between the metal and an unsaturated slag has frequently been assumed. These assumptions imply that the different desulphurising powers of the oxides are of minor importance relative to the effect of the total concentration of basic oxides on the activity of the oxygen ions in the slag. (More recent data show that these oxides are not exactly equivalent, but this evidence is considered later.)

THE 'EXCESS BASE' PARAMETER

In one application of this approach the sulphur partitioning between the slag and the metal has been plotted against a function called the 'excess base' for basic slags with a wide range of compositions [15]. The 'excess base' (EB) is calculated from the relation:

$$EB = n\mathrm{CaO}+n\mathrm{MgO}+n\mathrm{MnO}-2n\mathrm{SiO_2}-4n\mathrm{P_2O_5}-2n\mathrm{Al_2O_3}-n\mathrm{Fe_2O_3}$$

where n is the number of moles of oxide per 100 grams of slag. Each mole of the basic oxides supplies one oxygen ion to the solution, whilst two oxygen ions are acquired by a mole of silica to form the saturated SiO_4^{4-} grouping, and so on. Hence the excess base can be regarded as a measure of the mole fraction of unattached or 'free' oxygen ions in the solution. When the data are plotted in this way the points are distributed uniformly about a straight line (Figure 14.2) and no systematic

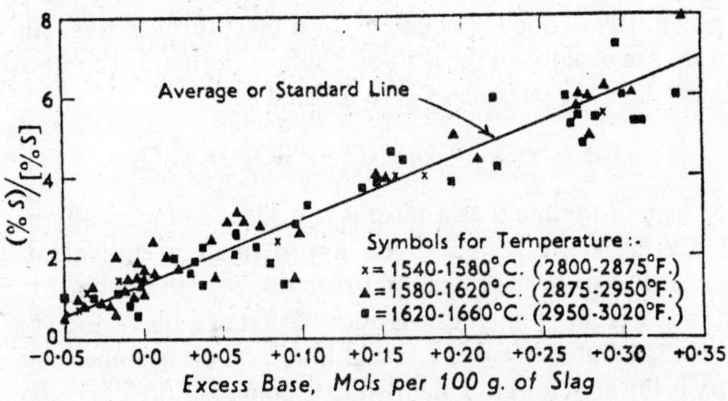

FIG. 14.2. Relation between sulphur partitioning and 'excess base' (see text) for basic open-hearth type slags at 1,540°–1,660°C. (After Grant and Chipman [15].)

trends are observed in the distribution of the points as the magnesia and manganese oxide contents of the melts are varied over a fairly wide range. These data have also been plotted in other ways [16, 17] but there is no indication that the basic oxides are not equivalent.

In the calculation of the 'excess base' quantities it is assumed that ferrous oxide is a neutral component of the slag and for a constant value of the excess base the sulphur distribution is found to be almost independent of the ferrous oxide content in the range 3–80 per cent. However, when these data are combined with those from other investigations and plotted on a pseudo-ternary composition diagram, it is found that the iso-partitioning lines are roughly parallel to the ortho-silicate–ferrous oxide join (Figure 9.3) [18] and this probably accounts for the apparent independence of the sulphur distribution of the ferrous oxide content of the slag when plotted against the excess base function. When the data are plotted to show the sulphur distribution versus the ratio of bases to acids (i.e., basicity ratio) in the slag, the points are fitted by a family of curves, each of which represents a constant ferrous oxide content of the slag. The curves intersect at compositions corresponding approximately to the ortho-silicate–ferrous oxide join and diverge widely at high basicity (Figure 9.4). This observation is more consistent with the previous discussion of the sulphur distribution between the metal and lime–ferrous oxide slags which established that at high basicity the partitioning is dependent on the ferrous oxide content of the melt.

A similar investigation has been made of the sulphur distribution between slag and metal compositions typical of those which are encountered in blast furnace practice and good correlation is obtained between the distribution ratio and the 'excess base' [19]. However, the 'excess base' as previously defined is negative for a blast furnace slag, since the compositions are usually on the acid-rich side of the neutral composition, and for these slags the parameter is re-defined as:

$$\mathrm{EB} = n\mathrm{CaO} + \tfrac{2}{3}n\mathrm{MgO} - n\mathrm{SiO_2} - n\mathrm{Al_2O_3}$$

The factor of $\frac{2}{3}$ in the magnesia term is introduced empirically because lime and magnesia are found to be approximately equivalent on a weight-per cent basis in their desulphurising powers in the blast furnace and the atomic weight of magnesia is about $\frac{2}{3}$ that of lime. At high basicity the data are distributed about a straight line, but this becomes curved as the basicity is lowered and the meta-silicate composition (corresponding to EB = 0) is approached (Figure 14.3). In contrast to the data for steelmaking slags, the distribution in the blast-furnace type heats is found to increase as the temperature is raised.

These observations can be examined in terms of the equilibrium constant for equation 14.6 which can be written as:

$$K_S = \frac{(N_{S^{--}} \cdot \gamma_{S^{--}})[\%O \cdot f_O]}{(N_{O^{--}} \cdot \gamma_{O^{--}})[\%S \cdot f_S]}$$

In Figures 14.2, 9.3 and 9.4 (%S)/[%S] is plotted against some function of the mole fraction of oxygen ions ($N_{O^{--}}$) in the slag and so the slopes of the lines are related to the quantity:

$$\frac{(\gamma_{O^{--}})[f_S]}{(\gamma_{S^{--}})[\%O \cdot f_O]}$$

The activity coefficient of the sulphur in the metal can be regarded as independent of the composition of the slag but, if the slag and metal are in equilibrium, the quantity [%O $.f_O$] is determined by the ferrous oxide activity in the slag, so that the slopes of the plots are also related to:

$$\frac{(\gamma_{O^{--}})}{(\gamma_{S^{--}})(a_{FeO})}$$

In more recent investigations the number of variables affecting the slope has been reduced by plotting the ratio {(%S)/[%S]}. [%O] against

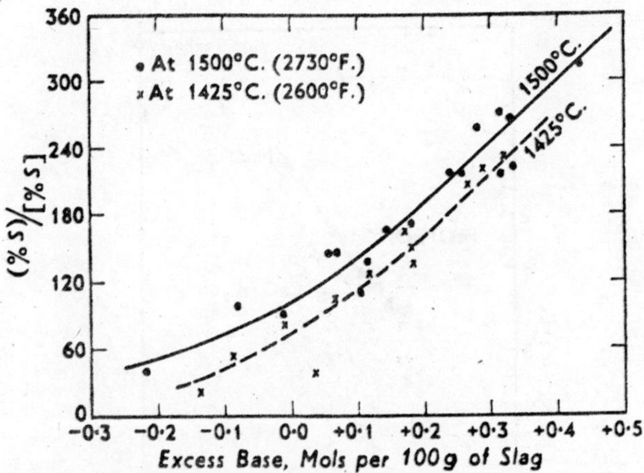

FIG. 14.3. Relation between sulphur partitioning and 'excess base' (see text) for blast furnace type slags at 1,425° and 1,500°C. (After Hatch and Chipman [19].)

the mole fraction of acid or basic oxides in the slag. This ratio is commonly called the S–O ratio and it is equal to:

$$K_S . \frac{(a_{O^{--}})[f_S]}{(\gamma_{S^{--}})[f_O]}$$

If it is again assumed that $[f_S]$ and $[f_O]$ in the metal do not change markedly with the composition of the slag, then the gradient of the plots is proportional to the ratio $(a_{O^{--}})/(\gamma_{S^{--}})$. The experimental data show that at high basicities in the slag this ratio decreases almost linearly with decrease in the basicity, but the rate of change is less rapid when the ratio of bases/acids is less than unity [17, 18]. Similar deductions can be drawn by plotting the sulphide capacity of the slag $C_S(=(\%S).p_{O_2}^{1/2}/p_{S_2}^{1/2})$ against the mole fraction of basic oxides. The C_S relations are examined in Chapter 9.4 and it is readily shown that the slope of these plots is again proportional to the ratio $(a_{O^{--}})/(\gamma_{S^{--}})$. Figure 14.4 shows the relationship between C_S and the basicity for a number of binary and ternary slag systems [20] and from this it is clearly evident that the basic oxides CaO, FeO, MnO and MgO are not equivalent in so far as their effects on the ratio $(a_{O^{--}})/(\gamma_{S^{--}})$ are concerned.

Similarly the data for the acid oxide–lime melts show that alumina, silica and phosphorus pentoxide have markedly different effects on the ratio. When this method of plotting is adopted, the data for open-hearth

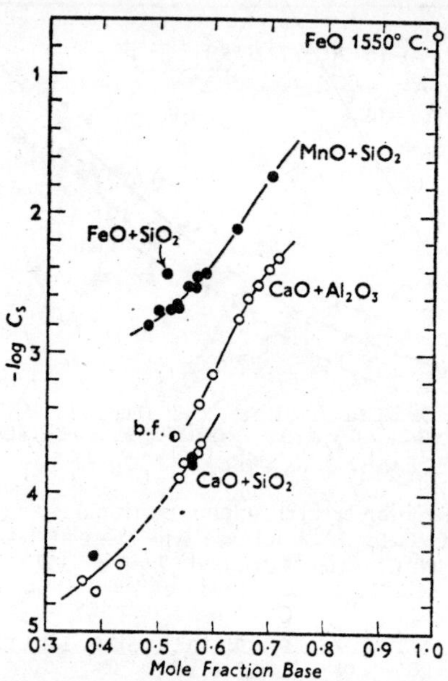

FIG. 14.4*a*. The sulphide capacities of various aluminate and silicate melts at 1,550°C. b.f. = blast furnace. (After Abraham and Richardson [20].)

type and blast furnace type melts are consistent with each other and also with the sulphur capacities of the simple slag mixtures as shown in Figure 14.4.

EFFECT OF STRUCTURE AND BOND TYPES

A general interpretation of these observations can be deduced from consideration of the effects of the structure and bonding in the slag on the thermodynamic properties of the various species present. Using a radioactive tracer technique the heat of transfer of sulphur from liquid iron into a pure ferrous oxide slag has been measured as – 12,000 cal g atom^{-1}, but with a ferrous silicate slag this is increased to $-27{,}000$ cal g atom^{-1} [21]. It has been suggested that this change is caused by the strong polarisation of the oxygen ions by the Si^{4+} ions which lowers the repulsive force between the negatively-charged oxygen and sulphide ions and increases the attractive force between the sulphide and the iron ions. This hypothesis is supported by data for the surface tension of slags, for the addition of a constant amount of sulphur to liquid silicates of

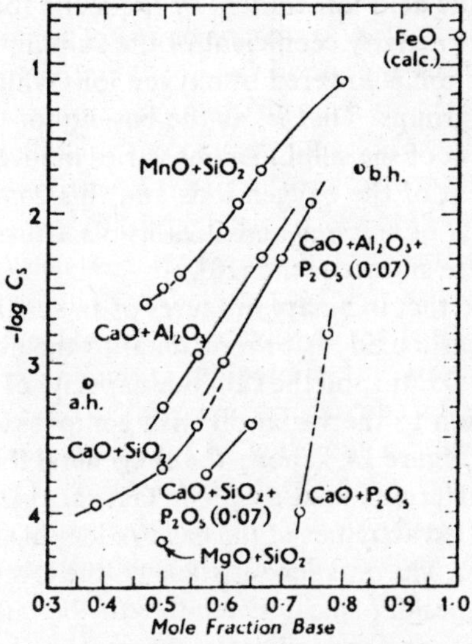

FIG. 14.4*b*. The sulphide capacities of various aluminate, phosphate and silicate melts at 1,650°C. a.h. = acid open-hearth, b.h. = basic open hearth. (After Abrahan and Richardson [20].)

barium, calcium and magnesium lowers their surface tension increasingly in that order [22]. As the polarisation of the cation-anion bond is increased the linkage with other ion groups is weakened and the surface tension is lowered. Because of their greater ionic size, sulphide ions are more polarisable than oxygen ions and, as magnesium ions are fairly strong polarisers, the Mg^{2+}—S bond is more strongly polarised than the Mg^{2+}—O^{--} bond. Barium and calcium on the other hand are comparatively weak polarisers and the difference between the metal–oxygen and metal–sulphur bonds is less marked. All this evidence suggests that sulphur ions in the slag are more strongly polarisable than the oxygen ions.

If a sulphur ion is substituted for an oxygen ion in the co-ordination shell of a strongly polarising silicon ion, the electron field of the silicate group is distorted to the extent that the structure becomes unstable and breaks down. From this it is concluded that sulphur ions can exchange with oxygen ions which are bonded to the basic cations but not with the oxygen ions in the oxy-acid (i.e. SiO_4^{4-}, PO_4^{3-}, etc.) groups. A similar conclusion is reached by reasoning from the change in the sulphide capacities with the acid-ion content of the melts [23]. On this basis it is expected that the activity coefficient of the sulphur in the slag increases as the concentration is lowered of oxygen ions which remain unbonded to the oxy-acid groups. That is, as the basicity of the slag is varied the activity coefficient of the sulphide ions varies in inverse relation with the activity coefficient of the oxygen ions and this conclusion is confirmed by measurements of sulphur activities in blast furnace type [4] and in binary manganese silicate melts [20].

If it is assumed that in binary mixtures of an acid and a basic oxide the activity of the basic oxide is proportional to the activity of the oxygen ions, then the S–O ratio and the sulphide capacity of the melt should vary directly in relation to the variation with composition of the activity of the basic oxide. Figure 14.4 shows that in general the sulphide capacities of the binary and ternary mixtures do in fact vary in the same order as the known or predicted activities of the basic oxides in aluminate, phosphate and silicate melts. The only inconsistency is the low value of the sulphide capacity in magnesium silicates, relative to the value at the same silica concentration in the calcium silicates; the reverse effect would be expected from the predicted activities of magnesia in the binary silicate melts, but this behaviour is confirmed by measurements of the sulphide capacities in pseudo-binary, meta–silicate mixtures of these two oxides [20]. The apparent anomaly may be due to the marked polarisation of the Mg^{2+}—S^{--} bond, for it is to be expected that this would raise the activity coefficient of the sulphide ions relative to the value at the same silica content in the calcium silicates. If the activities of lime and magnesia do not vary too markedly from each other in the binary silicate liquids then at constant silica concentration the relative values of the sulphide capacities of the two systems may well be determined by the activity coefficients of the sulphide ions.

THE EFFECT OF FERROUS OXIDE IN THE SLAG

The foregoing considerations have established that at constant composition and temperature the sulphide capacities of blast-furnace and steel-making slags are identical, but comparison of Figures 14.2 and 14.4

shows that for equivalent basicities the sulphur distribution in the blast furnace is about 80 to 100 times as great as in the open-hearth. It has already been shown that part of this difference can be accounted for in terms of the different values for the activity coefficients for sulphur and oxygen in the two processes. The only other major difference in the compositions of the bath is the concentration of ferrous oxide, which is less than 1 weight-per cent in the blast furnace slag and rarely below 10 per cent in the basic open-hearth slag. For slags of constant ratio:

$$(CaO+MgO+MnO)/(SiO_2+Al_2O_3+P_2O_5)$$

an increase in the ferrous oxide content raises the activity of the ferrous oxide and increases the activity of the oxygen in the metal at equilibrium with the slag. This should lower the sulphur partitioning, which has in fact been confirmed for slags of the blast furnace type by varying the ferrous oxide content over the range from 0·1 to 3·0 mole per cent [11]. When the ratio $(\%S)/[a_S]$ is plotted against the logarithm of the mole per cent of ferrous oxide, it is found that the family of points for fixed values of the 'excess-base' parameter can be fitted by a series of straight lines which are approximately parallel over the range from 0·1 to 1·0 mole per cent of ferrous oxide. At higher ferrous oxide contents the slope decreases rapidly and, when these data are combined with those from other investigations to extend the composition range up to 100 mole per cent of ferrous oxide, it is found that the distribution ratio passes through a minimum at about 5 per cent of ferrous oxide (Figure 14.5). As the basicity of the slag is raised the minima in the curves become less pronounced until, with basicities typical of those encountered in basic open-hearth refining slags, the distribution ratio is almost independent of the ferrous oxide content when this is greater than about 3 mole per cent.

It follows that as the ferrous oxide content in the open-hearth type slags is raised the increase in the metal oxygen activity is approximately balanced by some opposing effect, which is also dependent on the ferrous oxide concentration. From inspection of the equilibrium constant for the sulphur transfer reaction, it has been suggested [11, 24] that, when the concentration of ferrous oxide is very low, the rate of change of the metal oxygen activity with the ferrous oxide content is greater than the corresponding change in the activity of the oxygen ions in the slag. However, when the ferrous oxide concentration is greater than about 3 mole per cent these two activities change at about the same rate and the sulphur distribution ratio becomes approximately constant. This argument apparently is supported by the observation that the S–O ratio is approximately independent of the ratio FeO/CaO in open-hearth slags of constant silica content [18]. It is however assumed that the activity coefficients of the sulphide ions in the slag are almost independent of the composition of the slag, but the more recent measurements of slag activities do not justify this assumption and it is probably more correct to say that the ratio $(a_{O^{--}})/(\gamma_{S^{--}})$ changes at about the same rate as the metal oxygen activity for the slags of higher ferrous oxide content.

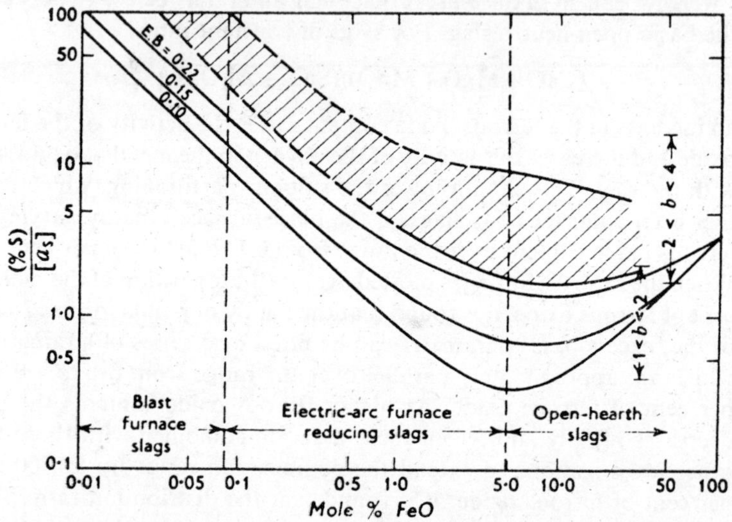

FIG. 14.5. Relation between sulphur distribution and ferrous oxide content of the slag for slags of various basicities. E.B. is 'excess base' (see text). The bands at the right-hand side indicate the range of sulphur distributions for open-hearth type slags with basicities between 1 and 2 and between 2 and 4. (After Chipman, *Metal Progr.*, 1952, Dec. p., 97, and modified to include the data of Bishop, Lander, Grant and Chipman [18].)

The intermediate range of ferrous oxide contents between 1·0 and 3·0 mole per cent is of interest, since these compositions fall in the range encountered in the reducing slags of the electric-arc furnace process. Figure 14.5 shows that within this range the equilibrium sulphur distribution is markedly dependent on the ferrous oxide content of the slag and this dependence explains why more sulphur can be removed from the metal when a slag is formed which is strongly reducing than when only mildly reducing conditions are maintained.

To summarise, it can be stated that the factors which favour high equilibrium sulphur partitioning are as follow:

(1) High activity of the oxygen ions in the slag. Generally, this is increased as the basicity of the slag is raised, when silica is replaced by alumina, or lime is replaced by manganese oxide, etc. (Figure 14.4).

(2) High activity coefficient of the sulphur in the metal. This is raised by elements such as aluminium, carbon, phosphorus and sulphur. It is lowered by chromium, copper and manganese and also as the concentration of sulphur is increased.

(3) Low activity of oxygen in the metal. Since the transfer of sulphur from metal to slag is an interface reaction the relevant metal oxygen activity is that in equilibrium with the slag and this increases as the activity of the ferrous oxide in the slag is raised. Because of the apparent balance of the change in factors 1 and 3 with change in the ferrous oxide content, this factor is significant only with slags of very low oxygen potential.

(4) Low activity coefficient of the sulphur ions in the slag. The available evidence suggests that this is increased as the basicity of the slag is lowered, or as magnesia and lime are replaced by manganese and ferrous oxides.

14.4 The Kinetics of Sulphur Transfer between Slag and Metal

In Part 2 of this text evidence is presented to show that the equilibrium sulphur distribution is not attained in the blast furnace and that in some way this is related to the non-attainment of the equilibrium silicon distribution. The determination of the equilibrium condition is difficult in the open-hearth process, because an exact sulphur balance can be obtained from the slag and metal only after they are removed from the furnace; during melting and refining the balance is complicated by the continuous transfer of sulphur between the slag and the gas phase. However, the available evidence seems to suggest that the equilibrium sulphur distribution between the slag and the metal is approached closely soon after the start of refining in the open-hearth and the equilibrium is rapidly restored after any adjustment to the composition or the temperature of the bath [17, 25].

A large number of investigations have been designed to yield information which might assist in the interpretation of the mechanism of the sulphur transfer reaction under the conditions which are encountered in the blast furnace and explain the inherent slow rate of the reaction under these conditions. It is fairly well established that the transfer of sulphur between the metal and the slag is a first-order reaction [26–29], in which case it is adequately represented by an equation such as 14.4 and it does not involve the formation of complex molecular entities. In one investigation at 1,550°C the overall reaction rate is stated to be first order in basic slags and 0·75 in acid slags [30], but other reports suggest that the order of the overall reaction is temperature dependent and increases from 0·9 at 1,410°C to 1·3 at 1,750°C [31] or from unity at about 1,400°C to second order at 1,700°C [32]. In the last case the effect of temperature on the order of the reaction is expressed as follows:

$$-\frac{d[\%S]}{dt} = \kappa[\%S]^{(1+B)} \qquad 14.13$$

where $-d[\%S]/dt$ is the rate of sulphur transfer from metal to slag and κ is the rate constant. B is a 'turbulence factor' which increases from zero to unity as the temperature is raised and which is regarded as a correction to the increase in the reaction rate to allow for the decrease in the viscosity of the slag and the increased turbulence of the metal caused by carbon monoxide evolution as the temperature of the melt increases.

It has been reported that the rate of transfer of sulphur is directly proportional to the area of the slag–metal interface [31–33]. This relation suggests that the reaction rate is controlled by diffusion across a thin static film of metal or slag at the interface and it has commonly been assumed that the slow step is the diffusion of oxygen across a stagnant film on the metal side of the interface. Substitution of the experimental data into the boundary layer equations leads to the conclusion that the rate of sulphur transfer from metal to slag should be about equal to the rate of transfer of iron atoms in the same direction, but this equivalence is not supported by the experimental data [28, 34]. However, the concentration of iron oxides is very low in a blast furnace slag and it is probably more correct to consider the diffusion of oxygen ions across the boundary layer on the slag side of the interface [35], in which case this is not conclusive evidence against a diffusion-controlled mechanism. In fact when the silicon in the metal is adjusted to the concentration in equilibrium with the slag, the rate of the sulphur transfer reaction can be described satisfactorily in terms of the metal boundary-layer diffusion theory [36].

If the reaction is diffusion controlled it would normally be expected that the reaction rate is increased by agitation of the melt which increases the area of the slag–metal interface and reduces or eliminates diffusion gradients within both the slag and the metal. The available evidence suggests that the sulphur transfer rate is rapid in the steel-refining processes where the bath is stirred by the carbon boil, but when the metal is quiet, as for example during the reducing-slag stage of the electric-arc process, the reaction rate can be increased by induced agitation of the bath with electro-magnetic stirrers.

With normal blast furnace-type metal and slag the rate is not altered substantially by mechanical agitation [37, 38]. The rate has been shown to increase markedly when the metal is disseminated into small globules which are mixed intimately with the slag [39], but in this experiment the metal solidified within a few seconds after it entered the slag and it is not known whether or not some sulphur would revert to the metal if the liquid phases had been in contact for a longer period of time. Mechanical agitation does increase the transfer rate when the silicon content of the metal is in equilibrium with the slag [36]; this behaviour suggests that the sulphur reaction in the blast-furnace is diffusion-controlled only when the reduction of silicon has reached the equilibrium condition.

A number of alternative control mechanisms have been proposed from time to time, but the bulk of the available data point strongly to the fact that the slow step which controls sulphur transfer under blast furnace conditions is the reduction of the silica in the slag. Some of the experimental and factual evidence supporting this conclusion is reviewed in Chapter 7 and it is widely recognised that sulphur transfer is more rapid under a basic slag, for which the equilibrium silicon content of the metal is low, than under an acid slag. It is significant that with an acid slag the activation energy for sulphur transfer is equal to the activation energy for silicon transfer. This activation energy falls slowly with increasing basicity in silica-rich slags, but drops suddenly at about the meta-silicate composition [29] where the activity of the silica in the slag is also changing rapidly.

THE EFFECT OF FLUORIDE IN THE SLAG

It has long been recognised that the rate of sulphur removal from the metal and into the slag is raised when calcium fluoride is added to the slag [40, 41], but the equilibrium sulphur distribution ratio is apparently unchanged by the presence of the fluorine ions [15, 42]. In a basic slag it is probable that the beneficial effect of fluorspar additions is in the fluxing of undissolved lime, since solid lime in lump form makes negligible contribution to the sulphide capacity or the S–O ratio of the slag, but raises the viscosity and thus retards the transfer reaction. The last effect probably accounts for the optimum basicities which are occasionally reported in the literature for sulphur removal with open-hearth type slags which are not saturated with basic oxides. The addition of fluorspar to finely-powdered lime is reported to increase the rate of sulphur transfer from the metal in the ladle-injection techniques [43]. Similar improvement in the rate of transfer is obtained by adding small amounts of alumina (bauxite) to basic slags [44, 45] and this addition is preferable to fluorspar when the slag is sold as a basic fertiliser.

The rate of sulphur transfer is also raised when fluorspar is added to an acid slag [29], but it is improbable that this effect is due to either the fluxing of solid lime or to the increase in the basicity of the slag which is caused by small additions of fluorspar. A more logical explanation is that the single-charged fluoride ions separate the chain- or ring-type silicate complexes into smaller structural units, thereby lowering the viscosity of the melt and increasing the reaction rate.

14.5 Ionic Treatments of the Sulphur Partitioning

Although the factors which control the rate and the equilibrium conditions of the sulphur reaction are now fairly well understood, there are at present insufficient data to allow the calculation of the sulphur distribution between a metal and a slag of any chosen composition in terms of the equilibrium constant for equation 14.6. Some success has been obtained by fitting equations to the variation with composition of the S–O ratio, but in each case empirical terms are introduced and these treatments cannot be regarded as satisfactory. The exchange of sulphur between slag and metal is most readily discussed in terms of the ionic species in the slag and it is understandable, therefore, that attempts have been made to formulate the equilibrium relations in terms of the ionic theories of solutions.

One of the earliest attempts assumes completely random mixing of the anions and cations in the slag and the activities of the ions are assumed to be directly proportional to their ionic fractions, but the equilibrium 'constant' is found to be a function of the slag composition [42, 46]. For slags containing less than 10 weight-per cent of silica a more satisfactory correlation is obtained using the Temkin model (see Chapter 4.1) and it has been suggested that these slags behave ideally with respect to the exchange of sulphur and oxygen ions [47, 48]. This does not apply at higher silica concentrations and it is necessary to include the activity coefficients of the oxygen and sulphur ions.

The most successful ionic treatments to date have been obtained by application of the interaction-energy relations of Flood et al to sulphur transfer. The equilibrium constant is given by [49]:

$$\log K'_S = \log \frac{[\%S].(N_{O^{--}})}{[\%O].(N_{S^{--}})}$$
$$= N_{Ca^{++}}.\log K'_{Ca^{++}} + N_{Fe^{++}}.\log K'_{Fe^{++}} + + N_{Mn^{++}}.\log K'_{Mn^{++}} + N_{Mg^{++}}.\log K'_{Mg^{++}} + \dots \qquad 14.14$$

where the apparent constants on the right-hand side of the equation are those for each of the exchange reactions, i.e.,

$$K'_{Ca^{++}} = \frac{[\%S](CaO)}{[\%O](CaS)} \tag{14.15}$$

$$K'_{Fe^{++}} = \frac{[\%S](FeO)}{[\%O](FeS)} \quad \text{etc.} \tag{14.16}$$

and the values of these constants at 1,600°C can be obtained from equations 14.9 to 14.12. From the form of equation 14.14 it is evident that the calculated sulphur distribution is dependent not only on the basicity but also on the composition of the slag and this conforms to the experimental evidence. It has been shown that the calculated distributions compare favourably with those found experimentally for slags within a limited range of compositions and it is unfortunate that attempts have not been made to apply the equations to a wider range of compositions. However, the treatment is restricted to basic slags, since the calculated oxygen ion fraction is zero for compositions at, or more acid than, the orthosilicate join. No satisfactory ionic treatment has been derived so far for neutral or acid slags.

14.6 The Removal of Sulphur by Vacuum Treatment of the Metal

Since elemental sulphur exists as a gas at liquid steel temperatures, it is natural to enquire whether or not it can be removed from the metal along with the other gases, oxygen, hydrogen and nitrogen, during vacuum melting and/or pouring. The partial pressure of sulphur gas which is in equilibrium with sulphur in the concentrations normally found in steels is very low and is of similar magnitude to the corresponding partial pressure of oxygen gas in equilibrium with the dissolved oxygen. Thus there is little probability of removing sulphur as the elemental gas at pressures of the order of one micron, which are the lowest pressures normally attained in the large-scale vacuum processes. However, there is some experimental evidence to show that the sulphur content of the metal can be lowered by a vacuum treatment and the extent of the removal is increased as the concentrations in the metal of carbon, silicon and aluminium are increased [50, 51]. These elements raise the activity coefficient of sulphur and thus increase the 'escape tendency' for the removal of the element into either the slag or the gas phase.

The effect of carbon on the sulphur content can be demonstrated by reference to a reaction such as:

$$S_{(in\,Fe)} + C_{(in\,Fe)} + CaO_{(s)} \rightleftharpoons CaS_{(s)} + CO_{(g)} \tag{14.17}$$

which could occur for example when the metal is contained in a lime crucible or when powdered lime is injected into the melt. The equilibrium for this reaction is pressure dependent and, if the reaction products are pure solid calcium sulphide and carbon monoxide at a pressure of one micron, for a steel containing 0·25 weight-per cent of carbon the equilibrium sulphur content is 4×10^{-7} weight-per cent [52]. If the activity coefficients of the carbon and sulphur are assumed to be constant, then the sulphur concentration is halved when the carbon content is doubled, and so on. According to this calculation it is theoretically possible to lower considerably the residual sulphur content by a vacuum treatment, but in Chapter 13.7 it is shown that the effective partial pressure of carbon monoxide is considerably greater than the total pressure in the vacuum chamber. This means that the equilibrium condition is not attained in practice and in vacuum-treated, iron–carbon–sulphur melts the final sulphur content is about 1,000 times as great as the equilibrium value [52].

The effects of silicon and aluminium can be illustrated by examination of the reaction:

$$2S_{(\text{in Fe})} + Si_{(\text{in Fe})} + 3CaO_{(s)} \rightleftharpoons CaS_{(s)} + 2CaO\,.\,SiO_{2(s)} \qquad 14.18$$

If the solids are pure and hence at unit activity the equilibrium sulphur content is about 0·003 weight-per cent for a 0·3 weight-per cent silicon steel. This reaction is not pressure dependent and the equilibrium is not affected by the partial pressure of oxygen in the atmosphere, but it is found experimentally that the lime is fluxed by the iron oxides which are formed when melting in air and the calculated sulphur content is obtained only when the metal is exposed to vacuum or to an oxygen-free atmosphere [52]. The free energy of formation of the calcium aluminates is less negative than for the corresponding calcium silicates. Thus it is expected that the equilibrium sulphur content is increased when silicon is replaced by aluminium in equation 14.18 and it has been shown experimentally that, weight for weight, silicon is more effective than aluminium [51], but an alternative explanation of this effect is also possible. Marked loss of silicon and sulphur has been noticed from melts containing these elements in an oxygen-free atmosphere and it has been suggested that a volatile silicon sulphide SiS or SiS_2 is formed under these conditions [53]. No evidence for the volatilisation of an aluminium sulphide at these temperatures has so far been reported.

Those elements which have the highest affinity for sulphur such as calcium [54] and cerium [55], have rather high vapour pressures and are not so effective desulphurisers in vacuum as in air, but these elements combine with oxygen in preference to sulphur; hence there may be an advantage in vacuum deoxidising and then adding these elements to the metal under a non-oxidising atmosphere.

There seems to be little probability of any significant change in the sulphur content of the metal during vacuum-casting treatments and, except for high silicon or high carbon steels, the residual sulphur contents obtained with the ladle injection techniques are at least as low as those which have been obtained so far with vacuum melting.

REFERENCES

1. K. G. SPEITH, H. VOM ENDE, F. BARDENHAUER and G. MAHN, *Stahl u. Eisen, Düsseldorf*, 1957, Vol. 79, p. 926.
2. K. M. GOLDMAN, G. DERGE and W. O. PHILBROOK, *Trans. Amer. Inst. min. (metall.) Engrs.*, 1954, Vol. 200, p. 534.
3. H. L. BISHOP, N. J. GRANT and J. CHIPMAN, ibid., 1958, Vol. 212, p. 185.
4. J. CHIPMAN, *Physical Chemistry of Process Metallurgy*, Interscience, 1961, p. 27.
5. G. G. HATCH and J. CHIPMAN, *Trans. Amer. Inst. min. (metall.) Engrs.*, 1946, Vol. 167, p. 134.
6. C. W. SHERMAN, H. I. ELVANDER and J. CHIPMAN, ibid., 1950, Vol. 188, p. 334.
7. C. B. ALCOCK and L. L. CHENG, *J. Iron St. Inst.*, 1960, Vol. 195, p. 169.
8. P. BARDENHEUER and W. GELLER, *Mitt. K.-Wilh. Inst. Eisenforsch.*, 1934, Vol. 16, p. 77.
9. W. A. FISCHER and H. SPITZER, *Arch. Eisenhüttenw.*, 1958, Vol. 29, p. 611.
10. G. R. ST PIERRE and J. CHIPMAN, *Trans. Amer. Inst. min. (metall.) Engrs.*, 1956, Vol. 206, p. 1474.
11. R. ROCCA, N. J. GRANT and J. CHIPMAN, ibid., 1951, Vol. 191, p. 319.
12. F. HARDERS, H. GREWE and W. OELSEN, *Stahl u. Eisen, Düsseldorf*, 1951, Vol. 71, p. 973.
13. H. FLOOD, T. FORLAND and K. GRJOTHEIM, *Physical Chemistry of Melts*, London, Inst. Min. Met., 1953, p. 46.
14. E. W. DEWING and F. D. RICHARDSON, *Trans. Faraday Soc.*, 1958, Vol. 54, p. 679; *J. Iron St. Inst.*, 1960, Vol. 194, p. 446; 1960, Vol. 195, p. 56.
15. N. J. GRANT and J. CHIPMAN, *Trans. Amer. Inst. min. (metall.) Engrs.*, 1946, Vol. 167, p. 134.
16. P. T. CARTER, *Disc. Faraday Soc.*, 1948, No. 4, p. 307.
17. E. T. TURKDOGAN, *J. Iron St. Inst.*, 1955, Vol. 179, p. 147.
18. H. L. BISHOP, H. N. LANDER, N. J. GRANT and J. CHIPMAN, *Trans. Amer. Inst. min. (metall.) Engrs.*, 1956, Vol. 206, p. 862.
19. G. G. HATCH and J. CHIPMAN, *Trans. Amer. Inst. min. (metall.) Engrs.*, 1949, Vol. 185, p. 274.
20. K. P. ABRAHAM and F. D. RICHARDSON, *J. Iron St. Inst.*, 1960, Vol. 196, pp. 309, 313.
21. L. A. SHVARTSMANN, *Hutn. List.*, 1957, Vol. 12, p. 965; 'Radio-isotopes in Scientific Research' *Proc. First UNESCO Internat. Conf. 1957*, Paris, 1958.

22. R. E. BONI and G. DERGE, *Trans. Amer. Inst. min. (metall.) Engrs.*, 1956, Vol. 206, p. 59.
23. F. D. RICHARDSON and C. J. B. FINCHAM, *J. Iron St. Inst.*, 1954, Vol. 178, p 4.
24. C. E. A. SHANAHAN, *Iron & Steel, Lond.*, 1950, Vol. 23, p. 313.
25. B. TRENTINI, *J. Iron St. Inst.*, 1957, Vol. 185, p. 522.
26. G. DERGE and W. O. PHILBROOKE, *Rev. Met.*, 1955, Vol. 52, p. 995.
27. T. SAITO and Y. KAWAI, *Sci. Reps. Res. Inst. Tohoku Univ.*, 1955, Series A, Vol. 7, p. 402.
28. S. RAMACHANDRAN, T. B. KING and N. J. GRANT, *Trans. Amer. Inst. min. (metall.) Engrs.*, 1956, Vol. 206, p. 1549.
29. R. G. WARD and K. A. SALMON, *J. Iron St. Inst.*, 1960, Vol. 196, p. 393.
30. O. A. ESIN and V. N. SHIKHOV, *Izv. Ak. Nauk Otdelenie Tekn.*, 1955, (2), p. 105.
31. J. SKALA and O. V. TRAVIN, *Hutn. List.*, 1957, Vol. 12, p. 1000.
32. O. V. TRAVIN and L. A. SHVARTSMANN, *Sessiya Akad Nauk*, 1955, p. 48; *Zhur. Fizi. Khimii*, 1955, Vol. 29, p. 2031.
33. G. DERGE, W. O. PHILBROOKE and K. M. GOLDMAN, *Trans. Amer. Inst. min. (metall.) Engrs.*, 1950, Vol. 188, p. 1111.
34. T. B. KING and S. RAMACHANDRAN, *Physical Chemistry of Steel-making*, New York, John Wiley, 1958, p. 125.
35. L. S. DARKEN, ibid., p. 133.
36. J. C. FULTON and J. CHIPMAN, ibid., p. 113.
37. G. G. HATCH and J. CHIPMAN, *Trans. Amer. Inst. min. (metall.) Engrs.*, 1949, Vol. 185, p. 249.
38. N. J. GRANT, B. KALLING and J. CHIPMAN, ibid., 1951, Vol. 191, p. 666.
39. C. E. A. SHANAHAN, *Physical Chemistry of Steel-making*, New York, John Wiley, 1958, p. 136.
40. T. P. COLCLOUGH, *J. Iron St. Inst.*, 1923, Vol. 107, p. 267.
41. W. B. LAURIE, ibid., 1939, Vol. 139, p. 257.
42. P. HERASYMENKO and G. SPEIGHT, ibid., 1950, Vol. 166, p. 169.
43. W. A. FISCHER and H. SPITZER, *Arch. Eisenhüttenw.*, 1958, Vol. 29, p. 535.
44. L. COOK, *Blast Furn.*, 1949, Vol. 37, p. 1067.
45. N. I. BEDA and N. K. KOTOV, *Stal'*, 1959, (7), p. 616.
46. P. HERASYMENKO and G. SPEIGHT, *J. Iron St. Inst.*, 1950, Vol. 166, p. 289.
47. A. M. SAMARIN, L. A. SHVARTSMANN and M. TEMKIN, *Acta Physicochim., URSS*, 1945, Vol. 20, p. 421.
48. A. M. SAMARIN and L. A. SHVARTSMANN, *Izv. Ak. Nauk, S.S.S.R.*, 1948, (9), p. 1457.
49. H. FLOOD and K. GRJOTHEIM, *J. Iron St. Inst.*, 1952, Vol. 171, p. 64.
50. G. A. GARNYK and A. M. SAMARIN, *Izv. Ak. Nauk Otdelenie Tekn.*, 1957, (5), p. 77.
51. T. P. FLORIDIS, *Trans. Amer. Inst. min. (metall.) Engrs.*, 1959, Vol. 215, p. 870.
52. R. G. WARD and R. HALL, *J. Iron St. Inst.*, 1960, Vol. 195, p. 75.
53. C. B. ALCOCK and F. D. RICHARDSON, *Acta Met.*, 1958, Vol. 6, p. 385.
54. J. CHIPMAN and TA LI, *Trans. Amer. Soc. Metals*, 1937, Vol. 25, p. 435.
55. F. C. LANGENBERG and J. CHIPMAN, *Trans. Amer. Inst. min. (metall.) Engrs.*, 1958, Vol. 212, p. 290.

CHAPTER 15

The Removal of Phosphorus from the Metal

15.1 Equilibrium Conditions

In contrast to the detailed and systematic investigations into the distribution between slag and metal of the elements which have been considered in previous chapters, the partitioning of phosphorus has received little attention. This is due largely to the difficulty in devising a suitable technique for measuring the activity of phosphorus in the slag and in the metal. The activities of oxygen, carbon, sulphur and some other alloy elements which may be present in the refining bath can be determined by equilibration of the melt with a suitable gas mixture, but no similar technique in which phosphorus is a constituent of the gas mixture has so far proved suitable for use in the range of temperatures and oxygen potentials encountered in the extraction and refining processes. Thermodynamic progress has also been hampered by the continued adoption of the traditional approach in which the phosphorus is assumed to be present in the slag as undissociated calcium phosphate compounds. These compounds may form in a liquid slag which is saturated with lime, but saturated slags are deliberately avoided in most practices and, except in the basic converter practice, it is extremely unlikely that the compounds are present in the slag before it solidifies. It is well recognised that the transfer of phosphorus into the slag is facilitated by a high basicity, a high oxygen potential and a low bath temperature but, with the limited data at present available, it is only possible to interrelate these variables by empirical equations.

The product of the oxidation of phosphorus in iron is the oxide P_2O_5 which is a gas at iron and steel-making temperatures. The free energy of formation of this oxide from phosphorus and oxygen in iron, relative to the dilute weight-per cent standard state, is obtained as follows.

For the reaction:

$$2P_{(g)} + 2\tfrac{1}{2}O_{2(g)} \rightleftharpoons P_2O_{5(l)} \qquad 15.1$$

the free-energy change is calculated from data for standard enthalpies, standard entropies and heat capacities [1] as being:

$$\Delta G^0_{(1843°K)} = -361{,}625 + 125{\cdot}58T \text{ cal mol}^{-1} \qquad 15.2$$

The heat capacity of liquid and gaseous P_2O_5 is unknown, but assuming no change from that of the solid and adding the free energy of vaporisation as given by:

$$P_2O_{5(l)} \rightleftharpoons P_2O_{5(g)} \tag{15.3}$$

$$\Delta G^0 = 13{,}000 - 14{\cdot}89T \text{ cal mol}^{-1} \tag{15.4}$$

we obtain for the free energy of formation of P_2O_5 gas:

$$P_{2(g)} + 2\tfrac{1}{2}O_{2(g)} \rightleftharpoons P_2O_{5(g)} \tag{15.5}$$

$$\Delta G^0 = -348{,}625 + 110{\cdot}69T \text{ cal mol}^{-1} \tag{15.6}$$

The activity of phosphorus in liquid iron has been determined by partitioning between liquid iron and liquid silver [2] and shows a positive deviation from ideality which is very large at high concentration. At 1,600°C the activity coefficient increases from 1·90 at 0·117 mole fraction to 4·67 at 0·177 mole fraction of phosphorus. The experimental technique does not yield accurate values for the activities in dilute solution in liquid iron but by comparison with other investigations [3, 4] it is concluded that, for concentrations up to at least 1 weight-per cent, phosphorus in liquid iron obeys Henry's law and the activities are close to ideal. Assuming ideal behaviour, the free energy of solution of phosphorus in iron relative to the dilute weight-per cent standard state is obtained by use of equation 2.30.

$$2P_{(\%\,\text{in Fe})} \rightleftharpoons P_{2(g)} \tag{15.7}$$

$$\Delta G^0 = 58{,}500 + 9{\cdot}1T \text{ cal mol}^{-1} \text{ [3]} \tag{15.8}$$

Combining equations 15.6 and 15.8 with equation 13.60 for the free-energy change accompanying the solution of oxygen in iron gives the standard free energy for the reaction:

$$2P_{(\%\,\text{in Fe})} + 5O_{(\%\,\text{in Fe})} \rightleftharpoons P_2O_{5(g)} \tag{15.9}$$

as

$$\Delta G^0 = -151{,}175 + 123{\cdot}74T \text{ cal mol}^{-1} \tag{15.10}$$

and

$$\log K_P = \log \frac{p_{P_2O_5}}{[a_P]^2[a_O]^5} = \frac{33{,}050}{T} - 27{\cdot}00 \tag{15.11}$$

At 1,600°C the equilibrium constant is $4{\cdot}47 \times 10^{-10}$. This value is approximate but, even if it is in error by several orders of magnitude, it is evident that the partial pressure of phosphorus pentoxide gas in equilibrium with the dissolved oxygen and phosphorus is very low and for all practical purposes the removal of phosphorus from the metal as a gas can be neglected.

The free energy of formation of liquid phosphorus pentoxide dissolved in the slag is obtained similarly by combining equations 15.2, 15.8 and 13.60 to obtain:

$$2P_{(\%\,\text{in Fe})} + 5O_{(\%\,\text{in Fe})} \rightleftharpoons P_2O_{5(l)} \qquad 15.12$$

$$\Delta G^0 = -164{,}175 + 138{\cdot}63T \text{ cal mol}^{-1} \qquad 15.13$$

and $$\log K_P = \log \frac{(a_{P_2O_5})}{[a_P]^2[a_O]^5} = \frac{35{,}880}{T} - 30{\cdot}31 \qquad 15.14$$

The equilibrium constant is equal to $7{\cdot}08 \times 10^{-12}$ at 1,600°C and $7{\cdot}6 \times 10^{-13}$ at 1,700°C. These values are again approximate, but they confirm that the concentration of phosphorus in the slag is negligible unless the activity of the pentoxide is extremely small and numerically is less than the partial pressure of the pentoxide gas in the equilibrium atmosphere. If the ratio $(a_{P_2O_5})/[a_O]^5$ is approximately independent of the temperature then from equation 15.14 the second power of the activity of the phosphorus in the metal is raised approximately tenfold as the temperature is increased by 100°C. A temperature effect of this order of magnitude is commonly found in practice; for example the residual phosphorus content in equilibrium with a normal basic Bessemer slag is increased from 0·01 weight-per cent in the temperature range 1,530 to 1,570°C to 0·03 weight-per cent at 1,630 to 1,700°C [5].

The metal–oxygen activity in the equilibrium relation is that for equilibrium with the slag since, in common with all the oxidation reactions except those for carbon, phosphorus is oxidised mainly at the slag–metal interface. It is therefore expected that the rate of transfer of phosphorus from metal to slag is controlled by diffusion across an interface boundary layer and, in support of this contention, it has been found that the rate is directly proportional to the area of the interface but is independent of the thickness of the slag layer [6].

THE EFFECTS OF OXYGEN POTENTIAL, TEMPERATURE AND SLAG BASICITY

The values of the equilibrium constant for the phosphorus reaction provide a clear explanation why the phosphorus contents of blast furnace slags are negligible. The metal oxygen activity of a blast furnace metal is very low and a rough calculation shows that in the relevant temperature range the activity coefficient of the pentoxide in the slag must be very much less than 10^{-20} if a significant amount of phosphorus is to enter the slag. A pictorial representation of this has been devised by Goodeve [7], who compared the oxygen potentials for each of the principal iron- and steel-making processes with those for the formation of calcium phosphate at 1 per cent concentration of P_2O_5 in the slag from a 1 per cent concentration of phosphorus in the metal. The diagram is reproduced in Figure 15.1 where the broken line represents the oxygen potentials in equilibrium with phosphorus pentoxide at this particular activity in the slag. The line is raised to higher oxygen potentials if the activity of the pentoxide is increased, and vice versa, so for phosphorus transfer into the slag the activity of the pentoxide must be very much

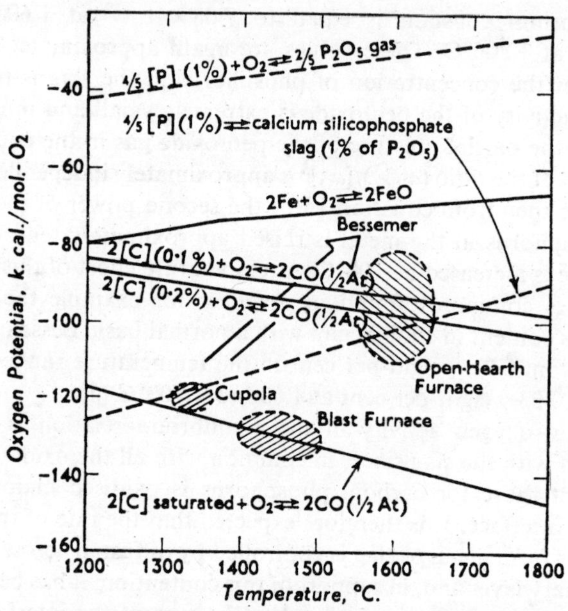

FIG. 15.1. Comparison of the oxygen potentials at which phosphorus can transfer from metal to slag with the oxygen potentials of various iron and steel-making processes. (After Goodeve [7].)

smaller in a low oxygen potential process, such as the blast furnace, than in the higher oxygen potential refining processes.

The diagram can be used also to make other qualitative predictions. Since the oxygen potentials for the formation of pure silica are very much lower than for pure phosphorus pentoxide, at equal mole fractions of these oxides in a basic slag the activity of the phosphorus pentoxide must be very much smaller than the activity of the silica. The oxygen potential lines for the oxidation of carbon and phosphorus intersect at a temperature which depends on the activities of the species in the metal and in the slag, and phosphorus may be oxidised in preference to carbon at all temperatures below this intersection, whilst the converse applies at higher temperatures. Since phosphorus exists in the slag as phosphorus pentoxide, or more correctly as the oxy-acid radical PO_4^{3-}, it is evident that the activity is lowered as the basicity of the slag is increased, with the result that the oxygen potential line for phosphorus oxidation is lowered and the temperature range over which phosphorus is oxidised in preference to carbon is increased. This feature is demonstrated clearly in the various basic converter processes. In the traditional Bessemer process the solution of lime in the slag is comparatively slow when the air blast is introduced from the bottom of the bath, the activity coefficient of the phosphorus pentoxide is high and does not fall to the level at which a significant amount of phosphorus can transfer into the slag until the carbon blow is almost completed. With the top-blown processes the lime is fluxed more rapidly into the slag and phosphorus can be removed prior to, or along with, the carbon depending on the composition of the slag and the temperature of the bath. Similarly, the relative amounts of carbon and phosphorus which are removed from the metal with a flush-slag process in the open-hearth or Rotor processes are critically dependent on the slag basicity and on the temperature to which the metal is heated before the first slag is removed.

15.2 The Activities of Phosphorus in the Metal and in the Slag

It is stated earlier in this chapter that phosphorus obeys Henry's law within the range of concentrations normally found in steels. The activity coefficient of phosphorus in liquid iron is raised by oxygen, nitrogen, silicon and sulphur (see Appendix 1), but the effects of other solutes at low concentration have not been investigated. The activity coefficient is raised by carbon in graphite-saturated solutions [8] and it is probable that dilute solutions of carbon in iron have a similar effect. Estimated values for carbon and some other solutes are given in the Appendix.

The activities of the phosphorus in the slag have not been examined experimentally, but it is evident that at least in basic slags the activity coefficient is very much less than unity. The phase diagram for the ternary system FeO–CaO–P_2O_5 shows an extensive region of liquid immiscibility [9] where the molten slag separates into two layers, one of which is rich in ferrous oxide and the other is rich in lime and phosphorus pentoxide, and it follows from this that the activities of the oxides are far from ideal in the ternary melts. The immiscibility region is reduced when silica is added to the melts and disappears when about 15 per cent of silica is present [10, 11] but, in view of the marked deviations from ideality in the simple melts, it is expected that the activity of the pentoxide is also less than ideal in the more complex steel-making slags.

In the previous chapter (see p. 428) it is suggested that the effect of the common basic oxides on the activity coefficient of the sulphide ions in the slag varies in the same order as the free energies of formation of their sulphide compounds and this applies also to the relation between the activities of silica in the binary silicate melts and the free energies of formation of the appropriate silicate compounds. Assuming that a similar relationship holds with phosphorus, the relative effectiveness of the basic oxides in lowering the activity coefficient of the pentoxide can be obtained from comparison of the free energies of formation of the corresponding phosphate compounds. By combining the data for the free energies of formation of these phosphates from gaseous phosphorus and oxygen [12–14] with equations 13.60 and 15.8 for the free energies of solution of oxygen and phosphorus in liquid iron, relative to the dilute weight-per cent standard state, we obtain the following relations:

$$3CaO_{(s)} + 2P_{(\%\,in\,Fe)} + 5O_{(\%\,in\,Fe)} \rightleftharpoons 3CaO.P_2O_{5(s)} \qquad 15.15$$

$$\Delta G^0 = -355{,}900 + 127T \text{ cal mol}^{-1} \qquad 15.16$$

and

$$\Delta G^0_{(1{,}873^\circ K)} = -119 \text{ kcal mol}^{-1}$$

$$3MgO_{(s)} + 2P_{(\%\,in\,Fe)} + 5O_{(\%\,in\,Fe)} \rightleftharpoons 3MgO.P_2O_{5(s)} \qquad 15.17$$

$$\Delta G^0 = -304{,}160 + 131T \text{ cal mol}^{-1} \qquad 15.18$$

and

$$\Delta G^0_{(1{,}873^\circ K)} = -59 \text{ kcal mol}^{-1}$$

$$3MnO_{(s)} + 2P_{(\%\,in\,Fe)} + 5O_{(\%\,in\,Fe)} \rightleftharpoons 3MnO.P_2O_{5(s)} \qquad 15.19$$

$$\Delta G^0 = -298{,}450 + 143T \text{ cal mol}^{-1} \qquad 15.20$$

and

$$\Delta G^0_{(1{,}873^\circ K)} = -31 \text{ kcal mol}^{-1}$$

The stability of the calcium phosphate is considerably greater than that of either the magnesium or the manganese phosphates and it is therefore suggested that the activities of phosphorus in the slag are lowered to a greater extent by lime than by magnesia or manganese oxide. The equations are not strictly applicable to steel-making conditions, since they apply to the formation of the solid phosphates from the solid basic oxides, whereas the liquid states are more appropriate to the refining slags, but this fact does not alter the relative order of effectiveness of the oxides. The free energies of formation of the ferrous phosphates have not been measured, but from other considerations it is probable that the effect of ferrous oxide on the activities of phosphorus in the slag is similar to the effect of manganese oxide.

The free energies of formation of these phosphate compounds are roughly 3 times as great (i.e., more negative) as those for the formation of the corresponding ortho-silicate compounds, hence at constant basicity the activities of phosphorus pentoxide are probably lower than the activities of silica. At 1,600°C the activity coefficient of silica varies from about 1·3 for compositions typical of acid refining slags to between 10^{-2} and 10^{-4} for basic refining slags [15]. The activity coefficient of phosphorus pentoxide is probably also close to unity for acid melts, the actual value depending on the extent of the liquid immiscibility in the phosphate-rich solutions, but in the basic melts the activity coefficients should be lower than the coefficients for silica by several orders of magnitude.

Generally, the standard enthalpies and free energies of formation are more negative for the phosphates, silicates and similar compounds of the alkali–metal oxides Na_2O and K_2O than for the corresponding compounds with the alkaline-earth oxides such as lime and magnesia. For the reaction:

$$3CaO.P_2O_5+3Na_2O \rightleftharpoons 3Na_2O.P_2O_5+3CaO \qquad 15.21$$

the heat of reaction is negative and is about 80 kcal per mole, which shows that calcium can be displaced from the phosphate by sodium oxide. Hence it is expected that the activity coefficient of phosphorus pentoxide is lowered when sodium or potassium oxide is added to the slag, while the basicity is held constant. This contention is supported by the fact that soda-ash treatments are used for ladle dephosphorisation.

The distribution of phosphorus between liquid iron and slags consisting of ferrous oxide, sodium oxide and silica has been investigated for a wide range of slag compositions [16] and it has been shown that more than 90 per cent of the phosphorus in the metal can be transferred to the slag when the silica content is less than 25 per cent. The distribution ratio is lowered as the silica content of the slag is increased, but between 50 and 80 per cent of the phosphorus is removed when the slag contains 35 to 40 per cent of silica and between 15 and 40 per cent of the phosphorus can be transferred into a slag containing 50 to 60 per cent of silica (Figure 15.2). This is considerably better than the extent to which the phosphorus is removed by normal steel-making slags and particularly with the more acid compositions. It is possible that part of the improvement with the acid slags is caused by an increase in the activity coefficient of the ferrous oxide, which would increase the equilibrium metal oxygen activity. It is improbable, however, that this accounts for the improved performance with the more basic compositions and the most satisfactory explanation is that the activity coefficient of the phosphorus pentoxide is lowered more markedly by sodium oxide than by lime.

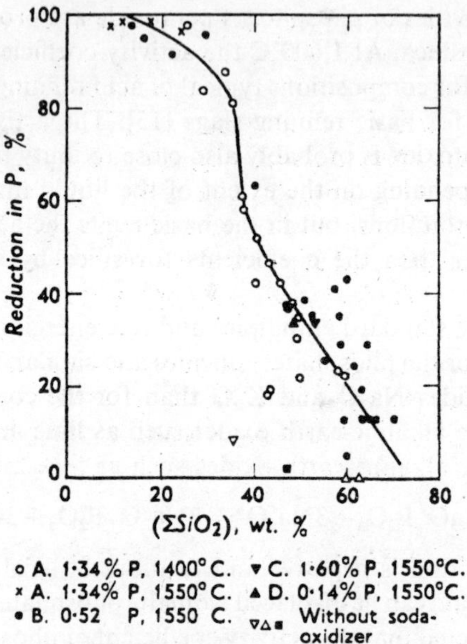

FIG. 15.2. Relation between the reduction in the phosphorus content of the metal and the silica content of soda-slags. (After Turkdogan and Maddocks [16].)

Unfortunately this information is of limited value to the steel-maker, for at high temperatures sodium oxide is reduced by molten iron to form sodium metal vapour:

$$Na_2O_{(l)} + Fe_{(l)} \rightleftharpoons FeO_{(l)} + 2Na_{(g)} \qquad 15.22$$

and the equilibrium moves further to the right as the temperature is raised [17]. The equilibrium constant for this reaction is given by:

$$K_{Na} = \frac{(a_{FeO})\, p_{Na}}{(a_{Na_2O})} \qquad 15.23$$

which shows that the loss of sodium as a vapour is increased as either the activity of the ferrous oxide is lowered or the activity of sodium oxide is raised. The activities of sodium oxide in these melts are unknown, but the activities calculated from the liquidus curves for the binary system Na_2O–SiO_2, assuming regular-solution behaviour [18], are very low for silica-rich melts. Thus as the basicity is increased the activity of sodium oxide must change more rapidly with composition in the binary system than the corresponding change in the activity of ferrous oxide in binary ferrous silicate melts. If similar behaviour holds for the more complex slags, then the loss of sodium should be comparatively small from acid slags, but should increase rapidly as the basicity is raised beyond about the ortho-silicate composition, and this is confirmed by observations in practice. The volatilised sodium metal is re-oxidised in the furnace atmosphere and the gaseous oxide is a very severe flux for any acid-refractory lining with which it comes into contact. Thus apart from the ladle treatments, it appears that the most promising field for the application of soda slags is for the removal of small amounts of phosphorus in the acid steel-making processes, or in a hot metal mixer using only mildly acid slags and operating at a comparatively low temperature.

15.3 Ionic and Empirical Treatments of the Phosphorus Partitioning

A true thermodynamic treatment of the phosphorus distribution between the slag and the metal is not possible at present because of the non-existence of reliable free energy and activity data for the reaction. It is therefore useful to examine some of the treatments in which the equilibrium is expressed in terms of the ionic theories of liquid slags. In most of these treatments the reaction is written as:

$$2P_{(\text{in Fe})} + 5O_{(\text{in Fe})} + 3O^{--}_{(\text{slag})} \rightleftharpoons 2PO_4^{3-}{}_{(\text{slag})} \qquad 15.24$$

and the equilibrium constant is expressed in terms of the ion fractions for the species in the slag and in weight-per cent or mole fractions for the solutes in the metal:

$$K'_P = \frac{(PO_4^{3-})^2}{[\%P]^2[\%O]^5(O^{--})^3} \qquad 15.25$$

Assuming that at low concentrations the activities of phosphorus and oxygen in the metal vary linearly with their concentrations, the apparent constant K'_P for the reaction is a function only of the temperature and the composition of the slag.

Flood et al. [19] adopted a procedure similar to that described in the previous chapter (see p. 440) for determining the sulphur distribution; that is, the value of K'_P is calculated from the relation:

$$\log K'_P = \sum N_{M^{2+}} \log K_{M^{2+}} \qquad 15.26$$

where $N_{M^{2+}}$ is the ionic fraction of a basic oxide cation and $K_{M^{2+}}$ is the equilibrium constant for reaction between phosphorus and the cation of the type:

$$2P_{(\text{in Fe})}+5O_{(\text{in Fe})}+3CaO_{(\text{slag})} \rightleftharpoons 3CaO.P_2O_{5(\text{slag})}$$

$$2P_{(\text{in Fe})}+5O_{(\text{in Fe})}+3MgO_{(\text{slag})} \rightleftharpoons 3MgO.P_2O_{5(\text{slag})}$$

and so on.

At the time this treatment was derived the values of these equilibrium constants had not been determined for liquid steel temperatures; hence values were obtained indirectly by fitting the equation to experimental data for the phosphorus distribution under simple slags and the following relation was proposed:

$$\log K_P' = 21N_{Ca^{2+}}+18N_{Mg^{2+}}+13N_{Mn^{2+}}+12N_{Fe^{2+}} \qquad 15.27$$

Comparing these constants with the experimental values which are given in equations 15.16, 15.18 and 15.20, it is found that lime, magnesia and manganese oxide are placed in the correct order by the relation, but the magnitude of the variation among the values of the constants differs considerably from those obtained experimentally. There have been a few reports of the application of this treatment to slags covering a narrow range of compositions, but no report has been made as to whether the fit is good or otherwise for the entire range of basic slag compositions.

A similar approach was later made by Turkdogan and Pearson [20] who related the equilibrium constant to the mole fraction of phosphorus pentoxide in the slag and derived data for the reaction:

$$2P_{(\text{in Fe})}+5O_{(\text{in Fe})} \rightleftharpoons P_2O_{5(\text{slag})}$$

$$K_P' = \frac{(\gamma_{P_2O_5})(N_{P_2O_5})}{[\%P]^2[\%O]^5} = \gamma_{P_2O_5}K_P'' \qquad 15.28$$

and

$$\log \gamma_{P_2O_5} = -\Sigma N_{MO}\log K_{MO}+\log K_P' \qquad 15.29$$

where N_{MO} is the mole fraction of each basic oxide in the slag. The constants for the interchange reactions differ only slightly from those given previously:

$$\log K_P'' = 22N_{CaO}+15N_{MgO}+13N_{MnO}+12N_{FeO}-2N_{SiO_2} \qquad 15.30$$

the main difference being the lower value for the magnesia constant, which brings the series more closely into line with the experimental values for the constants. The negative term for silica in equation 15.30 is introduced empirically to allow for the association between the basic oxide cations and the silicate anions, since this association must increase the activities of the phosphorus pentoxide in the slag. (In the ionic equations this association is allowed for in the calculation of the oxygen ion fraction.)

Using this relation the activity coefficients of phosphorus pentoxide, as calculated from actual distribution data for basic steel-making slags covering a wide range of compositions, vary from 10^{-14} to 10^{-18} and are very considerably lower than the values for the activity coefficient of silica at similar basicities. When the values of the activity coefficient at constant

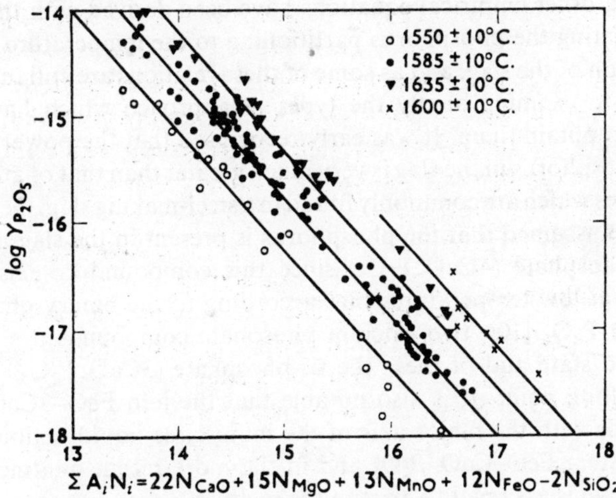

FIG. 15.3. The effect of slag composition and temperature on the activity coefficients of phosphorus pentoxide in open-hearth-type slags. (After Turkdogan and Pearson [20].)

temperature are plotted against the corresponding values of $\log K_p''$ the points are distributed with only a small amount of scatter about a line with a negative slope of 1.12 (Figure 15.3). The distribution of the points shows that the calculated activity coefficients are temperature dependent and the relation is given by:

$$\log \gamma_{P_2O_5} = -1{\cdot}12 \sum N_{MO} \log K_{MO} - \frac{4{,}200}{T} + 23{\cdot}58 \qquad 15.31$$

This temperature dependence could be real, since it indicates the expected decrease in the deviation from ideality as the temperature is raised, but it may arise partly from the effect of temperature on the values of the K_{MO} constants.

Since the constants in equation 15.31 are obtained by fitting the equation to the distribution data obtained with open-hearth type slags, it is reasonable to assume that the treatment can be used to predict the equilibrium phosphorus distribution for any slag within the range of compositions and temperatures from which the relations are derived. However, the constants are determined empirically and there is no *a priori* reason for assuming that the treatment holds for conditions which are outside these ranges.

From experimental studies, or from the combination of these with plant data, other empirical equations have been derived with the intention of relating the phosphorus partitioning to the temperature and the composition of the slag, and as some of these relations are still in use it is advisable to examine briefly the types of approach which have been adopted to obtain them. It was early recognised that the power of lime to hold phosphorus in the slag is very much greater than that of any of the other oxides which are commonly found in a steel-making slag. Generally, it has been assumed that the phosphorus is present in the slag as tetracalcium phosphate ($4CaO.P_2O_5$) since this compound co-exists with solid lime at low temperatures but, according to the binary phase diagram $CaO–P_2O_5$ [10], two calcium phosphate compounds are formed in the solid state and of these the tri-phosphate ($3CaO.P_2O_5$) has the higher melting point. It is also notable that the join $FeO–3CaO.P_2O_5$ corresponds with the major axis of the immiscible liquid region of the ternary system $FeO–CaO–P_2O_5$ and in the more recent treatments the tri-phosphate is assumed to exist in the melts.

A selection of the formulae which were proposed prior to 1940 for evaluating the phosphorus distribution has been examined by comparison with the actual distributions obtained in a number of open-hearth plants [21], and it is concluded that the relation derived by Schenck and Reiss [22] gives the most satisfactory correlation. This relation is expressed as:

$$4CaO.P_2O_{5(slag)} + 5Fe_{(l)} \rightleftharpoons 2P_{(in\,Fe)} + 5FeO_{(slag)} + 4CaO_{(slag)} \qquad 15.32$$

$$\log K'_P = \log\left(\frac{[\%P](FeO')^5(CaO')^4}{(\Sigma\, P_2O_5)}\right) + 0{\cdot}060\,(\Sigma\, P_2O_5)$$

$$= -\frac{51{,}800}{T} + 35{\cdot}05 \qquad 15.33$$

where $\Sigma\, P_2O_5$ signifies the total concentration of the pentoxide, in whatever form it is combined, in the slag. A similar relation has been derived in terms of the tri-phosphate. The primes denote the 'free' concentrations of ferrous oxide and lime, and these are obtained from a series of diagrams which interrelate the effects of temperature and the concentrations of various species in the slag on the concentrations of the oxides which are 'available' to take part in the reaction. With this adjustment it is claimed that K'_P is dependent only on the temperature and is independent of the compositions of the metal and the slag [22].

Equation 15.32 requires that the concentration of phosphorus is raised to the second power in the apparent equilibrium constant, but the power is unity in equation 15.33. It is found necessary to do this in order to obtain a fit with plant data and the second term, which includes the total concentration of the pentoxide in the slag, is also introduced empirically to allow for the 'non-ideal behaviour' of the calcium phosphates in the slag.

A different method for calculating the 'free' lime concentration has been adopted by Winkler and Chipman [23] who estimate this as the difference between the total mole fraction of lime in the slag and the quantity of lime which is combined to form the compounds $4CaO.2SiO_2$, $4CaO.P_2O_5$, $CaO.Fe_2O_3$, $2CaO.Al_2O_3$ and $4CaO.CaF_2.P_2O_5$. For this purpose the oxides of manganese and magnesium are assumed to be equivalent at low concentrations to an equal molar concentration of lime. However, a difficulty is encountered with this approach in explaining the transfer of phosphorus into neutral or slightly acid slags where, according to the definition, the 'free' lime concentration is zero, and it is found necessary to introduce into the computation a dissociation constant for the reaction:

$$4CaO.2SiO_2 \rightleftharpoons 2(CaO.SiO_2)+2CaO$$

It is further assumed that the activity of ferrous oxide is equal to its mole fraction when the slag constitution is determined in this way and that at low concentrations the activity of the phosphorus in liquid iron is proportional to the weight-per cent of phosphorus. The apparent equilibrium constant for equation 15.32 is then expressed as:

$$\log K'_P = \log \frac{[\%P]^2 (N_{FeO})^5 (N'_{CaO})^4}{(N_{4CaO.P_2O_5})}$$

$$= -\frac{40{,}067}{T} + 15{\cdot}06 \qquad 15.34$$

and the relation has been shown to give reasonably close agreement with the phosphorus distributions which are actually obtained in open-hearth practice [24].

A similar approach has been adopted by Balajiva et al. [25–27] who show a linear relation between the ratio $K' = (\%P_2O_5)/[\%P]^2(\%FeO)^5$ and the total lime content of the slag. When the concentrations are expressed in weight-per cent the ratio is given by:

$$\log K' = 1{,}180(\%CaO) - C \qquad 15.35$$

where C is a temperature dependent constant, but the data show a fairly considerable scatter, which is increased when K' is expressed in terms of the 'free' lime concentration in the slag. It was concluded that for constant ratios of CaO/SiO_2 there is an optimum concentration of ferrous oxide for phosphorus transfer into the slag and this seems reasonable from thermodynamic considerations. As the concentration of ferrous oxide is increased along a path of constant CaO/SiO_2 ratio, the activity of the ferrous oxide is raised and, because the mole fraction of lime varies inversely in proportion to the ferrous oxide content, it is probable that the activity coefficient of phosphorus pentoxide is also increased. The activity of ferrous oxide is not a linear function of the concentration and it is probable that ($\gamma_{P_2O_5}$) also varies non-linearly with the ferrous oxide content, so that as the ferrous oxide content is raised the phosphorus distribution may indeed pass through a maximum. However, the distribution is controlled primarily by the activity of the phosphorus pentoxide in the slag and if a maximum in the distribution does exist it is expected that it will not be very pronounced.

One interesting feature of these empirical treatments is that in each case a better correlation is obtained when the phosphorus partitioning is related to the ferrous oxide content of the slag than when it is expressed in terms of the oxygen content of the bulk metal. This is strong evidence in support of the contention that the oxidation of phosphorus is mainly, if not entirely, a slag–metal interface reaction.

15.4 Reversion of Phosphorus into the Metal

It is occasionally found that a small amount of phosphorus is transferred back from the slag into the metal following the addition of the deoxidisers to the metal bath. A more extensive reversion can occur while the metal is held in the ladle and particularly when about 80 or 90 per cent of the metal has been poured into the moulds. The extent of this reversion is usually greater when the metal temperature is high and/or when the basicity of the slag is low, and these facts form a guide to the probable causes of the reversal in the direction of the reaction.

The addition of the deoxidisers lowers the metal oxygen activity both in the bulk phase and at the slag–metal interface, so that if the phosphorus reaction is at, or very close to, equilibrium before the additions are made the equilibrium is displaced in the direction which causes the return of phosphorus to the metal. However, the oxygen activity is raised to the fifth power in the equilibrium relation and it follows that the phosphorus distribution is not changed appreciably by a moderate change in the oxygen content of the metal. The absorption into the slag of the reaction products from the deoxidation causes a negligible change in the composition of the slag and the activity coefficient of the phosphorus pentoxide is not affected appreciably, so the phosphorus reversion at this stage should be very small.

When the metal is poured into a ladle with an acid lining there is a slow but steady solution of the lining which lowers the basicity of the slag and the rate of solution is expected to increase with the temperature of the slag. Provided that the slag is saturated with lime an increase in the silica content has no effect on the activity of the phosphorus pentoxide, but generally the slag is not saturated and the activity is raised as the concentration of silica is increased. The increase in the rate of phosphorus reversion into the last metal in the ladle can be ascribed to two causes. Firstly, the activity of the phosphorus pentoxide is not a linear function of the silica concentration and it probably increases at a more rapid rate as the neutral composition is approached than in a highly basic slag. Thus the rate of phosphorus reversion should increase more rapidly than the rate at which the acid oxides are dissolving from the lining into the slag, and this effect will become more pronounced as the basicity of the tapping slag is lowered. Secondly, for a constant rate of change of the equilibrium partitioning, the amount of phosphorus which reverts from the slag is proportional to the volume ratio of metal to slag. If the volume ratio is changing more rapidly than the rate of the phosphorus reversion, then it is possible that the equilibrium partitioning is approached more closely towards the end of teeming, when the volume of metal remaining in the ladle is comparatively small. It is probable that both of these factors contribute to the reversion which is found in practice.

From these considerations it is evident that the extent of the phosphorus reversion is reduced as the time the metal is held in contact with the slag in the transfer ladle is lowered and as the basicity of the initial ladle slag is raised. Phosphorus reversion is not encountered when the ladle is lined with a basic refractory.

REFERENCES

1. O. KUBASCHEWSKI and E. LL. EVANS, *Metallurgical Thermochemistry*, third edition, Pergamon, 1958.
2. G. URBAIN, *Rev. Met.*, 1959, Vol. 56, p. 529.
3. J. B. BOOKEY, F. D. RICHARDSON and A. J. E. WELCH, *J. Iron St. Inst.*, 1952, Vol. 171, p. 404.
4. A. A. GRANOVKOYA and A. P. LISUBINOV, *Publ. Stalin Inst. for Steel*, 1954, Vol. 32, p. 79.
5. W. A. FISCHER and H. VOM ENDE, *Stahl u. Eisen, Düsseldorf*, 1952, Vol. 72, p. 1398.
6. O. A. ESIN and V. N. SHIKHOV, *Izv. Ak. Nauk Otdelenie Tekn.*, 1955, (3), p. 79.
7. C. GOODEVE, *J. Iron St. Inst.*, 1951, Vol. 166, p. 379.
8. M. OHTANI and N. A. GOKCEN, *Trans. Amer. Inst. min. (metall.) Engrs.*, 1960, Vol. 218, p. 533.
9. W. OELSEN and H. MAETZ, *Mitt. K.-Wilh. Inst. Eisenforsch.*, 1941, Vol. 23, p. 195.
10. G. TROMEL, *Stahl u. Eisen, Düsseldorf*, 1943, Vol. 63, p. 21.
11. W. OELSEN and H. MAETZ, *Arch. Eisenhüttenw.*, 1948, Vol. 19, p. 111.
12. J. B. BOOKEY, *J. Iron St. Inst.*, 1952, Vol. 172, pp. 61, 66.
13. C. G. STEVENS and E. T. TURKDOGAN, *Trans. Faraday Soc.*, 1954, Vol. 50, p. 370.
14. J. PEARSON, E. T. TURKDOGAN and E. M. FENN, *J. Iron St. Inst.*, 1954, Vol. 176, p. 441.
15. J. F. ELLIOTT, *Trans. Amer. Inst. min. (metall.) Engrs.*, 1955, Vol. 203, p. 485.
16. W. R. MADDOCKS and E. T. TURKDOGAN, *J. Iron St. Inst.*, 1949, Vol. 162, p. 249; 1952, Vol. 171, p. 128.
17. P. T. CARTER, ibid., 1950, Vol. 166, p. 321.
18. M. REY, *Disc. Faraday Soc.*, 1948, No. 4, p. 257.
19. H. FLOOD and K. GRJOTHEIM, *J. Iron St. Inst.*, 1950, Vol. 166, p. 169; 1952, Vol. 171, p. 64.
20. E. T. TURKDOGAN and J. PEARSON, ibid., 1953, Vol. 175, p. 398.
21. Y. K. ZEA, ibid., 1945, Vol. 151, p. 459.
22. H. SCHENCK and W. REISS, *Arch. Eisenhüttenw.*, 1939, Vol. 9, p. 589.
23. T. B. WINKLER and J. CHIPMAN, *Trans. Amer. Inst. min. (metall.) Engrs.*, 1946, Vol. 167, p. 111.
24. G. DERGE, *Basic Open-Hearth Steel-making*, second edition, A.I.M.E., p. 691.
25. K. BALAJIVA, A. G. QUARRELL and P. VAJRAGUPTA, *J. Iron St. Inst.*, 1946, Vol. 153, p. 115.
26. K. BALAJIVA and P. VAJRAGUPTA, ibid., 1947, Vol. 155, p. 563.
27. P. VAJRAGUPTA, ibid., 1948, Vol. 158, p. 494.

Appendix 1

Interaction Coefficients

A number of summaries have recently been published [1–4] of the effects of solute elements on the activity coefficients of certain elements at infinite dilution in liquid iron and extensive use has been made of these summaries in compiling the following Tables.

The following notation is used:

$$\epsilon_Y^X = \frac{\partial \ln f_Y}{\partial N_X}$$

$$e_Y^X = \frac{\partial \log f_Y}{\partial (\text{wt.-}\%X)}$$

Where only one of these is given in the original reference, the other is calculated from the relation:

$$e_Y^X = \frac{0{\cdot}2425}{M_X} . \epsilon_Y^X$$

where M_x is the atomic weight of X. Use is also made of the relations:

$$e_X^Y = \frac{M_X}{M_Y} . e_Y^X$$

and

$$\epsilon_X^Y = \epsilon_Y^X$$

which are discussed in Chapter 2.

Following the method used by Ohtani [3] and others, the interaction coefficients are plotted against the atomic numbers of the X solutes on the accompanying diagrams and some values (designated with an È in the Tables) have been obtained by interpolation or extrapolation on the graphs.

In many cases there is very poor agreement among the values given for a particular interaction coefficient, and those marked with a star in the Tables are either those which are most consistent with the periodic relations or those which, from other considerations, are probably more reliable than the other values for the coefficient.

Appendix Table 1 *The Activity Coefficients of Solute Elements in Dilute Binary Fe–X melts at 1,600°C*

Solute X	ϵ_X^X	e_X^X	*References*
Al	6·0	0·054	[5, 6]
B	22	0·49	E
C	9·5	0·19	[7]
	11·4	0·23*	[8, 9]
Cr	0	0	[10]
Co	0	0	E
Cu	− 5·5	−0·021	[11]
	− 6·9	−0·026	[12]
Mn	0	0	[13, 14]
Ni	− 0·01	0	[15, 16]
N	0	0	[4]
Nb	3·2	0·008	[17]
O	−13·4	−0·20*	[18]
	0 to −4·3	0 to −0·05	[3]
P	0	0	E
Si	37·0	0·32	[19]
S	− 3·7	−0·028	[20, 21, 22]
Ti	9·0	0·048	[23]
V	4·5	0·022	E
Zn	−12	−0·04	E

These data are presented in Figure A.1. and show very good agreement for the elements of the fourth period in the Periodic Table.

The following Tables give the interaction parameters at 1,600°C for the effects of dilute concentrations of various solutes on the activity coefficient at infinite dilution in iron of the solute named at the head of each Table.

Appendix Table 2 *Aluminium*

Solute X	ϵ_{Al}^X	e_{Al}^X	*References*
C	6·7	0·13	[5]
Co	2·1	0·008	E
Cu	1·5	0·006	E from Table 6
Mn	2·8	0·012	E from Table 7
Ni	1·8	0·008	E from Table 8
N	0·28	0·005	[4]
O	−1,340	−20	[1, 24, 25]
P	6·5	0·050	E
Si	6·9	0·058	[5]
S	6·1	0·042	[38]

The values for ϵ_{Al}^{P} and ϵ_{Al}^{Co} are interpolated by assuming a linear relationship for the parameters of the elements of the third and fourth periods respectively.

Appendix Table 3 *Carbon*

Solute	ϵ_C^x	e_C^x	*References*
Al	6·7	0·058	[5]
Cr	−4·3	−0·020	[1, 24]
	−5·1	−0·024	[2]
Co	2·9	0·012	[2]
Cu	4·2	0·016	[2, 11]
Mn	−0·5	−0·002	[1]
Mo	−3·5	−0·009	[2]
Ni	2·9	0·012	[2]
	3·6	0·015	E
N	12·1	0·21	[4]
	6·7	0·116	[26]
	6·35	0·11	[27]
	6·18	0·107	[28]
Nb	−23·0	−0·06	[2]
	−53·6	−0·14	[18]
O	−24·0	−0·364	[8]
	−22·4	−0·34	[29]
	−20·5	−0·31	[1]
	−15·8	−0·24*	[30]
	−8·6	−0·10*	[2]
P	6·0	0·047	E
Si	10·2	0·088	[1]
	6·4	0·055	[29]
S	13·3	0·10	[24]
	12·0	0·09	[2]
	5·4	0·045	[1]
Sn	0	0	[2]
Ti	−12·8	−0·07	E
V	−8·0	−0·038	[2]
W	−2·3	−0·003	E
Zn	4·5	0·017	E

The relations between the interaction coefficients and the atomic number of the elements are shown in Figure A.2. The data for the elements of the third period show wide scatter and the line is drawn through the points which are considered to be the most reliable. The data for Group IV elements are fitted by a smooth curve with the exception of either nickel or copper. It is probable that the value for copper is the more reliable of these two and the parameter for nickel is re-estimated on this

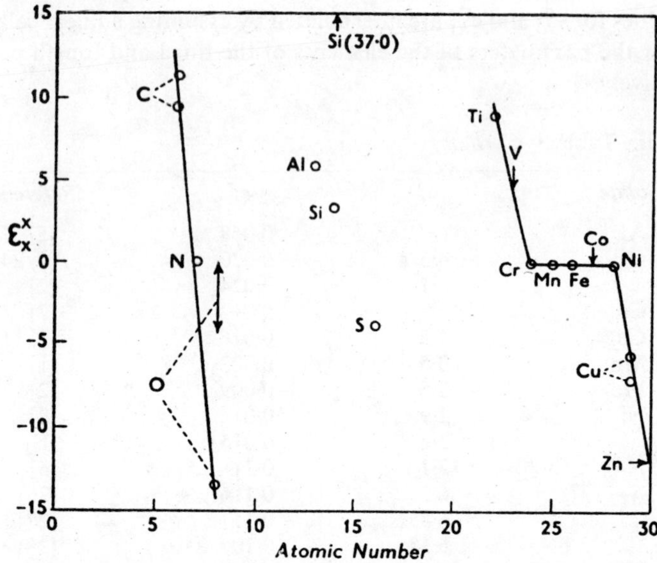

FIG. A.1. Relation between ϵ_X^X and atomic number of X at 1,600°C.

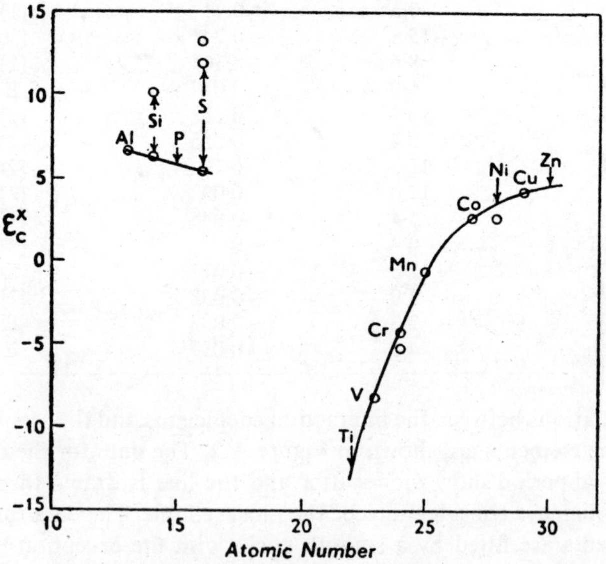

FIG. A.2. Relation between ϵ_C^X and atomic number of X at 1,600°C.

basis. The points for elements of the second period show very wide scatter and it is not possible to establish a graphical relation with the atomic numbers.

Appendix Table 4 *Chromium*

Solute X	ϵ_{Cr}^{X}	e_{Cr}^{X}	*References*
C	− 5·1	−0·103	[2]
	− 4·3	−0·087	[1, 24]
N	−89	−1·54	[26]
	− 9·9	−0·171*	[31]
	− 9·7	−0·168*	[4, 32]
	− 8·8	−0·152	[27]
	− 8·3	−0·144	[33]
O	− 8·8	−0·133	[1]
	−13·7	−0·208*	[10]
S	− 4·1	−0·031	[34]

Appendix Table 5 *Cobalt*

Solute X	ϵ_{Co}^{X}	e_{Co}^{X}	*References*
Al	2·1	0·017	E from Table 2
C	2·9	0·054	[2]
N	2·7	0·043	[4]
	1·7	0·027	[28]
O	1·7	0·024	[18]
P	0·9	0·007	E
Si	1·5	0·013	E
S	0·3	0·002	E

The activities of sulphur in liquid iron–cobalt alloys show a small positive deviation from ideality at 1,540°C [35], but the measurements do not extend to the dilute solutions and ϵ_{Co}^{S} cannot be determined accurately from the results. The heats of formation of the most stable sulphides of manganese, nickel and copper vary in the same order as the effects of these elements on the activity coefficient of sulphur and ϵ_{Ni}^{S} is zero. The heat of formation of cobalt sulphide is a little less negative than that for nickel sulphide and this suggests that ϵ_{Co}^{S} has a small positive value.

The interaction coefficients for the effects of manganese, cobalt, nickel and copper on the activity coefficient of aluminium are related linearly to the atomic numbers of the elements, but the linear relation does not apply to their effects on the activity coefficients of silicon, phosphorus and sulphur which are in the same short period as aluminium (Figure A.3).

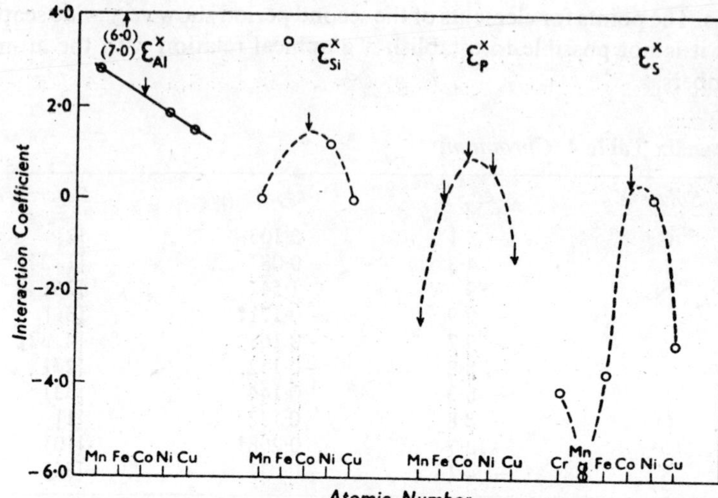

FIG. A.3. Variation of the interaction coefficients for elements of the third period with the atomic numbers of elements in the first transition series.

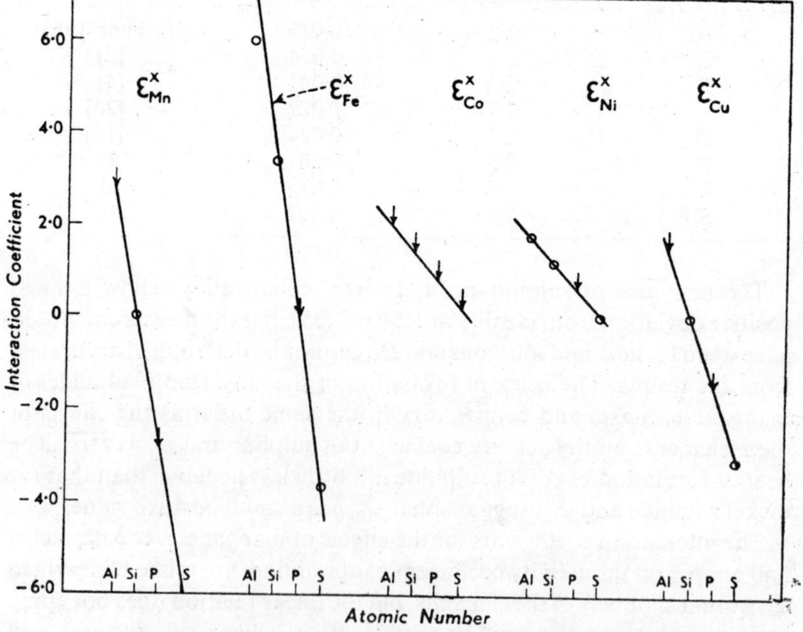

FIG. A.4. Variation of the interaction coefficients for elements of the first transition series with the atomic numbers of elements of the third period.

Appendix Table 6 *Copper*

Solute X	ϵ_{Cu}^{X}	e_{Cu}^{X}	*References*
Al	1·5	0·014	E
C	4·2	0·085	[2, 11]
N	2·4	0·042*	[4]
	0·52	0·009	[28]
O	−2·5	−0·038	[18]
P	−1·5	−0·012	E
Si	0	0	[11]
S	−3.1	−0·023	[36]

Appendix Table 7 *Manganese*

Solute X	ϵ_{Mn}^{X}	e_{Mn}^{X}	*References*
Al	2·8	0·027	E
C	− 0·5	−0·009	[1]
N	−21	−0·363	[26]
	− 5·6	−0·097	[33]
	− 4·5	−0·078*	[4]
	0	0	[3]
O	− 8·2 to +2·0	−0·124 to +0·030	[37]
	0	0	[1, 24]
P	− 2·8	−0·022	E
Si	0	0	[1]
S	− 5·9	−0·045	[24]
	− 5·7	−0·043	[1, 38]

Appendix Table 8 *Nickel*

Solute X	ϵ_{Ni}^{X}	e_{Ni}^{X}	*References*
Al	1·8	0·016	E
C	2·9	0·059	[2]
	3·6	0·073	E from Table 3
H	0	0	[39]
N	2·7	0·047	[32]
	2·4	0·042	[4, 26, 28]
	1·7	0·029	[40]
	0	0	[33]
O	1·4	0·022	[41]
P	0·6	0·005	E
Si	1·2	0·011	[1]
S	0	0	[54]

Appendix Table 9 *Nitrogen*

Solute X	ϵ_N^X	e_N^X	*References*
Al	0·28	0·0025	[4]
As	5·6	0·018	[28]
C	12·1	0·25	[4]
	6·7	0·135	[26]
	6·35	0·13	[27]
	6·18	0·125	[28]
Cr	−89	−0·415	[26]
	− 9·9	−0·046*	[31]
	− 9·7	−0·045*	[4, 32]
	− 8·8	−0·041	[27]
	− 8·3	−0·0385	[33]
Co	2·7	0·011	[4]
	1·7	0·007	[28]
Cu	0·52	0·002	[28]
	2·4	0·009*	[4]
Mn	−21	−0·094	[26]
	− 5·6	−0·0245	[33]
	− 5·2	−0·023*	[42]
	− 4·5	−0·02*	[4]
	− 3·0	−0·013	[43]
	0	0	[3]
Mo	− 5·1	−0·013	[40]
	− 4·4	−0·011	[4]
	− 1·6	−0·004	[28]
Ni	2·7	0·011*	[32]
	2·4	0·010*	[4, 26, 28]
	1·7	0·007	[40]
	0	0	[33]
Nb	−25·6	−0·067	[4]
O	3·3	0·05	[4]
	0	0	[28]
P	6·4	0·05	[27]
Si	7·5	0·065	[28]
	5·4	0·047*	[4, 44]
S	1·7	0·013	[28]
	7·0	0·053	E
Sn	3·4	0·007	[4]
	0·98	0·002	[28]
Ta	−25	−0·034	[4]
V	−24	−0·12	[40]
	−21	−0·10*	[4]
	−19	−0·09*	[31]
W	− 1·5	−0·002	[4]

However, there is a linear relation between the atomic numbers and the interaction coefficients of aluminium, silicon, phosphorus and sulphur with each of the elements manganese, nickel and copper (Figure A.4). Tentative values for ϵ_{Co}^{Si}, ϵ_{Co}^{P} and ϵ_{Co}^{S} are therefore obtained by drawing a line parallel to the ϵ_{Ni}^{X} atomic number relation and through the estimated value for ϵ_{Co}^{Al}.

The relations shown in Figure A.3 suggest that the interaction coefficients for the effects of elements in the first transition series on the activity

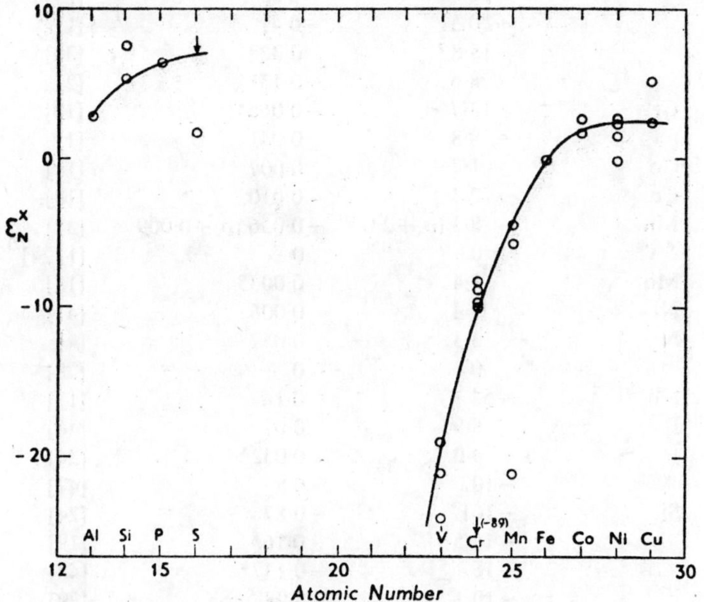

FIG. A.5. Relation between ϵ_N^X and atomic number of X at 1,600°C.

coefficients of silicon, phosphorus and sulphur show a maximum for cobalt, but the relations are not sufficiently well established to justify further comment.

The variation of the interaction coefficients for nitrogen with the atomic numbers of the second solutes is shown in Figure A.5. The relation holds for the elements of the first transition series, but the experimental value for sulphur implies that a maximum occurs in the relation for the elements of the third period. Assuming a continuous increase through this period in the interaction coefficients, a value for ϵ_N^S is estimated which is more consistent with the values for ϵ_C^S.

The relations between the interaction coefficients ϵ_O^x and the atomic numbers are shown in Figure A.6. The points show considerable scatter,

Appendix Table 10 *Oxygen*

Solute X	ϵ_O^x	e_O^x	*References*
Al	−1,340	−12·0	[1, 24, 25]
B	−27·7	−0·62	[45]
C	−24·0	−0·485	[8]
	−22·4	−0·45	[29]
	−20·5	−0·41	[1]
	−15·8	−0·32*	[30]
	− 8·6	−0·13*	[2]
Cr	−13·7	−0·064	[10]
	− 8·8	−0·041	[1]
Co	1·7	0·007	[18]
Cu	− 2·5	−0·010	[18]
Mn	− 8·3 to +2·0	−0·036 to +0·009	[37]
	0	0	[1, 24]
Mo	1·4	0·0035	[18]
Ni	1·4	0·006	[41]
N	3·3	0·057	[4]
	0	0	[28]
Nb	−54	−0·14	[17]
P	8·9	0·07	[46]
	− 4·0	−0·032*	[24]
	−102	−0·8	[47]
Si	−101	−0·87	[24]
	−18·5	−0·16*	[19]
	−16	−0·137*	[48]
	−10·1	−0·087	[29]
	− 2·3	−0·02	[1]
S	132	1·0	[49]
	0	0*	[50, 51]
Ti	−100	−0·51	E
	−37	−0·187	[23]
V	−57	−0·27	[52]
	−53	−0·25	[24]
W	6·4	0·009	[18]

particularly with the solute elements from the second and third periods, and the lines are drawn to join the points which are considered to be the most reliable.

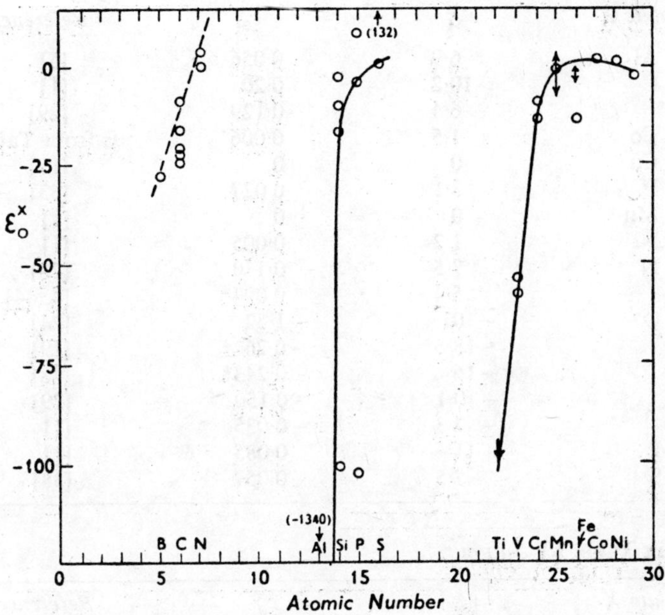

FIG. A.6. Relation between ϵ_O^X and the atomic number of X at 1,600°C.

Appendix Table 11 *Phosphorus*

Solute X	ϵ_P^X	e_P^X	*References*
Al	6·5	0·056	E from Table 2
C	6·0	0·12	E from Table 3
Co	0·9	0·004	E from Table 5
Cu	− 1·5	−0·006	E from Table 6
Mn	− 2·8	−0·012	E from Table 7
Ni	0·6	0·003	E from Table 8
N	6·4	0·111	[27]
O	−102	−1·5	[47]
	− 4·0	−0·062	[24]
	8·9	0·135	[46]
Si	11·0	0·094	[1]
S	5·4	0·041	[38]

Appendix Table 12 *Silicon*

Solute X	ϵ_{Si}^{X}	e_{Si}^{X}	*References*
Al	6·9	0·056	[5]
C	10·2	0·20	[1]
	6·4	0·129	[29]
Co	1·5	0·006	E from Table 5
Cu	0	0	[11]
H	3·1	0·027	[53]
Mn	0	0	[1]
Ni	1·2	0·005	[1]
N	7·5	0·130	[28]
	5·4	0·094*	[4, 44]
O	−101	−1·52	[24]
	−18·5	−0·262*	[19]
	−16	−0·243*	[48]
	−10·1	−0·150	[29]
	− 2·3	−0·035	[1]
P	11	0·085	[1]
S	7·5	0·057	[38]

Appendix Table 13 *Sulphur*

Solute X	ϵ_{S}^{X}	e_{S}^{X}	*References*
Al	6·1	0·055	[38]
C	13·3	0·27	[24]
	12·0	0·24	[2]
	5·4	0·110	[1]
Cr	− 4·1	−0·019	[34]
Co	0·3	0·001	E from Table 5
Cu	− 3·1	−0·012	[36]
Mn	− 5·9	−0·026	[24]
	− 5·7	−0·025	[1, 38]
Ni	0	0	[54]
N	1·7	0·030	[28]
	7·0	0·12	E from Table 9
O	132	2·0	[49]
	0	0*	[50, 51]
P	5·4	0·042	[38]
Si	7·5	0·065	[38]

The Effects of Solute Elements on the Activity Coefficient of Carbon in Carbon-Saturated Fe–C–X melts

The available data have recently been reviewed by Fuwa and Chipman [2] and by Ohtani and Gokcen [3] and the values for $(\partial \ln N_C)/(\partial N_X)_{a_C}$

recommended by them are given in Table 4.1. Values of $(\partial \ln f_C)/(\partial N_X)_{N_C}$ obtained from these data by use of equation 13.41 are given in Table 14 and Figure 7.

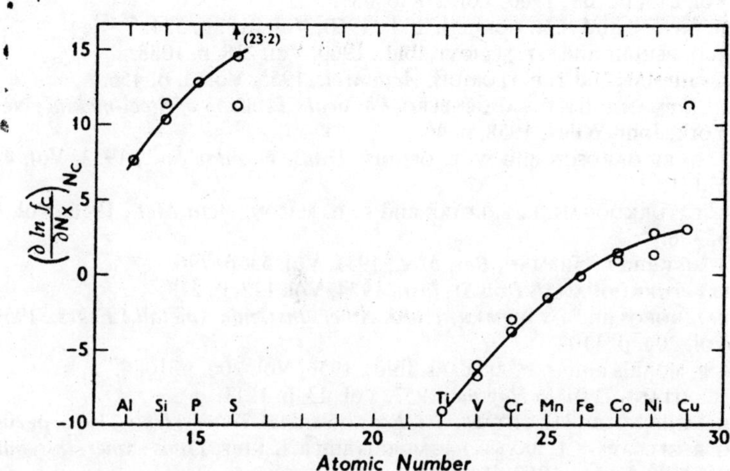

FIG. A.7. Variation of $(\partial \ln f_C/\partial N_X)_{N_C}$ with atomic number of X for carbon-saturated solutions.

Appendix Table 14 *Interaction Coefficients for Carbon in Carbon-Saturated Solutions*

Solute X	$\left(\frac{\partial \ln f_C}{\partial N_X}\right)_{N_C}$ Reference [2]	Reference [3]
Al	7·5	7·5
Cr	− 2·7	− 3·6
Co	1·4	1·1
Cu	11·5	3·1
Mn	− 1·4	− 1·4
Ni	2·8	1·4
Nb	− 11·5	—
P	12·8	12·8
Si	11·2	10·3
S	11·2	14·5, 23·2
Ti	—	− 9·2
V	− 6·1	− 6·4

REFERENCES

1. J. CHIPMAN, *J. Iron St. Inst.*, 1955, Vol. 180, p. 97.
2. T. FUWA and J. CHIPMAN, *Trans. Amer. Inst. min. (metall.) Engrs.*, 1959, Vol. 215, p. 708; 1960, Vol. 218, p. 887.
3. M. OHTANI and N. A. GOKCEN, ibid., 1960, Vol. 218, p. 534.
4. R. D. PEHLKE and J. F. ELLIOTT, ibid., 1960, Vol. 218, p. 1088.
5. J. CHIPMAN and T. P. FLORIDIS, *Acta Met.*, 1955, Vol. 3, p. 456.
6. J. CHIPMAN and F. C. LANGENBERG, *Physical Chemistry of Steel-making*, New York, John Wiley, 1958, p. 46.
7. F. D. RICHARDSON and W. E. DENNIS, *Trans. Faraday Soc.*, 1953, Vol. 49, p. 171.
8. E. T. TURKDOGAN, L. E. LEAKE and C. R. MASON, *Acta Met.*, 1956, Vol. 4, p. 396.
9. A. RIST and J. CHIPMAN, *Rev. Met.*, 1957, Vol. 53, p. 796.
10. E. T. TURKDOGAN, *J. Iron St. Inst.*, 1954, Vol. 178, p. 278.
11. P. J. KOROS and J. CHIPMAN, *Trans. Amer. Inst. min. (metall.) Engrs.*, 1956, Vol. 206, p. 1102.
12. J. P. MORRIS and G. R. ZELLARS, ibid., 1956, Vol. 206, p. 1086.
13. M. OHTANI, *Tetsu to Hagane*, 1957, Vol. 43, p. 1211.
14. K. SANBONGI and M. OHTANI, *Sci. Reps. Res. Inst. Tohoku Univ.*, 1955, p. 204.
15. G. R. ZELLARS, S. L. PAYNE, J. P. MORRIS and R. L. KIPP, *Trans. Amer. Inst. min. (metall.) Engrs.*, 1959, Vol. 215, p. 181.
16. R. SPEISER, A. J. JACOBS and J. W. SPRETNAK, ibid., 1959, Vol. 215, p. 185.
17. M. ELLE and J. CHIPMAN, ibid., 1961, Vol. 221, p. 701.
18. T. P. FLORIDIS and J. CHIPMAN, ibid., 1958, Vol. 212, p. 549.
19. J. CHIPMAN and T. C. M. PILLAY, ibid., 1961, Vol. 221, p. 1277.
20. J. P. MORRIS and R. C. BUEHL, ibid., 1950, Vol. 188, p. 317.
21. C. W. SHERMAN, H. I. ELVANDER and J. CHIPMAN, ibid., 1950, Vol. 188, p. 1389.
22. A. ADACHI and Z. MORITA, *Tetsu to Hagane*, 1958, Vol. 44, p. 637.
23. J. CHIPMAN, *Trans. Amer. Inst. min. (metall.) Engrs.*, 1960, Vol. 218, p. 767.
24. J. PEARSON and E. T. TURKDOGAN, *J. Iron St. Inst.*, 1954, Vol. 176, p. 19.
25. N. A. GOKCEN and J. CHIPMAN, *Trans. Amer. Inst. min. (metall.) Engrs.*, 1953, Vol. 197, p. 173.
26. T. SAITO, *Sci. Reps. Res. Inst. Tohoku Univ.*, 1950, Vol. 2, p. 909.
27. T. KOOTZ, *Arch. Eisenhüttenw.*, 1941, Vol. 15, p. 77.
28. H. SCHENCK, M. FROHBERG and H. GRAF, ibid., 1958, Vol. 29, p. 673; 1959, Vol. 30, p. 533.
29. H. SCHENCK and G. H. GERDOM, ibid., 1959, Vol. 30, p. 451.
30. S. BANYA and S. MATOBA, *Tetsu to Hagane*, 1958, Vol. 44, p. 643.
31. R. M. BRICK and J. A. CREEVY, *Metals Technology*, A.I.M.E. Techn. Publn. No. 1165, April, 1940.
32. J. C. HUMBERT and J. F. ELLIOTT, *Trans. Amer. Inst. min. (metall.) Engrs.*, 1960, Vol. 218, p. 1076.
33. H. WENTRUP and O. REIF, *Arch. Eisenhüttenw.*, 1949, Vol. 20, p. 359.
34. N. R. GRIFFING and G. W HEALY, *Trans. Amer. Inst. min. (metall.) Engrs.*, 1960, Vol. 218, p. 849.
35. C. B. ALCOCK and L. L. CHENG, *J. Iron St. Inst.*, 1960, Vol. 195, p. 169.

36. T. ROSENQVIST and E. M. COX, *Trans. Amer. Inst. min. (metall.) Engrs.*, 1950, Vol. 188, p. 1389.
37. B. V. LINCHIVSKII and A. A. SAMARIN, *Izv. Ak. Nauk Otdelenie Tekn.*, 1957, (2), p. 9.
38. C. W. SHERMAN and J. CHIPMAN, *Trans. Amer. Inst. min. (metall.) Engrs.*, 1952, Vol. 194, p. 597.
39. T. BUSCH and R. A. DODD, ibid., 1960, Vol. 218, p. 488.
40. V. KASHYAP and N. PARLEE, ibid., 1958, Vol. 212, p. 86.
41. H. A. WRIEDT and J. CHIPMAN, ibid., 1956, Vol. 206, p. 1195.
42. S. Z. BEER, ibid., 1961, Vol. 221, p. 1.
43. R. A. DODD and N. A. GOKCEN, ibid., 1961, Vol. 221, p. 233.
44. M. M. KARNAUKHOV and A. M. MOREZOV, *Bull. Acad. Sci. SSSR*, 1947, Vol. 6, p. 735.
45. C. YUAN HSI, *Acta Met. Sinica*, 1958, Vol. 3, p. 252.
46. D. DUTILLOY and J. CHIPMAN, *Trans. Amer. Inst. min. (metall.) Engrs.*, 1960, Vol. 218, p. 428.
47. J. B. BOOKEY, F. D. RICHARDSON and A. J. E. WELCH, *J. Iron St. Inst.*, 1952, Vol. 171, p. 404.
48. S. MATOBA, K. GUNJI and T. KUWANA, *Tetsu to Hagane*, 1959, Vol. 45, p. 1328.
49. D. C. HILTY and W. CRAFTS, *Trans. Amer. Inst. min. (metall.) Engrs.*, 1952, Vol. 196, p. 1307.
50. S. MATOBA and T. UNO, *Tetsu to Hagane*, 1942, Vol. 28, p. 651.
51. Z. BUZHEK and A. M. SAMARIN, *Dokl. obsch. Sobr. Ak. Nauk, S.S.S.R.*, 1957, Vol. 114, p. 97; *Hutn. List.*, 1959, Vol. 14, p. 926.
52. M. N. DASTUR and J. CHIPMAN, *Trans. Amer. Inst. min. (metall.) Engrs.*, 1949, Vol. 185, p. 441.
53. H. LIANG, M. B. BEVER and C. F. FLOE, ibid., 1946, Vol. 167, p. 395.
54. J. A. CORDIER and J. CHIPMAN, ibid., 1955, Vol. 203, p. 905.

Appendix 2

Slag and Metal Analyses for Basic Open-Hearth Heats

Minutes after Clear Melt	*Metal Temperature °C*	*Metal Composition Weight-per cent* C	Mn	Si	S	P	O
07 Carbon							
0	1,580	0·52	0·21	0·027	0·054	0·080	*
65	1,580	0·24	0·20	0·025	0·054	0·055	0·019
115	1,590	0·14	0·19	0·023	0·052	0·049	*
150	*	0·12	0·18	0·022	0·048	0·040	0·038
220	*	0·07	0·14	0·022	0·045	0·022	*
255	1,590	0·065	0·14	0·022	0·040	0·024	0·047
08 Carbon							
0	1,536	0·62	0·32	*	0·030	0·063	*
93	1,530	0·36	0·13	*	0·020	0·010	*
168	1,593	0·12	0·11	*	0·020	0·010	*
223	1,616	0·08	0·11	0·03	0·018	0·012	0·042
11 Carbon							
120	1,565	0·33	0·24	0·019	0·056	0·033	*
150	1,565	0·21	0·16	0·019	0·040	0·028	*
185	1,580	0·11	0·15	0·018	0·038	0·025	0·035
14 Carbon							
0	1,520	1·29	0·25	0·02	0·043	0·031	*
125	1,574	0·48	0·165	0·045	0·045	0·007	*
175	1,606	0·275	0·175	0·03	0·034	0·007	*
220	1,619	0·135	0·14	0·045	0·029	0·008	*
17 Carbon							
0	1,560	0·48	0·17	*	0·042	0·011	*
44	1,575	0·31	0·16	*	0·031	0·018	*
65	1,585	0·25	0·17	*	0·031	0·018	*
88	1,590	0·17	0·15	*	0·029	0·010	*
24 Carbon							
120	1,590	0·40	0·22	0·037	0·044	0·040	*
145	1,580	0·30	0·19	0·028	0·042	0·037	0·015
175	1,580	0·24	0·18	0·026	0·040	0·034	0·013
28 Carbon							
0	1,540	0·535	0·24	0·03	0·042	0·19	*
57	1,568	0·485	0·25	0·03	0·030	0·064	*
83	1,592	0·445	0·20	0·025	0·026	0·043	*
129	*	0·35	0·135	0·03	0·023	0·025	*
162	1,592	0·28	0·135	0·03	0·023	0·024	*

Slag Composition Weight-per cent								
CaO	SiO_2	FeO	Fe_2O_3	MgO	Al_2O_3	P_2O_5	MnO	S
42·0	12·0	11·5	1·76	9·41	3·44	13·17	5·28	0·28
45·0	11·5	8·77	3·12	8·55	3·98	13·48	3·91	0·34
46·0	10·6	9·72	2·48	9·56	3·03	12·55	4·65	0·34
45·0	9·6	11·37	3·52	9·63	3·12	11·50	3·91	0·35
43·8	8·8	14·30	5·04	9·78	3·00	9·22	4·46	0·40
43·2	8·24	16·06	7·20	9·65	2·86	8·78	4·00	0·39
52·25	14·9	4·34	2·54	5·32	2·55	8·14	4·12	0·281
50·5	10·0	13·51	4·75	5·10	1·98	6·47	4·80	0·226
47·0	8·75	17·46	5·03	6·30	1·80	5·49	4·10	0·224
44·68	8·86	19·78	5·85	7·29	1·82	4·83	3·45	0·119
40·6	12·30	11·02	3·12	5·14	3·55	10·49	3·53	0·42
48·2	10·40	13·46	4·80	5·58	2·99	9·69	3·13	0·38
40·7	9·36	14·26	4·96	5·72	2·59	8·42	3·36	0·37
50·0	22·6	3·7	8·6	4·70	1·55	1·04	6·4	0·123
53·5	16·6	8·7	4·15	5·80	2·50	1·39	4·8	0·222
54·5	16·1	7·8	3·81	6·67	2·38	1·32	4·46	0·229
55·0	15·8	11·4	2·29	8·02	2·06	1·17	4·75	0·212
48·8	12·2	9·3	5·4	7·9	2·13	7·47	5·94	0·248
52·0	11·7	10·1	4·4	6·0	2·12	6·83	5·22	0·322
51·5	11·1	10·2	4·3	6·9	2·15	6·51	4·9	0·310
48·0	11·2	12·4	4·15	6·1	2·0	6·46	4·7	0·308
55·4	9·9	9·58	3·44	5·36	3·64	8·08	3·91	0·48
55·4	9·6	10·8	3·52	5·00	3·05	8·80	3·72	0·50
57·2	8·8	11·09	3·20	4·85	2·62	7·16	3·55	0·41
51·82	18·15	2·73	1·6	5·45	2·48	8·09	4·0	0·172
55·0	13·4	4·9	2·0	5·8	2·4	8·19	3·75	0·229
55·0	13·5	5·9	2·9	6·1	2·4	8·30	3·85	0·195
50·5	11·4	7·5	4·4	5·5	1·95	7·48	3·3	0·215
51·25	12·3	8·7	4·0	5·78	1·94	7·35	3·42	0·206

Appendix 2 (*continued*)

Minutes after Clear Melt	*Metal Temperature °C*	*Metal Composition Weight-per cent*					
		C	Mn	Si	S	P	O
38 Carbon							
0	1,475	1·31	0·20	0·02	0·045	0·011	*
128	1,510	0·69	0·18	0·04	0·033	0·009	*
163	1,570	0·535	0·17	0·04	0·026	0·010	*
198	1,610	0·375	0·18	0·04	0·024	0·015	*
55 Carbon							
0	1,528	0·76	0·31	*	0·040	0·105	*
78	1,562	0·61	0·22	*	0·026	0·030	*
99	1,576	0·60	0·23	*	0·027	0·021	*
119	1,590	0·55	0·13	*	0·026	0·021	*
80 Carbon							
0	1,584	1·14	0·54	0·08	0·043	0·098	*
57	1,550	1·06	0·265	0·035	0·030	0·027	*
111	1,554	0·97	0·26	0·025	0·030	0·024	*
155	1,580	0·80	0·23	0·04	0·021	0·022	*

* Not determined.

The last analysis given for each heat shows the metal and slag compositions just prior

Slag Composition Weight-per cent								
CaO	SiO_2	FeO	Fe_2O_3	MgO	Al_2O_3	P_2O_5	MnO	S
43·0	16·3	12·6	3·5	5·25	1·0	3·38	8·9	0·248
52·0	15·0	10·9	3·3	5·2	1·2	2·70	6·05	0·243
51·8	14·3	9·7	5·8	7·4	1·0	3·01	6·1	0·200
53·0	14·2	9·5	6·9	7·7	1·32	2·96	5·7	0·166
49·7	15·5	3·2	1·4	3·8	1·98	11·25	5·4	0·201
50·0	12·1	8·2	3·9	4·08	1·48	9·22	4·89	0·249
51·2	10·8	9·1	3·3	4·2	1·5	8·80	5·1	0·241
52·0	11·3	10·2	3·3	4·2	1·5	8·61	5·1	0·234
55·0	15·2	2·66	1·91	4·77	2·2	10·56	6·1	0·203
48·2	12·8	6·71	3·65	4·4	1·95	8·33	5·9	0·223
54·0	12·0	7·0	3·1	4·9	1·84	10·16	6·1	0·196
53·6	12·5	7·2	3·4	5·33	2·0	9·63	5·6	0·176

to the addition of the deoxidisers and/or tapping the furnace.

Index